Dog–Human Relationships: Behavior, Physiology, and Wellbeing

Dog–Human Relationships: Behavior, Physiology, and Wellbeing

Editor

Betty McGuire

MDPI

Basel • Beijing • Wuhan • Barcelona • Belgrade • Novi Sad • Cluj • Manchester

Editor
Betty McGuire
Cornell University
Ithaca
USA

Editorial Office
MDPI
Grosspeteranlage 5
4052 Basel, Switzerland

This is a reprint of articles from the Special Issue published online in the open access journal *Animals* (ISSN 2076-2615) (available at: https://www.mdpi.com/journal/animals/special_issues/dog_human_relationships).

For citation purposes, cite each article independently as indicated on the article page online and as indicated below:

Lastname, A.A.; Lastname, B.B. Article Title. *Journal Name* **Year**, *Volume Number*, Page Range.

ISBN 978-3-7258-1693-4 (Hbk)
ISBN 978-3-7258-1694-1 (PDF)
doi.org/10.3390/books978-3-7258-1694-1

Cover image courtesy of Betty McGuire

Contents

About the Editor

Betty McGuire

Betty McGuire is a Senior Lecturer (Retired) in the Department of Ecology and Evolutionary Biology at Cornell University, Ithaca, NY. She received her undergraduate training at Pennsylvania State University, completed her doctoral degree at University of Massachusetts Amherst, and was a Postdoctoral Fellow at the University of Illinois. For many years, she studied social behavior, reproduction, and ecology of small mammals, including voles, rats, and shrews, using both laboratory and field approaches. More recently, her research has focused on behavior of shelter dogs, particularly scent-marking and food-guarding behaviors, and dog-human interactions. She has been recognized for training research undergraduates, with whom she publishes. Betty has coauthored textbooks in human biology, vertebrate biology, and animal behavior.

Preface

This Special Issue includes 22 papers on dogs interacting with humans in many different capacities, environments, and countries. With a focus on canine and human behavior, physiology, and wellbeing, there are papers concerning interactions between humans and shelter dogs, pet dogs, therapy dogs, service dogs, working dogs, and free-living dogs. As such, this collection of papers will be of interest to researchers, veterinarians, animal shelter administrators and staff, as well as people training and handling therapy, service, and working dogs. I thank authors for their contributions, the many reviewers who helped to improve all papers, as well as staff members at Animals who carefully managed reviews, revisions, acceptances, and proofs. I especially wish to thank Iulia Chinan, who helped start this Special Issue and saw it through its first year, and Sandra Spatariu, who came on in its second year and saw the project through to completion. Finally, I thank Cyndi Cai, who first introduced me to Animals and its Special Issues.

Regarding dogs living in shelters, Lisa Gunter and colleagues provide evidence that brief outings away from the shelter, and especially temporary stays in foster homes, increased a dog's likelihood of adoption when compared to control shelter dogs without such interventions. Additionally, these away-from-shelter programs were more successful at shelters with significant resources and when a greater percentage of participants were community members than volunteers and staff. Another canine enrichment activity at shelters is leash walking. Building on studies by other researchers showing that mature dogs display greater uneasiness around unfamiliar men than unfamiliar women and our previous study showing sex of an unfamiliar walker influenced scent-marking behavior of mature shelter dogs during leash walks, my students and I examined this same question while also including juvenile dogs and subsequent walks when dogs were familiar with walkers. As in our earlier study, we found that mature male dogs urinated less frequently when walked by an unfamiliar man than by an unfamiliar woman, and mature dogs of both sexes were less likely to defecate when walked by an unfamiliar man than by an unfamiliar woman. Unlike mature dogs, juvenile dogs were generally unaffected by sex of walker. These patterns for mature and juvenile dogs did not change over walks as dogs became familiar with walkers. Our results indicate that sex of human and sex and maturity of dog influence interactions between people and dogs in shelters, and potentially elsewhere. Familiarity did not influence scent-marking behavior of shelter dogs, perhaps because they are exposed daily to so many different people. A few days after entering most shelters, dogs are behaviorally evaluated by staff and one part of many evaluations concerns a dog's reaction to a stranger. Andrew Song and I examined whether sex of the stranger influenced dog reactions in this testing situation. We found that dogs responded significantly more strongly to male strangers than female strangers, but even responses to male strangers fell within the range of behaviors considered not concerning by staff. Thus, from a practical standpoint, stranger tests on shelter behavioral evaluations probably do not have to consider sex of the stranger. Future studies should examine whether sex of an unfamiliar person influences stranger-directed aggression in dogs in homes.

Living in a shelter is stressful for most dogs. JoAnna Platzer and colleagues investigated the domestication syndrome hypothesis by exploring potential connections between pigmentation, behavior, and the physiological stress response in shelter dogs. If pigmentation predicts susceptibility to stress, then shelters could employ interventions based on coat color and pattern to reduce stress in specific dogs. The authors found that dogs with different coat colors and patterns did not differ in a commonly used physiological measure of stress (urinary cortisol:creatinine ratio), thereby not supporting assumptions of the domestication syndrome hypothesis. Animal shelters also are challenging work environments for staff, and most literature on sheltering reflects Western shelters and perspectives. Deyvika Srinivasa and colleagues conducted semi-structured interviews with staff at three shelters in India to identify challenges and resiliency factors. Interview analyses revealed two main challenges and three resiliency factors, some of which differed from those of Western shelters. The first challenge was high intake numbers due to abandoned pet dogs and large numbers of free-ranging dogs. The second was inadequate funding reflecting lack of government support and difficulty obtaining donations from community members due to religious and cultural beliefs favoring larger animals, such as cows and donkeys, over dogs. Resiliency factors included staff adaptability, strong co-worker relationships, and a deep commitment to caring for animals. In Iran, free-ranging dogs also are common, especially in and around cities. Farshad Amiraslani accessed news articles published online on an Iranian news site to examine spatial distribution of free-ranging dogs, views on their impacts, and the emergence of dog clinics. He found that cities in most large provinces had free-ranging dogs, and all impacts were negative, including bites, rabies, and costs to municipalities of capturing and caring for such dogs. Although in the past, interested individuals largely took responsibility for free-ranging dogs, some dog clinics now exist in Iranian cities. Such clinics typically capture, sterilize, vaccinate, and care for injured or ill dogs. However, space limitations usually result in temporary housing, with most dogs released in natural areas after treatment.

Most papers in this Special Issue focus on dogs living in homes. Two of these papers examined dog adoption and another considered issues encountered by owners of dogs with behavioral problems. Rebecca Mead and colleagues investigated factors that influence the extent to which dog owners and prospective dog owners in the UK engaged in preacquisition research. About half of dog owners and two-thirds of prospective dog owners conducted preacquisition research, and likelihood of doing so was influenced by source of dog, previous dog ownership, and demographic factors, such as a person's age. This information may inform efforts by welfare organizations to better prepare dog adopters. Ian Dinwoodie and colleagues used a self-reporting questionnaire to examine the relationship between factors used when selecting a dog to adopt and eventual owner satisfaction. Owner satisfaction was positively influenced when time spent considering an adoption was short, companionship was the primary motive for adoption, and dog behavior and personality, rather than physical characteristics, were selection factors. Canine behavioral problems can negatively impact the bond between dogs and their owners. Although validated tools are available to assess caregiver burnout in humans caring for other humans, Carmen Luz Barrios and colleagues identified lack of such a tool for owners caring for dogs with behavioral issues. The authors developed a questionnaire by modifying a scale used to evaluate human caregiver syndrome, adding new questions, and then validating the questionnaire. This new tool may improve the wellbeing of both dogs and owners.

Several studies examined the impact of dog ownership on owner behavior, and one examined the impact on owner physiology. Benedikt Hielscher-Zdzieblik and colleagues examined the effects of dog breed and age on physical activity of owners in Germany. Whereas most previous studies on this topic were cross-sectional in nature, these authors conducted a 3-year longitudinal study. At baseline (time zero), owners of different breeds engaged in different types of activities with their dogs. When data from participants who completed follow-up questionnaires at years 1, 2, and 3 were analyzed, owner physical activity was stable over time and not associated with dog characteristics such as age, size, and activity level. Ana Junça-Silva and colleagues conducted five different surveys of teleworkers (= remote workers) in Portugal during the COVID-19 pandemic to examine the relationships between attitude toward telework, positive affect, self-reported job performance, and emotional attachment and physical closeness to pet dogs. Some teleworkers owned dogs while others did not have any pets. Dog owners viewed telework more positively, and reported higher levels of positive affect and job performance, with positive affect mediating the relationship between teleworking and job performance. Additionally, this mediating relationship was strengthened when owners' emotional and physical closeness to their dog were high. Anamarie Johnson and colleagues analyzed in-home videos made by American owners when presenting their dog with eight different types of dental chews, and then compared dog behaviors, scored by trained coders, to owner reports of their own and their dog's preference for particular chews. The relationship between owner-perceived dog preferences and dog behavior from videotapes was stronger than that between owner preference and dog behavior from videotapes. The results suggest that researchers interested in factors affecting dog and owner preferences for pet products should consider the use of in-home video recordings because such videos capture dog behavior in a setting where dogs are most comfortable. Finally, Justyna Wojtaś and colleagues examined whether correlations exist between cortisol levels in hair samples from female owners and their cats and dogs living in the same households in Poland. Each owner also completed questionnaires concerning the frequency of interactions with each of her cats and dogs as well as rating her emotional relationship with each pet. The authors did not find significant correlations between cortisol levels of female owners and their dogs, female owners and their cats, or between cats and dogs living in the same household. However, some significant correlations were discovered when considering frequency of interaction and rating of emotional relationship. For example, hair cortisol levels were significantly positively correlated in owners and dogs if owners groomed the dog once a week as well as when owners described the dog as giving them a reason to get up each morning.

The two remaining papers on pet dogs evaluated dog-owner attachment and communication. Emily Richards and colleagues tested dogs on a social evaluation task in which an experimenter either helped (prosocial condition) or refused to help (antisocial condition) the dog's owner open a container; in the control condition, the owner did not turn to either experimenter to ask for help. The dogs were then tested for which individual (prosocial, antisocial, or control) they preferred. Owners also completed a subset of questions from the Canine Behavioral Assessment & Research Questionnaire (C-BARQ) focused on assessing each dog's level of attachment to its owner. The main goal was to determine whether level of attachment to owner predicted a dog's performance on the social evaluation task. Results indicated that attachment was a significant predictor of dogs' preference, but only in the prosocial condition. These results suggest one reason - individual differences in level of attachment to owner - that might explain the mixed results characteristic of studies on canine social evaluation. Courtney Sexton and colleagues examined how dog facial features affect communication between dogs and their owners. The authors developed a standardized method to code dog facial patterns and coloration and used the Dog Facial Action Coding System to measure expressivity of dogs' faces in four conditions: 1) dog at rest without eye contact from owner; 2) eye contact only from owner; 3) eye contact plus unfamiliar words from owner; and 4) eye contact plus familiar words from owner. The authors found dogs with plainer faces tended to be more expressive, and owners more accurately characterized rates of expressivity in adult dogs with plainer faces. Also, whereas movements of muscles in the upper face of dogs were higher in conditions 2 and 3 (ambiguous cues from owners), movements of muscles were more equally distributed across facial regions in condition 4 (clear social cues from owners).

Two studies examined use of therapy dogs. Lisa Townsend and colleagues studied reactivation of a hospital-based dog therapy program during the COVID-19 pandemic, when many such programs were temporarily stopped. The authors collected information, such as type and age of participant, length of interaction, and adherence of dog handlers to safety protocols for humans (using personal protective equipment and hand sanitizer) and dogs (limiting length of visit). Healthcare workers made up the largest percentage of visit recipients (72%), with the remainder split equally among pediatric patients, adult patients, and public adults, such as visitors. Visit durations were longer with patients than healthcare workers, and handler adherence to human and dog safety protocols was high. These findings should inform development of policies and procedures for other visitation programs using therapy dogs. Although use of therapy dogs in medical settings is somewhat common and associated with positive effects, such as reducing patient anxiety and perception of pain, the practice is less common in dental settings. Using a pre-tested and validated survey, Katelyn Cass and colleagues investigated the perceptions and concerns of orthodontic patients and their caregivers about therapy dogs. The authors did not perform dog therapy or examine its effectiveness in reducing anxiety. More than one-third of patients experienced at least moderate anxiety related to orthodontic care, and the majority of participants thought therapy dogs would reduce anxiety and make dental visits more enjoyable. A minority of participants expressed concerns about safety, cleanliness, and allergens, and nearly half would select an orthodontic office with therapy dogs over an office without therapy dogs. These data indicate that most patients and caregivers support the use of therapy dogs in orthodontic offices, and few have concerns.

Two papers focused on service dogs trained to guide visually impaired people. Yana Bender and colleagues conducted semi-structured interviews with German owners of guide dogs to determine factors influencing dog-handler compatibility. The authors found that similar activity levels, hobbies, and personalities characterized a good match between owners and guide dogs, and such owners reported smaller impacts of previous guide dog ownership. These findings may inform the process by which people are matched with guide dogs. Given the long hours of demanding work performed by guide dogs, Enrique De la Fuente-Moreno and colleagues examined cortisol levels in a group of guide dogs compared to another group of dogs of the same two breeds that also underwent guide dog training but were living as companion dogs. The authors collected four saliva samples from each dog over a 1-hour period of isolation (before isolation; 15 minutes after isolation and the sound of a gunshot; 30 minutes after isolation; 45 minutes after isolation). Guide dogs had higher levels of cortisol than companion dogs at all four points in time. Although it is unclear whether the higher cortisol levels in guide dogs reflected stress resulting from their work, these findings at least suggest that guide dogs may face demanding working conditions.

Working dogs include police and military dogs as well as those involved in scent detection, herding, hunting, and search and rescue. Ariella Moser and colleagues examined characteristics that influence performance of Australian biosecurity detection dogs working at airports, docks, and mail centers to find organic material that could pose a risk to agriculture or native species. The authors conducted focus group interviews with biosecurity dog handlers, trainers, and supervisors to develop a questionnaire, which was then completed by handlers when assessing dogs currently used in biosecurity scent detection. Search motivation and arousal were positively associated with ratings of detection performance. A dog's emotional stability did not predict detection ratings, but was emphasized during focus groups because it can influence dog welfare, safety, and ease of handling by different people. Working dogs and companion dogs are not always in close contact with their handlers and owners when communication is desirable. Liliana Rodríguez-Vizzuett and colleagues conducted a scoping review of literature on digital technology that supports remote intentional communication between dogs and humans. For both dogs and humans, some technologies allow the generation of messages (e.g., touch screens for dogs and mobile apps for humans), receipt of messages (e.g., audio and/or vibrational commands for dogs and mobile apps for humans), or both (video chat for both dogs and humans). During remote dog-human communication, the most studied devices were wearable devices for dogs, such as harnesses and collars, and mobile apps for people. Finally, the authors found that most investigations evaluating these digital technologies were pilot studies, and concluded that although research on remote communication between dogs and humans has great potential, it is in very early stages.

To complete our collection of papers, Peter Verbeek and colleagues conducted a scoping review of literature published from 2012 to 2022 on the dog-human bond. Their focus was to assess to what degree and in which ways research on the dog-human bond considered welfare of the dogs. They found that number of publications on the dog-human bond increased over time as did number with a primary focus on dog welfare. Overall, about one-third of papers had a primary focus on dog welfare, and such papers typically concerned pet dogs in western, educated, industrialized, rich, democratic societies. Publications with a focus on dog welfare most commonly addressed behavioral issues, particularly problematic behavior, as compared to other aspects of dog welfare, such as nutrition, environment, and health. The authors concluded by suggesting more research in this area is needed on therapy dogs, service dogs, working dogs, and free-living dogs, and in geographic regions other than Europe and North America. Finally, they suggested that Tinbergen's four questions–proximate causation, development, function, and ultimate causation–could be a useful guide in developing future research questions on dog welfare and the dog-human bond.

Betty McGuire
Editor

animals

MDPI

Article

The Influence of Brief Outing and Temporary Fostering Programs on Shelter Dog Welfare

Lisa M. Gunter [1,*], Emily M. Blade [1], Rachel J. Gilchrist [1], Betsy J. Nixon [1], Jenifer L. Reed [2], JoAnna M. Platzer [2], Ingrid C. Wurpts [3], Erica N. Feuerbacher [2] and Clive D. L. Wynne [1]

1 Department of Psychology, Arizona State University, Tempe, AZ 85281, USA; emily@shelteranimalscount.org (E.M.B.); rgilchri@asu.edu (R.J.G.); cwynne1@asu.edu (C.D.L.W.)
2 School of Animal Sciences, Virginia Polytechnic Institute and State University (Virginia Tech), Blacksburg, VA 24061, USA; jtay6366@gmail.com (J.L.R.); jplatzer@vt.edu (J.M.P.); enf007@vt.edu (E.N.F.)
3 State Farm Insurance, Phoenix, AZ 85016, USA; ingrid.wurpts@gmail.com
* Correspondence: lisagunter@vt.edu

Simple Summary: Animal shelters can be stressful for dogs, but human interaction can improve their experience. While at the shelter, dogs' stress can be reduced by spending time with a person outside of their kennel as can leaving the shelter for an overnight or longer stay in a foster home. In this study, we analyzed data of 1955 dogs from 51 animal shelters that went on an outing of a few hours or fostering stay of 1–2 nights, and 25,946 dogs that resided at these shelters but did not experience these interventions (controls). We found that outings and temporary fostering stays increased dogs' likelihood of adoption by five and over 14 times, respectively. While dogs that experienced these interventions spent longer in the shelter awaiting adoption as compared to non-intervention dogs, this difference in length of stay was present prior to the dogs' outings and fostering stays. We found that shelters' intervention programs were more successful when members of the community were more involved in providing these experiences (in contrast to volunteers and staff) as well as when these organizations had more resources. Animal shelters should consider implementing brief outing and temporary fostering programs to improve the welfare of shelter-living dogs.

Abstract: Human interaction is one of the most consistently effective interventions that can improve the welfare of shelter-living dogs. Time out of the kennel with a person has been shown to reduce physiological measures of stress as can leaving the shelter for a night or more in a foster home. In this study, we assessed the effects of brief outings and temporary fostering stays on dogs' length of stay and outcomes. In total, we analyzed data of 1955 dogs from 51 animal shelters that received these interventions as well as 25,946 dogs residing at these shelters that served as our controls. We found that brief outings and temporary fostering stays increased dogs' likelihood of adoption by 5.0 and 14.3 times, respectively. While their lengths of stay were longer in comparison to control dogs, this difference was present prior to the intervention. Additionally, we found that these programs were more successful when greater percentages of community members (as compared to volunteers and staff) were involved in caregiving as well as when programs were implemented by better-resourced shelters. As such, animal welfare organizations should consider implementing these fostering programs as evidence-based best practices that can positively impact the outcomes of shelter dogs.

Keywords: dogs; animal shelter; human–animal interaction; welfare; adoption

Citation: Gunter, L.M.; Blade, E.M.; Gilchrist, R.J.; Nixon, B.J.; Reed, J.L.; Platzer, J.M.; Wurpts, I.C.; Feuerbacher, E.N.; Wynne, C.D.L. The Influence of Brief Outing and Temporary Fostering Programs on Shelter Dog Welfare. *Animals* **2023**, *13*, 3528. https://doi.org/10.3390/ani13223528

Academic Editor: Betty McGuire

Received: 3 October 2023
Revised: 4 November 2023
Accepted: 6 November 2023
Published: 15 November 2023

1. Introduction

Millions of dogs enter animal sheltering facilities across the United States each year [1]. While a dog's temporary stay in the shelter is likely stressful when compared to life in a home [2–4], the outcomes of dogs that experience this fate have improved considerably over the past two decades [5,6]. Overall, dogs are more often reclaimed by their owners, adopted, or transferred to other shelters for placement, while canine euthanasia is occurring less often [7–9].

This improvement in live outcomes provides an opportunity to further explore the welfare of dogs as they reside in animal shelters. Mellor and Reid [10] describe animal welfare as a state of being, both mental and physical, that an animal experiences. In total, these experiences are subjective within the animal, and an animal's welfare or well-being is the integration of these experiences. Within the literature, scientists have focused their evaluations of welfare on the animal's biological functioning, its affective state, and how closely its living situation matches its natural state of being [11].

Because a dog's residency in the animal shelter is temporary, we can measure its welfare in two ways: proximally and distally. The proximal evaluation of a dog's welfare is concerned with what the dog is currently experiencing. This approach is closest in perspective to how applied scientists measure the welfare of other captive animals in order to provide them with optimal care [12]. Here, measurements of health, physiology, behavior, and cognition inform welfare assessment.

Animal-based measurements, such as a dog's body condition score, skin condition, and overall cleanliness can inform welfare assessment in the shelter [13,14]. Cortisol, a hormone involved in animals' stress response system, is elevated when dogs are living in the shelter as compared to a home [3,4,15–17]. In the animal shelter, dogs rest and sleep less compared to when they are temporarily staying or living in a home [3,16,18]. Furthermore, when shelter dogs sleep more during the daytime in the shelter, they demonstrate a positive bias during a spatial cognitive bias task, an indication of better welfare [19].

Human interaction is one of the most well-studied interventions in animal sheltering. Specifically, dogs spending time with a person outside of the kennel has been consistently shown to improve dogs' proximal welfare by reducing measures of stress and improving their behavior (for an in-depth review, see Gunter and Feuerbacher [20]). Gunter et al. [16] found that stays of one or two nights in a home reduce dogs' cortisol levels and increase their longest bouts of rest. Ferhinger [15] found a similar effect on cortisol when dogs were provided with three days of fostering. Conversely, brief outings with a person into the community have been shown to increase dogs' cortisol levels, even after accounting for their activity [19]. Regardless of their direction of impact, the effects of out-of-shelter human-interaction interventions are short-lived. Upon return to the shelter, dogs typically return to baseline cortisol and activity levels [16,21].

Measuring dogs' distal welfare involves the ultimate goal of animal sheltering: dogs permanently leaving the shelter and living in a human home. From this perspective, dogs' lengths of stay and outcomes are evaluated to assess welfare. Many characteristics of the dog can influence how long it stays in the shelter and its adoption likelihood, but these qualities are often immutable, such as the dogs' morphology or how it arrived to the shelter [22]. Only a handful of empirically evaluated adoption interventions have been shown to reduce dogs' time in the shelter or increase the possibility of a positive outcome. These include the facilitation of a dog's adoption by a foster caregiver [23], altering the dog's behavior with potential adopters [24], and removing labels used to describe a dog's visually-identified breed [25,26].

Thus, while the proximal effects of brief outings and temporary fostering on shelter dog welfare have been explored, what is less understood is whether dogs' length of stay in the care of the shelter or their likelihood of adoption are altered by these interventions. Generally, canine foster caregiving has been shown to improve dogs' distal welfare [27,28]. Thus, in the present study, we hypothesized that both brief outings and temporary fostering would result in reduced lengths of stay and better outcomes for shelter dogs, as compared to dogs living in the shelter during the same time period that did not experience these interventions.

2. Materials and Methods

2.1. Shelter Recruitment and Enrollment

Animal welfare organizations operating in the United States were contacted for enrollment in this study via websites, social media announcements, and targeted email invitations. After confirming that an organization's dogs lived in a facility for most of their sheltering stay (as opposed to a foster caregiver's home), we conducted interviews with these organizations. Prospective organizations needed to collect data about their program in order to participate in the study.

Additionally, information about the shelters' existing brief outing and temporary fostering programs was used as criteria for enrollment. For shelters participating in the brief outing component of the study, these organizations needed to be (1) without a brief outing program or (2) their program was experienced by an average of less than 10% of their canine population. For shelters enrolling in the study's temporary fostering component, these organizations needed to already have a brief outing program in place that provided at least 10% or more of their canine population with this activity in order to participate. Additionally, they either had (1) no temporary fostering program in place or (2) an existing program that, on average, served less than 10% of their canine population. Shelters with existing brief outing and temporary fostering programs, in which at least 10% or more of their dog populations participated, were not eligible to enroll in the study.

During the interview process, demographic information about the shelters was collected, including their admission type (i.e., open, managed, or limited) and organization type (i.e., municipal, private nonprofit, or private nonprofit with municipal contracts). Open admission was defined as shelters with unrestricted animal intake from the areas they served, while those with managed admission controlled the arrival of incoming animals. Limited-admission shelters restricted the animals accepted into their care [29]. Information was also gathered about the types of fostering opportunities offered by each organization and their adoption policies.

Shelters also provided animal intake and budgetary data for the year prior to the study. Using these data, we calculated each shelter's live release rate (dividing the sum of all live outcomes for dogs by total dog outcomes [30]) as well as the shelter's resource level (as previously described in Gunter et al. [28]).

2.2. Programmatic Training and Support

After study enrollment, staff at participating animal shelters attended a training program, provided by Maddie's Fund (Pleasanton, CA, USA), which discussed procedures for implementing and operating brief outing and temporary fostering programs. Once shelter staff completed this training, members of the research team met every other week with staff in video calls and engaged in email correspondence, assisting the shelter in the development of either a brief outing or temporary fostering program.

This initial support culminated in the launch of the shelter's program, which occurred no later than 45 days after attending the training program. During the program's launch, a member of the research team provided on-site assistance to the shelter for 3 days, and afterward, the team continued remote bi-weekly contact and email correspondence with each organization until data collection was complete. Following the shelter's final brief outing or temporary fostering stay, data collection continued for an additional 7 days to record the dogs' outcomes.

2.3. Dogs

Shelter staff determined which dogs would participate in their shelter's intervention; however, outings and stays needed to be with dogs six months of age or older and residing at the shelter at the time of participation. Shelters provided data about each dog that experienced the intervention, including their age, weight, sex, intake date and type, outcome date and type, and adoption status (i.e., available for adoption or not available due to medical, behavioral, or other reasons) on the date of their intervention experience.

Using dogs' intake and outcome data, as well as the duration of their outing or stay, we were able to calculate their lengths of stay prior to and after the intervention.

2.4. Brief Outings and Temporary Fostering Stays

Brief outings were conducted off the property of the animal shelter with at least one person (e.g., staff, volunteer, or community member). Similar to the brief outings described by Gunter et al. [21], outings lasted approximately 1–4 h. Off-site excursions to shelter-facilitated adoption events were not included. Temporary fostering stays were defined as dogs spending 1–2 nights in the home of a shelter staff person, volunteer, or member of the community, which has previously been described by Gunter et al. [16]. Although shelter dogs were sometimes placed into foster homes that included one or more resident dogs, shelter dogs were not placed together, except in instances where dogs were considered a bonded pair by shelter staff.

The duration of the intervention experiences and whether a dog bite occurred, either to a person or another dog, were also recorded.

2.5. Foster Caregivers

Shelter staff also collected data about the foster caregiver and their brief outing or temporary fostering experience, including the caregiver's age, their relationship to the organization, the date and time that their outing or stay began and ended, and if they adopted their dog. From these data, we were able to calculate the total number of caregivers that participated at each shelter as well as the total number of foster experiences provided.

Additional information about the caregivers providing temporary foster care was gathered, including whether the person was previously involved with the shelter's brief outing or temporary foster programs and the number of dogs residing in their home. If the caregiver had resident dogs, the shelter recorded the method of introduction between the resident and shelter dog (e.g., conducted at the shelter, at the caregiver's home, or the dogs were separated during the fostering stay).

2.6. Intervention Impact and Program Performance

In order to evaluate the impacts of these interventions on shelter dog outcomes and length of stay, inventory reports were gathered about the dogs that resided in their shelters but did not participate in the intervention during the data collection period. These non-intervention dogs served as our study's controls. The reports included dogs' intake date and type, age, weight, sex, and outcome date and type and were obtained from shelter data management programs and other sources (e.g., cloud-based spreadsheets and paper records). Dogs in both the intervention and control conditions were either residing in the shelter prior to the launch of their intervention program or arrived during the data collection period. Due to data tracking and reporting limitations, comparison data on non-participating dogs were not available from Mendocino County Animal Control.

In an effort to evaluate the performance of these interventions amongst our participating shelters while accounting for differences in their canine intake, we ranked shelters on (1) the number of foster experiences provided during the data collection period, (2) the number of foster caregivers providing those experiences, and (3) the number of days needed for the shelter to enroll 40 dogs in the study, regardless of intervention type. This 40-dog-per-shelter sample size benchmark was used in order to reach an adequate number of participants based on previous research [16,21].

As such, programs that provided the most intervention experiences to dogs and had the most caregivers participating in their programs were ranked highest on those variables, while a shelter that needed the fewest days to collect their data was ranked higher than a shelter that took longer to reach the study's sample size. Using each shelter's rankings on these three variables, a summed ranking value was calculated. This overall rank was used to assess program performance in relation to characteristics of the shelter and its foster caregivers.

2.7. Data Analytic Approach

In the estimation of descriptive statistics during data analysis, we identified that many continuous variables (related to the shelters, dogs, interventions, caregivers, and program performance) were non-normally distributed. As such, means, measures of data variability (i.e., standard deviation, standard error, and confidence intervals), medians, and ranges are reported.

To estimate the difference in adoption rates between dogs who experienced brief outings versus temporary fostering stays, we used a chi-square goodness of fit test.

To understand the impacts of these interventions on adoption versus other outcomes (i.e., remained in care, transfer out, or euthanasia), we used two multinomial logistic regression models comparing dogs that received an intervention (outing or fostering stay) with control dogs (those that did not receive an intervention). Adoption served as the reference category for the dependent variable. We attempted to include shelter as a random effect in these models, but they failed to converge, so only models with fixed effects were employed.

In these regression analyses, relative risk ratios (RR) are reported for brief outings (BO/RR) and temporary fostering (TF/RR), which indicates the probability of an outcome for the intervention dogs compared to the probability of the same outcome for dogs that did not receive a brief outing or temporary fostering stay. As such, an RR value greater than one indicates how many times more likely that particular outcome is to occur for intervention dogs than dogs in the comparison group. With ratios of probabilities less than one, those RR values can be used as the divisor with one as the dividend to yield the outcome that is X times more likely to occur. Confidence intervals for relative risk ratios are reported alongside these values.

We used two linear regression models to estimate the effect of the intervention (brief outing or temporary fostering stay) on length of stay. Our dependent variable, length of stay in days, was log-transformed to account for its positive skew. In these analyses, dogs that were returned to their owners were excluded.

In both types of regression models, we entered dog-level covariates including their sex (i.e., male or female), age (in months), weight (in kilograms), number of times the intervention was experienced, and intake type (i.e., stray, cruelty/neglect, transfer in, owner surrender/return). Among dogs that were temporarily fostered, we estimated the additional effect of an intervention-level covariate: number of resident dogs in the caregiver's home.

To explore factors related to the performance of intervention programs among our study shelters, we used an ordinary least squares regression model with intervention type (brief outing or temporary fostering) and shelter-related characteristics (i.e., organization and admission types, resource level, and proportions of volunteer and community caregivers). Our dependent variable, program performance, was a summed ranking value based on foster experiences, caregivers, and days of data collection.

All models were evaluated for data sparsity by cross-tabulating categorical independent variables with the categorical dependent variable. In instances where cell counts were at or near zero, groups (e.g., dogs remaining in the organization's care at the shelter or in foster care) were combined when appropriate. For models utilizing continuous covariates (i.e., dog age and weight), we screened for outliers and capped or floored these variables. When dogs had more than one outing or fostering stay during data collection, associated logistic and OLS regression models were estimated using the dog's earliest intervention experience in order to avoid an individual dog contributing multiple cases to our model estimations. All analyses were conducted in R.

3. Results

3.1. Descriptive Statistics

3.1.1. Shelters

Between February 2019 and March 2020, we collected data with 51 US animal shelters about their brief outing and temporary fostering programs with each shelter. In total, 60 foster programs are represented in this sample (nine animal shelters participated in both interventions investigated in this study).

Live release rates (LRRs) for dogs varied across our shelters but were relatively high with an average rate of 91.9% (*SD* = 9.4%), ranging from 63 to 100% with a median of 95.6%. Shelters' annual operating budgets for the year prior to the study varied considerably from USD 200,000 to over USD 23 million (*M* = USD 3,983,677, *Mdn* = USD 2,155,613, *SD* = USD 4,394,998).

The number of animals that shelters brought into their facilities each year differed by several fold. While the average number was 6569 animals (*Mdn* = 4879, *SD* = 7051), the intake of shelters ranged from 241 to 32,788 animals. On average, the proportion of dogs in these shelters accounted for roughly half of all animals (*M* = 50.4%, *SD* = 16.5) with a range of 27 to 100%, and a median of 47.8%.

Table 1 describes the average, median, and range of operating budgets and annual intakes by resource level (as previously described in Gunter et al. [28]), including the count of shelters in our sample that were included at each resource level.

Table 1. Shelter resource levels and associated annual budgets, animal intake numbers, and resources per animal.

Resource Level	*M*, *Mdn* Annual Budget	Annual Budget Range (Min–Max)	*M*, *Mdn* Animal Intake	Animal Intake Range (Min–Max)	*M*, *Mdn* Resources per Animal	Resources per Animal Range (Min–Max)	Shelters in This Dataset
Very Low	1.9 M, 965 K	200 K–4.3 M	10,867, 7215	826–29,595	209, 242	118–276	11
Low	4.2 M, 3 M	1.7 M–13 M	10,016, 8118	3873–32,788	406, 403	370–439	9
Moderately Low	2.2 M, 1.8 M	730 K–4.3 M	4077, 3241	1256–8916	562, 572	473–648	10
Moderate	2.3 M, 1.2 M	300 K–6.9 M	3021, 1494	408–8834	759, 749	699–845	11
High	8.4 M, 6.6 M	1.5 M–23 M	8244, 6377	1498–23,093	1028, 1022	918–1105	6
Very High	7.4 M, 9.9 M	386 K–14 M	4299, 3478	241–8117	1849, 1647	1480–2852	11

Note. Abbreviations: Millions (M), Thousands (K). Average Annual Budget and Average Resources Per Animal and their associated ranges are in USD. Resources Per Animal is an estimated value calculated by dividing a shelter's annual budget by the previous year's animal intake. Resource Level calculations are not included for two organizations that were unable to provide yearly animal intake numbers.

Organizationally, most shelters were either private nonprofits (45%), private nonprofits with municipal contracts (31.7%), or public municipal agencies (23.3%). More often, shelters were open intake (46.7%), but managed-intake facilities were also common (36.7%). A smaller proportion of shelters in our study were limited admission (16.7%).

When describing the performances of shelters' intervention programs as a summed ranking value of three performance metrics (i.e., number of intervention experiences, caregivers participating, and days of data collection), the average performance value for these programs was 62.2 (*Mdn* = 59.0, *SD* = 23.6, Range: 12–116).

Overall, 2327 dogs had a brief outing (1728) or temporary fostering stay (599) as part of this study. Because dogs could have more than one outing or stay, 3481 intervention experiences occurred: 2786 brief outings and 695 temporary fostering stays. Overall, 2367 caregivers participated, either as brief outing (1842) or temporary fostering (525) caregivers. These data represent all recorded experiences. A small portion of these experiences failed to meet study criteria (e.g., the outing duration was less than 1 hour or the fostering stay exceeded three days) and were removed from subsequent descriptive and statistical analyses.

3.1.2. Dogs

The data of 27,901 dogs were used in our analyses: 1955 that participated in the brief outing and temporary fostering interventions, and 25,946 dogs that resided in the study shelters at the same time as the intervention dogs and served as controls. Over half of dogs in this dataset entered the shelter as strays (57.7%). Almost one-quarter of dogs were surrendered by their owner (16.9%) or were a failed adoption (6.4%), 14.1% were transferred from another facility, and 4.9% were brought to the shelter as part of a cruelty or neglect case. Males and females were relatively equally represented (males: 52.2%). Dogs were, on average, just over three years of age at the time of entering the study (M = 39.7 months, SD = 35.7, Mdn = 24, Range: 5.6–267.9) and weighed a mean of 18.4 kg (SD = 10.3, Mdn = 18.6, Range: 0.45–77.0).

A majority of dogs (87.8%) that received an outing or foster stay were available for adoption at the time of the study; however, 12.2% were not available, due to behavioral (4.2%), medical (4.0%), or other (4.0%) reasons (e.g., stray hold or awaiting transfer). The average length of stay for intervention dogs, excluding time out of the shelter during their brief outing or temporary fostering stay, was 35.1 days (SD = 42.3) with a median of 21.0 days. Dogs' average length of stay pre-intervention was 32.7 days (Mdn = 14, SD = 52.4, Range: 0–623), and 9.9 days (Mdn = 5, SD = 14.6, Range: 0–157) after study participation. Dogs in our control condition had an average length of stay of 9.5 days (Mdn = 5, SD = 14.0, Range: 0–267).

At the end of the study, we found that outcomes for dogs in the interventions were mostly positive, although nearly a quarter (23.6%) remained in the care of their organization at the end of data collection (i.e., seven days after the final dog participated in the shelter's intervention). Of those dogs still in the organization's care, nearly all (98.7%) were residing in the shelter (as compared to a foster home). Almost two-thirds (65.2%) of dogs had been adopted into a home, 8.2% were transferred to another animal welfare organization, and less than one percent (0.9%) were returned to their owner. Less than two percent of dogs that participated in our interventions were euthanized for behavioral (1.1%), medical (0.3%), or capacity reasons (0.5%). Dogs that were returned to their owners were not included in the statistical analyses.

3.1.3. Intervention

The average duration of an outing was 3.0 h (SD = 1.3, Mdn = 2.6, Range: 1–10) and 1.6 days (SD = 0.6, Mdn = 1.9, Range: 0.5–3 days) for a temporary fostering stay. Over three-quarters of dogs (77.1%) had only one outing or stay during the study, but 22.9% of dogs had two or more experiences. Overall, a total of 2437 brief outings and 496 temporary fostering stays were eligible for inclusion in these analyses.

During the 2934 intervention experiences that occurred as part of this study, a total of six bites were reported, representing <1% of all experiences. Most often, these bites were inflicted upon a person (five) while one incident was with another dog during a brief outing.

3.1.4. Foster Caregivers

In total, 1842 brief outing and 408 temporary fostering caregivers were included in the statistical analyses. Caregivers were, on average, 39.0 years old (SD = 15.0) with a median age of 35.7 years. We found that members of the community, with no prior relationship to the shelter, were most often providing brief outings (47.5%) while shelter volunteers provided another 42.7% of outings. Conversely, volunteers more often temporarily fostered (45.4%) while community members provided 37.1% of temporary foster experiences. Additionally, shelter staff provided 7.7% of the study's brief outings, and 11.3% of the temporary fostering stays. A small portion of outings (2.1%) and foster experiences (6.4%) were provided by caregivers who were not categorized by our study shelters.

Just over half (50.9%) of temporary foster caregivers had no resident dog in their home; with community members, this occurred much more often (70.1%; Table 2). When caregivers had a dog(s) living in their home, they most often introduced the dogs at the shelter prior to fostering (37.9%), followed closely by caregivers electing to keep the dogs separated during fostering (35.1%). Over a quarter of caregivers (27.0%) carried out the introduction between the dogs at their home.

Table 2. Temporary fostering caregivers and number of resident dogs living in their homes.

		Percent of Caregivers (%)				
		Number of Resident Dogs in Home				
Type of Caregiver	***n***	**0**	**1**	**2**	**3**	**Not Reported**
Community Member	174	70.1	21.3	5.8	1.7	0.6
Volunteer	173	53.8	15.6	11.0	2.9	19.7
Staff	42	28.6	26.2	23.8	19.1	2.4
Not Reported	19	0	0	0	0	100
Total	408	50.9	19.1	12.5	3.8	13.7

For most intervention experiences analyzed in this study, caregivers did not adopt their dogs. Only 4.2% of outings and 12.0% of fostering stays resulted in an adoption by the caregiver providing the experience; however, this difference in caregiver adoptions between brief outings and temporary fostering stays was statistically significant: $\chi^2 = 46.9$, $p < 0.001$.

3.2. *Intervention and Non-Intervention Dogs*

3.2.1. Intervention Impact on Shelter Outcomes

To better understand how brief outings and temporary fostering influenced dogs' outcomes, we employed a series of multinomial logistic regression models. These models included the intervention impact (brief outing or temporary fostering versus controls) as well as the covariates of dog age, weight, sex, and intake type. To reduce outlier influence, weight was restricted from 1.36–45.36 kg, with values outside this range set to the described minimum or maximum value, and age capped at 150 months. These restrictions and cap for weight and age, respectively, were utilized in subsequent models in which these variables were included as covariates.

Our categorical dependent variable in these multinomial logistic regression models was adoption versus transfer to another agency, remaining at the shelter, euthanasia, or becoming lost or unexpectedly dying in the shelter. Because no dogs that received an intervention were lost or died in the shelter, this outcome was grouped with euthanasia.

As represented in Figure 1, we found that dogs that experienced either intervention were less likely to be euthanized, become lost or die at the shelter (BO/RR = 0.20, 95% CI [0.14, 0.28]; TF/RR = 0.07, 95% CI [0.02, 0.21]), or be transferred to another animal welfare organization (BO/RR = 0.53, 95% CI [0.43, 0.64]; TF/RR = 0.24, 95% CI [0.15, 0.40]) than be adopted when compared to non-intervention dogs, controlling for other factors. As such, when dogs experienced a brief outing or a temporary fostering stay, they were 5.0 and 14.29 times more likely, respectively, to be adopted than euthanized as compared to dogs that did not receive these interventions. Intervention dogs were also more likely to remain in the care of their shelter at the end of the study than be adopted when compared to non-intervention dogs (BO/RR = 1.97, 95% CI [1.70, 2.26]; TF/RR = 2.30, 95% CI [1.72, 3.08]). Full model results are reported in Table S1 (Supplementary Materials).

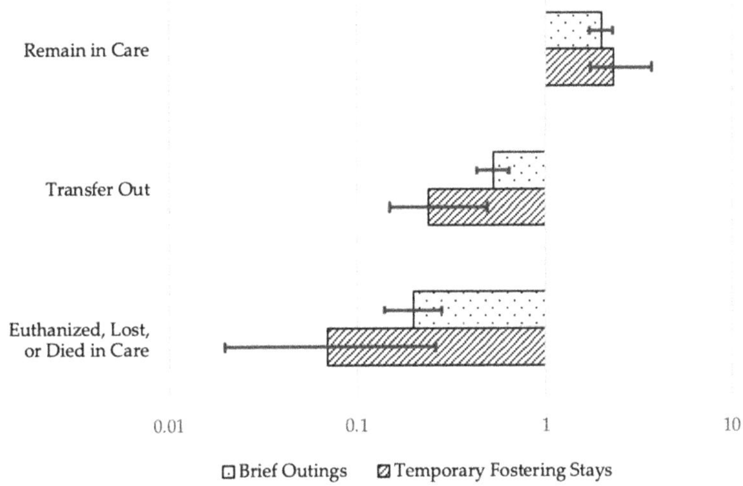

Figure 1. Relative risk (RR) ratios of adoption versus other outcomes for intervention dogs as compared to controls, adjusted for dog age, weight, sex, and intake type.

3.2.2. Intervention Impact on Length of Stay

To assess the impact of brief outings and temporary fostering on dogs' time living in the shelter, we carried out two multilevel model analyses. These models included fixed effects for intervention and characteristics about the dog (i.e., sex, weight (restricted), age (capped), and intake type) as well as a random intercept for the shelter to estimate the intervention's effect on length of stay (log-transformed) amongst intervention and non-intervention dogs that were adopted.

In the intervention models described in Table 3, we found that dogs that were heavier ($p < 0.001$) or older ($p = 0.003$) had longer lengths of stay. For dogs that experienced either intervention, they had, on average, longer lengths of stay than non-intervention dogs ($p < 0.001$). Furthermore, intervention dogs that arrived as a stray, part of a cruelty or neglect case, or were transferred from another organization had longer lengths of stay than non-intervention dogs that were owner-surrendered or returned after a failed adoption ($p < 0.001$). In our temporary fostering model, we found that female dogs had shorter lengths of stay than males, even when accounting for the other covariates ($p = 0.002$).

Table 3. Fixed effects of brief outing and temporary fostering interventions and model covariates on dogs' length of stay.

	Brief Outing					Temporary Fostering				
Fixed Effect	**Est**	*SE*	*df*	*t*	*p*	**Est**	*SE*	*df*	*t*	*p*
(Intercept)	1.46	0.08	53	18.07	<0.001	1.60	0.05	139	29.72	<0.001
Intervention vs. Population	0.88	0.03	6235	26.81	<0.001	0.89	0.05	4123	16.30	<0.001
Female vs. Male	−0.01	0.02	6291	−0.28	0.781	−0.11	0.03	4154	−3.71	0.002
Dog Weight	0.01	0.00	6301	17.34	<0.001	0.01	0.00	4156	9.83	<0.001
Dog Age	0.00	0.00	6298	6.87	<0.001	0.00	0.00	4154	2.98	0.003
Stray vs. Owner Surrender	0.48	0.03	6316	16.72	<0.001	0.53	0.04	4098	14.99	<0.001
Cruelty/Neglect vs. Owner Surrender	0.56	0.06	6308	9.72	<0.001	0.94	0.11	4128	8.38	<0.001
Transfer In vs. Owner Surrender	0.13	0.03	6309	4.18	<0.001	0.18	0.04	4118	4.57	<0.001

3.3. *Intervention Dogs*

3.3.1. Duration of Brief Outings and Temporary Fostering Stays

To identify factors associated with outing duration, we utilized a multilevel model to examine the effects of dog characteristics (i.e., sex, weight (restricted), age (capped), intake type), caregiver characteristics (i.e., age, type), a dog bite to a human or other dog during the outing as well as a random intercept for shelter on the duration of outings. We found significant effects of dog weight and caregiver type. Dogs of greater weight had significantly shorter outings than dogs of lesser weight, $t(2203) = -2.0$, $p = 0.042$, while caregivers who were volunteers at the shelter were more likely to take dogs on longer outings than caregivers from the community, $t(1625) = 2.7$, $p = 0.006$, accounting for all other covariates.

To identify factors associated with the duration of temporary foster care, we utilized a multilevel model to examine the effects of dog characteristics (i.e., sex, weight (restricted), age (capped), intake type), caregiver characteristics (i.e., type, age, previous fostering experience), a bite to a person during the experience, and number of resident dogs in the caregiver's home as well as a random intercept for shelter on fostering duration. Only one variable, the occurrence of a human bite, predicted significantly shorter temporary fostering stays, $t(397) = -3.06$, $p = 0.002$, accounting for all other covariates. Full model results are reported in Table S2 (Supplementary Materials).

3.3.2. Shelter Outcomes by Intake Type

Using only the data from our intervention dogs, we conducted two multinomial logistic regression analyses with covariates to assess the effect of intake type on dog outcomes. Adoption was our reference category for the outcome variable, and dogs that were surrendered by their owners or returned by their adopters was our reference category for the intake type predictor variable. Other possible outcomes for intervention dogs include transfer to another organization, remaining in care (at the shelter or in foster care), euthanasia, or becoming lost or dying in the shelter. The covariates of dog age (capped), weight (restricted), and sex were used. In our temporary fostering analysis, the number of resident dogs in the caregiver's home was included. Dog counts by intervention, intake type, and shelter outcome are provided in Table 4.

Table 4. Counts of dogs by intervention, intake type, and outcome.

	Outcome Counts of Brief Outing Dogs				Outcome Counts of Temporary Fostering Dogs			
Intake Type	**Adoption**	**Remain in Care**	**Transfer Out**	**Euthanized, Lost, or Died in Care**	**Adoption**	**Remain in Care**	**Transfer Out**	**Euthanized, Lost, or Died in Care**
Owner Surrender	341	147	40	6	122	27	1	2
Stray	319	143	93	25	83	21	13	0
Transfer In	275	97	7	3	134	19	3	1
Cruelty/Neglect	11	7	5	2	1	3	0	0

Among dogs that had a brief outing, we found that stray dogs were more likely to be transferred out to another agency (BO/RR = 2.44), or euthanized (BO/RR = 3.86), than be adopted as compared to owner-surrendered and returned dogs, controlling for other factors (Table 5). Dogs that were transferred into the shelter and had a brief outing were less likely to be transferred out again (BO/RR = 0.22) than be adopted as compared to dogs that were owner-surrendered or returned, controlling for other factors.

Table 5. Dog outcome relative risk (RR) ratios with 95% lower and upper confident limits (LCLs and UCLs) for dogs that experienced a brief outing.

	Shelter Outcomes for Brief Outing Dogs								
	Remain in Care vs. Adopted			Transfer Out vs. Adopted			Euthanized, Lost, or Died in Care vs. Adopted		
Covariates	**RR**	**LCL**	**UCL**	**RR**	**LCL**	**UCL**	**RR**	**LCL**	**UCL**
(Intercept)	0.19	0.12	0.29	0.18	0.09	0.35	0.01	0.001	0.03
Female vs. Male Dogs	1.06	0.83	1.34	0.86	0.59	1.26	0.56	0.26	1.22
Dog Weight (kg)	1.02	1.01	1.04	0.97	0.95	0.99	1.07	1.02	1.12
Dog Age (months)	1.01	1.00	1.01	1.01	1.00	1.01	0.99	0.97	1.01
Stray vs. Owner Surrender	1.16	0.87	1.54	2.44	1.59	3.74	3.86	1.52	9.84
Cruelty/Neglect vs. Owner Surrender	1.60	0.60	4.23	4.49	1.46	13.81	11.98	2.12	67.78
Transfer In vs. Owner Surrender	0.98	0.71	1.34	0.22	0.09	0.50	0.77	0.19	3.16

Note. Adoption is the outcome reference category in this analysis, and dogs that were surrendered by their owners or returned by their adopter is the comparison group for intake type. An RR value > 1 indicates that the comparison outcome is that many times more likely to occur instead of adoption as the predictor value increases (or for the comparison intake type than for owner-surrendered or returned dogs). With an RR value < 1, divide 1 by the RR value to calculate how many times more likely adoption is to occur as the predictor value increases (or for the comparison intake type than for owner-surrendered or returned dogs).

For cruelty and neglect dogs, those that experienced a brief outing during their shelter stay were more likely to be transferred to another facility (BO/RR = 4.49) and more likely to be euthanized (BO/RR = 11.98) than adopted as compared to dogs that were surrendered by their owner or returned, controlling for other factors.

The relative risk ratios and confidence intervals estimated with the multinomial logistic regression model for temporarily fostered dogs (as described in Table 6) should be interpreted with caution. As shown in Table 4, some predictor and dependent variable categories were rare. Although the model met convergence criteria, the presence of extremely small cell counts limits our abilities to properly estimate parameter variability. However, further grouping of shelter outcome categories to address low counts (e.g., grouping dogs that were transferred out with those that were lost, died in care, or euthanized) would have created unmeaningful groups. As such, these categories remain as described, despite their low prevalence. Nevertheless, these multinomial logistic regression results provide directional understanding of the relationships between intake type and outcome.

Table 6. Dog outcome relative risk (RR) ratios with 95% lower and upper confident limits (LCLs and UCLs) for dogs who experienced a temporary fostering stay.

	Shelter Outcomes for Temporarily Fostered Dogs								
	Remain in Care vs. Adopted			Transfer Out vs. Adopted			Euthanized, Lost, or Died in Care vs. Adopted		
Covariates	**RR**	**LCL**	**UCL**	**RR**	**LCL**	**UCL**	**RR**	**LCL**	**UCL**
(Intercept)	0.17	0.06	0.51	0.002	0	0.03	0.02	0.001	0.99
Number of Resident Dogs	0.82	0.56	1.21	1.94	1.18	3.19	<0.001	<0.001	<0.001
Female vs. Male Dogs	0.92	0.50	1.68	1.31	0.44	3.94	<0.001	<0.001	<0.001
Dog Weight (kg)	1.01	0.98	1.04	1.06	0.99	1.13	1.04	0.92	1.18
Dog Age (months)	1.00	0.10	1.01	0.99	0.97	1.01	0.10	0.97	1.03
Stray vs. Owner Surrender	1.10	0.51	2.34	22.20	2.72	180.90	<0.001	<0.001	<0.001
Cruelty/Neglect vs. Owner Surrender	6.35	0.34	117.10	<0.001	<0.001	<0.001	0.83	0.83	0.83
Transfer In vs. Owner Surrender	0.50	0.23	1.10	2.41	0.24	24.58	0.52	0.04	6.42

Note. Adoption is the outcome reference category in this analysis, and dogs that were surrendered by their owners or returned by their adopter is the comparison group for intake type. An RR value > 1 indicates that the comparison outcome is that many times more likely to occur instead of adoption as the predictor value increases (or for the comparison intake type than for owner-surrendered or returned dogs). With an RR value < 1, divide 1 by the RR value to calculate how many times more likely adoption is to occur as the predictor value increases (or for the comparison intake type than for owner-surrendered or returned dogs).

We found that stray dogs that were temporarily fostered were more likely to be transferred (TF/RR = 22.20) than adopted and less likely to be euthanized (TF/RR < 0.001) as compared to owner-surrendered and returned dogs, controlling for other factors. Dogs from cruelty and neglect cases that experienced a temporary fostering stay were less likely to be euthanized (TF/RR = 0.83) or transferred out (TF/RR < 0.001) than be adopted as compared to owner-surrendered/returned dogs, controlling for other factors.

For dogs that were temporarily fostered by a caregiver, we found that more resident dogs in their home corresponded to a higher likelihood of the fostered dog being transferred out of the shelter (TF/RR = 1.94) and a much lower likelihood of being euthanized (TF/RR < 0.001).

3.3.3. Post-Intervention Length of Stay

To better appreciate how a brief outing or temporary fostering stay may have influenced dogs' length of stay in the shelter after receiving the intervention, we employed multilevel modeling with fixed effects for dog sex, weight (restricted), age (capped), and intake type, and number of resident dogs in the caregiver's home (for the temporary fostering model only) as well as a random intercept for the shelter to estimate the effect that these interventions had on post-intervention length of stay (log-transformed) among dogs that had a shelter outcome in our intervention groups. Log-transformed length of stay was our dependent variable.

For these dogs, we found that a dog's weight was significantly related to longer lengths of stay post-intervention for dogs that experienced a brief outing, $t(833) = 3.19$, $p = 0.001$, or temporary fostering stay, $t(220) = 3.60$, $p = 0.001$. That is, as the weight of the dog increased, so did their time in the shelter post-intervention. For dogs that were temporarily fostered, their age also positively predicted longer lengths of stay after fostering, $t(215) = 3.7$, $p = 0.0003$. Full results of this model are reported in Table S3 (Supplementary Materials).

3.4. *Overall Intervention Performance*

We found that shelters with higher percentages of caregivers who were community members were more likely to have higher performing programs, $t(45) = 4.27$, $p < 0.001$, controlling for intervention type and other covariates. Additionally, public municipal agencies were more likely to have lower performing programs when compared to private, nonprofit organizations with municipal contracts, $t(45) = -2.08$, $p = 0.044$. Shelters with more resources were likely to have higher program performances, $t(45) = 2.27$, $p = 0.028$.

No other variables included in the model significantly predicted intervention performance. We also tested two interactions in our model, intervention type by proportion of volunteers who were foster caregivers as well as intervention type by proportion of community caregivers, but neither interaction was statistically significant, indicating that the effect of the proportion of volunteers or community members on program performance did not differ by intervention type. Table 7 describes the main effects and interactions that were tested in the ordinary least squares regression model of program performance.

Table 7. Effects of shelter characteristics on the performance of intervention programs.

Effect	Est	*SE*	*t*	*p*
(Intercept)	37.04	13.68	2.71	0.010
Intervention (Temporary Fostering vs. Brief Outing)	−50.05	34.55	−1.45	0.154
Percent Volunteers	22.44	15.85	1.42	0.164
Percent Community Members	62.70	14.68	4.27	<0.001
Shelter Resources	0.01	0.00	2.27	0.028
Municipal vs. Nonprofit w/Municipal Contracts	−12.85	6.20	−2.08	0.044
Nonprofit vs. Nonprofit w/Municipal Contracts	−8.38	6.51	−1.29	0.205
Managed vs. Open Admission	−1.17	6.19	−0.19	0.851
Limited vs. Open Admission	−10.73	8.85	−1.21	0.232
Intervention Type by Percent Volunteers	29.77	40.99	0.73	0.471
Intervention Type by Percent Community Members	5.30	36.58	0.15	0.885

4. Discussion

Our investigation found that interventions consisting of either a brief outing or temporary stay in a caregiver's home resulted in shelter dogs being adopted more often as compared to dogs in shelters that did not receive these interventions. Dogs that participated in these interventions were also less likely to be transferred to another facility for placement. Adoptions by caregivers were infrequent but occurred more often after an overnight stay than an outing.

Our findings add to a growing body of fostering literature, including work by Ferhinger [15] and Gunter et al. [16,21], that has investigated the proximal effects of human interaction provided outside of the animal shelter on the welfare of shelter-living dogs. Previous studies found that overnight stays of any duration (one, two, or three nights) reduced dogs' cortisol levels and increased rest whereas brief outings did not, and as such, these interventions do not have the same effects on dogs' immediate welfare and recommendations for their usage have differed [20].

The current study provides evidence about the distal benefits of both brief outings and temporary fostering stays, most importantly their influence on shelter dog adoptions. Simply put, dogs leave animal shelters alive more often when they have an outing of just a few hours or stay in a home with a person, five or over 14 times so, respectively. Moreover, these dogs typically had longer shelter stays prior to experiencing these interventions. Previously, Patronek and Crowe [27] found a positive effect of canine foster caregiving, increasing the likelihood of live outcomes by five to over 20 times, depending on a dog's intake type into the shelter, as compared to dogs that did not enter foster care.

Nevertheless, dogs in our study that were surrendered by their owners or returned by adopters, and then temporarily fostered, were more likely to be euthanized than temporarily fostered dogs that arrived as strays or were part of cruelty or neglect investigations. During temporary fostering stays, it is possible that caregivers' observations coincided with behavioral concerns about these dogs that were expressed by their previous owners and played a role in the dogs' negative outcomes. Prior work by Duffy et al. [31] and Stephen and Ledger [32] found that relinquishing owners' reports about their dogs' aggression toward strangers were significantly correlated to reports by the dogs' new adopters about the same behavior. Nevertheless, future research with a larger sample size of temporarily fostered dogs would allow for more accurate parameter estimation concerning these intake types and outcomes.

Conversely, owner-surrendered and returned dogs that left the shelter on brief outings were more likely to be adopted as compared to their stray and cruelty/neglect counterparts. Additionally, we found that when dogs were temporarily fostered in homes with multiple resident dogs, they were more likely to be adopted and twice as likely to be transferred out of the shelter for placement. It is possible that a dog's friendliness with other dogs may be related to these outcomes, a factor that has been previously reported as influential in dog adoptions and relinquishments to the shelter [33–36].

With regards to caregivers adopting their fostered dogs, it is possible that such decisions could be influenced by the duration of the caregiving experience. Here, we found that brief outings had the lowest percentage of adoptions by a caregiver, 4%, while 12% of temporary fostering stays resulted in adoption. During the pandemic, Gunter et al. [28] found that adult dogs were adopted by their caregivers in 18% of foster experiences, which typically involved much more time in the caregiver's home. However, Gunter et al. [28] also found that intention to adopt (i.e., trial adoption programs), the type of foster caregiver, and number of dogs in a caregiver's home influenced adoption likelihood as well. Thus, it seems that while shelters will achieve better distal welfare for dogs by utilizing brief outing and temporary fostering interventions, the adoption of these dogs will likely not be by the caregivers themselves.

Another aspect of dogs' welfare in the shelter is the time spent in the organization's care awaiting an outcome. Few experimental interventions in the shelter have been shown to reduce dogs' time living in the shelter, while a greater number of interventions have been identified that increase adoption likelihood [23–27]. In the present study, we found that dogs that participated in either intervention had longer lengths of stay and more often remained in the care of the shelter at the end of study compared to non-intervention dogs. More specifically, we found that dogs that received a brief outing or fostering stay had lengths of stay between 32 and 34 days, respectively, prior to the intervention, while dogs that did not experience either intervention resided in the shelter for just 10 days.

It is conceivable that intervention dogs' longer lengths of stay before their brief outings or fostering stays may have been related to qualities about the dogs themselves; specifically, characteristics that negatively impact adoption likelihood, such as a dog's morphology [37] or social behavior during meet-and-greets [38]. Animal shelters are often encouraged to use brief outing or temporary fostering programs for adoption promotion, particularly those dogs that have resided in the shelter for extended periods of time [39]. As such, intervention dogs' longer lengths of stay may not be related to the intervention's effect, but, instead, were a consideration when shelter staff selected dogs for outings and fostering stays.

After the intervention, shelters would have had more information about these dogs and might have felt better informed about their viability as adoption candidates. Supportively, we found that placement of intervention dogs was more likely to occur through adoption, rather than transferring them to another organization for placement, which may be indicative of the shelter's continued investment or a lack of perceived attractiveness by other organizations. Additionally, no intervention dogs were lost or died unexpectedly in the shelter during the study. After their brief outing or temporary fostering stay, dogs waited an average of just 10 days to be adopted, which is considerably shorter than their lengths of stay beforehand. Such a finding suggests that ultimately, these dogs' distal welfare was positively impacted by the interventions.

Furthermore, highly desirable dogs may have not resided long enough in the shelter to participate in our interventions, which could account for the difference in length of stay observed between the two groups. As such, future studies may consider matching dogs on multiple morphological and behavioral variables to further understand the effects of these interventions.

Age has also been shown to influence time to adoption from foster care during the pandemic [28] as well as likelihood of return after adoption [35]. Across our dataset, we found that heavier and older dogs stayed longer in the shelter awaiting an outcome, and the effect of weight on length of stay persisted post-intervention. The effects of age and weight on shelter dog outcomes have been previously reported, most recently by Cain et al. [7,40]. In our study, we found that a dog's weight also influenced the duration of their brief outing, such that larger dogs received shorter outings than smaller dogs.

One possible explanation for this effect of weight on outing duration may be related to the force exerted by larger dogs while on-leash. Shih et al. [41] found that dogs of greater size and weight exhibit more tension on-leash, and increased tension on the leash has been shown to negatively impact volunteers' satisfaction walking shelter dogs [42]. Thus, it is possible that caregivers on outings with larger dogs may have been less satisfied due in part to an inability to handle their dogs, resulting in earlier returns to the shelter. In future studies, examining the effects of dog walking equipment on outing duration and caregiver satisfaction may elucidate ways that these experiences can be improved, which could increase the distal benefits of this intervention, particularly for larger dogs that often reside in shelters longer.

Caregivers in the present study were slightly older than those that fostered dogs during the pandemic [28], but both studies found that caregivers are usually early middle-aged adults. Previously, foster caregivers have been reported to more often be pet owners [43]. However, we found a greater proportion of temporary foster caregivers in this study and those that were pandemic caregivers did not own a dog [28]. While caregiving opportunities in these studies may have been more appealing to non-dog-owners, it is also possible that these studies' larger datasets, which captured all caregivers participating in the interventions versus only those caregivers willing to complete a survey, may have led to this difference in findings. As we did not collect the pet-owning status of brief outing caregivers (because the shelter dogs were not residing in their homes), we are unable to describe this attribute further. Worth noting, however, is that Ackermann et al. [43] did identify that a key difference in the motivation of early middle-aged foster caregivers, as opposed to younger and older caregivers, was not wanting the responsibility of pet ownership, which may explain the lack of dog ownership by caregivers that we observed in this study.

In their exploration of fostering during the pandemic, Gunter et al. [28] found that members of the community with no prior relationship to the shelter played the largest role in foster caregiving. In this study, we also found that community involvement was influential. Specifically, brief outing and temporary fostering programs that had greater proportions of community caregivers providing experiences were higher performing as defined here by the number of intervention experiences the shelter provided, foster caregivers they engaged, and days needed to carry out the study. Across our shelters, members of the community were more often engaged in brief outing programs whereas volunteers had a greater presence in shelters' temporary fostering efforts.

The importance and impact of community engagement in animal sheltering, such as strategies used by animal control and field service officers, have been previously described by Moss et al. [44]. Our findings suggest that not only do interventions that engage individuals beyond the shelter's volunteers and staff lead to more successful programs, but these shorter-duration fostering interventions can significantly impact outcomes for dogs. As such, we believe that removing barriers to community participation in these programs can save the lives of more dogs awaiting adoption in United States animal shelters.

Prior work has identified that individuals are hesitant to foster shelter animals because of the emotional attachment and time involved in caregiving as well as limitations caused by their own pets and housing status [45]. Brief outings address these concerns as they are of a minimal duration and do not require housing the dog. As evidenced with our data, this particular intervention may be a powerful engagement tool, particularly as animal shelters struggle to recruit and retain foster caregivers [46].

Social exchange theory, as described by Schafer [47] in relation to volunteers, may be a better way to understand the motivations of foster caregivers, and provide us insights into how brief outings and temporary foster care could shape greater community involvement. Foster caregiving is high-stakes volunteerism [28,48], and it is likely that as the duration of caregiving increases, so do the costs to the caregiver and risk (and reward) of emotional attachment. In order to address these concerns, we posit that shorter-duration foster care as studied here should be commonly practiced so that the rewards of caregiving easily exceed the costs, especially for first-time caregivers. As rewards are repeatedly experienced by caregivers through brief outings and temporary fostering stays, riskier fostering opportunities of longer durations could be embarked upon. Such an approach may address the emotional stress of this type of volunteerism, and retention issues that are often experienced by animal welfare organizations [43,46,49,50].

During this study's nearly three thousand intervention experiences, dog bites were exceedingly rare, but when they did occur, they more often involved the dog biting a person versus another dog. Not surprisingly, we found that such events were related to shorter fostering stays in homes, likely indicative of their negative effect on the caregiver's experience. Bites to humans and dogs were also rarely reported by Gunter et al. [28] in over 2500 fostering experiences of longer durations. With the relatively low risk to human safety associated with foster caregiving of varying durations and the benefits of these programs on shelter dog outcomes, it is not surprising that organizations with foster care programs have higher rates of live release and lower returns of adopted dogs [51].

With regards to the evaluation of the shelters' brief outing and temporary fostering programs, we did find that municipal shelters typically had lower performing programs. During the pandemic, Gunter et al. [28] also observed that municipal shelters' utilization of foster care was lower and more quickly returned to pre-pandemic levels compared to shelters that were either private nonprofits or private nonprofit organizations with municipal contracts. Furthermore, in this study, we found that shelters with greater resources had higher performing programs, highlighting the importance of human and financial resources in animal welfare. Thus, while brief outings and temporary fostering programs should be explored by a variety of organizations given their potential impact on dog welfare, it is likely that municipal agencies will need to provide additional support to staff while better-resourced shelters may be able to implement these programs more easily with greater effect.

When considering the limitations of our study, it is likely that the requirements of programmatic training, implementation, and data collection were barriers to participation for lower-resourced shelters, undermining our efforts to enroll a diverse sample of animal shelters operating in the United States. The average live release rate for shelters in the current study was above 90%, which is high, but comparable with data reported by industry organizations [52]. Additionally, it is unknown how our findings may generalize to animal shelters in other countries, particularly localities where foster caregiving is not as commonplace.

In their study, Gunter et al. [28] reported similar resource levels of participating shelters, and in that investigation, shelters were not required to attend training or implement a specific intervention beyond placing dogs in foster care. Nevertheless, the obligation of placing dogs in foster care in and of itself could have been an impediment to participation across both studies as nearly all participating shelters had existing canine foster care programs at the time of study enrollment.

Data collection about dogs receiving these interventions was overseen by our research team; however, we relied upon shelter management systems for data about non-participating dogs. While such systems are routinely utilized in research about dogs in animal shelters, the availability of data and completeness of dogs' records likely differed between our control and intervention conditions. Moreover, longer data collection periods following the interventions would have likely resulted in fewer intervention dogs remaining in care at the end of the study, improving our outcome predictions.

It is possible that not all dogs at these shelters were eligible to participate in the interventions that were tested. The majority of dogs that experienced a brief outing or temporary fostering stay were available for adoption with a small proportion that were unavailable due to behavioral or medical concerns. Nevertheless, dogs with greater safety concerns related to their behavior were likely not selected for participation in our interventions but remained available for adoption or were euthanized soon after intake, which may have contributed to the higher likelihood of adoption in our intervention groups and shorter lengths of stay for control dogs. As described in previous studies about brief outings and temporary fostering interventions [16,21], shelter staff often do not enroll dogs with histories of aggression in these types of programs, which may explain the low incidents of human and dog bites that were reported.

5. Conclusions

This study demonstrates that brief outings and temporary fostering stays result in a greater likelihood of adoption for dogs in animal shelters when compared to dogs that do not experience these interventions. Adoptions were seldom by the caregivers themselves, although when this did occur, it was more often after a temporary fostering stay. Nevertheless, dogs that participated in either intervention had longer lengths of stay and were more often awaiting adoption at the end of the study as compared to non-intervention dogs, although this difference in length of stay was present prior to study enrollment and may be related to morphological and behavioral qualities of the intervention dogs.

When intervention programs of either type had greater percentages of community members participating, these programs were higher performing. Brief outing caregivers were more often individuals from the community whereas shelter volunteers were more involved in temporary foster care. As such, brief outings may be a means to address the caregiver recruitment issues faced by animal welfare organizations.

In all, shorter-duration fostering interventions as studied here may better balance the costs and rewards involved with this type of high-stakes volunteerism and assist in the retention of foster caregivers. However, shelter resources play a role in the programmatic success of these interventions. Organizations need to provide the human and financial means necessary to operate these programs in order to positively impact the dogs in their care.

Supplementary Materials: The following supporting information can be downloaded at: https://www.mdpi.com/article/10.3390/ani13223528/s1.

Author Contributions: Conceptualization, L.M.G., E.N.F. and C.D.L.W.; methodology, L.M.G., E.N.F., R.J.G., J.M.P. and E.M.B.; validation, E.M.B., R.J.G., B.J.N., J.L.R. and J.M.P.; formal analysis, L.M.G., E.M.B., B.J.N. and I.C.W.; investigation, E.M.B., R.J.G., J.L.R. and J.M.P.; original draft preparation, L.M.G.; writing—review and editing, L.M.G., R.J.G., B.J.N., J.L.R., J.M.P., I.C.W., E.N.F. and C.D.L.W.; visualization, L.M.G. and I.C.W.; supervision, L.M.G., E.N.F. and C.D.L.W.; funding acquisition, C.D.L.W., E.N.F. and L.M.G. All authors have read and agreed to the published version of the manuscript.

Funding: This research was made possible by a grant from Maddie's Fund, Thanks to Maddie.

Institutional Review Board Statement: This study did not require ethical approval.

Informed Consent Statement: Not applicable.

Data Availability Statement: The data presented in this study are openly available in the Virginia Tech Data Repository.

Acknowledgments: Thank you to the staff and volunteers of Animal Care and Control Team of Philadelphia, Animal Care Centers of New York City, Animal Defense League of Texas, Asheville Humane Society, Baltimore Animal Rescue & Care Shelter, Charleston Animal Society, Chautauqua County Humane Society, Chesapeake Animal Services, City of Bakersfield Animal Care Center, City of Bloomington Animal Care and Control, City of San Antonio Animal Care Services, Companion Animal Alliance, Compassion Without Borders, Contra Costa County Animal Services, Davis County Animal Care and Control, Denison Animal Welfare Group, Detroit Animal Care and Control, East Bay Society for the Prevention of Cruelty to Animals, Fort Worth Animal Care and Control, Fredericksburg Society for the Prevention of Cruelty to Animals, Friends of Strays, Great Plains Society for Prevention of Cruelty to Animals, Humane Rescue Alliance, Humane Society Naples, Humane Society of Northeast Georgia, Humane Society of Western Montana, Humane Society Waterville Area, Jacksonville Humane Society, Kentucky Humane Society, Kitsap Humane Society, Lake Humane Society, Fulton County Animal Services, Mendocino County Animal Care Services, Michiana Humane Society, Palm Valley Animal Center, Panhandle Animal Shelter, Pasadena Humane, Pasco County Animal Services, PAWS & More Animal Shelter, Paws Humane Society, Pennsylvania Society for the Prevention of Cruelty to Animals Lancaster Center, Regional Center for Animal Care and Protection, Ridgefield Operation for Animal Rescue, Saving Grace Pet Adoption Center, Second Chance Animal Services, Society for the Improvement of Conditions for Stray Animals, Society for the Prevention of Cruelty to Animals Brazoria County, Society for the Prevention of Cruelty to Animals Florida, Spokane County Regional Animal Protection Service, City of Stockton Animal Services, and Wayside Waifs. Additional thanks to graduate student, Lindsay Isernia; implementation coordinator, Alysha Verba; and research assistants, Michelle Dwyer and Caroline Hudgens.

Conflicts of Interest: The authors declare no conflict of interest. The funders had no role in the design of the study; in the collection, analyses, or interpretation of data; in the writing of this manuscript; or in the decision to publish the results.

References

1. Pet Statistics. Available online: https://www.aspca.org/animal-homelessness/shelter-intake-and-surrender/pet-statistics (accessed on 17 May 2023).
2. Sandri, M.; Colussi, A.; Perrotta, M.G.; Stefanon, B. Salivary cortisol concentration in healthy dogs is affected by size, sex, and housing context. *J. Vet. Behav.* **2015**, *10*, 302–306. [CrossRef]

3. van der Laan, J.E.; Vinke, C.M.; van der Borg JA, M.; Arndt, S.S. Restless nights? Nocturnal activity as a useful indicator of adaptability of shelter housed dogs. *Appl. Anim. Behav. Sci.* **2021**, *241*, 105377. [CrossRef]
4. Rooney, N.J.; Gaines, S.A.; Bradshaw, J.W. Behavioural and glucocorticoid responses of dogs (*Canis familiaris*) to kennelling: Investigating mitigation of stress by prior habituation. *Physiol. Behav.* **2007**, *92*, 847–854. [CrossRef]
5. Wenstrup, J.; Dowidchuk, A. Pet overpopulation: Data and measurement issues in shelters. *J. Appl. Anim. Welf. Sci.* **1999**, *2*, 303–319. [CrossRef]
6. Rodriguez, J.R.; Davis, J.; Hill, S.; Wolf, P.J.; Hawes, S.M.; Morris, K.N. Trends in intake and outcome data from US animal shelters from 2016 to 2020. *Front. Vet. Sci.* **2022**, *9*, 863990. [CrossRef]
7. Cain, C.J.; Woodruff, K.A.; Smith, D.R. Factors associated with shelter dog euthanasia versus live release by adoption or transfer in the United States. *Animals* **2021**, *11*, 927. [CrossRef]
8. Woodruff, K.; Smith, D.R. An estimate of the number of dogs in US shelters in 2015 and the factors affecting their fate. *J. Appl. Anim. Welf. Sci.* **2020**, *23*, 302–314. [CrossRef]
9. Rowan, A.; Kartal, T. Dog population & dog sheltering trends in the United States of America. *Animals* **2018**, *8*, 68.
10. Mellor, D.J.; Reid, C.S.W. Concepts of animal well-being and predicting the impact of procedures on experimental animals. *Improv. Well-Being Anim. Res. Environ.* **1994**, 3–18.
11. Mellor, D.; Patterson-Kane, E.; Stafford, K.J. *The Sciences of Animal Welfare*, 1st ed.; Wiley-Blackwell: Hoboken, NJ, USA, 2009.
12. Ward, S.J.; Hosey, G. The need for a convergence of agricultural/laboratory and zoo based approaches to animal welfare. *J. Appl. Anim. Welf. Sci.* **2020**, *23*, 484–492. [CrossRef]
13. Kiddie, J.L.; Collins, L.M. Development and validation of a quality of life assessment tool for use in kennelled dogs (*Canis familiaris*). *Appl. Anim. Behav. Sci.* **2014**, *158*, 57–68. [CrossRef]
14. Barnard, S.; Pedernera, C.; Candeloro, L.; Ferri, N.; Velarde, A.; Dalla Villa, P. Development of a new welfare assessment protocol for practical application in long-term dog shelters. *Vet. Rec.* **2016**, *178*, 18. [CrossRef] [PubMed]
15. Fehringer, A.A. Stress in Shelter Dogs and the Use of Foster Care to Improve Animal Welfare. Bachelor's Thesis, Pennsylvania State University, State College, PA, USA, 2014.
16. Gunter, L.M.; Feuerbacher, E.N.; Gilchrist, R.; Wynne, C.D.L. Evaluating the effects of a temporary fostering program on shelter dog welfare. *PeerJ* **2019**, *7*, e6620. [CrossRef] [PubMed]
17. van der Laan, J.E.; Vinke, C.M.; Arndt, S.S. Evaluation of hair cortisol as an indicator of long-term stress responses in dogs in an animal shelter and after subsequent adoption. *Sci. Rep.* **2022**, *12*, 5117. [CrossRef]
18. Hoffman, C.L.; Ladha, C.; Wilcox, S. An actigraphy-based comparison of shelter dog and owned dog activity patterns. *J. Vet. Behav.* **2019**, *34*, 30–36. [CrossRef]
19. Owczarczak-Garstecka, S.C.; Burman, O.H. Can sleep and resting behaviours be used as indicators of welfare in shelter dogs (*Canis lupus familiaris*)? *PLoS ONE* **2016**, *11*, e0163620. [CrossRef]
20. Gunter, L.; Feuerbacher, E. Canine enrichment. In *Animal Behavior for Shelter Veterinarian and Staff*, 2nd ed.; John Wiley & Sons: Hoboken, NJ, USA, 2022; pp. 263–287.
21. Gunter, L.M.; Gilchrist, R.J.; Blade, E.M.; Barber, R.T.; Feuerbacher, E.N.; Platzer, J.M.; Wynne, C.D. Investigating the impact of brief outings on the welfare of dogs living in US shelters. *Animals* **2021**, *11*, 548. [CrossRef]
22. Protopopova, A.; Gunter, L.M. Adoption and relinquishment interventions at the animal shelter: A review. *Anim. Welf.* **2017**, *26*, 35–48. [CrossRef]
23. Mohan-Gibbons, H.; Weiss, E.; Garrison, L.; Allison, M. Evaluation of a novel dog adoption program in two US communities. *PLoS ONE* **2014**, *9*, e91959. [CrossRef]
24. Protopopova, A.; Brandifino, M.; Wynne, C.D. Preference assessments and structured potential adopter-dog interactions increase adoptions. *Appl. Anim. Behav. Sci.* **2016**, *176*, 87–95. [CrossRef]
25. Gunter, L.M.; Barber, R.T.; Wynne, C.D. What's in a name? Effect of breed perceptions & labeling on attractiveness, adoptions & length of stay for pit-bull-type dogs. *PLoS ONE* **2016**, *11*, e0146857.
26. Cohen, N.P.; Chodorow, M.; Byosiere, S.E. A label'sa label, no matter the dog: Evaluating the generalizability of the removal of breed labels from adoption cards. *PLoS ONE* **2020**, *15*, e0238176. [CrossRef]
27. Patronek, G.J.; Crowe, A. Factors associated with high live release for dogs at a large, open-admission, municipal shelter. *Animals* **2018**, *8*, 45. [CrossRef]
28. Gunter, L.M.; Gilchrist, R.J.; Blade, E.M.; Reed, J.L.; Isernia, L.T.; Barber, R.T.; Foster, A.M.; Feuerbacher, E.N.; Wynne, C.D. Emergency fostering of dogs from animal shelters during the COVID-19 pandemic: Shelter practices, foster caregiver engagement, and dog outcomes. *Front. Vet. Sci.* **2022**, *9*, 862590. [CrossRef]
29. Shelter Terminology. Available online: https://higherlogicdownload.s3.amazonaws.com/ASV/fa11b6a5-ea22-45cc-9b33-416a24d44499/UploadedImages/Policy_Stmts/Shelter_Terminology.pdf (accessed on 6 November 2023).
30. Calculations. Available online: https://www.shelteranimalscount.org/animal-welfare-glossary/calculations/ (accessed on 24 October 2023).
31. Duffy, D.L.; Kruger, K.A.; Serpell, J.A. Evaluation of a behavioral assessment tool for dogs relinquished to shelters. *Prev. Vet. Med.* **2014**, *117*, 601–609. [CrossRef] [PubMed]
32. Stephen, J.; Ledger, R. Relinquishing dog owners' ability to predict behavioural problems in shelter dogs post adoption. *Appl. Anim. Behav. Sci.* **2007**, *107*, 88–99. [CrossRef]

33. Luescher, A.U.; Medlock, R.T. The effects of training and environmental alterations on adoption success of shelter dogs. *Appl. Anim. Behav. Sci.* **2009**, *117*, 63–68. [CrossRef]
34. Siettou, C.; Fraser, I.M.; Fraser, R.W. Investigating some of the factors that influence "consumer" choice when adopting a shelter dog in the United Kingdom. *J. Appl. Anim. Welf. Sci.* **2014**, *17*, 136–147. [CrossRef] [PubMed]
35. Powell, L.; Reinhard, C.; Satriale, D.; Morris, M.; Serpell, J.; Watson, B. Characterizing unsuccessful animal adoptions: Age and breed predict the likelihood of return, reasons for return and post-return outcomes. *Sci. Rep.* **2021**, *11*, 8018. [CrossRef]
36. Friend, J.R.; Bench, C.J. Evaluating factors influencing dog post-adoptive return in a Canadian animal shelter. *Anim. Welf.* **2020**, *29*, 399–410. [CrossRef]
37. Protopopova, A.; Wynne, C.D. Judging a dog by its cover: Morphology but not training influences visitor behavior toward kenneled dogs at animal shelters. *Anthrozoös* **2016**, *29*, 469–487. [CrossRef]
38. Protopopova, A.; Wynne, C.D.L. Adopter-dog interactions at the shelter: Behavioral and contextual predictors of adoption. *Appl. Anim. Behav. Sci.* **2014**, *157*, 109–116. [CrossRef]
39. Get 'em Home Challenge Winning Strategies for Helping Long-Stay Pets in Animal Shelters. Available online: https://chewonthis.maddiesfund.org/2019/08/get-em-home-challenge-winning-strategies-for-helping-long-stay-pets-in-animal-shelters/ (accessed on 11 August 2023).
40. Cain, C.J.; Woodruff, K.A.; Smith, D.R. Phenotypic characteristics associated with shelter dog adoption in the United States. *Animals* **2020**, *10*, 1959. [CrossRef] [PubMed]
41. Shih, H.Y.; Georgiou, F.; Curtis, R.A.; Paterson, M.B.; Phillips, C.J. Behavioural evaluation of a leash tension meter which measures pull direction and force during human–dog on-leash walks. *Animals* **2020**, *10*, 1382. [CrossRef] [PubMed]
42. Shih, H.Y.; Phillips, C.J.; Mills, D.S.; Yang, Y.; Georgiou, F.; Paterson, M. Dog pulling on the leash: Effects of restraint by a neck collar vs. a chest harness. *Front. Vet. Sci.* **2021**, *8*, 1002. [CrossRef]
43. Ackerman, R.; Watson, B.; Serpell, J.; Reinhard, C.L.; Powell, L. Understanding the Motivations of Foster Caregivers at Animal Shelters. *Animals* **2023**, *13*, 2694. [CrossRef]
44. Moss, L.R.; Hawes, S.M.; Connolly, K.; Bergstrom, M.; O'Reilly, K.; Morris, K.N. Animal control and field services officers' perspectives on community engagement: A qualitative phenomenology study. *Animals* **2022**, *13*, 68. [CrossRef]
45. Foster Caregiver Research Report. Available online: https://www.maddiesfund.org/assets/research/maddies-fund-foster-caregiver-research-report-june-2017.pdf (accessed on 26 August 2023).
46. Exploring Foster Programs and Their People, Management, and Leadership Challenges. Available online: https://www.maddiesfund.org/assets/grants/exploring-foster-programs-rogelberg-report.pdf (accessed on 4 September 2023).
47. Schafer, R.B. Equity in a relationship between individuals and a fraternal organization. *J. Volunt. Action Res.* **1979**, *8*, 12–20. [CrossRef]
48. Reese, L.A. Shelter and rescue programmes associated with higher live release and lower return rates for dogs. *Anim. Welf.* **2021**, *30*, 419–430. [CrossRef]
49. Reese, L.A.; Jacobs, J.; Grebey, T. Factors contributing to the satisfaction of animal shelter volunteers: The importance of voice. *J. Appl. Anim. Welf. Sci.* **2023**, *26*, 132–146. [CrossRef]
50. Reese, L.A.; Jacobs, J.; Seelenbinder, B.; Stedhouwer, T.; Velychko, N.; Wathen, L. The Emotional Aspects of Dog Fostering: Both Ends of the Leash. *Anthrozoös* **2023**, *36*, 369–387. [CrossRef]
51. Reese, L.A. Community factors and animal shelter outcomes. *J. Appl. Anim. Welf. Sci.* **2022**, 1–19. [CrossRef] [PubMed]
52. Q1 2023 Analysis. Available online: https://www.shelteranimalscount.org/wp-content/uploads/2023/05/Q1-Report.pdf (accessed on 3 September 2023).

animals

MDPI

Article

Influence of Walker Sex and Familiarity on Scent-Marking Behavior of Juvenile and Mature Shelter Dogs

Betty McGuire [1,*], Philippa Kok [2], Miles Garland [1], Bailey Guy [2], Alexandra Jackson [2] and Scott Haber [3]

1 Department of Ecology and Evolutionary Biology, Cornell University, Ithaca, NY 14853, USA; msg284@cornell.edu
2 Department of Animal Science, Cornell University, Ithaca, NY 14853, USA; pvk8@cornell.edu (P.K.); bjg224@cornell.edu (B.G.); aej34@cornell.edu (A.J.)
3 College of Arts and Sciences Communications Office, Cornell University, Ithaca, NY 14853, USA; sah67@cornell.edu
* Correspondence: bam65@cornell.edu

Simple Summary: Many mammals behave differently with familiar people versus strangers, and sometimes the sex of the person is another important influence. We studied shelter dogs during walks to determine the effects on behavior of the dog's sex and maturity and the walker's sex and familiarity. In Study 1, unfamiliar men and unfamiliar women walked dogs. In Study 2, after walks with unfamiliar men and women, dogs were walked again when walkers were familiar. In both studies, mature males urinated at higher rates when walked by a woman than by a man, whereas mature females urinated at similar rates with women and men. Mature males and mature females were less likely to defecate when walked by a man than by a woman. Juvenile dogs were less affected than mature dogs by the walker's sex, suggesting experience influenced responses in mature dogs. In Study 2, the effects on urination and defecation of a dog's sex and maturity and the walker's sex did not change over walks as dogs became familiar with walkers. Shelter dogs may be less responsive to the degree of familiarity with people than other mammals because they are directly exposed to so many people. Our results indicate that dog sex and maturity and human sex influence dog–human interactions.

Citation: McGuire, B.; Kok, P.; Garland, M.; Guy, B.; Jackson, A.; Haber, S. Influence of Walker Sex and Familiarity on Scent-Marking Behavior of Juvenile and Mature Shelter Dogs. *Animals* **2023**, *13*, 3649. https://doi.org/10.3390/ani13233649

Academic Editor: Angelo Gazzano

Received: 30 September 2023
Revised: 3 November 2023
Accepted: 22 November 2023
Published: 25 November 2023

Abstract: Many mammals living on farms, in zoos, and in research settings behave differently with familiar people versus strangers, and the sex of the person can also influence interactions. We conducted two studies to examine the influence of a dog's sex and maturity and a walker's sex and familiarity on the behavior of shelter dogs during leash walks. In Study 1 with unfamiliar walkers (n = 113 dogs), we found that mature males urinated at higher rates when walked by a woman than by a man, whereas mature females urinated at similar rates. Mature males and mature females were less likely to defecate when walked by a man than by a woman. Juvenile dogs were generally less affected than mature dogs by a walker's sex, suggesting a role for experience in mature dogs' responses. In Study 2, when dogs were walked more than once by a man and a woman (n = 81 dogs), we found patterns of urination and defecation like those in Study 1. Importantly, the effects of the dog's sex and maturity and the walker's sex on dogs' patterns of urination and defecation did not change over walks as dogs became familiar with walkers. Dogs in shelters are directly exposed to so many people that they may be less responsive to differing degrees of familiarity than mammals living in other settings. Our data indicate that dog maturity and sex and human sex influence dog–human interactions.

Keywords: urination; defecation; experience; environment; dog-human interaction

1. Introduction

Familiarity can influence interactions between humans and other mammals, including those kept on farms [1], in zoos [2], in research settings [3], and as companion animals [4].

For instance, handled piglets (*Sus scrofa*) interacted longer with their handler than with a stranger and showed less agitation and avoidance when caught by their handler [5]. Observations of several species housed at a zoo—African bush elephant, *Loxodonta africana*, Rothschild's giraffe, *Giraffa camelopardalis rothschildi*, South American tapir, *Tapirus terrestris*, and meerkat, *Suricata suricatta*—indicated less avoidance of familiar keepers than unfamiliar keepers [6]. Laboratory rats, *Rattus norvegicus*, preferred familiar individuals to strangers when given the choice of which person to crawl up on [7] and spent more time near familiar than unfamiliar caretakers in an open field test [8]. However, preferential behavior toward familiar humans has not been found in all studies and sometimes different results are found within a species. In studies with pet cats, *Felis catus*, familiarity with humans has been shown to have either no effect [9] or a negative influence on sociability behaviors [10] and a positive influence on the duration of contact [11]. Similarly, in one study domestic horses (*Equus ferus caballus*) groomed by a familiar handler showed a lower stress response than when groomed by an unfamiliar handler [12], while in another study, horses handled by their owners and an unfamiliar handler showed similar behavioral compliance and physiological measures of stress [13].

Domestic dogs (*Canis lupus familiaris*) have been a popular model for studies of human–animal relationships. For dogs, the effects of familiarity on behavior typically have been tested using paradigms involving exposure to the owner and a stranger, although some studies have used a familiar person (not the owner) and a stranger and more rarely, all three types of people are included (owner, familiar person, and stranger). Dogs tested in the owner versus stranger paradigm exhibited more play and exploration in the presence of their owner than a stranger and following brief separations from their owner, displayed more contact-seeking behavior toward the returning owner than the stranger [14,15]. Dogs tied out in the yard of their home and approached by either their owner or a stranger gazed more, spent more time near, wagged their tails more, and barked less during the owner's approach than the stranger's approach [16]. When choosing to interact with either their owner or a stranger, dogs preferred their owner when tested in an unfamiliar setting but preferred the stranger in a familiar setting [4]. Although dogs preferentially attended to pointing cues given by their owners as compared to strangers [17,18], owners and strangers were equally effective as demonstrators when dogs were tested in a detour task [19]. Finally, when tested with owners, a familiar woman, and an unfamiliar woman in eight different situations, dogs always preferred their owner to the unfamiliar woman and their owner to the familiar woman during stressful situations; minor differences were found in the responses of dogs to familiar and unfamiliar women (familiar and unfamiliar men were not included [20]).

Given the absence of owners in settings such as animal shelters and research facilities, familiarity studies conducted in these environments instead test dogs with only familiar versus unfamiliar people, and most such studies have found few or no differences in dog behavioral responses. In one study, even though shelter dogs spent more time within arm's reach facing (but not in direct contact with) a familiar person than an unfamiliar person in the first 2 min of a 10 min test, this effect did not characterize the remaining 8 min of the test [21]. Similarly, familiarity had little or no effect on approach/withdrawal responses of shelter dogs and time spent at different distances from people who were either familiar or unfamiliar [22]. When observed during encounter tests with either a familiar caretaker or an unfamiliar person, laboratory Beagles responded in the same friendly manner to both people [23]. However, other dog breeds (Labrador retrievers, miniature schnauzers, and cocker spaniels) kept in a long-stay enriched kennel environment, where they experienced high levels of daily contact with people, preferred unfamiliar to familiar people [21]. Despite the many diverse situations and settings in which dogs have been studied, we could find no studies that examined how familiarity with a person might affect canine scent-marking behavior during leash walks.

Several studies suggest that canine scent-marking behavior is sensitive to environmental conditions. For example, adult male dogs that used the raised-leg urinary posture characteristic of mature males temporarily switched back to the juvenile lean-forward posture, with all feet remaining on the ground, in fearful situations involving loud noises [24,25] Consistent with this finding relating changes in urinary behavior to stressful conditions, we previously reported that under the challenging conditions of shelter life, the percent of urinations in which mature male dogs used the raised-leg posture was 73% [26] as compared to 94–97% reported for mature male dogs living in other situations [27–30]. Further, in another study, we showed that some scent-marking behaviors exhibited during leash walks by mature shelter dogs (adults and seniors) differed when the dogs were walked by an unfamiliar man versus an unfamiliar woman [31]. When walked by an unfamiliar man, male dogs were more likely to use the juvenile lean-forward posture and urinate less frequently (another characteristic of juvenile males) than when walked by an unfamiliar woman. In contrast, the sex of an unfamiliar walker did not influence urinary posture or frequency of urination in female dogs. However, both male and female dogs were less likely to defecate when walked by an unfamiliar man than by an unfamiliar woman. Previous studies of dogs in kennels, shelters, and guide dog programs also reported behavioral responses consistent with greater unease in dogs exposed to an unfamiliar man than an unfamiliar woman (e.g., less time spent near, more time spent barking at, more time with tail in the low position, more frequent lip-licking, and presence of warning behaviors such as growling and raised hackles [32–37]). Thus, based on previous scent-marking studies and other studies that assessed a wide range of different behaviors, we interpreted the behavioral differences we observed in scent-marking in our previous study [31] to reflect greater unease with unfamiliar men than unfamiliar women.

Here, we present two studies on the scent-marking behavior of shelter dogs during walks by men and women. In Study 1, we extend our previous findings [31] by including juvenile dogs along with mature dogs walked by an unfamiliar man and an unfamiliar woman. Our goal was to determine whether the behavioral responses of juveniles toward unfamiliar male and unfamiliar female walkers would differ from those of mature dogs. To our knowledge, age-related differences in how dogs respond to unfamiliar men and unfamiliar women have not been reported, and any differences found between juvenile and mature dogs could shed light on when such responses develop. In Study 2, an unfamiliar man and an unfamiliar woman walked shelter dogs, and this was followed by subsequent walks when dogs were familiar with the male and female walkers. Our goal in Study 2 was to determine whether responses of juvenile and mature dogs to unfamiliar male and unfamiliar female walkers would be maintained on subsequent walks when dogs were familiar with walkers. If uneasiness decreased over walks, this could inform shelters looking for ways to reduce fear and stress in dogs in their care.

As mentioned, we could find no reports of age-related differences in the response to an unfamiliar man as compared to an unfamiliar woman; this is because most previous studies used either mature dogs only [32–35] or mostly mature dogs (94 of 111 dogs studied) and did not examine the effects of age [36] or did not examine the interaction between the age of the dog and sex of the stranger [37]. However, there are data indicating that reactivity toward unfamiliar humans (not differentiated by sex) increases with age in dogs. In several studies, increased age was associated with an increased risk of stranger-directed aggression in pet dogs, as assessed by owner-completed questionnaires [38–40]. Additionally, an observational study of free-roaming village dogs in Mexico tested whether dogs of three age classes would approach a female stranger: most puppies completely approached, most juveniles partially approached, and most adults did not approach at all [41]. Given that reactivity toward unfamiliar humans seems to increase with age in dogs, in Study 1 we expected that juvenile dogs would show little or no difference in scent-marking behavior when walked by an unfamiliar man and an unfamiliar woman and that mature dogs would show responses like those found in our earlier study [31]. For Study 2, because previous studies with shelter dogs [21,22] and pet dogs [20] found little difference in their behavior when with familiar people (not the owner) and unfamiliar people, we predicted that scent-marking behavior during walks with an unfamiliar man and an unfamiliar woman would not change when dogs were familiar with their walkers, and this would characterize both juvenile and mature dogs.

2. Materials and Methods

These studies were carried out under protocol 2012-0150, which was approved by Cornell University's Institutional Animal Care and Use Committee.

2.1. Study Shelter

We conducted behavioral observations of dogs during walks at the Tompkins County SPCA in Ithaca NY, USA. The shelter is no-kill, open-admission with scheduled intake and uses a conversation-based approach for adoptions. Active volunteer programs exist for both cats and dogs. Dog volunteers walk, socialize, train, and sometimes groom the dogs. As previously reported, women outnumber men as dog volunteers and in most staff positions at the shelter [31]. The behavioral observations included here began in September 2018 and ran through mid-March of 2020, when the shelter closed to volunteers due to the COVID-19 pandemic (all members of our walking team volunteer at the shelter). The shelter reactivated its dog volunteer program in July 2021. We restarted our observations in September 2021 and completed them in June 2023.

2.2. Care and Housing of Dogs

Dogs are admitted to the rescue building where they are housed in chain link cages with an indoor space (2.2 m^2) and an outdoor run (3.5 m^2). They undergo a veterinary examination at intake, which includes vaccinations, a fecal exam, deworming, flea control, heartworm testing, and, in older dogs, a complete blood count/chemistry profile. A urinalysis is performed for dogs of any age if surrendering owners report urinary issues or if symptoms, such as frequent urination, are observed by shelter staff or volunteers. Dogs without a microchip receive one. Most dogs admitted to the shelter are mixed breeds; the number of purebred dogs is unknown because DNA testing is not routinely performed.

A few days after intake, dogs are behaviorally evaluated in the Pet Adoption Center, which is adjacent to the rescue building [42–44]. Following evaluation, they are moved to the adoption floor where they are housed in cubicles ranging in size from 5.2 to 7.3 m^2. In both the Rescue Center and the Pet Adoption Center, dogs are almost always individually housed; exceptions include puppies from the same litter and dogs surrendered from the same household that staff judge need to be housed together; none of the dogs in our study were housed together. Each cubicle has a bed, blanket, water bowl, toys, and often a crate. Staff feed the dogs between 08:00 and 09:00 h and between 14:30 and 15:00 h. Several times a day, staff and volunteers either walk the dogs or take them to a large outdoor play yard; each day, start and end times of walks or visits to the play yard are recorded on a dry-erase board in the dog wing. Other forms of enrichment include day trips or overnight stays with volunteers and play groups of compatible dogs. All dogs wear either a buckle or martingale collar; a harness (previously fitted by staff) and leash (at least 1.8 m long) hang on a hook outside each dog's cubicle. Most harnesses are the PetSafe® Easy Walk® brand (Radio Systems® Corporation, Knoxville, TN, USA).

2.3. Data Collection

Our walking team consisted of six males (four 20–22 years of age and the remaining two, 31 and 37) and six females (five 20–22 years of age and one 64). Walks began on shelter grounds and continued across the street into a large field (16.6 ha; 42°28′20″ N, 76°26′22″ W), the substrate of which was mostly grass. A creek, forest, and other fields bordered the walking area. Each member of the team individually walked dogs at the shelter from one to three times a week. During walks, we let dogs freely investigate their surroundings and set the pace of the walk; however, they were not allowed to directly interact with other dogs or people other than their walker. There was no set walking route because adjustments had to be made during each walk, given that other volunteers with shelter dogs (and more rarely members of the public) were walking in the fields as well.

Upon arrival at the shelter, a walker checked the dry-erase board in the dog wing and chose a dog that had not been outside (either walked or in the play yard) for at least 2 h. Dog walking shifts (2 h in duration) at the shelter are scheduled for 1200, 1430, and on one day each week, there is an additional shift at 1700 h; thus, we walked the dogs about 2–3 h after their previous walk or time spent in the play yard. A walker entered the dog's cubicle, greeted and harnessed the dog, and walked the dog out of the shelter. Behavioral observations began once outside and ended outside 20 min later, precisely timed using cell phones. During this 20 min period, we verbally recorded scent marking behaviors—each urination and defecation—using our cell phones (e.g., the voice memo app on an iPhone 12, model MN9G2LL/A, Apple Inc., Cupertino, CA, USA). We repeated these same procedures on the second walk for each dog, which had to occur at least 1 day after the first walk. If dogs remained on the adoption floor after our second walk, we walked them additional times using the same methods described for the first walks.

We estimate that we interacted with each dog for about 30 min: (1) time spent with each dog before walks (greeting and harnessing the dog in its cubicle); (2) the 20 min of data collection during the walk; and (3) time spent after data collection had ended (walking back to the shelter, removing the harness, and again briefly interacting with the dog before leaving its cubicle). Gácsi et al. [45] found that shelter dogs form attachments to humans after 30 min of total contact (10 min over three days); these data informed our decision to categorize dogs as familiar with us after we had walked them once. We chose a 20 min observation period to be consistent with our previous study on how the sex of an unfamiliar walker influences the scent-marking behavior of dogs at this shelter [31].

After walks, we transferred the data from verbal recordings onto paper check sheets. In addition to the behavioral data collected, we photographed each dog and retrieved their demographic information from door signs and shelter records (e.g., intake date, source, identification number, and age). We used intake date to calculate time at the shelter, defined as the number of days from intake to the day of each walk with us; we also recorded days elapsed between the first walk and subsequent walks. All data were uploaded to Box, a service for data and document sharing and storage.

Over the course of data collection, we observed dogs between the ages of 4 months and 13 years. We classified dogs as juveniles (4 months to <1 year) and mature (1 year and older). To understand whether familiarity would influence dog responses to walkers, our goal was to have each dog walked at least two times by at least one male walker (when unfamiliar and familiar) and at least two times by at least one female walker (when unfamiliar and familiar); we described these dogs as having complete data for Study 2. Ideally, the second walks occurred within a few days of the first walks. However, it was not always possible in the shelter environment to collect complete data or for the second walks to occur within a few days of the first walks. Causes of incomplete data collection or longer periods between first and second walks included dogs being adopted throughout our research or transferred to other shelters or rescues; dog meets with potential adopters (and also with their resident dogs) that understandably took priority over our walks; behavioral or medical issues requiring removal of dogs from the adoption floor for a period of time, thereby lengthening the days elapsed between first and subsequent walks; and dogs temporarily removed from the shelter for day trips, overnight stays, or fosters with volunteers. Although necessary, our collecting data only on dogs that had not been out for at least 2 h also limited the dogs we could observe at the shelter on any day when other dog volunteers were present. Finally, we excluded data from two dogs with health conditions (one was obese and the other arthritic) and two dogs that were very fearful of strangers and not released for walks with volunteers until more than 2 months after their admission to the shelter (most dogs arrive on the adoption floor about 10–14 days after admission to the shelter and are immediately available for walking by volunteers).

By the end of data collection, we had 32 dogs with incomplete data (4 juvenile and 6 mature males and 3 juvenile and 19 mature females), defined as walked once by at least one unfamiliar man and at least one unfamiliar woman but without subsequent walks by both the same man and the same woman. We had complete data for 81 dogs (9 juvenile and 30 mature males and 12 juvenile and 30 mature females). For these 81 dogs, the second walk typically occurred within a few days of the first walk but there were exceptions because of the challenges described when studying dogs in a shelter environment (median = 3 days after first walk; 67% of second walks occurred from 1 to 7 days after the first walk; 26% from 8 to 14 days after the first walk and 6% from 15 to 21 days; and three walks occurred 23, 28, and 35 days after the first walk). The three second walks with the longest times elapsed from the first walk involved dogs that were walked additional times after the second walk. Thirteen of the 81 dogs were walked twice by the same man and twice by the same woman (i.e., no additional walks). Thirty-one dogs were walked more than two times by both the same man and the same woman, and thirty-seven were walked more than two times by either the same man or the same woman but not by both. Overall, number of additional walks per dog ranged from one to thirty-two, with most dogs being walked 1–3 times after their second walk. We used first walks from dogs with incomplete data (n = 32) and complete data (n = 81) to extend our previous study [31]; thus, we had 113 dogs in Study 1. We used first and subsequent walks from the 81 dogs with complete data to examine the effects of familiarity on dog behavior in Study 2. Note that data from three mature females and six mature males used in our previous study [31] were included in Studies 1 and 2; data from their second and additional walks were not included in our previous study, which focused only on first walks when walkers were unfamiliar. These nine were the only dogs from our previous study that had some second walks and sometimes additional walks as well. Two other dogs (Bru Bru and Mega) from this same period (2018–2019) were not part of our previous study and were included here. Table S1 contains all data from Studies 1 and 2 (Table S1: Data on Urination and Defecation by Shelter Dogs).

Most dogs were spayed or neutered before arrival on the adoption floor, and all were spayed or neutered before adoption. In an earlier study of scent-marking behavior displayed by mature dogs during walks at this shelter and another local shelter, one of us (BM) found that female urination rates did not change after spaying, but male urination rates decreased after neutering [46]. Thus, we also recorded the reproductive condition of dogs at the time of each walk. Of the 113 dogs with either incomplete or complete data, 45 of the 49 males were neutered for all their walks; one juvenile male was intact for all his walks and three juvenile males were intact for their first walks and either intact or neutered for subsequent walks. Of the 64 female dogs with either incomplete or complete data, 61 were spayed for all their walks; one adult was intact for all her walks, and one juvenile and one adult were intact for some of their first walks and spayed for all other walks.

2.4. Statistical Analyses

In Studies 1 and 2, we summarized data into means and standard deviations for urination rate and into percentages for defecation. For all analyses, we used R version 4.3.0 [47] and the following packages: glmmTMB [48] and emmeans [49].

2.4.1. Study 1

We used a linear mixed effects model to model urination rate (total number of urinations/20 min) as a function of the sex of dog, sex of walker, maturity status of dog (juvenile versus mature), and all two-way and three-way interactions of those variables, along with the additional fixed effect of weeks at shelter (we converted days to weeks); we included dog ID and walker ID as random effects. We used Cohen's d to estimate effect sizes. A generalized linear mixed effects model with a binomial distribution and a logit link was used to model defecation with the same variables described for urination rate.

2.4.2. Study 2

For urination rate, we used a linear mixed effects model with the sex of dog, sex of walker, maturity status of dog, and all two-way and three-way interactions of those variables, along with additional fixed effects of weeks at shelter and walk number. Random effects of dog ID, dog ID interacted with sex of the walker, and walker ID, as well as a random slope of the walk number and weeks at the shelter for each dog were included in the model. We also examined two-way interactions between the walk number and sex of dog, sex of walker, and maturity status of dog to examine whether the effects of these variables changed over walks as dogs became increasingly familiar with walkers. We used Cohen's d to estimate effect sizes.

We used a generalized linear mixed effects model with a binomial distribution and a logit link to model defecation with the same variables described for urination rate except that random effects only included dog ID, dog ID interacted with sex of the walker, and walker ID. As with urination rate, we also examined two-way interactions between walk number and the sex of dog, sex of walker, and maturity status of dog.

3. Results

3.1. Study 1: First Walks

Table 1 shows descriptive statistics for the two scent-marking behaviors in relation to the sex of dog, sex of unfamiliar walker, and maturity status of dog; this information is meant to provide a general overview of the raw data.

Table 1. Descriptive statistics based on raw data for rate of urination (mean ± *SD*) by male and female shelter dogs ($n = 113$), categorized as either juvenile or mature, during a 20 min walk by either an unfamiliar male or an unfamiliar female. Percentages of walks in which dogs defecated also are shown.

Dog's Sex	Walker's Sex	Dog's Maturity [1]	Urination Rate [2]	% Walks with Defecation
Male	Male	Juvenile	0.07 ± 0.03	50.0
Male	Male	Mature	0.18 ± 0.13	41.7
Male	Female	Juvenile	0.10 ± 0.07	71.4
Male	Female	Mature	0.38 ± 0.32	54.2
Female	Male	Juvenile	0.04 ± 0.02	28.6
Female	Male	Mature	0.09 ± 0.07	23.5
Female	Female	Juvenile	0.09 ± 0.05	50.0
Female	Female	Mature	0.12 ± 0.12	68.8

[1] Juveniles (4 months to <1 year); mature (≥1 year). [2] Total number of urinations/20 min.

3.1.1. Urination

We found a significant three-way interaction between the sex of the dog, the sex of the walker, and the maturity status of the dog for rate of urination (total number of urinations/20 min; $F = 7.98$, $d.f. = 1, 107.83$, $p < 0.01$). Mature male dogs walked by an unfamiliar woman had higher rates of urination than when walked by an unfamiliar man ($d = 2.12$); in contrast, mature female dogs had similar rates of urination when walked by an unfamiliar woman and by an unfamiliar man ($d = 0.09$; Figure 1a). The sex of an unfamiliar walker did not influence the rates of urination in either juvenile male ($d = 0.05$) or juvenile female dogs ($d = 0.20$; Figure 1a).

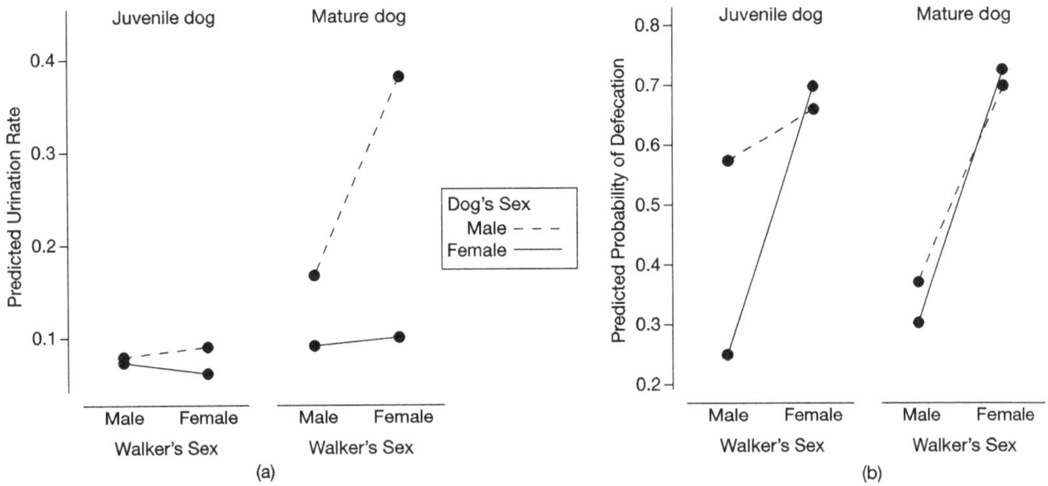

Figure 1. Scent-marking behaviors of dogs in relation to sex of dog, sex of unfamiliar walker, and maturity status of dog; data represent estimated marginal means from models. (**a**) Predicted rates of urination by male and female juvenile dogs and male and female mature dogs when walked by male and female walkers; (**b**) Predicted probabilities of defecation by male and female juvenile dogs and male and female mature dogs when walked by male and female walkers. Walks were 20 min in duration (n = 113 dogs).

When considering sex differences in the rates of urination, mature male dogs had higher rates of urination than mature female dogs, when walked by both an unfamiliar man (d = 0.78) and an unfamiliar woman (d = 2.81; Figure 1a). For juvenile dogs, there was no sex difference in the rates of urination when walked by either an unfamiliar man (d = 0.09) or by an unfamiliar woman (d = 0.34; Figure 1a).

For the dog's maturity status, mature male dogs had higher rates of urination than juvenile male dogs when walked by an unfamiliar man (d = 0.97) and an unfamiliar woman (d = 3.04; Figure 1a). Maturity status did not affect the rates of urination by female dogs when walked by either an unfamiliar man (d = 0.28) or an unfamiliar woman (d = 0.56; Figure 1a).

3.1.2. Defecation

We found a significant effect of the sex of the walker on the likelihood that dogs would defecate during a walk ($X^2 = 7.71$, $d.f. = 1$, $p < 0.01$; Figure 1b). A dog had a 0.699 probability of defecation when walked by a woman and a 0.374 probability when walked by a man. The odds that a dog would defecate with a female walker were 3.9 times higher than with a male walker ($p < 0.01$). Contrasts revealed that almost all groups–juvenile females, mature females, and mature males–had a significantly greater likelihood of defecation when walked by a woman than by a man; the juvenile males did not have a significantly greater likelihood, although their pattern was in the same direction as the other groups (Figure 1b).

3.2. *Study 2: All Walks*

3.2.1. Urination

Table 2 shows descriptive statistics for the rates of urination for walks one through four, as a sample of the raw data collected for all walks. None of the two-way interactions between the sex of the dog, sex of the walker, and maturity status of the dog with walk number was significant, indicating that the effects of these variables on the rate of urination did not change over walks as dogs became more familiar with individual walkers. We dropped these two-way interactions from the final model.

Table 2. Descriptive statistics based on raw data for rate of urination (Mean ± *SD*) by male and female shelter dogs, categorized as either juvenile or mature, during 20 min walks by either a male or a female. Results for the first four walks are included (n = 81 dogs for the first two walks; n = 68 dogs for walk 3; n = 49 dogs for walk 4).

Dog's Sex	Walker's Sex	Dog's Maturity [1]	Walk 1	Walk 2	Walk 3	Walk 4
Male	Male	Juvenile	0.07 ± 0.04	0.06 ± 0.03	0.07 ± 0.06	0.08 ± 0.08
Male	Male	Mature	0.17 ± 0.11	0.14 ± 0.10	0.16 ± 0.08	0.11 ± 0.05
Male	Female	Juvenile	0.08 ± 0.08	0.09 ± 0.06	0.11 ± 0.08	0.09 ± 0.02
Male	Female	Mature	0.36 ± 0.23	0.39 ± 0.25	0.37 ± 0.22	0.38 ± 0.22
Female	Male	Juvenile	0.05 ± 0.02	0.05 ± 0.03	0.07 ± 0.04	0.03 ± 0.04
Female	Male	Mature	0.08 ± 0.06	0.09 ± 0.06	0.10 ± 0.06	0.09 ± 0.06
Female	Female	Juvenile	0.06 ± 0.04	0.06 ± 0.03	0.05 ± 0.03	0.09 ± 0.06
Female	Female	Mature	0.13 ± 0.10	0.15 ± 0.13	0.16 ± 0.16	0.18 ± 0.15

[1] Juveniles (4 months to <1 year); mature (≥1 year).

As in Study 1, we found a significant three-way interaction between the sex of the dog, sex of the walker, and maturity status of the dog for the rate of urination (F = 5.31, *d.f.* = 1, 74.7, $p < 0.05$). Mature male dogs walked by a woman had higher rates of urination than when walked by a man (d = 3.30); in contrast, mature female dogs had similar rates of urination when walked by a woman and by a man (d = 0.59; Figure 2a). The sex of walker did not influence the rates of urination in either juvenile males (d = 0.49) or juvenile females (d = 0.10; Figure 2a).

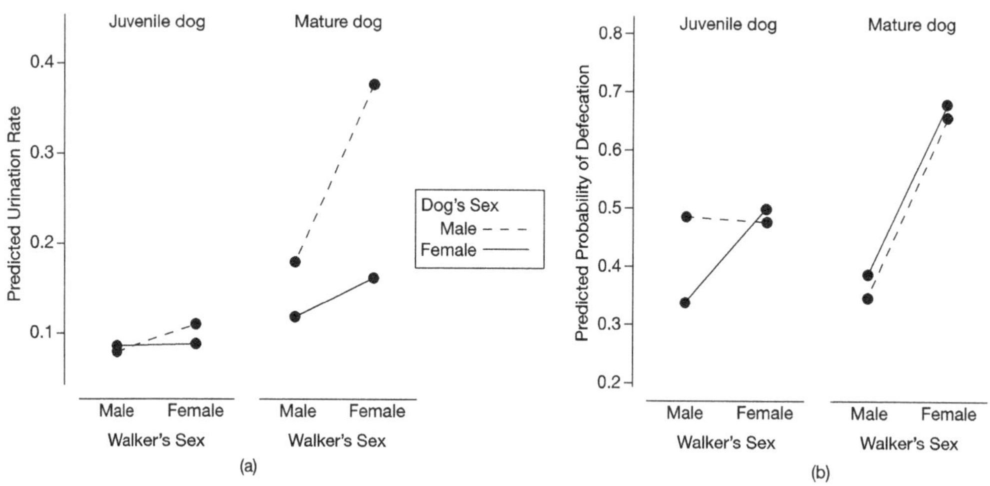

Figure 2. Scent-marking behaviors of dogs in relation to sex of dog, sex of walker, and maturity status of dog; data represent estimated marginal means from models. (**a**) Predicted rates of urination by male and female juvenile dogs and male and female mature dogs when walked by male and female walkers; (**b**) Predicted probabilities of defecation by male and female juvenile dogs and male and female mature dogs when walked by male and female walkers. Walks were 20 min in duration and data are from at least the first and second walk of each dog (n = 81 dogs); many dogs had additional walks. Given we found no significant interaction between walk number and sex of dog, sex of walker, or maturity of dog, these interactions were dropped from models; thus, the predicted values in figure are averaged across all walks of each dog.

When considering sex differences in rates of urination, mature male dogs walked by a woman had higher rates of urination than mature female dogs walked by a woman (d = 3.61); there was a trend ($p < 0.06$) for mature male dogs walked by a man to have higher rates of urination than mature female dogs walked by a man (d = 0.90; Figure 2a). For juvenile dogs, there was no sex difference in the rates of urination when walked by either a man (d = 0.19) or a woman (d = 0.40; Figure 2a).

Regarding the dog's maturity status, mature male dogs walked by a man had higher rates of urination than juvenile males walked by a man (d = 1.49; Figure 2a). Similarly, mature male dogs walked by a woman had higher rates of urination than juvenile males walked by a woman (d = 4.30; Figure 2a). Maturity status did not affect rates of urination by female dogs when walked by either a man (d = 0.40) or by a woman (d = 1.09; Figure 2a).

Rate of urination decreased with weeks at shelter ($F = 6.33$, $d.f. = 1, 15.2$, $p < 0.05$) but increased with walk number ($F = 18.24$, $d.f. = 1, 38.7$, $p < 0.001$).

3.2.2. Defecation

To provide a sample of the raw data collected during all walks, Table 3 shows the percentage of walks in which defecation occurred for the first four walks. None of the two-way interactions between the sex of the dog, sex of the walker, and maturity status of the dog with walk number was significant, indicating that the effects of these variables on the likelihood of defecation did not change over walks as dogs became more familiar with walkers. We dropped these two-way interactions from the final model.

Table 3. Percentages of 20 min walks based on raw data in which male and female shelter dogs, categorized as either juvenile or mature, defecated. Results for the first four walks are included (n = 81 dogs for the first two walks; n = 68 dogs for walk 3; n = 49 dogs for walk 4).

Dog's Sex	Walker's Sex	Dog's Maturity [1]	Walk 1	Walk 2	Walk 3	Walk 4
Male	Male	Juvenile	63.6	44.4	33.3	66.7
Male	Male	Mature	37.5	29.4	37.5	0.00
Male	Female	Juvenile	60.0	64.3	44.4	60.0
Male	Female	Mature	61.4	74.4	69.2	70.0
Female	Male	Juvenile	20.0	40.0	60.0	0.00
Female	Male	Mature	36.6	42.9	36.8	46.7
Female	Female	Juvenile	54.2	60.0	75.0	50.0
Female	Female	Mature	63.2	77.3	68.8	81.8

[1] Juveniles (4 months to <1 year); mature (≥1 year).

For likelihood that dogs would defecate during a walk, we found a significant interaction between the sex of the walker and maturity status of the dog ($X^2 = 4.17$, $d.f. = 1$, $p < 0.05$; Figure 2b). The probability of defecation by a juvenile dog was similar when walked by either a woman (0.470) or a man (0.401). In contrast, the probability of defecation by a mature dog was higher when walked by a woman (0.676) than by a man (0.363). The odds that a mature dog would defecate when walked by a woman were 3.66 times larger than when walked by a man ($p < 0.05$).

4. Discussion

4.1. Study 1: First Walks

Consistent with our previous findings at this shelter [31], mature male dogs had higher rates of urination when walked by an unfamiliar woman than by an unfamiliar man, and mature female dogs had similar rates of urination when walked by an unfamiliar woman and by an unfamiliar man. As before, we interpret the reduced rates of urination by mature male dogs to reflect greater uneasiness with unfamiliar men than with unfamiliar women. A pattern of greater unease with unfamiliar men, sometimes displayed by both male and female dogs, has been reported in several studies using diverse behavioral measures, which included less time spent near, more time spent barking at, more time with tail in the low position, more frequent lip-licking, and the presence of warning behaviors such as raising hackles and growling [32–37]. Additionally, the urinary behavior of male dogs—both posture and rate of urination—appears generally sensitive to fearful or stressful conditions [24–26,31].

Our inclusion of juvenile dogs in the present study revealed that the sex of an unfamiliar walker did not influence rates of urination by either juvenile males or juvenile females. Some studies on stranger-directed aggression in dogs found that avoidance and warning or aggressive responses to unfamiliar people increase with age [38–41]. Age-related increases in reactivity to unfamiliar people may reflect the cumulative effect of experiences with strangers perceived by dogs as threatening [39]. In the same way, cumulative experiences with unfamiliar men perceived as threatening might explain why we found that mature male dogs, but not juvenile male dogs, urinated at lower rates when walked by an unfamiliar man than by an unfamiliar woman. However, we did not directly test this possibility. We also found sex differences in the rates of urination for mature dogs (males had higher rates than females) but not for juveniles. Regardless of the sex of the walker, mature males had higher urination rates than juvenile males, whereas mature females and juvenile females had similar rates of urination. These findings regarding sex differences in and effects of maturation on rates of urination have been well documented for dogs [30,50,51].

Dogs were more likely to defecate when walked by an unfamiliar woman (0.699 probability) than by an unfamiliar man (0.374 probability), again suggesting greater uneasiness with unfamiliar men. The present results agree with our previous findings on the likelihood of defecation for mature dogs [31]. A closer look at our current data revealed that the observed pattern of a greater likelihood of defecation with a female walker was significant for mature males, mature females, and juvenile females but not for juvenile males, although their response was in the same direction as the other groups.

We did not determine the specific cues used by dogs to discriminate the sex of unfamiliar walkers. Potential cues include tactile, auditory, visual, and olfactory stimuli. Subtle sex differences in petting techniques appeared responsible for lower cortisol levels found in shelter dogs petted by women than in those petted by men [52]. In subsequent studies when men and women were trained to use a standardized petting technique, reductions in cortisol levels were similar in dogs petted by men and by women [53,54]. Shih et al. [36] found that male walkers made more frequent physical contact with shelter dogs than did female walkers, who were more likely to talk to dogs during walks and use high-pitched voices. Pet dogs were better at matching human male voices to images of male faces in comparison to human female voices and faces; the authors suggested that dogs' general wariness of men might promote the learning of male facial and vocal characteristics [55]. Finally, given the keen olfactory sense of dogs [56], it would not be surprising if they used olfactory cues to distinguish male and female walkers. Indeed, laboratory rats and mice discriminate experimenter sex using androgen-based olfactory cues and exhibit a strong physiological stress response in the presence of male, but not female, stimuli [57].

4.2. Study 2: All Walks

For both rate of urination and likelihood of defecation, none of the two-way interactions between the walk number and sex of the dog, sex of the walker, and maturity status of the dog was significant, indicating that the effects of these variables on urination and defecation did not change over walks as the dogs become familiar with walkers. Our findings agree with those of other studies reporting little difference in the response of dogs to familiar people (not the owner) and unfamiliar people [20–23] and stand in contrast to the preferential behavior typically shown by pet dogs to their owners over strangers [14–18]. Even more so than dogs in homes, dogs in shelters might be expected to display similar behavior in the presence of unfamiliar and familiar people because of their novel and challenging environment. While in the shelter, dogs have direct physical contact with many different people of varying degrees of familiarity (e.g., at our study shelter, veterinarians, veterinary students, shelter staff, volunteers, and members of the public), which might reduce their likelihood of behaving differently when with unfamiliar and familiar people.

For comparison, most pet dogs living in a community in Cheshire England were estimated by their owners to interact with 3–5 people outside their household each week [58].

The patterns found for urination in Study 2, using data from all the walks, were very similar to those found in Study 1, using data only from the first walks. The single exception concerned sex differences in the rates of urination. In both Study 1 and Study 2, mature male dogs walked by a woman had higher rates of urination than mature female dogs walked by a woman. However, the significant finding in Study 1 that mature male dogs walked by a man had higher rates of urination than mature female dogs walked by a man was instead a nonsignificant trend in Study 2 ($p < 0.06$), perhaps reflecting smaller sample sizes of mature dogs in Study 2 ($n = 30$ males; $n = 30$ females) than in Study 1 ($n = 36$ males; $n = 49$ females).

For likelihood of defecation during walks in Study 1, we detected a significant effect of the sex of the walker, with dogs more likely to defecate when walked by an unfamiliar woman than by an unfamiliar man. In Study 2, we found a significant interaction between the sex of the walker and maturity status of the dog. The probability of defecating by a juvenile dog was similar when walked by a woman (0.470) and by a man (0.401), whereas the probability of defecating by a mature dog was significantly higher when walked by a woman (0.676) than by a man (0.363). Indications that juveniles might behave differently than mature dogs were evident in Study 1, but only for juvenile males, who were the one group not to show a significantly higher probability of defecation when walked by a woman than by a man, although their pattern was the same as in other groups. It is unclear why juvenile females showed a significantly higher likelihood of defecation when walked by a woman than by a man in Study 1 but not in Study 2; their overall pattern, however, was similar across the two studies and sample sizes were somewhat smaller in Study 2 ($n = 12$) than in Study 1 ($n = 15$).

4.3. Limitations

Our studies have several limitations. First, we did not control for the age of the walker, which ranged from low-twenties to mid-sixties. Some mammals can discriminate human age as evidenced by their behavioral and physiological responses to certain stimuli. African bush elephants (*Loxodonta africana*) use auditory cues to differentiate men, who pose a significant hunting threat, from boys, who do not and show more defensive bunching and investigative sniffing after playbacks of men's voices [59]. Domestic horses (*Equus ferus caballus*) use both auditory and visual cues to differentiate adults from children and show increased heart rates during children's vocalizations [60]. However, Koda and Shimoju [33] found that dogs behaved in a similar way toward unfamiliar women (from 20 to 40 years old) and unfamiliar girls (from 8 to 13 years old); they did not include a comparison of unfamiliar men and unfamiliar boys in their study. These data on dogs suggest that at least in the case of human females, dogs respond in a similar way to individuals of different ages. Other limitations relate to the challenges of collecting data over time on individual dogs in shelters. Throughout our studies, dogs were adopted and sometimes returned, or temporarily unavailable to us due to in-shelter activities (e.g., meeting potential adopters and their dogs, and either medical or behavioral issues requiring removal from the adoption floor for 10 or more days) or out-of-shelter activities (e.g., short-term fosters and day trips with volunteers). Longer times than ideal between first and second walks for some dogs in Study 2 resulted from both in-shelter and out-of-shelter activities. Nevertheless, by including all walks for each dog we hoped to reduce the impact of those initial delays. Finally, whereas our sample sizes for mature dogs might be considered moderate (Study 1, 36 males and 49 females; Study 2, 30 males and 30 females), our sample sizes for juvenile dogs (Study 1, 13 males and 15 females; Study 2, 9 males and 12 females) were less than half those of mature dogs. The smaller number of juveniles likely reflects the shorter lengths of stay for juvenile dogs than mature dogs at our study shelter [44].

5. Conclusions

In Study 1 (first walks), we extended findings from our previous study showing that the sex of an unfamiliar person influences the scent-marking behaviors of mature shelter dogs [31], further strengthening our suggestion that researchers studying mature shelter dogs (and perhaps all mature dogs) should report the sex of all personnel involved in handling and data collection. We also extended our previous findings by including juvenile dogs, whose rate of urination was unaffected by the sex of an unfamiliar walker and the likelihood of defecation was unaffected in males but not females. The generally greater responsivity of mature dogs to the sex of an unfamiliar person may reflect cumulative experiences with male strangers perceived as threatening, although this was not tested here.

In Study 2 (all walks), we found that the effects on urination and defecation of the dogs' sex and maturity and the walkers' sex did not change over walks as dogs became familiar with walkers. We suggest that shelter dogs have direct contact with so many different people of varying degrees of familiarity that they may be less responsive than other mammals–for example, those on farms, in zoos, and in research environments–to familiar versus unfamiliar people.

Our data indicate that the maturity of the dog, as well as the sex of the dog and sex of the human, influence dog–human interactions. Future studies should examine the specific cues used by dogs to discriminate men and women, perhaps starting with androgen-based olfactory cues, and determine the cause(s) of the change in response to men as the dogs mature.

Supplementary Materials: The following supporting information can be downloaded at: https://www.mdpi.com/article/10.3390/ani13233649/s1, Table S1: Data on Urination and Defecation by Shelter Dogs.

Author Contributions: Conceptualization, B.M.; methodology, B.M.; coordination and review of statistical analyses, B.M.; data collection, B.M., M.G., B.G., P.K., A.J. and S.H.; data curation, B.M., M.G., B.G., P.K., A.J. and S.H.; writing—original draft preparation, B.M.; writing—review and editing, B.M., M.G., B.G., P.K., A.J. and S.H.; supervision, B.M. All authors have read and agreed to the published version of the manuscript.

Funding: This research was funded in part by the Laidlaw Foundation.

Institutional Review Board Statement: This research was conducted under the protocol Behavior of Shelter Dogs 2012-0150, which was approved again on 16 September 2021, by the Institutional Animal Care and Use Committee of Cornell University. Institutional Review Board Statement not applicable.

Informed Consent Statement: Not applicable.

Data Availability Statement: Data are available online as Supplementary Material.

Acknowledgments: We thank Jim Bouderau for permission to walk dogs at the shelter. Stephen Parry, Cornell Statistical Consulting Unit, helped with statistics and Willy Bemis helped with figures. We thank three anonymous Reviewers for their helpful comments.

Conflicts of Interest: The authors declare no conflict of interest.

References

1. Waiblinger, S.; Boivin, X.; Pedersen, V.; Tosi, M.-V.; Janczak, A.M.; Visser, E.K.; Jones, R.B. Assessing the human-animal relationship in farmed species: A critical review. *Appl. Anim. Behav. Sci.* **2006**, *101*, 185–242. [CrossRef]
2. Cole, J.; Fraser, D. Zoo animal welfare: The human dimension. *J. Appl. Anim. Welf. Sci.* **2018**, *21*, 49–58. [CrossRef]
3. van Driel, K.S.; Talling, J.C. Familiarity increases consistency in animal tests. *Behav. Brain Res.* **2005**, *159*, 243–245. [CrossRef] [PubMed]
4. Feuerbacher, E.N.; Wynne, C.D.L. Dogs don't always prefer their owners and can quickly form strong preferences for certain strangers over others. *J. Exp. Anal. Behav.* **2017**, *108*, 305–317. [CrossRef] [PubMed]
5. Tanida, H.; Miura, A.; Tanaka, T.; Yoshimoto, T. Behavioral response to humans in individually handled weanling pigs. *Appl. Anim. Behav. Sci.* **1995**, *42*, 249–259. [CrossRef]
6. Martin, R.A.; Melfi, V. A comparison of zoo animal behavior in the presence of familiar and unfamiliar people. *J. Appl. Anim. Welf. Sci.* **2016**, *19*, 234–244. [CrossRef]

7. Davis, H.; Taylor, A.A.; Norris, C. Preference for familiar humans by rats. *Psychon. Bull. Rev.* **1997**, *4*, 118–120. [CrossRef]
8. McCall, R.B.; Lester, M.L.; Corter, C.M. Caretaker effect in rats. *Dev. Psychobiol.* **1969**, *1*, 771. [CrossRef]
9. Vitale, K.R.; Udell, M.A.R. The quality of being sociable: The influence of human attentional state, population, and human familiarity on domestic cat sociability. *Behav. Proc.* **2019**, *158*, 11–17. [CrossRef]
10. Ellis, S.L.H.; Thompson, H.; Guijarro, C.; Zulch, H.E. The influence of body region, handler familiarity and order of region handled on the domestic cat's response to being stroked. *Appl. Anim. Behav. Sci.* **2015**, *173*, 60–67. [CrossRef]
11. Edwards, C.; Heiblum, M.; Tejeda, A.; Galindo, F. Experimental evaluation of attachment behaviors in owned cats. *J. Vet. Behav.* **2007**, *2*, 119–125. [CrossRef]
12. Scopa, C.; Greco, A.; Contalbrigo, L.; Fratini, E.; Lanatà, A.; Scilingo, E.P.; Baragli, P. Inside the interaction: Contact with familiar humans modulates heart rate variability in horses. *Front. Vet. Sci.* **2020**, *7*, 582759. [CrossRef] [PubMed]
13. Ijichi, C.; Griffin, K.; Squibb, K.; Favier, R. Stranger danger? An investigation into the influence of human-horse bond on stress and behaviour. *Appl. Anim. Behav. Sci.* **2018**, *206*, 59–63. [CrossRef]
14. Topál, J.; Miklósi, Á.; Csányi, V.; Dóka, A. Attachment behavior in dogs (*Canis familiaris*): A new application of Ainsworth's (1969) Strange Situation Test. *J. Comp. Psychol.* **1998**, *112*, 219–229. [CrossRef] [PubMed]
15. Prato-Previde, E.; Custance, D.M.; Spiezio, C.; Sabatini, F. Is the dog-human relationship an attachment bond? An observational study using Ainsworth's Strange Situation Test. *Behaviour* **2003**, *140*, 225–254. [CrossRef]
16. Rappolt, G.A.; John, J.; Thompson, N.S. Canine responses to familiar and unfamiliar humans. *Aggress. Behav.* **1979**, *5*, 155–161. [CrossRef]
17. Cook, A.; Arter, J.; Jacobs, L.F. My owner, right or wrong: The effect of familiarity on the domestic dog's behavior in a food choice task. *Anim. Cogn.* **2014**, *17*, 461–470. [CrossRef] [PubMed]
18. Cunningham, C.L.; Ramos, M.F. Effect of training and familiarity on responsiveness to human cues in domestic dogs (*Canis familiaris*). *Anim. Cogn.* **2014**, *17*, 805–814. [CrossRef]
19. Pongrácz, P.; Miklósi, Á.; Kubinyi, E.; Gurobi, K.; Topál, J.; Csányi, V. Social learning in dogs: The effect of a human demonstrator on the performance of dogs in a detour test. *Anim. Behav.* **2001**, *62*, 1109–1117. [CrossRef]
20. Kerepesi, A.; Dóka, A.; Miklósi, Á. Dogs and their human companions: The effects of familiarity on dog-human interactions. *Behav. Proc.* **2015**, *110*, 27–36. [CrossRef]
21. Pullen, A.J.; Merrill, R.J.N.; Bradshaw, J.W.S. The effect of familiarity on behaviour of kennel housed dogs during interactions with humans. *Appl. Anim. Behav. Sci.* **2012**, *137*, 66–73. [CrossRef]
22. Conley, M.J.; Fisher, A.D.; Hemsworth, P.H. Effects of human contact and toys on the fear responses to humans of shelter-housed dogs. *Appl. Anim. Behav. Sci.* **2014**, *156*, 62–69. [CrossRef]
23. Döring, D.; Haberland, B.E.; Ossig, A.; Küchenhoff, H.; Dobenecker, B.; Hack, R.; Schmidt, J.; Erhard, M.H. Behavior of laboratory beagles towards humans: Assessment in an encounter test and a simulation of experimental situations. *J. Vet. Behav.* **2014**, *9*, 295–303. [CrossRef]
24. Berg, I.A. Development of behavior: The micturition pattern in the dog. *J. Exp. Psych.* **1944**, *34*, 343–368. [CrossRef]
25. Martins, T.; Valle, J.R. Hormonal regulation of the micturition behavior of the dog. *J. Comp. Physiol. Psych.* **1948**, *41*, 301–311. [CrossRef]
26. Gough, W.; McGuire, B. Urinary posture and motor laterality in dogs (*Canis lupus familiaris*) at two shelters. *Appl. Anim. Behav. Sci.* **2015**, *168*, 61–70. [CrossRef]
27. Bekoff, M. Scent-marking by free-ranging domestic dogs–olfactory and visual components. *Biol. Behav.* **1979**, *4*, 123–139.
28. Cafazzo, S.; Natoli, E.; Valsecchi, P. Scent-marking behaviour in a pack of free-ranging domestic dogs. *Ethology* **2012**, *118*, 955–966. [CrossRef]
29. Pal, S.K. Urine marking by free-ranging dogs (*Canis familiaris*) in relation to sex, season, place and posture. *Appl. Anim. Behav. Sci.* **2003**, *80*, 45–59. [CrossRef]
30. Sprague, R.H.; Anisko, J.J. Elimination patterns in the laboratory beagle. *Behaviour* **1973**, *47*, 257–267. [CrossRef]
31. McGuire, B.; Fry, K.; Orantes, D.; Underkofler, L.; Parry, S. Sex of walker influences scent-marking behavior of shelter dogs. *Animals* **2020**, *10*, 6–32. [CrossRef]
32. Lore, R.K.; Eisenberg, F.B. Avoidance reactions of domestic dogs to unfamiliar male and female humans in a kennel setting. *Appl. Anim. Behav. Sci.* **1986**, *15*, 261–266. [CrossRef]
33. Koda, N.; Shimoju, S. Human-dog interactions in a guide-dog training program. *Psychol. Rep.* **1999**, *84*, 1115–1121. [CrossRef] [PubMed]
34. Wells, D.L.; Hepper, P.G. Male and female dogs respond differently to men and women. *Appl. Anim. Behav. Sci.* **1999**, *61*, 341–349. [CrossRef]
35. Bergamasco, L.; Osella, M.C.; Savarino, P.; Larosa, G.; Ozella, L.; Manassero, M.; Badino, P.; Odore, R.; Barbero, R.; Re, G. Heart rate variability and saliva cortisol assessment in shelter dog: Human–animal interaction effects. *Appl. Anim. Behav. Sci.* **2010**, *125*, 56–68. [CrossRef]
36. Shih, H.-Y.; Paterson, M.B.A.; Georgiou, F.; Pachana, N.A.; Phillips, C.J.C. Who is pulling the leash? Effects of human gender and dog sex on human–dog dyads when walking on-leash. *Animals* **2020**, *10*, 1894. [CrossRef]
37. McGuire, B.; Song, A. Influence of sex of stranger on responses of shelter dogs during canine behavioral evaluations. *Animals* **2023**, *13*, 2461. [CrossRef]

38. Blackwell, E.J.; Twells, C.; Seawright, A.; Casey, R.A. The relationship between training methods and the occurrence of behavior problems, as reported by owners, in a population of domestic dogs. *J. Vet. Behav.* **2008**, *3*, 207–217. [CrossRef]
39. Casey, R.A.; Loftus, B.; Bolster, C.; Richards, G.J.; Blackwell, E.J. Human directed aggression in domestic dogs (*Canis familiaris*): Occurrence in different contexts and risk factors. *Appl. Anim. Behav. Sci.* **2014**, *152*, 52–63. [CrossRef]
40. Didehban, N.; Borujeni, M.P.; Avizeh, R.; Mosallanejad, B. Problematic behaviors in companion dogs: A survey of their prevalence and associated factors. *J. Vet. Behav.* **2020**, *39*, 6–13. [CrossRef]
41. Ruiz-Izaguirre, E.; Eilers, K.C.H.A.M.; Bokkers, E.A.M.; Ortolani, A.; Ortega-Pacheco, A.; de Boer, I.J.M. Human–dog interactions and behavioural responses of village dogs in coastal villages in Michoacán, Mexico. *Appl. Anim. Behav. Sci.* **2014**, *154*, 57–65. [CrossRef]
42. Sternberg, S. *Assess-A-Pet: The Manual*; Assess-A-Pet: New York, NY, USA, 2006.
43. Bollen, K.S.; Horowitz, J. Behavioral evaluation and demographic information in the assessment of aggressiveness in shelter dogs. *Appl. Anim. Behav. Sci.* **2008**, *112*, 120–135. [CrossRef]
44. McGuire, B.; Chan, J.; Jean-Baptiste, K.; Kok, P.; Rosenbaum, E. Results of behavioral evaluations predict length of stay for shelter dogs. *Animals* **2021**, *11*, 3272. [CrossRef] [PubMed]
45. Gácsi, M.; Topál, J.; Miklósi, Á.; Dóka, A.; Csányi, V. Attachment behavior of adult dogs (*Canis familiaris*) living at rescue centers: Forming new bonds. *J. Comp. Psychol.* **2001**, *11*, 423–431. [CrossRef] [PubMed]
46. McGuire, B. Effects of gonadectomy on scent-marking behavior of shelter dogs. *J. Vet. Behav.* **2019**, *30*, 16–24. [CrossRef]
47. R Core Team. *R: A Language and Environment for Statistical Computing*; R Foundation for Statistical Computing: Vienna, Austria, 2023; Available online: https://www.R-project.org/ (accessed on 24 July 2023).
48. Brooks, M.E.; Kasper, K.; van Benthem, K.J.; Magnusson, A.; Berg, C.W.; Nielsen, A.; Skaug, H.J.; Maechler, M.; Bolker, B.M. glmmTMB Balances Speed and Flexibility Among Packages for Zero-inflated Generalized Linear Mixed Modeling. *R J.* **2017**, *9*, 378–400. [CrossRef]
49. Lenth, R.V. Emmeans: Estimated Marginal Means, Aka Least-Squares Means. R Package Version 1.8.5. 2023. Available online: https://CRAN.R-project.org/package=emmeans (accessed on 24 July 2023).
50. Beach, F.A. Effects of gonadal hormones on urinary behavior in dogs. *Physiol. Behav.* **1974**, *12*, 1005–1013. [CrossRef]
51. Ranson, E.; Beach, F.A. Effects of testosterone on ontogeny of urinary behavior in male and female dogs. *Horm. Behav.* **1985**, *19*, 36–51. [CrossRef]
52. Hennessy, M.B.; Davis, H.N.; Williams, M.T.; Mellott, C.; Douglas, C.W. Plasma cortisol levels of dogs at a county animal shelter. *Physiol. Behav.* **1997**, *62*, 485–490. [CrossRef]
53. Hennessy, M.B.; Williams, M.T.; Miller, D.D.; Douglas, C.W.; Voith, V.L. Influence of male and female petters on plasma cortisol and behaviour: Can human interaction reduce the stress of dogs in a public animal shelter? *Appl. Anim. Behav. Sci.* **1998**, *61*, 63–77. [CrossRef]
54. Hennessy, M.B.; Voith, V.L.; Hawke, J.L.; Young, T.L.; Centrone, J.; McDowell, A.L.; Linden, F.; Davenport, G.M. Effects of a program of human interaction and alterations in diet composition on activity of the hypothalamic-pituitary-adrenal axis in dogs housed in a public animal shelter. *J. Am. Vet. Med. Assoc.* **2002**, *221*, 65–71. [CrossRef] [PubMed]
55. Yong, M.H.; Ruffman, T. Domestic dogs match human male voices to faces, but not for females. *Behaviour* **2015**, *152*, 1585–1600. [CrossRef]
56. Bern, G.S.; Brooks, A.M.; Spivak, M. Scent of the familiar: An fMRI study of canine brain responses to familiar and unfamiliar human and dog odors. *Behav. Proc.* **2015**, *110*, 37–46. [CrossRef] [PubMed]
57. Sorge, R.E.; Martin, L.J.; Isbester, K.A.; Sotocinal, S.G.; Rosen, S.; Tuttle, A.H.; Wieskopf, J.S.; Acland, E.L.; Dokova, A.; Kadoura, B.; et al. Olfactory exposure to males, including men, causes stress and related analgesia in rodents. *Nat. Methods* **2014**, *11*, 629–632. [CrossRef]
58. Westgarth, C.; Pinchbeck, G.L.; Bradshaw, J.W.S.; Dawson, S.; Gaskell, R.M.; Christley, R.M. Dog-human and dog-dog interactions of 260 dog-owning households in a community in Cheshire. *Vet. Rec.* **2008**, *162*, 436–442. [CrossRef]
59. McComb, K.; Shannon, G.; Sayialel, K.N.; Moss, C. Elephants can determine ethnicity, gender, and age from acoustic cues in human voices. *Proc. Natl. Acad. Sci. USA* **2014**, *111*, 5433–5438. [CrossRef]
60. Jardat, P.; Ringhofer, M.; Yamamoto, S.; Gouyet, C.; Degrande, R.; Parias, C.; Reigner, F.; Calandreau, L.; Lansade, L. Horses form cross-modal representations of adults and children. *Anim. Cogn.* **2023**, *26*, 369–377. [CrossRef]

animals

MDPI

Article

Influence of Sex of Stranger on Responses of Shelter Dogs during Canine Behavioral Evaluations

Betty McGuire [1,*] and Andrew Song [2]

[1] Department of Ecology and Evolutionary Biology, Cornell University, Ithaca, NY 14853, USA
[2] Department of Information Science, Cornell University, Ithaca, NY 14853, USA; ajs557@cornell.edu
* Correspondence: bam65@cornell.edu

Simple Summary: Dogs are often less comfortable around unfamiliar men than unfamiliar women, yet most studies of stranger-directed aggression have not examined whether the sex of the stranger is a risk factor. We analyzed data collected by staff at a NY shelter to determine whether dogs responded differently to unfamiliar men and unfamiliar women during the Stranger test of the behavioral evaluation. Of the 283 dogs tested, 26 were undersocialized and from the same home; 19.2% were assessed as showing no concerning behavior during the Stranger test. Of the 257 remaining dogs, 89.9% were assessed as showing no concerning behavior. Dogs tested with a male stranger (n = 55) had significantly higher scores on the test, indicating greater uneasiness, than dogs tested with a female stranger (n = 202). However, the mean score for dogs tested with a male stranger (2.2) fell within the range of scores considered not concerning by the shelter (1–3). In summary, we found that the sex of a stranger influenced dog responses during the Stranger test, but in practice, our findings do not indicate that changes are needed in how shelters conduct or interpret tests for stranger-directed aggression. Our findings also underscore the importance of socialization for dogs.

Abstract: In many situations, domestic dogs display greater uneasiness with unfamiliar men than unfamiliar women. However, little is known about whether the sex of an unfamiliar person is a risk factor for stranger-directed aggression, especially with respect to behaviors less intense than biting. We analyzed data collected by behavioral staff over a 27-month period (n = 283 dogs) at a New York shelter to determine whether the sex of an unfamiliar person influenced behaviors assessed during the Stranger test of the canine behavioral evaluation. Scores ranged from 1 (calm and friendly) to 5 (will not approach stranger or unsafe to allow an approach). No concerning behaviors (scores 1–3) were assessed for 19.2% of 26 undersocialized dogs from one home and 89.9% of the remaining 257 dogs. Within the group of 257, those tested with a male stranger had significantly higher scores than those tested with a female stranger; the effect size was small to moderate. Thus, we found that dogs responded differently to male and female strangers during this testing situation, but from a practical standpoint, our findings do not warrant adjustments in how shelters conduct or interpret tests for stranger-directed aggression. Our findings also highlight the importance of early exposure to different people and situations for dogs.

Keywords: aggression; familiarity; risk factor; temperament test; undersocialized dogs

Citation: McGuire, B.; Song, A. Influence of Sex of Stranger on Responses of Shelter Dogs during Canine Behavioral Evaluations. *Animals* **2023**, *13*, 2461. https://doi.org/10.3390/ani13152461

Academic Editor: Lynette A. Hart

Received: 24 June 2023
Revised: 21 July 2023
Accepted: 28 July 2023
Published: 30 July 2023

1. Introduction

Domestic dogs—in shelters and other environments—respond differently to unfamiliar men and unfamiliar women, with the general pattern being greater uneasiness with unfamiliar men, sometimes more pronounced in male dogs than female dogs. Shelter dogs enrolled in an enhanced human interaction program showed improved sociability toward unfamiliar women but not unfamiliar men [1]. At another shelter, when an unfamiliar person stood in front of their cage, dogs watched and barked longer when the person was a man rather than a woman [2]. The sex of an unfamiliar walker influenced scent-marking behavior during leash walks of mature shelter dogs: when walked by an unfamiliar man, male dogs urinated less frequently, were more likely to revert to the juvenile urinary posture (no hindlimb raised), and were less likely to defecate than when walked by an unfamiliar woman [3]. The sex of an unfamiliar walker did not influence frequency of urination or urinary posture in female dogs; however, like male dogs, female dogs were less likely to defecate when walked by an unfamiliar man than an unfamiliar woman. At a different shelter, dogs displayed behaviors associated with vigilance (gazing at the handler) and stress (lip-licking) more frequently when walked by men than women and spent less time with their tail in the high position when walked by men than by women [4]. Dogs living in other environments show similar responses to unfamiliar men; those enrolled in a training program for guide dogs made less frequent physical contact with unfamiliar men than unfamiliar women [5]. Finally, in a commercial kennel environment, male dogs made less frequent direct physical contact with and spent less time near an unfamiliar man than an unfamiliar woman; in contrast, female dogs did not differ in their behavior toward an unfamiliar man and an unfamiliar woman [6].

In the same way, another dog-human interaction that has been examined with respect to familiarity is aggression directed by dogs toward people; in most studies, aggression directed at strangers is considered separately from aggression directed at household members. Stranger-directed aggression is commonly assessed in animal shelters that conduct canine behavioral evaluations before making dogs available for adoption; about 6–8% of dogs were assessed as showing concerning or dangerous behaviors in the presence of strangers [7,8]. Aggression toward unfamiliar people has also been studied in dogs living in homes. Prevalence measures based on owner-completed questionnaires vary widely, perhaps reflecting different scoring systems and dog populations (5–10%, 9–11; 24–26%, 12,13; 78%, 14). Research has focused also on identifying risk factors for stranger-directed aggression, including characteristics of the dog, the owner, and the home environment [9–18]. However, with the exception of literature on victim risk factors for dog bites, which typically reports a predominance of male victims [19–22], we could find no information on whether the sex of an unfamiliar person influences other dog behaviors associated with stranger-directed aggression, such as barking, raising hackles, and growling. Sex of the stranger was not reported in previous studies of shelter dogs [7,8] or included in those using owner-completed questionnaires, which asked how dogs respond to a "stranger" or an "unfamiliar person," without specifying sex [9–18].

In the present study, we examined whether dogs at a New York animal shelter reacted differently to an unfamiliar man or an unfamiliar woman during the Stranger test of the canine behavioral evaluation. During the Stranger test at this shelter, dogs are assessed as showing responses to the stranger ranging from calm and friendly to not being safe enough to allow the dog to approach the stranger. Based on findings from previous studies of dog reactions to unfamiliar men and unfamiliar women [1–6], we predicted that dogs tested with an unfamiliar man would display greater uneasiness than dogs tested with an unfamiliar woman. Given that studies have found uneasiness toward men to be more pronounced in male dogs than female dogs [3,6], whereas others have not found this pattern [1,2,4,5], we considered it equally likely for us to find either more pronounced uneasiness toward male strangers in male dogs or similar levels of uneasiness with male strangers in male and female dogs. In addition to providing information on an understudied risk factor for stranger-directed aggression generally, our findings may inform how shelters conduct and interpret tests with strangers. These shelter-related applications are important because results from behavioral evaluations influence the development of in-shelter handling plans (e.g., whether a dog has a limited team of volunteers who must be introduced by staff) and special adoption efforts to make the best possible match between dog and adopter (e.g., whether a potential adopter's household has frequent visitors).

2. Materials and Methods

This research was conducted under protocol 2012-0150, which was reapproved on 16 September 2021 by the Institutional Animal Care and Use Committee of Cornell University.

2.1. Study Shelter and Canine Records

We obtained canine records from the Tompkins County SPCA in Ithaca, NY, USA. The shelter is no-kill, open-admission, and uses scheduled intake and a conversation-based approach to adoptions. The records, entered by shelter staff into the PetPoint data management system, included behavioral results from the Stranger test of the shelter's canine behavioral evaluation (Section 2.3) as well as intake data, from which we retrieved demographic information for each dog evaluated (sex, reproductive status, and age). Our study began 1 January 2021 (the time when, upon our request, the name of the stranger began being routinely recorded by staff for each test) and ended 2 April 2023. This time period yielded records for 283 dogs, which represented all dogs behaviorally evaluated. Some dogs were admitted to the shelter but not behaviorally evaluated, and therefore, were not part of our sample; these dogs included those retrieved by owners shortly after their arrival at the shelter (behavioral evaluations typically occur 3 days after intake), those euthanized at their owner's request, and very rare instances in which dogs were identified at intake as unsafe to make available for adoption and euthanized. Behavioral evaluations might also be skipped for extremely undersocialized dogs, such as those not used to living inside a home and having routine contact with people; however, this, too, was a very rare occurrence (n = 1 dog during our study).

2.2. Dogs, Care, and Housing

The 283 dogs included juveniles (n = 55; from 4 months to <1 year), adults (n = 195; from 1 year to <8 years), and seniors (n = 33; ≥8 years); we did not include puppies because their behavioral evaluations differed from those of older dogs and their results were not entered into PetPoint. Included in the 283 dogs were 26 Chihuahuas or Chihuahua mixes (14 males and 12 females) from the same home. These dogs had very little exposure to people other than the owner (hereafter, described as undersocialized dogs). Most had been spayed or neutered before admission to the shelter (88.5%; 23/26) and all but one were either adults (57.7%; 15/26) or seniors (38.5%; 10/26). Given their unusual living situation before admission to the shelter, we excluded them from formal data analyses (Section 2.4). Chihuahuas and Chihuahua mixes from other homes were not excluded. Most dogs were mixed breeds; the number of purebred dogs in our study population was unknown (DNA testing was not performed).

We describe the housing and care of dogs from intake to behavioral testing, which is the relevant time period for the present study. Intake occurred in the Rescue building, where veterinary staff examined dogs, weighed them, and checked their teeth to estimate age. The intake exam also included vaccinations, flea control, a fecal exam, deworming, and a heartworm test. Following the intake exam, dogs were housed in the Rescue building in chain-link cages (indoor space, 2.2 m^2, and outdoor run, 3.5 m^2). Almost all dogs were individually housed; exceptions were made for dogs surrendered from the same household when staff judged the dogs should remain together (e.g., the undersocialized dogs were typically housed in pairs). Each cage contained a raised bed, blanket, water bowl, and toys; staff fed dogs between 08:00 and 09:00 h and between 14:00 and 15:00 h. Several times a day, staff either walked dogs or brought them to an outdoor enclosure. Typically, three days after intake, dogs were behaviorally evaluated and subsequently housed in the Pet Adoption Center, adjacent to the Rescue building.

2.3. Behavioral Evaluation

The canine behavioral evaluation at our study shelter is based on Sternberg's Assess-a-Pet [23], subsequently modified by Bollen and Horowitz [24]. Present at each test were an evaluator from the shelter's Behavior Program and a scribe, also a member of shelter staff. The evaluator was female, and three other female staff members rotated through the role of scribe, with one exception when a different female staff member (an animal control officer) served as scribe. The same female evaluator conducted all behavioral evaluations except two during our study period (n = 281 of 283 dogs); two of the female staff members typically serving as scribes conducted the other two evaluations. The evaluation includes nine tests, some of which have subtests. The first test, Cage presentation, takes place in the Rescue building, and then the dog is leashed and brought to a conference room in the Pet Adoption Center, where all subsequent tests are conducted. In Appendix A, we provide brief descriptions of all tests and subtests that are not the focus of the present study; below is a more detailed description of the Stranger test.

The Stranger test is the eighth test in the nine-test sequence of the canine behavioral evaluation (Appendix A). The stranger is at least 18 years old, unfamiliar to the dog, and may be either a male or female staff member or dog volunteer, wearing ordinary clothing (not scrubs). With the evaluator holding the dog's leash, the test begins when the stranger knocks on the door of the conference room and the evaluator says, "Come in!" The stranger enters the room, stops at least 2 m away from the dog, and stares at the dog for 3 s. Next, the stranger takes a step forward, stops, and reaches toward the dog. Finally, the stranger squats down, facing sideways, and speaks to the dog in a friendly voice. The dog's behavioral response to the stranger is scored as follows:

1. Remains calm and is friendly upon solicitation;
2. Is nervous about the stranger (ears back, tail tucked), but is friendly upon solicitation;
3. Alarm barks or growls and backs off completely, but is friendly upon solicitation;
4. Alarm barks, hackles up, growls, does not calm readily, but eventually is friendly in a cautious way upon solicitation;
5. Alarm barks, hackles up, growls, cannot settle, and either will not approach upon solicitation or is not safe to allow an approach.

The scribe writes her score on the evaluation form, and the evaluator reviews the score with reference to her own score. Any difference in score is discussed, and an agreement is reached regarding the score entered into PetPoint. Note that a particular score does not mean that a dog showed every behavior listed for that score. For example, dogs receiving a score of 5 may not alarm bark, raise hackles, or growl but simply avoid any interaction with the stranger; this information is recorded in the notes section of PetPoint. In the shelter scoring system, scores 1–3 are classified as "no concerning behavior", a score of 4 is considered "concerning behavior", and a score of 5 is "dangerous behavior". The behavioral responses of dogs that receive a score of 5 and will not approach the stranger are considered dangerous because of the potential for defensive aggression should someone attempt to force interaction despite a dog's attempts to get away or warn them away. Responses of dogs that receive a score of 5 and are assessed as unsafe to allow them to approach the stranger are considered dangerous because of the potential for offensive aggression; this assessment of unsafe to allow an approach is rare at our study shelter (e.g., of the 283 dogs in our sample, only one was considered unsafe to allow an approach; Section 3.1.1).

2.4. Data and Statistical Analyses

From each PetPoint record of the Stranger test, we retrieved the dog's name, identification number, score (1–5), date of test, evaluator's name, and all notes about the test, which included the name of the person serving as the stranger. From the name of the stranger, we assigned the sex of the stranger. BM has conducted research at this shelter for over 10 years, so she is very familiar with both dog volunteers and staff members, making the assignment of the sex of a stranger based on the recorded name straightforward. We checked paper copies of the evaluations to retrieve the name of the scribe for each test. Dogs that are adopted and returned to the shelter are behaviorally evaluated again; for dogs with two or more behavioral evaluations during our study period (n = 23), we used results from their first evaluation to make testing conditions across dogs as similar as possible.

For the 26 undersocialized dogs, we simply report descriptive information about their performance on the Stranger test; two were evaluated by staff members other than the main evaluator. Having excluded these 26 undersocialized dogs, we were left with 257 dogs for data analyses; all were assessed by the same evaluator.

Of the 257 tests, 55 (21.4%) involved a male stranger and 202 (78.6%) involved a female stranger; the much larger number of female strangers reflects the female bias in both dog volunteers and staff members (in most positions) at our study shelter [3]. Table 1 shows the demographic characteristics of the 257 dogs tested with either a male or female stranger. We used chi-square tests to evaluate whether dogs tested with a male stranger had generally similar demographic characteristics as those tested with a female stranger; we found no significant differences (sex: $X^2 = 0.05$, *d.f.* = 1, $p = 0.82$; reproductive status: $X^2 = 0.66$, *d.f.* = 1, $p = 0.42$; age class: $X^2 = 1.37$, *d.f.* = 2, $p = 0.51$).

Table 1. Demographic characteristics of shelter dogs in relation to whether they were tested with either a male stranger ($n = 55$) or a female stranger ($n = 202$) during the Stranger test of the canine behavioral evaluation.

Demographic Characteristics of Dogs	Tested with Male Stranger	Tested with Female Stranger
Sex		
Male	41.8% (23/55)	43.6% (88/202)
Female	58.2% (32/55)	56.4% (114/202)
Reproductive status		
Intact	47.3% (26/55)	53.5% (108/202)
Neutered/spayed	52.7% (29/55)	46.5% (94/202)
Age class [1]		
Juvenile	25.5% (14/55)	19.8% (40/202)
Adult	63.6% (35/55)	71.8% (145/202)
Senior	10.9% (6/55)	8.4% (17/202)

[1] Juveniles, from 4 months to <1 year; adults, from 1 year to <8 years; and seniors, ≥8 years.

We calculated the prevalence of concerning or dangerous behavior for the Stranger test, defined as the number of dogs assessed as showing either concerning or dangerous behavior/number of dogs tested. Separate prevalence measures are provided for the 26 undersocialized dogs and the 257 dogs. We used a standard least squares model to examine whether the sex of the stranger, along with the demographic characteristics of the 257 dogs—sex, reproductive status, and age class—predicted a dog's score on the Stranger test. We examined the main effects and the one two-way interaction we were interested in, sex of dog by sex of stranger. This interaction was not significant, so we dropped it from the final model ($F = 1.89$, *d.f.* = 1, $p = 0.17$). We used Student's *t*-tests when comparing means for both sex and reproductive status and Tukey's HSD to correct for multiple comparisons when comparing means for age class. Because of sample size differences, especially with regard to the number of dogs tested with either a male stranger or a female stranger, we used Hedges' g to calculate effect sizes. We set the p value threshold at $p < 0.05$ and used JMP Pro (version 15.0.0) for all statistical analyses except the calculation of effect sizes, which was performed using Social Sciences Statistics (https://www.socscistatistics.com (accessed on 28 May 2023).

3. Results

3.1. Main Study with 257 Dogs

3.1.1. Prevalence

The vast majority of dogs (89.9%; 231/257) were assessed as showing no concerning behavior during the Stranger test. The prevalence of concerning behavior was 6.6% (17/257), and the prevalence of dangerous behavior was 3.5% (9/257). During our study period, only one dog was assessed as unsafe to allow him to approach the stranger, but this was during his second behavioral evaluation, and we only used results from first evaluations.

He was tested upon first entering the shelter (score of 3 on the Stranger test), adopted and returned a few months later, and evaluated again (score of 5, unsafe to allow an approach). Nevertheless, he was successfully adopted after being on an in-shelter handling plan for 1 month (limited team of volunteers who must be introduced by staff). Thus, for the first behavioral evaluations, none of the dogs with a score of 5 was assessed as unsafe to allow them to approach the stranger; all would not approach upon solicitation.

Of the 26 dogs with a score of either 4 or 5, only one was described as trembling and frozen in place; this dog did not bark, raise hackles, or growl. The other 25 dogs with scores of 4 or 5 displayed at least one of these three behaviors.

3.1.2. Response to a Stranger in Relation to the Sex of the Stranger and Demographic Characteristics of Dogs

Descriptive statistics for scores on the Stranger test in relation to the sex of the stranger and dog demographic characteristics are shown in Table 2. Dogs tested with a male stranger had significantly higher scores on the Stranger test than dogs tested with a female stranger, indicating greater uneasiness with male strangers ($F = 6.81$, *d.f.* = 1, $p < 0.01$; Table 2). The effect size was in the small to moderate range (Hedges' $g = 0.36$).

Table 2. Score on the Stranger test of the canine behavioral evaluation in relation to the sex of the stranger and demographic characteristics of shelter dogs ($n = 257$). Scores ranged from 1 (calm and friendly) to 5 (alarm barks, hackles up, growls, cannot settle, and will not approach the stranger upon solicitation). Sample sizes of dogs are in parentheses. Within specific variables, values with different superscript letters are significantly different ($p < 0.05$).

Variables	Score on the Stranger Test (Mean ± *SD*)
Sex of stranger	
Male	2.22 ± 1.29 (55) [a]
Female	1.81 ± 1.09 (202) [b]
Sex of dog	
Male	1.92 ± 1.14 (111)
Female	1.89 ± 1.15 (146)
Reproductive status of dog	
Intact	2.06 ± 1.21 (134) [a]
Neutered/spayed	1.73 ± 1.05 (123) [b]
Age class of dog [1]	
Juvenile	1.80 ± 1.12 (54)
Adult	1.98 ± 1.18 (180)
Senior	1.52 ± 0.79 (23)

[1] Juveniles, from 4 months to <1 year; adults, from 1 year to <8 years; and seniors, ≥8 years.

Reproductive status significantly influenced scores on the Stranger test, with intact dogs having higher scores than spayed/neutered dogs ($F = 5.34$, *d.f.* = 1, $p < 0.03$; Table 2); the effect size was small ($g = 0.29$). Sex and age class of dogs did not influence scores on the Stranger test (sex: $F = 0.18$, *d.f.* = 1, $p = 0.67$; age class: $F = 1.98$, *d.f.* = 2, $p = 0.14$; Table 2).

3.2. *Undersocialized Dogs*

Scores on the Stranger test for the 26 undersocialized dogs from one household were as follows: 1 ($n = 1$); 2 ($n = 4$); 3 ($n = 0$); 4 ($n = 5$); and 5 ($n = 16$). Thus, 19.2% (5/26) of these dogs were assessed as showing no concerning behavior (scores 1–3), 19.2% as showing concerning behavior (score of 4), and 61.6% as showing dangerous behavior (score of 5). None of the dogs with a score of 5 was assessed as unsafe to allow them to approach the stranger; all would not approach the stranger upon solicitation. For all 21 dogs with a score of either 4 or 5, notes in the PetPoint files described them as showing the following responses: trembling, hiding, freezing, and shutting down. Notes further indicated these dogs did not engage in alarm barking, raising hackles, or growling. When considering all

26 undersocialized dogs, the mean score ($\pm SD$) for those tested with a male stranger was 4.83 ± 0.41 (n = 6) and for those tested with a female stranger, 4.00 ± 1.38 (n = 20).

All of the undersocialized dogs were released for adoption. All 26 were adopted, and three were returned (11.5%): one for biting when the adopter tried to remove him from a hiding place, one for warning behaviors but no biting, and the other after a year, when the adopter could no longer care for her (this dog was the only one of the 26 to have a score of 1 on the Stranger test). All three dogs were adopted again and not returned to the shelter.

4. Discussion

The vast majority of the 257 dogs behaviorally evaluated during our study period (2021–2023) were assessed as showing no concerning behavior on the Stranger test (89.9%), with the prevalence of concerning or dangerous behavior at 10.1% (6.6% concerning + 3.5% dangerous). A prevalence of 10.1% is slightly higher than the 5.8% previously reported for dogs tested at this shelter from 2014 to 2019 [7]. This likely reflects that the current study included all dogs tested, whereas McGuire et al. [7] were studying length of stay at the shelter, so they only included dogs that were tested, released for adoption, and housed at the shelter until adoption (i.e., not in foster homes; typically, dogs placed in foster homes are either especially fearful or have medical conditions such that a home environment is more suitable than the shelter environment). Thus, in contrast to McGuire et al. [7], our study population included dogs that were tested and either returned to their owner, transferred to a rescue, or placed in a foster home; some of the dogs transferred to a rescue or placed in a foster home likely had more intense responses during the behavioral evaluation, possibly towards strangers, than those tested, released for adoption, and housed at the shelter. Our prevalence measure of 10.1% is generally similar to those from another shelter [8] and some studies using owner-completed questionnaires [9–11]. Other studies based on owner-completed questionnaires reported higher prevalence measures for stranger-directed aggression [12–14].

Dogs tested with a male stranger had higher scores on the Stranger test, which indicates greater uneasiness, than dogs tested with a female stranger. Although this difference was statistically significant, the effect size was in the small to moderate range, and both means (male stranger, 2.22, and female stranger, 1.81) fell within the range of scores considered no concerning behavior on the test (1–3). We did not find a significant interaction between the sex of the dog and the sex of the stranger, indicating that male and female dogs responded in a similar manner to male strangers. Our findings from the Stranger test are consistent with results from other studies indicating greater uneasiness with unfamiliar men than unfamiliar women in both male and female shelter dogs [1,2,4] and dogs in a guide dog training program [5]. In an earlier study at the Tompkins County SPCA, both male and female dogs were less likely to defecate when walked by an unfamiliar man than an unfamiliar woman, again indicating similar levels of uneasiness with male strangers [3]. However, male but not female dogs displayed lower rates of urination and were more likely to revert to the juvenile urinary posture when walked by an unfamiliar man than by an unfamiliar woman [3]. The sex of the dog seems to modulate the response differently based on the specific behavior under study.

In regard to the greater uneasiness shown by dogs toward unfamiliar men than unfamiliar women in our study, it is possible that—from a dog's perspective—unfamiliarity with men has two components. First, male staff members and male dog volunteers serving as strangers are unknown to the dogs as individuals, thus "unfamiliar". Second, given the female bias in staff and dog volunteers at our study shelter [3], male strangers are also members of a less familiar group, human males. The possibility that unfamiliarity could have two components might suggest that shelters make efforts to recruit more males when filling staff and volunteer positions; this could ease the first interactions between dogs and potential adopters who are male.

Of the three demographic characteristics examined, only reproductive status was a significant predictor of score on the Stranger test, with scores of intact dogs higher than those of neutered and spayed dogs. However, the effect size was small, and the means were within the range considered to be no concerning behavior (intact, 2.06; neutered/spayed, 1.73). In an earlier study based on nearly 5 years of data at our study shelter, reproductive status was not significantly associated with the likelihood of concerning or dangerous behavior on the Stranger test [25]. Studies using owner-completed questionnaires reported no effect of reproductive status on stranger-directed aggression [15,18], and a study based on phone interviews found no decrease in stranger-directed aggression in male dogs following neutering [26]. Farhoody et al. [17] reported a significant increase in the odds of stranger-directed aggression in neutered or spayed dogs when compared to intact dogs, but this result was driven by dogs whose surgeries were performed between 7 and 12 months of age. The reasons for these divergent findings are unclear.

We found that the sex of the dog did not influence scores on the Stranger test. Our finding of no effect of dog sex on stranger-directed aggression is consistent with those based on owner-completed questionnaires [12,15,18]. In contrast, Takeuchi et al. [27] found that male dogs tended to be over-represented in cases of stranger-directed aggression at a behavior clinic. Finally, we found that the age of the dog did not influence scores on the Stranger test, which agrees with findings from one study based on an owner-completed questionnaire [18]. However, other studies have found either an increase in stranger-directed aggression with age [9,12,15] or mixed results, with adults more likely to be aggressive than adolescents and seniors [14]. Our present findings that neither sex nor age class influenced scores on the Stranger test are consistent with findings from earlier research at our study shelter [25].

The prevalence of concerning or dangerous behavior during the Stranger test for the 26 undersocialized dogs from one home was 80.8% (19.2% concerning + 61.6% dangerous). None of these dogs with concerning or dangerous behavior displayed alarm barking, raising hackles, or growling. Instead, they hid, trembled, and froze, and those with a score of 5 would not approach the stranger, leaving open the possibility for defensive aggression should their signals be ignored and interaction forced. The conditions of the Stranger test (e.g., an unfamiliar room in the shelter, two new people with whom they have only spent a few minutes, then a knock on the door, followed by a third person they have never met) seemed especially stressful for these dogs, on top of the already challenging conditions of shelter life (e.g., loud sounds, new smells, many unfamiliar people and dogs). The responses displayed by the undersocialized dogs in our study are typical of moderately to extremely fearful dogs in shelters, of which about one-third come from hoarding situations [28], the source of the dogs in our study. The high prevalence of fearful behaviors in these dogs highlights the importance of exposing dogs, especially when young, to different people and situations [29]. Despite their fearfulness in the shelter, all were adopted, and only three were returned (11.5%); this likelihood of return is similar to the 13.7% reported for small dogs at this shelter from 2014 to 2019 [30]. The three returned dogs were subsequently adopted and not returned to the shelter. Collins et al. [28] reported similar positive outcomes from a much larger sample of fearful dogs that went through a behavioral rehabilitation program: 99% of dogs that completed the program were adopted, with 88% of adopters very satisfied and 8% somewhat satisfied with their dog.

Our study has several limitations. First, canine behavioral evaluations at shelters have been criticized for their lack of rigorous scientific validation and poor predictability of many post-adoption behaviors [8,31–33]. However, proponents view these evaluations as one of several sources of information about dogs in shelters [24,34]. Second, although the Stranger test attempts to mimic a stranger knocking on the door of a dog's home, the contexts and settings of the two situations are very different. Whereas dogs in their homes may be defending household members and/or their home area, dogs in shelters have typically been housed there for only a few days before testing and have interacted with the evaluator and scribe for only a few minutes before someone the dog has never met knocks on the conference room door. Additionally, the dogs are leashed during the Stranger test, whereas dogs are likely not leashed in their home when a stranger approaches.

The fact that our measure of prevalence of concerning or dangerous behavior during the Stranger test (10.1% for the 257 dogs) is similar to measures found in some studies based on owner-completed questionnaires [9–11] suggests that perhaps the sound of knocking signals to dogs the arrival of a stranger, even in a setting very different from a home [25]. Third, shelters are challenging environments for dogs, and certain aspects of the Stranger test (e.g., the door opening following knocking and possible changes in leash tension) might induce fear or arousal in the dogs right before they see the stranger, potentially resulting in false positives. Fourth, because we did not videotape the Stranger tests, we could not assess intra-rater reliability, which would be useful given the same evaluator conducted all of the Stranger tests for the 257 dogs formally analyzed. Fifth, we did not determine the particular stimuli used by the dogs to discriminate the sex of the stranger. In the setting of the Stranger test, potential stimuli include visual (e.g., size and appearance of the stranger), olfactory (e.g., androgen-based cues), and auditory cues (e.g., voice of the stranger). Sixth, because our data came from a single shelter, our findings may not apply to other shelters with different dog populations and behavioral evaluations [35,36]. Finally, given the female bias in the two sources of strangers at our study shelter—dog volunteers and shelter staff [3]—our sample of dogs tested with a male stranger (n = 55) was much smaller than our sample tested with a female stranger (n = 202). Herzog [37] recently highlighted the large numbers of women in professions related to animal care and research on human-animal relationships; he also noted that studies of the human-animal bond often have too few male participants. Herzog [37] suggested authors make the gender breakdown of study participants clear, which we have done here for those administering the Stranger tests and serving as the stranger. We also checked the demographic characteristics of dogs tested with male strangers versus female strangers to make sure the two groups of dogs were similar in terms of sex, reproductive status, and age class.

The present study used a between-subjects experimental design. It would be interesting to examine dogs' reactions to male and female strangers using a within-subject design. Given questions about whether Stranger tests with shelter dogs adequately mimic situations in homes, it would be best to conduct this research with dogs in their own homes. Such research might further contribute to our understanding of victim risk factors for canine aggression directed at strangers.

5. Conclusions

We found that dogs tested with an unfamiliar male had significantly higher scores than dogs tested with an unfamiliar female on the Stranger test of the canine behavioral evaluation; this finding provides another example of dogs responding differently to unfamiliar men and unfamiliar women. However, the effect size was small to moderate, and the mean score for dogs tested with a male stranger fell within the range of scores grouped as no concerning behavior on this particular test of the evaluation. From a practical standpoint, our findings do not indicate that changes are needed in how shelters conduct or interpret tests for stranger-directed aggression with regard to the sex of the stranger.

Although unanticipated and descriptive, our data from the undersocialized Chihuahuas and Chihuahua mixes revealed that most were assessed as showing concerning or dangerous behavior during the Stranger test, yet all were adopted, and the likelihood of return for them was similar to that previously recorded for all small dogs at our study shelter over a 5-year period [30]. These data from 26 undersocialized small dogs from a single home, together with data from a much larger number of dogs of all sizes displaying moderate to extreme fear in the shelter environment [28], show that positive outcomes for fearful dogs are not only possible but likely when given time and behavioral support.

Supplementary Materials: The following are available online at https://www.mdpi.com/article/10.3390/ani13152461/s1, File S1 McGuire and Song Stranger Test Data.

Author Contributions: Conceptualization, B.M.; supervision, B.M.; methodology, B.M.; data collection and curation, B.M. and A.S.; statistical analyses, B.M. and A.S.; writing—original draft preparation, B.M. and A.S.; writing—review and editing, B.M. and A.S. All authors have read and agreed to the published version of the manuscript.

Funding: This research received no external funding.

Data Availability Statement: All data from this study are included in Supplementary Materials.

Acknowledgments: Jim Bouderau, Executive Director of the Tompkins County SPCA, gave us permission to analyze dog records. Emme Hones, Behavior Program Manager at the shelter, agreed to record the name of the stranger for us beginning in January 2021, provided PetPoint files with dog intake data and Stranger test results, and shared her knowledge and expertise regarding canine behavioral evaluations. Stephen Parry provided statistical advice. We thank four anonymous reviewers for their helpful comments; one suggested the idea that, from a dog's perspective, there may be two components to unfamiliarity with regard to human males at shelters where the majority of staff members and dog volunteers are female.

Conflicts of Interest: The authors declare no conflict of interest.

Appendix A

Table A1 provides brief descriptions of the tests and subtests of the canine behavioral evaluation at our study shelter. Details of the Stranger test are presented in Section 2.3.

Table A1. Tests and subtests of the Tompkins County SPCA's canine behavioral evaluation.

Test	Subtest	Evaluator's Actions
Cage presentation	Confrontational	Faces dog, bends at the waist, and makes direct eye contact
	Friendly	Faces sideways, bends down, and speaks to the dog in a friendly manner
Sociability	Stand and ignore	Stands and ignores the dog for 60 s
	Stroke three times	Slowly strokes dog from neck to tail three times
	Sit and ignore	Sits and ignores dog for 5 s
	Pet and talk	Sits, pets, and talks to the dog for 20 s in a friendly manner
Teeth exam		Makes five attempts to lift the dog's upper lip and hold for 5 s
Handling	Far side	Stands perpendicular to the dog and strokes its far side
	Hind foot	Runs hand down the dog's back and picks up hind foot
	Tail tug	Runs hand down the dog's tail and tugs slightly
	Check ears	Touches and looks inside both ears
	Press shoulders	Applies slight pressure to the dog's shoulders
	Lead by collar	Moves the dog around by collar
	Wipe with towel	Wipes the dog's body with a towel
	Hug	Gives the dog a hug
Arousal		Initiates play with the dog with toys for 30 s and stops
Food bowl		Gives the dog a mix of kibble and canned food in a bowl, and using the Assess-a-Hand, strokes the dog's back and attempts to pull the bowl away
Possession		Gives the dog a raw hide chew or pig's ear, and using the Assess-a-Hand, attempts to take the item away
Stranger		See Section 2.3
Dog-to-dog [1]	See a dog	Holds leash of the dog being tested while the scribe brings into the conference room a leashed, unfamiliar dog from the adoption floor
	Meet a dog	Allows dogs to interact while leashed

[1] Previously tested dogs that did not show aggression toward other dogs during their own evaluation serve as stimulus dogs in this test.

References

1. Bergamasco, L.; Osella, M.C.; Savarino, P.; Larosa, G.; Ozella, L.; Manassero, M.; Badino, P.; Odore, R.; Barbero, R.; Re, G. Heart rate variability and saliva cortisol assessment in shelter dog: Human–animal interaction effects. *Appl. Anim. Behav. Sci.* **2010**, *125*, 56–68. [CrossRef]
2. Wells, D.L.; Hepper, P.G. Male and female dogs respond differently to men and women. *Appl. Anim. Behav. Sci.* **1999**, *61*, 341–349. [CrossRef]
3. McGuire, B.; Fry, K.; Orantes, D.; Underkofler, L.; Parry, S. Sex of walker influences scent-marking behavior of shelter dogs. *Animals* **2020**, *10*, 632. [CrossRef] [PubMed]
4. Shih, H.-Y.; Paterson, M.B.A.; Georgiou, F.; Pachana, N.A.; Phillips, C.J.C. Who is pulling the leash? Effects of human gender and dog sex on human–dog dyads when walking on-leash. *Animals* **2020**, *10*, 1894. [CrossRef]
5. Koda, N.; Shimoju, S. Human-dog interactions in a guide-dog training program. *Psychol. Rep.* **1999**, *84*, 1115–1121. [CrossRef]
6. Lore, R.K.; Eisenberg, F.B. Avoidance reactions of domestic dogs to unfamiliar male and female humans in a kennel setting. *Appl. Anim. Behav. Sci.* **1986**, *15*, 261–266. [CrossRef]
7. McGuire, B.; Chan, J.; Jean-Baptiste, K.; Kok, P.; Rosenbaum, E. Results of behavioral evaluations predict length of stay for shelter dogs. *Animals* **2021**, *11*, 3272. [CrossRef]
8. Clay, L.; Paterson, M.B.A.; Bennett, P.; Perry, G.; Phillips, C.J.C. Do behaviour assessments in a shelter predict the behaviour of dogs post-adoption? *Animals* **2020**, *10*, 1225. [CrossRef]
9. Casey, R.A.; Loftus, B.; Bolster, C.; Richards, G.J.; Blackwell, E.J. Human directed aggression in domestic dogs (*Canis familiaris*): Occurrence in different contexts and risk factors. *Appl. Anim. Behav. Sci.* **2014**, *152*, 52–63. [CrossRef]
10. Salonen, M.; Sulkama, S.; Mikkola, S.; Puurunen, J.; Hakanen, E.; Tiira, K.; Araujo, C.; Lohi, H. Prevalence, comorbidity, and breed differences in canine anxiety in 13,700 Finnish pet dogs. *Sci. Rep.* **2020**, *10*, 2962. [CrossRef]
11. Dinwoodie, I.R.; Dwyer, B.; Zottola, V.; Gleason, D.; Dodman, N.H. Demographics and comorbidity of behavior problems in dogs. *J. Vet. Behav.* **2019**, *32*, 62–71. [CrossRef]
12. Didehban, N.; Borujeni, M.P.; Avizeh, R.; Mosallanejad, B. Problematic behaviors in companion dogs: A survey of their prevalence and associated factors. *J. Vet. Behav.* **2020**, *39*, 6–13. [CrossRef]
13. Khoshnegah, J.; Azizzadeh, M.; Gharaie, A.M. Risk factors for the development of behavior problems in a population of Iranian domestic dogs: Results of a pilot study. *Appl. Anim. Behav. Sci.* **2011**, *131*, 123–130. [CrossRef]
14. Flint, H.E.; Coe, J.B.; Serpell, J.A.; Pearl, D.L.; Niel, L. Risk factors associated with stranger-directed aggression in domestic dogs. *Appl. Anim. Behav. Sci.* **2017**, *197*, 45–54. [CrossRef]
15. Blackwell, E.J.; Twells, C.; Seawright, A.; Casey, R.A. The relationship between training methods and the occurrence of behavior problems, as reported by owners, in a population of domestic dogs. *J. Vet. Behav.* **2008**, *3*, 207–217. [CrossRef]
16. Duffy, D.L.; Hsu, Y.; Serpell, J.A. Breed differences in canine aggression. *Appl. Anim. Behav. Sci.* **2008**, *114*, 441–460. [CrossRef]
17. Farhoody, P.; Mallawaarachchi, I.; Tarwater, P.M.; Serpell, J.A.; Duffy, D.L.; Zink, C. Aggression toward familiar people, strangers, and conspecifics in gonadectomized and intact dogs. *Front. Vet. Sci.* **2018**, *5*, 18. [CrossRef]
18. Hsu, Y.; Sun, L. Factors associated with aggressive responses in pet dogs. *Appl. Anim. Behav. Sci.* **2010**, *123*, 108–123. [CrossRef]
19. Overall, K.L.; Love, M. Dog bites to humans—Demography, epidemiology, injury, and risk. *JAVMA* **2001**, *218*, 1923–1934. [CrossRef]
20. Westgarth, C.; Brooke, M.; Christley, R.M. How many people have been bitten by dogs? A cross-sectional survey of prevalence, incidence and factors associated with dog bites in a UK community. *J. Epidemiol. Community Health* **2018**, *72*, 331–336. [CrossRef]
21. Ozanne-Smith, J.; Ashby, K.; Stathakis, V.Z. Dog bite and injury prevention—Analysis, critical review, and research agenda. *Inj. Prev.* **2001**, *7*, 321–326. [CrossRef] [PubMed]
22. Daigle, L.; Delesalle, L.; Ravel, A.; Ford, B.; Aenishaenslin, C. Occurrence and risk factors of dog bites in northern indigenous communities: A scoping review. *Front. Vet. Sci.* **2022**, *9*, 777640. [CrossRef] [PubMed]
23. Sternberg, S. *Assess-A-Pet: The Manual*; Assess-A-Pet: New York, NY, USA, 2006.
24. Bollen, K.S.; Horowitz, J. Behavioral evaluation and demographic information in the assessment of aggressiveness in shelter dogs. *Appl. Anim. Behav. Sci.* **2008**, *112*, 120–135. [CrossRef]
25. McGuire, B.; Jean-Baptiste, K. Relationships between demographic characteristics of shelter dogs and performance on tests of a behavioral evaluation and between performance and adoption success. *J. Vet. Behav.* 2023, *in press*. [CrossRef]
26. Neilson, J.C.; Eckstein, R.A.; Hart, B.L. Effects of castration on problem behaviors in male dogs with reference to age and duration of behavior. *JAVMA* **1997**, *211*, 180–182.
27. Takeuchi, Y.; Ogata, N.; Houpt, K.A.; Scarlett, J.M. Differences in background and outcome of three behavior problems of dogs. *Appl. Anim. Behav. Sci.* **2001**, *70*, 297–308. [CrossRef]
28. Collins, K.; Miller, K.; Zverina, L.; Patterson-Kane, E.; Cussen, V.; Reid, P. Behavioral rehabilitation of extremely fearful dogs: Report on the efficacy of a treatment protocol. *Appl. Anim. Behav. Sci.* **2022**, *254*, 105689. [CrossRef]
29. Serpell, J.; Duffy, D.L.; Jagoe, J.A. Becoming a dog: Early experience and the development of behavior. In *The Domestic Dog: Its Evolution, Behavior and Interactions with People*, 2nd ed.; Serpell, J., Ed.; Cambridge University Press: New York, NY, USA, 2016; pp. 93–117. [CrossRef]
30. McGuire, B. Characteristics and Adoption Success of Shelter Dogs Assessed as Resource Guarders. *Animals* **2019**, *9*, 982. [CrossRef]

31. Patronek, G.J.; Bradley, J. No better than flipping a coin: Reconsidering canine behavior evaluations in animal shelters. *J. Vet. Behav.* **2016**, *15*, 66–77. [CrossRef]
32. Patronek, G.J.; Bradley, J.; Arps, E. What is the evidence for reliability and validity of behavior evaluations for shelter dogs? A prequel to "No better than flipping a coin". *J. Vet. Behav.* **2019**, *31*, 43–58. [CrossRef]
33. McGuire, B.; Orantes, D.; Xue, S.; Parry, S. Abilities of canine shelter behavioral evaluations and owner surrender profiles to predict resource guarding in adoptive homes. *Animals* **2020**, *10*, 1702. [CrossRef]
34. Clay, L.; Paterson, M.; Bennett, P.; Perry, G.; Rohlf, V.; Phillips, C.J.C. In defense of canine behavioral assessments in shelters: Outlining their positive applications. *J. Vet. Behav.* **2020**, *38*, 74–81. [CrossRef]
35. Bradley, J.; Rajendran, S. Increasing adoption rates at animal shelters: A two-phase approach to predict length of stay and optimal shelter allocation. *BMC Vet. Res.* **2021**, *17*, 70. [CrossRef] [PubMed]
36. Cain, C.J.; Woodruff, K.A.; Smith, D.R. Phenotypic characteristics associated with shelter dog adoption in the United States. *Animals* **2020**, *10*, 1959. [CrossRef] [PubMed]
37. Herzog, H. Women Dominate Research on Human-Animal Bond. *Am. Psychol.* **2021**. Available online: https://www.psychologytoday.com/us/blog/animals-and-us/202105/women-dominate-research-the-human-animal-bond (accessed on 22 April 2023).

animals

MDPI

Article

Exploring the Domestication Syndrome Hypothesis in Dogs: Pigmentation Does Not Predict Cortisol Levels

JoAnna M. Platzer *, Lisa M. Gunter and Erica N. Feuerbacher

School of Animal Sciences, Virginia Polytechnic Institute and State University (Virginia Tech), Blacksburg, VA 24061, USA; lisagunter@vt.edu (L.M.G.); enf007@vt.edu (E.N.F.)
* Correspondence: jplatzer@vt.edu

Simple Summary: Domestic dogs have a wide variety of colorations, and previous research has found that, in certain breeds, coat color can be linked to behavior. However, it is unknown if coloration is connected to dogs' stress responses. To explore this question, we studied dogs living under stressful conditions: an animal shelter. We analyzed their urinary levels of cortisol, a stress hormone, to explore whether values from the shelter and on outings with people correlated with their coloration, specifically, their coat color/pattern, nose color, and extent of white spotting. In this preliminary study, we did not find a connection between their cortisol levels and coloration. While more research is needed, these initial findings do not suggest that dogs differ in their stress responses as a result of coloration alone.

Abstract: Previous research has found connections between pigmentation, behavior, and the physiological stress response in both wild and domestic animals; however, to date, no extensive research has been devoted to answering these questions in domestic dogs. Modern dogs are exposed to a variety of stressors; one well-studied stressor is residing in an animal shelter. To explore the possible relationships between dogs' responses to stress and their pigmentation, we conducted statistical analyses of the cortisol:creatinine ratios of 208 American shelter dogs as a function of their coat color/pattern, eumelanin pigmentation, or white spotting. These dogs had been enrolled in previous welfare studies investigating the effect of interventions during which they left the animal shelter and spent time with humans. In the current investigation, we visually phenotype dogs based on photographs in order to classify their pigmentation and then conduct post hoc analyses to examine whether they differentially experience stress as a function of pigmentation. We found that the dogs did not differ significantly in their urinary cortisol:creatinine ratios based on coat color/pattern, eumelanin pigmentation, or white spotting, either while they were residing in the animal shelter or during the human interaction intervention. These preliminary data suggest that pigmentation alone does not predict the stress responses of shelter dogs; however, due to the small sample size and retrospective nature of the study, more research is needed.

Keywords: shelter dogs; cortisol; stress; pigmentation; morphology; the domestication syndrome hypothesis

Citation: Platzer, J.M.; Gunter, L.M.; Feuerbacher, E.N. Exploring the Domestication Syndrome Hypothesis in Dogs: Pigmentation Does Not Predict Cortisol Levels. *Animals* **2023**, *13*, 3095. https://doi.org/10.3390/ani13193095

Academic Editor: Betty McGuire

Received: 28 July 2023
Revised: 19 September 2023
Accepted: 27 September 2023
Published: 4 October 2023

1. Introduction

1.1. The Domestication Syndrome Hypothesis

Domesticated animals differ considerably from their wild relatives and sometimes within the domesticated phenotypes seemingly unrelated traits correlate with each other, an observation dating back to Charles Darwin [1]. Behaviorally, domesticated animals have reduced reactivity to humans and, as such, are less likely than their wild counterparts to behave fearfully or aggressively towards humans [2]. Physiologically, their stress response systems are also less reactive to the same stressors [3,4]. Morphologically, domesticated animals are often depigmented, with floppier ears, curlier tails, and more neotenic craniofacial structures, among other traits [5,6]. The commonality of these traits and the possibility of a universal causative mechanism underlying this phenomenon is the basis of the domestication syndrome hypothesis [7].

Many research studies support the domestication syndrome hypothesis. Notably, in their investigation of experimental domestication, scientists from the Institute of Cytology and Genetics in Novosibirsk, Russia, found that breeding foxes strictly on the basis of their behavior towards humans, selecting the least fearful and aggressive animals to breed, also led to physiological and morphological changes. The selected line of tame foxes showed increased frequencies of many of the traits included in the domestication syndrome hypothesis; the first morphological change observed was coat depigmentation. This depigmentation occurred in the form of white spotting and brown mottling. Physiological changes occurred as well: foxes that were selected for decreased behavioral reactivity also displayed lower physiological reactivity to human handling. Tame foxes had lower baseline cortisol levels and showed smaller increases in cortisol in response to an acute stressor involving human handling [7]. These seminal results demonstrate that morphological and physiological changes can arise without direct selection for those traits.

Scientists have explored possible causal mechanisms of the domestication syndrome hypothesis. One plausible explanation is the neural crest hypothesis [8], which posits that animals with reduced behavioral reactivity towards humans likely have a less active hypothalamic–pituitary–adrenal (HPA) axis [3,4]. Consequently, selecting animals for breeding that exhibit reduced behavioral reactivity might produce animals with less responsive HPA axes, and a mildly deficient neural crest might explain this diminished HPA axis responsivity. The neural crest is a group of cells that crucially contribute to the development of various tissues implicated in traits associated with the domestication syndrome, such as pigment-producing melanocytes [8]. A correlation between morphological (e.g., pigmentation) and physiological (e.g., HPA axis reactivity) traits might support the possibility of a causative mechanism underlying the domestication syndrome hypothesis.

1.2. Dog Pigmentation Overview

Many domesticated mammals display a striking variety of colors and patterns, the most pronounced of which is the domestic dog (*Canis lupus familiaris*) [9]. The mutations leading to changes in mammalian pigmentation can occur early in the domestication process, and humans will actively select for novel variations for aesthetic, superstitious, and functional reasons [10–14]. Variation away from wild-type coloration in animals living in their native environments can reduce camouflage and thereby diminish fitness; however, these natural selection pressures are relaxed in the anthropogenic niche [10]. Even modern free-ranging dogs that primarily breed without human intervention, living on the fringes of human society, exhibit variation in pigmentation [15].

The wide variety of pigmentation in dogs is due to just two pigments, eumelanin and pheomelanin, and the depigmentation of both. Eumelanin is typically black, but various mutations have resulted in eumelanin being expressed in shades of silvery gray ("blue"), brown ("liver"), or silvery tan ("isabella") [9]. Pheomelanin is typically reddish yellow, but less-understood *intensity* mutations have resulted in pheomelanin expression in shades of red, orange, tan, yellow, cream, or nearly white [16]. This lightening of eumelanin and pheomelanin is a form of depigmentation, however depigmentation can also result in the total absence of pigment [7]. A very common form of depigmentation is white spotting [17]. White spotting can be thought of as "erasing" pigment: where white spotting is present, no pigment is produced [18]. White spotting causes pink skin and white fur. The expression of white spotting can range from relatively small areas (e.g., a white dot on the chest) to covering nearly the entire dog (e.g., the dog appears almost completely white).

The color of the eumelanin and pheomelanin and the depigmentation of these two pigments are just a few factors impacting the dog's appearance. Another factor is how these pigments are distributed across the coat [9]; it is this distribution that determines the dog's coat pattern. Eumelanin and pheomelanin can be distributed in different areas (e.g., alternating stripes of eumelanin and pheomelanin produce the coat pattern known as brindle) and even banded on the same hair together (e.g., agouti). Dogs can also have coats colored entirely by eumelanin or pheomelanin, resulting in, for example, black and yellow Labrador retrievers, respectively. Eumelanin also colors the nose and skin of dogs, such that the shade of a dog's eumelanin is visible on its nose (except in rare instances when the nose is fully depigmented) even if its coat has no eumelanin.

To summarize, a dog's coat pattern is determined by the distribution of eumelanin and pheomelanin. If a dog has only one of these pigments present in its fur, that dog will be solid-colored. In cases of phaeomelanistic fur, we can still gather information about the dog's eumelanin pigmentation from their nose color. Depigmentation further impacts the dog's appearance: various shades of eumelanin and phaeomelanin are possible and a dog might also display white spotting over any portion of its body. Just as dogs possess a variety of pigmentation phenotypes, they also display many behavioral phenotypes. Recent research suggests there might be a connection between these variables.

1.3. Dogs and Stress

Pet dogs experience a multitude of stressors in our anthropogenic world. Entering the animal sheltering system is a well-documented stressor for dogs, likely due in part to the social isolation [19], spatial restrictions [19], and excessive noise [20] they experience in this environment. Cortisol is commonly used to evaluate an animal's stress levels [21] and its concentration can be measured in several bodily samples, however urine and feces are often preferred as they are considered the least invasive to collect [22,23].

Dogs in shelters have higher cortisol levels than dogs in homes [24,25]; as such, those conducting research in animal shelters often focus on identifying interventions that reduce dogs' stress levels. Some of the most successful interventions involve interactions with humans. Even temporarily removing dogs from a shelter can reduce their stress levels, although spending time in foster homes [26] is a better intervention for decreasing dogs' cortisol levels than short outings into the community [27]. However, dogs' cortisol levels vary across individuals during their time in the shelter [28]. Despite evidence that pigmentation can be predictive of dogs' behavior [29–34], no study to our knowledge has investigated whether dogs' pigmentation correlates with cortisol levels, particularly those living in animal shelters.

Previous research has found connections between pigmentation and glucocorticoids in other species [35]. As mutations often have pleiotropic effects [36], it is possible that selection for certain pigmentation types has incidentally selected for changes in these animals' stress response systems. Furthermore, because stress can impact a dog's behavior [37], it is possible that differential sensitivity to stress can underlie the behavioral differences observed in differently pigmented dogs. Indeed, previous research has found correlations between dogs' behavior and their pigmentation supporting the plausibility of this hypothesis [29–34]. Thus, when we consider the range of morphological and behavioral variability that shelter dogs display, these animals are a useful population to explore questions about morphology, physiology, and behavior in present-day domestic dogs. More specifically, if pigmentation is a predictor of stress susceptibility in dogs, phenotype-based interventions designed to reduce their stress levels can be further explored. In the present study, we utilize the urinary cortisol:creatinine ratios of shelter dogs exposed to human–animal interventions outside of the animal shelter to uncover the relationships between our three pigmentation variables of interest: coat pattern, eumelanin pigmentation, and white spotting.

2. Materials and Methods

We visually phenotyped dogs aged six months and older living in American animal shelters that had been enrolled in previous research studies designed to evaluate the effects of short-term outings and weeklong fostering on shelter dog welfare [27,38]. We then utilized the cortisol:creatinine ratio data from these studies in order to analyze the morphological and physiological data to explore our research questions.

2.1. Visual Phenotyping

We classified the dogs according to the visual presentation of their pigmentation using photographic visual phenotyping. Photographs were sourced from animal shelter websites and social media, as well as taken in-person by the research team. Using these photographs, we categorized the dogs according to their coat pattern, eumelanin color, and amount of white spotting. All three variables were independent of one another, such that a dog's classification in one category had no bearing on its classification in another.

2.1.1. Coat Pattern

We classified the dogs as solid eumelanin (Figure 1A), brindle (Figure 1B), solid pheomelanin (Figure 1C), shaded yellow (Figure 1D), agouti (Figure 1E), black saddle (i.e., saddleback or creeping tan; Figure 1F), and black back (Figure 1G), based on Bannasch et al. [39] and Brancalion et al. [9], according to the apparent distribution of eumelanin and pheomelanin across their coats. Based on the small sample sizes, we pooled together the phenotypes of black saddle (n = 8) and black back (n = 16) into a single category called "black with tan" and removed the category agouti (n = 1) prior to analysis. Our sample included 88 dogs with coats of solid eumelanin, 34 with brindle coats, 42 with coats of solid pheomelanin, 20 with shaded yellow coats, and 24 with black with tan coats (see proportional breakdown in Figure 2). Thus, our coat pattern analysis included 207 dogs.

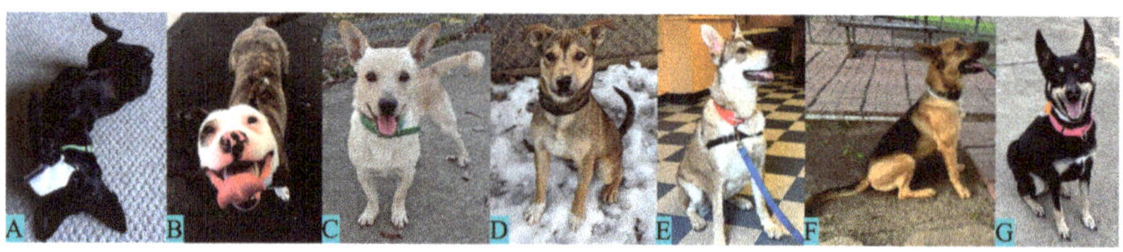

Figure 1. Coat pattern categorization: (**A**): solid eumelanin, (**B**): brindle, (**C**): solid pheomelanin, (**D**): shaded yellow, (**E**): agouti (not included in analysis due to small sample size), (**F**): black saddle, and (**G**): black back.

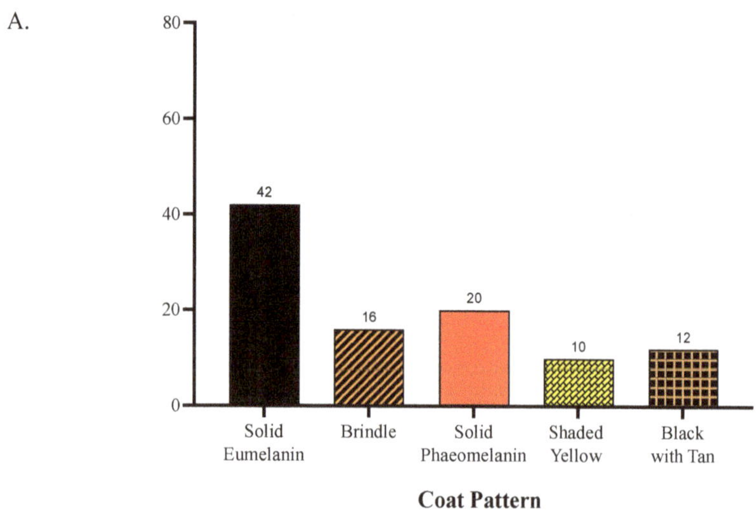

Figure 2. *Cont.*

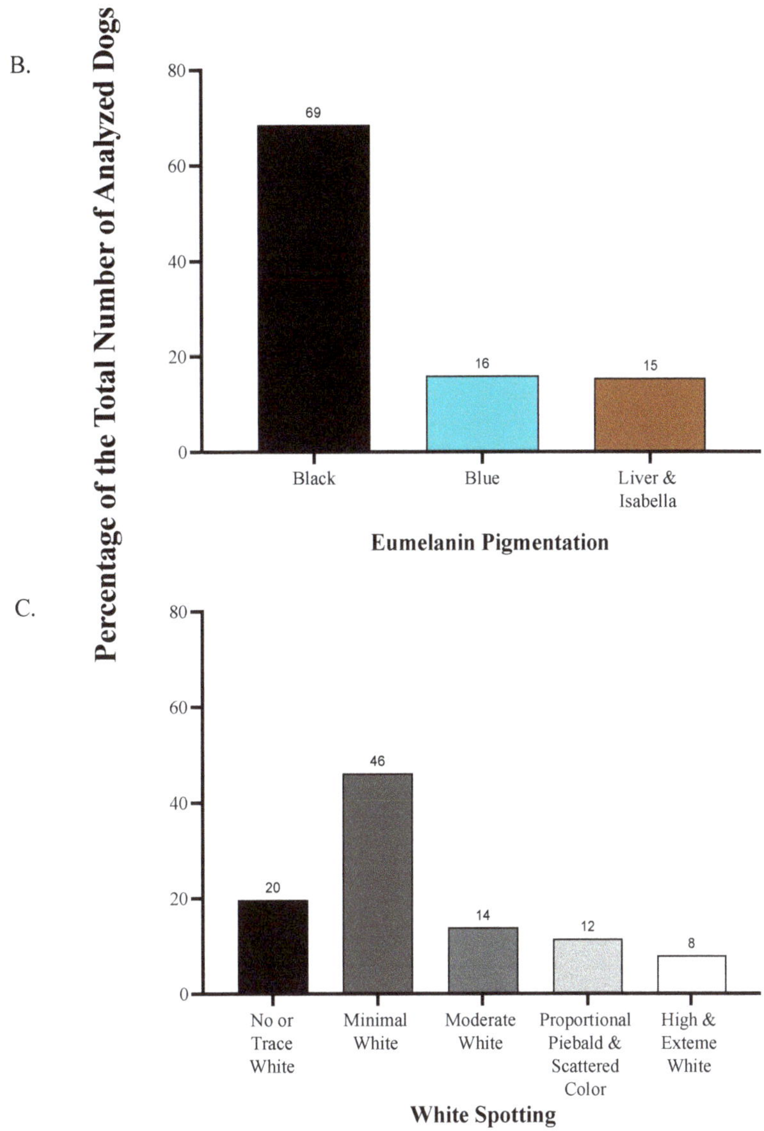

Figure 2. Pigmentation variable distributions in our study samples: (**A**): coat pattern, (**B**): eumelanin pigmentation, (**C**): white spotting.

2.1.2. Eumelanin Color

We classified the dogs as black (Figure 3A), blue (Figure 3B), liver (Figure 3C), or isabella (Figure 3D) according to the apparent color of eumelanin pigment present on their nose, skin around the eyes and muzzle, and any visible fur expressing eumelanin pigment. Due to the small sample sizes and difficulties in visual discernment between liver and isabella eumelanin (which are genotypically distinct but can overlap phenotypically), we pooled the dogs of liver and isabella eumelanin together into a single category prior to the analysis. Within our sample, 143 dogs had black eumelanin, 34 with blue eumelanin, and 31 dogs had either liver or isabella eumelanin (Figure 2). A total of 208 dogs were included in the eumelanin pigmentation analysis.

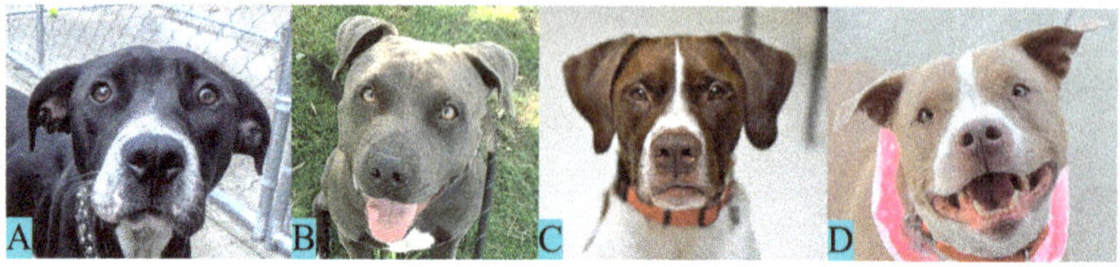

Figure 3. Eumelanin pigmentation categorization: (**A**): black eumelanin, (**B**): blue eumelanin, (**C**): liver eumelanin, and (**D**): isabella eumelanin.

2.1.3. White Spotting

We classified the dogs according to the apparent extent of white spotting on their bodies following the white spotting categorization scheme utilized by Morrill et al. [40] with one modification: the sixth white spotting category was subdivided into two categories due to observations that subjects placed into the single category within the original scheme showed considerable variations. Our two new categories were described as "no or trace white" and "minimal white." White spotting is a continuous variable; however, for our purposes, the dogs were placed into one of seven categories for analyses. In the order of increasing white spotting, the categories utilized in this study were labeled as no or trace white (Figure 4A), minimal white (Figure 4B), moderate white (Figure 4C), proportional piebald (Figure 4D), scattered color (Figure 4E), high white (Figure 4F), and extreme white (Figure 4G). Because of the relatively small sample sizes and similarities between phenotypes, we pooled together the dogs in the categories of proportional piebald (n = 12) and scattered color (n = 12), and the categories of high white (n = 13) and extreme white (n = 4) prior to the analysis. Due to difficulties in phenotyping the dogs with coats of extremely pale pheomelanin or graying fur, we were unable to categorize five dogs in their level of white spotting. Our sample included 40 dogs with no or trace white, 94 with minimal white, 28 with moderate white, 24 with proportional piebald or scattered color, and 17 with high white or extreme white (Figure 2). As such, our white spotting analysis included 203 dogs.

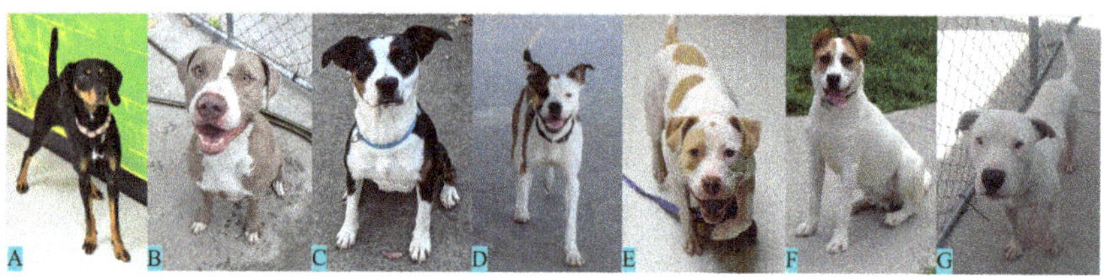

Figure 4. White spotting categorization: (**A**): no or trace white, (**B**): minimal white, (**C**): moderate white, (**D**): proportional piebald, (**E**): scattered color, (**F**): high white, and (**G**): extreme white.

2.2. *Cortisol Collection*

The dogs in the present study had previously participated in one of two welfare studies about foster caregiving [27,38]: therefore, there were three experimental phases in this study that aligned with those investigations. As such, we classified each dog's urine sample according to the study in which it was collected ("intervention type"), and, within that study, when the sample was collected relative to the intervention ("intervention phase"). Intervention type was either a brief outing [27] or weeklong fostering [38]. Intervention phase describes the time point from which the cortisol value was derived: pre-intervention (at the shelter), during the intervention (while spending time with humans outside of the shelter), or post-intervention (after the dog was returned to the shelter). We used the dogs' urinary cortisol:creatinine ratio as it described the dog's cortisol level without being impacted by its relative hydration.

2.2.1. Brief Outing Intervention

In the study by Gunter et al. [27], the physiological data from 40 dogs at Fulton County Animal Services (FCAS) in Atlanta, GA, 41 dogs from Detroit Animal Care and Control (DACC) in Detroit, MI, and 42 dogs from the Regional Center for Animal Care and Protection (RCACP) in Roanoke, VA, were collected, resulting in 123 subjects participating in the study. During the study, the dogs spent time with a human away from the shelter for approximately 2.5 h in order to investigate the effect of this intervention on the dogs' welfare. Dogs had varying levels of prior exposure to their caregiver, who was either a community member, shelter volunteer or staff person, or part of the research team. During the three-day study, the dogs experienced their brief outing in the late morning or early afternoon of the second day. Prior to the outing, the dogs' urine was collected for the first time in the afternoon of the day before and then in the morning prior to the outing. The third collection occurred in the afternoon of the second day, after the dogs had recently returned from their outing. This collection was reflective of the dogs' experiences during the outing and was our intervention collection time point. Following the outing, the dogs' urine was collected in the morning and afternoon of the study's third and final days. These urine samples were then used to measure the dogs' cortisol:creatinine ratios before, during, and after the intervention. We referred to Gunter et al. [27] for a full description of the intervention and methods used to collect and analyze the cortisol data.

2.2.2. Weeklong Fostering Intervention

The physiological data from 41 dogs at Charlottesville-Ablemarle SPCA (CASPCA) in Charlottesville, VA, and 44 dogs at the Pima Animal Care Center (PACC) in Tucson, AZ, were collected for the Gunter et al. [38] study resulting in a total of 85 subjects participating. For the study, the dogs lived in foster caregivers' homes for seven days to study the effects of weeklong fostering on the dogs' welfare. Foster caregivers were members of the public and shelter volunteers, and, generally, the dogs were unfamiliar with their foster caregivers and others in the household prior to their fostering stay. Dogs' urine was collected for 17 consecutive days to measure their cortisol:creatinine ratios: five mornings in the shelter prior to fostering, seven mornings in the caregiver's home, and then five mornings in the shelter after foster care. The urine samples were analyzed using the same methods as described in the brief outing study.

2.3. Analysis

To investigate whether the dogs' coat pattern, eumelanin pigmentation, or white spotting influenced their cortisol responses, we analyzed the dogs' cortisol:creatinine ratios from our five study shelters using three linear mixed models in IBM SPSS Statistics (Version 29). In order to utilize these data, despite the differing numbers of collection time points between the studies, the cortisol values were categorized into one of three phases corresponding to the collection time point. Those phases were either prior to the intervention in the shelter (Phase 1), during the intervention (Phase 2), or in the shelter after the intervention (Phase 3). Phase-level analyses of the dogs' cortisol:creatinine values were previously performed by Gunter et al. [26,27].

Based on this previous research [26,27], each linear mixed model included the following fixed effects as these variables were shown to affect cortisol levels and were entered into the models as covariates: dogs' weight (kg) and age (months) in addition to the intervention type (brief outing or weeklong fostering) and intervention phase (in-shelter pre-intervention, during the intervention, and in-shelter post-intervention).

With regard to our present research questions about the dogs' morphology, one categorical variable was entered into each model as a fixed effect. A five-level categorical variable (i.e., solid eumelanin, brindle, solid pheomelanin, shaded yellow, and black with tan) was employed to describe the dogs' coat patterns; a three-level categorical variable (i.e., black, blue, and liver or isabella) was employed to describe the dogs' eumelanin pigmentation; and a five-level categorical variable (i.e., no or trace white, minimal white, moderate white, proportional piebald and scattered color, and high and extreme white) was employed to describe the dogs' white spotting.

To disambiguate the known effects of the welfare interventions and changes in cortisol levels observed during the study's phases from our present research questions, two- and three-way interactions were entered into our linear mixed models as fixed effects. These included an intervention-type-by-phase interaction and an intervention-type-by-phase-by-morphology-variable interaction. While the intervention-type-by-phase interaction was included in our analyses for appropriate model specifications, the results were reported in other publications by Gunter et al. [27,38]. Additionally, dog and intercept were included as random effects and intervention phase was included as a repeated effect. A variance covariance matrix was employed, and a diagonal covariance matrix for the repeated measure of phase. The method of restricted maximum likelihood (REML) was used for estimating variance parameter values, and a statistical significance level of $p < 0.05$ was used throughout our statistical models.

3. Results

3.1. Descriptive Statistics

In total, 208 dogs from five study shelters (FCAS, DACC, RCACP, CASPCA, and PACC) participated in the study. The dogs had an average weight of 23.6 kg ($SD = 7.0$), average age of 38.2 months ($SD = 31.2$), and average cortisol:creatinine ratio of 20.1 $\frac{nmol}{L} : \frac{nmol}{L} \times 10^{-6}$ ($SD = 15.2$). Dogs included in our sample were more often female (53.5%).

3.2. Linear Mixed Models

The dogs included in the present study contributed 1994 cortisol:creatinine values for the analysis. To utilize the values from both the brief outing and weeklong fostering studies in our linear mixed models, we calculated the average cortisol:creatinine ratios for the intervention phases. Each dog contributed three values: the mean of its samples from the in-shelter period before the intervention, the mean value of sample(s) during the intervention, and the mean of in-shelter values after the intervention. This process yielded a total of 616 average cortisol:creatinine ratio values that were used in the analyses below.

To explore the effects of the coat pattern on the dogs' cortisol responses, the cortisol:creatinine ratio values were statistically analyzed to detect the possible effects of coat pattern, intervention type (i.e., brief outings or weeklong fostering), intervention phase (i.e., before, during, or after the intervention), and interaction of intervention type and phase, or a three-way interaction of intervention-type-by-phase-by-coat-pattern with the dogs' age and weight also entered into the model. We found that the dogs' cortisol:creatinine ratios differed significantly (at $p \leq 0.05$) as a function of the following variables: intervention type ($p < 0.001$), the interaction of intervention type and phase (p-value to be reported in [38]), dog weight ($p = 0.012$), and dog age ($p = 0.007$). However, the dogs' cortisol:creatinine ratios did not significantly differ as a function of coat pattern ($p = 0.591$), intervention phase ($p = 0.652$) or in an interaction between intervention type, intervention phase, and coat pattern ($p = 0.295$). Table 1 provides the estimated marginal means and standard errors of the dogs' cortisol:creatinine ratios as a function of coat pattern, intervention type, and intervention phase. Thus, after accounting for the interventions and their phases, we did not detect differences in the dogs' cortisol levels related to their coat patterns (Figure 5).

Table 1. Estimated marginal means and standard errors of urinary cortisol:creatinine ratios calculated using individual linear mixed models for sub-groups of dogs differing with respect to coat pattern, eumelanin pigmentation, and white spotting.

Pigmentation Variable	Brief Outing Intervention			Weeklong Fostering Intervention		
	Before	During	After	Before	During	After
	M (*SE*)	*M* (*SE*)	*M* (*SE*)	*M* (*SE*)	*M* (*SE*)	*M* (*SE*)
Coat Pattern						
Solid Eumelanin	28.3 (1.7)	30.5 (1.8)	27.9 (1.7)	19.0 (2.6)	11.9 (2.7)	18.1 (2.6)
Brindle	29.3 (2.8)	28.9 (3.0)	31.4 (2.8)	19.6 (3.9)	16.3 (4.1)	17.2 (4.0)
Solid Pheomelanin	32.2 (2.5)	36.1 (2.7)	36.2 (2.5)	20.3 (3.4)	11.7 (3.6)	16.7 (3.4)
Shaded Yellow	31.0 (4.8)	39.4 (5.0)	35.3 (4.8)	16.8 (3.8)	13.2 (4.0)	16.0 (3.8)
Black with Tan	38.4 (5.9)	46.7 (6.2)	34.1 (5.9)	18.3 (3.1)	9.7 (3.3)	16.1 (3.1)
Eumelanin Pigmentation						
Black	28.5 (1.5)	33.1 (1.6)	29.6 (1.5)	17.8 (1.6)	11.3 (1.7)	16.2 (1.6)
Blue	30.1 (2.7)	29.4 (2.9)	31.5 (2.8)	22.9 (4.0)	16.0 (4.3)	22.2 (4.1)
Liver or Isabella	34.6 (2.7)	34.6 (2.8)	35.7 (2.7)	22.2 (5.0)	13.5 (5.2)	16.3 (5.0)
White Spotting						
No or Trace White	35.3 (2.9)	39.2 (3.0)	35.0 (2.9)	14.0 (2.9)	10.0 (3.1)	13.0 (3.0)
Minimal	29.5 (1.7)	31.2 (1.8)	32.2 (1.7)	19.7 (2.1)	13.2 (2.2)	19.0 (2.1)
Moderate	26.7 (3.2)	31.2 (3.3)	26.2 (3.1)	19.4 (3.9)	12.6 (4.2)	15.7 (4.0)
Proportional Piebald or Scattered Color	27.5 (3.2)	30.0 (3.4)	25.9 (3.2)	27.6 (4.6)	12.2 (4.8)	20.2 (4.6)
High or Extreme White	25.9 (3.9)	27.7 (4.1)	28.5 (3.9)	17.6 (5.3)	12.2 (5.6)	16.5 (5.3)

Estimated marginal means (Ms) and standard errors (SEs) of values of urinary cortisol:creatinine ratios obtained for dogs classified into various pigmentation categories as a function of intervention type and phase.

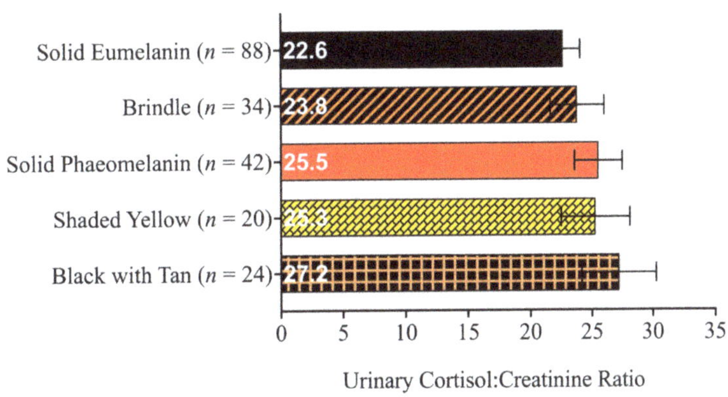

Figure 5. Estimated marginal means of urinary cortisol:creatinine ratios and standard errors of the dogs shown as a function of coat pattern categories. The sample size of each coat pattern category is given next to its label.

To investigate the effects of eumelanin pigmentation on the shelter dogs' cortisol levels, the dogs' urinary cortisol:creatinine ratios were analyzed to detect an effect of eumelanin pigmentation, intervention type, intervention phase, an interaction of intervention type and phase, or a three-way interaction of intervention-type-by-phase-by-eumelanin-pigmentation along with the dogs' age and weight. We found that the dogs' cortisol:creatinine ratios differed significantly (at $p \leq 0.05$) as a function of the following variables: intervention type ($p < 0.001$), intervention phase ($p = 0.02$), the interaction between intervention type and intervention phase (p-value to be reported in [38]), dog weight ($p < 0.001$), and dog age ($p < 0.001$). However, the dogs' cortisol:creatinine ratios did not significantly differ as a function of eumelanin pigmentation ($p = 0.322$) or in a three-way interaction between intervention type, intervention phase, and eumelanin pigmentation ($p = 0.387$). The estimated marginal means and standard errors of the dogs' cortisol:creatinine ratios as a function of eumelanin pigmentation, intervention type, and intervention phase are provided in Table 1. Thus, after accounting for the interventions and their phases, we did not find differences in the dogs' cortisol responses based on their eumelanin pigmentation (Figure 6).

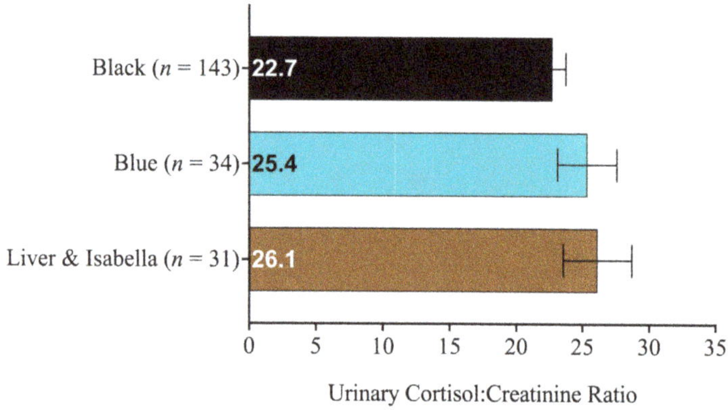

Figure 6. Estimated marginal means of urinary cortisol:creatinine ratios and standard errors of the dogs shown as a function of eumelanin pigmentation categories. The sample size of each category of eumelanin pigmentation is given next to its label.

To better understand the impact of white spotting on dogs' cortisol responses, their urinary cortisol:creatinine ratios were analyzed to detect the possible effects of white spotting, intervention type, intervention phase, an interaction of intervention type and phase, or a three-way interaction of intervention-type-by-phase-by-white-spotting along with the variables of dog age and weight. We found that the dogs' cortisol:creatinine ratios differed significantly (at $p \leq 0.05$) as a function of the following variables: intervention type ($p < 0.001$), intervention phase ($p = 0.049$), the interaction between intervention type and intervention phase (p-value to be reported in [38]), dog weight ($p = 0.011$), and dog age ($p < 0.001$). However, the dogs' cortisol:creatinine ratios did not significantly differ as a function of white spotting ($p = 0.830$) or in a three-way interaction between intervention type, phase, and white spotting ($p = 0.234$). The estimated marginal means and standard errors in the dogs' cortisol:creatinine ratios as a function of pigmentation, intervention, and phase included in this model are provided in Table 1. As such, we did not detect an effect of white spotting on the dogs' cortisol levels either in the shelter or during the human–animal interaction intervention (Figure 7).

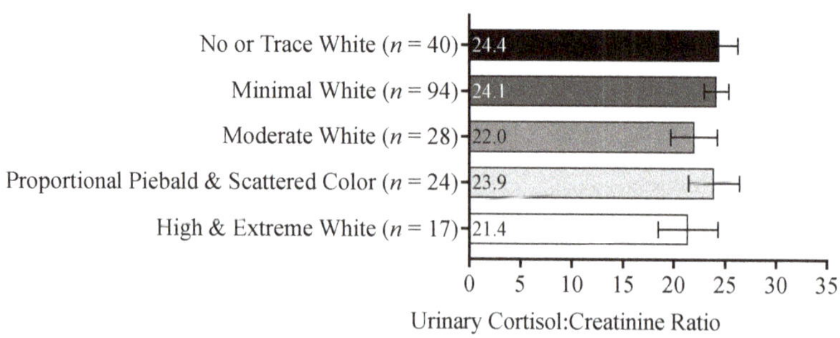

Figure 7. Estimated marginal means of urinary cortisol:creatinine ratios and standard errors of the dogs shown as a function of white spotting categories. The sample size of each category of white spotting is given next to its label.

4. Discussion

This translational study was designed to investigate questions of basic and applied relevance, including the applicability of the domestication syndrome hypothesis to the cortisol responsivity of domestic dogs, allowing for the detection of relationships between morphological variables and physiological stress that might offer insights into improving the welfare of shelter-living dogs. In order to answer these research questions, we utilized data from dogs living in American animal shelters that were exposed to a human–animal intervention during which they left the shelter. These dogs displayed a range of coat patterns, eumelanin pigmentation, and white spotting, and we examined their urinary cortisol:creatinine ratios, as a function of these three pigmentation characteristics, prior to, during, and after the interventions. While this study's retrospective design and limited sample size placed some limitations on the conclusions we can draw from the results, we found that none of our pigmentation variables predicted the dogs' urinary cortisol levels at any time during the welfare studies.

The domestication syndrome hypothesis purports that depigmentation in domesticated animals may be linked to an increased resilience to the stressors associated with living in an anthropomorphic niche. Very little research has been conducted about the domestication syndrome hypothesis in domestic dogs, despite their distinction of being the first domesticated species. Nevertheless, our results align with the existing literature. Hansen Wheat et al. [41] demonstrated that three morphological traits associated with the domestication syndrome hypothesis (i.e., white spotting, floppy ears, and curly tails) showed no covariation with expected behavioral traits on a breed level. However, a previous study by this research team [42] revealed that the expected behavioral correlations of domestication, such as reduced fear and aggression as well as increased sociability and playfulness, were less pronounced in modern dog breeds compared to those of ancient breeds. They hypothesized that these findings might be the result of the emphasis breeders have placed on selection for morphological traits in purebred dogs since the Victorian era, overriding the existing correlates of the domestication syndrome hypothesis.

Our study was designed to complement the existing literature [41,42]. While these investigations focused on breed-level analyses utilizing samples of purebred dogs [41,42], we investigated individual dogs of unknown origins, many of which were likely mixed breed. Furthermore, while studies by Hansen Wheat et al. [41,42] explored the possible correlations between morphology and behavior, our study examined potential relationships between morphology and the physiological stress response.

It is important to note that, while we lacked information on the likely complex breed heritages of the dogs in our study, previous research suggests that North American mixed-breed dogs often have modern-breed ancestry [40,43]. Therefore, it is likely that the dogs in our study may have been subjected to the same selection pressures prioritizing morphology, leading to a possible and previously proposed decoupling of domestication syndrome hypothesis-associated traits [42]. By utilizing a heterogeneous sample of mixed-breed dogs and exploring the interaction between dogs' morphological and physiological characteristics, our study offers further insights into the applicability of the domestication syndrome hypothesis to contemporary American dogs living in animal shelters.

Our results, however, do not readily align with the existing body of research examining the interrelationships between pigmentation and behavior in domestic dogs. In contrast to the domestication syndrome hypothesis, several studies (all using modern purebred dogs as subjects) have reported that reduced pigmentation is associated with undesirable behavioral traits. In English cocker spaniels, Korean jindos, and Labrador retrievers, the recessive red mutation has been shown to be associated with increased aggression (spaniels and retrievers; [29–32]) and fearfulness (jindos; [33]). In Labrador retrievers, the recessive eumelanin-lightening liver mutation was associated with lower trainability [32,34]. Notably, however, in English cocker spaniels, the presence of substantial white spotting aligns with the expectations of the domestication syndrome hypothesis: dogs with more white spotting reportedly exhibit lower levels of aggression compared to their counterparts with less white spotting [30,31].

While our study does not directly assess behavior, it is crucial to acknowledge the influence of the physiological stress response on behavior. A more reactive physiological stress response system can correlate with greater behavioral reactivity (e.g., [44]). Hence, we hypothesized that cortisol levels (or, more precisely, urinary cortisol:creatinine ratios), as an indicator of the stress response, might be higher in dogs with pigmentation phenotypes that prior literature has associated with greater behavioral reactivity. However, in these results, we did not find correlations between pigmentation and cortisol levels. Such findings suggest that the differences in behavior observed in these previous studies might have been driven by some factor(s) other than differences in cortisol production.

From the perspective of applied welfare, we hypothesized that if pigmentation was related to the cortisol levels of dogs living in shelters (or with changes in the cortisol levels arising in response to a human–animal interaction intervention), this knowledge might be useful for personnel involved in animal rescues and shelters. Considering these organizations' limited resources, the ability to visually triage dogs upon entry, in order to identify individuals that might be at particular risk for welfare impairments, could be beneficial. Nevertheless, as our results failed to demonstrate that dogs' stress responses differ as a function of their pigmentation phenotype, shelter staff should not focus on dog pigmentation when assessing dogs' stress levels in the shelter, and it is unlikely to aid adopters in predicting future behavior when choosing a dog (e.g., [40,45]).

This study was subject to several limitations that may have influenced our results. Firstly, while cortisol is a widely used physiological indicator of an animal's response to a stressor(s), it is simply a measure of arousal, regardless of its emotional valence (i.e., physiologically, excitement and anxiety can both be viewed as forms of stress) [46]. Thus, we were unable to differentiate between the eustress and distress the animal experienced, despite our specific interest in distress and its relevance to the domestication syndrome hypothesis. Secondly, this was a retrospective examination using the data previously collected in other studies, which led to some design weaknesses. Despite the significant number of urinary cortisol samples collected from shelter-living dogs that were utilized in our analyses, the number of subjects was likely not large enough to test our statistical models' interaction terms (which were needed in order to include the known effects of our human–animal interaction intervention). Furthermore, the sample sizes in each category of pigmentation type were unequal, and, in some cases, the total number of dogs within a specific category of a morphological variable, such as coat pattern and white spotting, were limited and required pooling together, which likely reduced our detection abilities.

The dogs in this study had diverse genetic backgrounds and often unknown life histories, introducing additional variability that may have impacted our cortisol values [47]. Because this investigation was envisaged as a translational study with an applied relevance to dogs living in animal shelters, the human–animal intervention was not standardized, which would have been the preferred approach for more basic explorations assessing the applicability of the domestication syndrome hypothesis with domestic dogs. Lastly, we employed visual phenotyping rather than genotyping to assess pigmentation, acknowledging that different genotypes could produce similar phenotypes [9]. Future studies using closely related dogs with similar life experiences, such as littermates, or dogs from populations that have not undergone recent intense artificial selection for morphological characteristics, such as free-ranging dogs, may be better able to address the possible connection between differential stress responding and morphology within the context of the domestication syndrome hypothesis.

5. Conclusions

Utilizing data from a sample of dogs living in animal shelters in the United States, we phenotyped subjects for three pigmentation variables: coat pattern, eumelanin pigmentation, and white spotting. We then examined whether these morphological characteristics predicted the dogs' urinary cortisol:creatinine ratios while they resided in the animal shelter and in response to a human–animal interaction intervention.

In our investigation, we found that neither coat pattern, eumelanin pigmentation, nor white spotting predicted the dogs' cortisol levels, suggesting that dogs may not differentially respond to the stressors of the shelter or human interactions as a function of their pigmentation. Thus, these preliminary results do not support assumptions of the domestication syndrome hypothesis related to pigmentation and physiological stress responses in this population of domestic dogs. Nevertheless, future studies should be conducted to determine if these null results were caused by study limitations.

Author Contributions: Conceptualization, J.M.P., E.N.F. and L.M.G.; methodology, J.M.P., E.N.F. and L.M.G.; validation, J.M.P. and L.M.G.; formal analysis, L.M.G.; investigation, J.M.P.; resources, E.N.F.; data curation, J.M.P. and L.M.G.; writing—original draft preparation, J.M.P.; writing—review and editing, J.M.P., E.N.F. and L.M.G.; visualization, J.M.P.; supervision, E.N.F.; project administration, J.M.P.; funding acquisition, E.N.F. and L.M.G. All authors have read and agreed to the published version of the manuscript.

Funding: This research was funded in part by Maddie's Fund, a nonprofit organization that provides funding and other resources for animal rescue and sheltering organizations and research initiatives.

Institutional Review Board Statement: The animal study protocol was approved by the Institutional Animal Care and Use Committee of Arizona State University (17-1552R approved on 10 February 2017 and 20-1760R approved on 19 December 2019).

Data Availability Statement: Publicly available datasets were analyzed in this study. These data can be found at: https://data.lib.vt.edu/articles/dataset/Morphology_Cortisol_Data_for_Repository/24101649/1 (accessed on 14 September 2023).

Acknowledgments: We would like to thank the staff, volunteers, public participants, and dogs at the Regional Center for Animal Care and Protection, Lifeline Animal Project at Fulton County Animal Services, Detroit Animal Care and Control, Charlottesville Albemarle SPCA, and Pima Animal Care Center, as well as the rest of our research team, for making this study possible. In particular, Emily Blade, Dana Eldred, Rachel Gilchrist, Meghann Ramsey, Jenifer Reed, and Meera Solomon for providing photos of the participating dogs in addition to their contributions in intervention implementation and cortisol sampling. We would like to dedicate this paper to those lovely dogs who contributed the data to this study but unfortunately were unable to find adoptive homes.

Conflicts of Interest: The authors declare no conflict of interest.

References

1. Darwin, C. The Variation of Animals and Plants under Domestication (Vol. 2). *J. Murray*. 1868. Available online: https://www.gutenberg.org/files/3332/3332-h/3332-h.htm (accessed on 10 February 2021).
2. Zeder, M.A. The Domestication of Animals. *J. Anthropol. Res.* **2012**, *68*, 161–190. [CrossRef]
3. Minton, J.E. Function of the hypothalamic-pituitary-adrenal axis and the sympathetic nervous system in models of acute stress in domestic farm animals2. *J. Anim. Sci.* **1994**, *72*, 1891–1898. [CrossRef] [PubMed]
4. Newberry, R. Behavioral, emotional, & cognitive effects of domestication. In *APA Handbook of Comparative Psychology: Vol. 1. Basic Concepts, Methods, Neural Substrate, and Behavior*; American Psychological Association: Washington, DC, USA, 2017; Chapter 16; pp. 315–329. [CrossRef]
5. Lord, K.A.; Larson, G.; Coppinger, R.P.; Karlsson, E.K. The History of Farm Foxes Undermines the Animal Domestication Syndrome. *Trends Ecol. Evol.* **2020**, *35*, 125–136. [CrossRef] [PubMed]
6. Sánchez-Villagra, M.R.; Geiger, M.; Schneider, R.A. The taming of the neural crest: A developmental perspective on the origins of morphological covariation in domesticated mammals. *R. Soc. Open Sci.* **2016**, *3*, 160107. [CrossRef]
7. Trut, L.; Oskina, I.; Kharlamova, A. Animal evolution during domestication: The domesticated fox as a model. *BioEssays News Rev. Mol. Cell. Dev. Biol.* **2009**, *31*, 349–360. [CrossRef]
8. Wilkins, A.S.; Wrangham, R.W.; Fitch, W.T. The "Domestication Syndrome" in Mammals: A Unified Explanation Based on Neural Crest Cell Behavior and Genetics. *Genetics* **2014**, *197*, 795–808. [CrossRef]
9. Brancalion, L.; Haase, B.; Wade, C.M. Canine coat pigmentation genetics: A review. *Anim. Genet.* **2022**, *53*, 3–34. [CrossRef]
10. Linderholm, A.; Larson, G. The Role of Humans in Facilitating and Sustaining Coat Colour Variation in Domestic Animals. *Semin. Cell Dev. Biol.* **2013**, *24*, 587–593. [CrossRef]
11. Rigg, R. *Livestock Guarding Dogs: Their Current Use World Wide*; IUCN/SSC Canid Specialist Group Occasional Paper; IUCN: Gland, Switzerland, 2001; Available online: http://www.canids.org/occasionalpapers/ (accessed on 21 July 2023).
12. Saif, R.; Iftekhar, A.; Asif, F.; Alghanem, M.S. Dog Coat Colour Genetics: A Review. *Adv. Life Sci.* **2020**, *7*, 215–224.

13. Grandin, T. *The Way I See It: The Dangers of Overselection*; Western Horseman: Fort Worth, TX, USA, 1998; pp. 120–124. Available online: https://mountainscholar.org/bitstream/handle/10217/4424/H105.pdf;sequence=1 (accessed on 10 February 2021).
14. Grandin, T.; Deesing, M. Genetics and Animal Welfare. Available online: https://www.grandin.com/welfare/genetics.animal.welfare.html (accessed on 6 April 2021).
15. Daniels, T.J.; Bekoff, M. Population and Social Biology of Free-Ranging Dogs, Canis familiaris. *J. Mammal.* **1989**, *70*, 754–762. [CrossRef]
16. Schmutz, S.M.; Berryere, T.G. Genes affecting coat colour and pattern in domestic dogs: A review. *Anim. Genet.* **2007**, *38*, 539–549. [CrossRef] [PubMed]
17. Schmutz, S.M.; Berryere, T.G.; Dreger, D.L. MITF and White Spotting in Dogs: A Population Study. *J. Hered.* **2009**, *100*, S66–S74. [CrossRef]
18. Baranowska Körberg, I.; Sundström, E.; Meadows, J.R.S.; Rosengren Pielberg, G.; Gustafson, U.; Hedhammar, Å.; Karlsson, E.K.; Seddon, J.; Söderberg, A.; Vilà, C.; et al. A Simple Repeat Polymorphism in the MITF-M Promoter Is a Key Regulator of White Spotting in Dogs. *PLoS ONE* **2014**, *9*, e104363. [CrossRef] [PubMed]
19. Beerda, B.; Schilder, M.B.H.; Van Hooff, J.A.R.A.M.; De Vries, H.W.; Mol, J.A. Chronic Stress in Dogs Subjected to Social and Spatial Restriction. I. Behavioral Responses. *Physiol. Behav.* **1999**, *66*, 233–242. [CrossRef]
20. Scheifele, P.; Martin, D.; Clark, J.G.; Kemper, D.; Wells, J. Effect of kennel noise on hearing in dogs. *Am. J. Vet. Res.* **2012**, *73*, 482–489. [CrossRef]
21. Hellhammer, D.H.; Wüst, S.; Kudielka, B.M. Salivary cortisol as a biomarker in stress research. *Psychoneuroendocrinology* **2009**, *34*, 163–171. [CrossRef]
22. Beerda, B.; Schilder, M.B.H.; Janssen, N.S.C.R.M.; Mol, J.A. The Use of Saliva Cortisol, Urinary Cortisol, and Catecholamine Measurements for a Noninvasive Assessment of Stress Responses in Dogs. *Horm. Behav.* **1996**, *30*, 272–279. [CrossRef]
23. Schatz, S.; Palme, R. Measurement of Faecal Cortisol Metabolites in Cats and Dogs: A Non-invasive Method for Evaluating Adrenocortical Function. *Vet. Res. Commun.* **2001**, *25*, 271–287. [CrossRef]
24. Sandri, M.; Colussi, A.; Perrotta, M.G.; Stefanon, B. Salivary cortisol concentration in healthy dogs is affected by size, sex, and housing context. *J. Vet. Behav.* **2015**, *10*, 302–306. [CrossRef]
25. Dudley, E.S.; Schiml, P.A.; Hennessy, M.B. Effects of repeated petting sessions on leukocyte counts, intestinal parasite prevalence, and plasma cortisol concentration of dogs housed in a county animal shelter. *J. Am. Vet. Med. Assoc.* **2015**, *247*, 1289–1298. [CrossRef]
26. Gunter, L.M.; Feuerbacher, E.N.; Gilchrist, R.J.; Wynne, C.D.L. Evaluating the effects of a temporary fostering program on shelter dog welfare. *PeerJ* **2019**, *7*, e6620. [CrossRef] [PubMed]
27. Gunter, L.M.; Gilchrist, R.J.; Blade, E.M.; Barber, R.T.; Feuerbacher, E.N.; Platzer, J.M.; Wynne, C.D.L. Investigating the Impact of Brief Outings on the Welfare of Dogs Living in US Shelters. *Animals* **2021**, *11*, 548. [CrossRef] [PubMed]
28. Hennessy, M.B.; Voith, V.L.; Mazzei, S.J.; Buttram, J.; Miller, D.D.; Linden, F. Behavior and cortisol levels of dogs in a public animal shelter, and an exploration of the ability of these measures to predict problem behavior after adoption. *Appl. Anim. Behav. Sci.* **2001**, *73*, 217–233. [CrossRef] [PubMed]
29. Houpt, K.A.; Willis, M.B. Genetics of behavior. In *The Genetics of the Dog*; Ruvinsky, A., Sampson, J., Eds.; CABI Pub: Wallingford, UK; New York, NY, USA, 2001; ISBN 978-0-85199-520-5.
30. Pérez-Guisado, J.; Lopez-Rodríguez, R.; Muñoz-Serrano, A. Heritability of dominant–aggressive behaviour in English Cocker Spaniels. *Appl. Anim. Behav. Sci.* **2006**, *100*, 219–227. [CrossRef]
31. Podberscek, A.L.; Serpell, J.A. The English Cocker Spaniel: Preliminary findings on aggressive behaviour. *Appl. Anim. Behav. Sci.* **1996**, *47*, 75–89. [CrossRef]
32. van Rooy, D.; Wade, C.M. Association between coat colour and the behaviour of Australian Labrador retrievers. *Canine Genet. Epidemiol.* **2019**, *6*, 10. [CrossRef]
33. Kim, Y.K.; Lee, S.S.; Oh, S.I.; Kim, J.S.; Suh, E.H.; Houpt, K.A.; Lee, H.C.; Lee, H.J.; Yeon, S.C. Behavioural reactivity of the Korean native Jindo dog varies with coat colour. *Behav. Process.* **2010**, *84*, 568–572. [CrossRef]
34. Lofgren, S.E.; Wiener, P.; Blott, S.C.; Sanchez-Molano, E.; Woolliams, J.A.; Clements, D.N.; Haskell, M.J. Management and personality in Labrador Retriever dogs. *Appl. Anim. Behav. Sci.* **2014**, *156*, 44–53. [CrossRef]
35. Roulin, A.; Ducrest, A.-L. Association between melanism, physiology and behaviour: A role for the melanocortin system. *Eur. J. Pharmacol.* **2011**, *660*, 226–233. [CrossRef]
36. Reissmann, M.; Ludwig, A. Pleiotropic effects of coat colour-associated mutations in humans, mice and other mammals. *Semin. Cell Dev. Biol.* **2013**, *24*, 576–586. [CrossRef]
37. Mills, D.; Karagiannis, C.; Zulch, H. Stress—its effects on health and behavior. *Vet. Clin. N. Am. Small Anim. Pract.* **2014**, *44*, 525–541. [CrossRef] [PubMed]
38. Gunter, L. *The Influences of Weeklong Fostering and Co-Housing on Shelter Dog Welfare*; [Manuscript in preparation]; Virginia Tech: Blacksburg, VA, USA, 2023.
39. Bannasch, D.L.; Kaelin, C.B.; Letko, A.; Loechel, R.; Hug, P.; Jagannathan, V.; Henkel, J.; Roosje, P.; Hytönen, M.K.; Lohi, H.; et al. Dog colour patterns explained by modular promoters of ancient canid origin. *Nat. Ecol. Evol.* **2021**, *5*, 1415–1423. [CrossRef] [PubMed]

40. Morrill, K.; Hekman, J.; Li, X.; McClure, J.; Logan, B.; Goodman, L.; Gao, M.; Dong, Y.; Alonso, M.; Carmichael, E.; et al. Ancestry-inclusive dog genomics challenges popular breed stereotypes. *Science* **2022**, *376*, eabk0639. [CrossRef]
41. Hansen Wheat, C.; van der Bijl, W.; Wheat, C.W. Morphology does not covary with predicted behavioral correlations of the domestication syndrome in dogs. *Evol. Lett.* **2020**, *4*, 189–199. [CrossRef]
42. Hansen Wheat, C.; Fitzpatrick, J.L.; Rogell, B.; Temrin, H. Behavioural correlations of the domestication syndrome are decoupled in modern dog breeds. *Nat. Commun.* **2019**, *10*, 2422. [CrossRef]
43. Padmabandu, M. Top 10 Most Common Breeds in Embark-Tested Dogs. Embark. 2022. Available online: https://embarkvet.com/resources/most-common-dog-breed-ancestry/ (accessed on 5 May 2023).
44. Trut, L.N.; Plyusnina, I.Z.; Oskina, I.N. An experiment on fox domestication and debatable issues of evolution of the dog. *Russ. J. Genet.* **2004**, *40*, 644–655. [CrossRef]
45. Protopopova, A.; Brandifino, M.; Wynne, C.D.L. Preference assessments and structured potential adopter-dog interactions increase adoptions. *Appl. Anim. Behav. Sci.* **2016**, *176*, 87–95. [CrossRef]
46. Iyasere, O.S.; Durosaro, S.O.; Oyeniran, V.J.; Daramola, J.O. Is an increase of glucocorticoid concentrations related to the degree of arousal or valence experienced by an animal to a stimulus? *Domest. Anim. Endocrinol.* **2022**, *81*, 106752. [CrossRef]
47. Novais, A.; Monteiro, S.; Roque, S.; Correia-Neves, M.; Sousa, N. How age, sex and genotype shape the stress response. *Neurobiology* **2017**, *Stress 6*, 44–56. [CrossRef]

animals

MDPI

Article

Interviews with Indian Animal Shelter Staff: Similarities and Differences in Challenges and Resiliency Factors Compared to Western Counterparts

Deyvika Srinivasa [1], Rubina Mondal [2], Kai Alain Von Rentzell [1] and Alexandra Protopopova [1,*]

1 The Animal Welfare Program, Faculty of Land and Food Systems, The University of British Columbia, Vancouver, BC V6T 1Z4, Canada
2 Department of Biological Sciences, Indian Institute of Science Education and Research Kolkata, Mohanpur 741246, India
* Correspondence: a.protopopova@ubc.ca

Simple Summary: Previous knowledge about animal sheltering systems and perspectives of staff working in animal shelters has been centered around Western countries. However, staff in Indian shelters must tackle different kinds of problems, such as care of free-ranging dogs. We conducted interviews with ten participants working in animal shelters in India to begin to gain an understanding of their experiences. Participants reported that inadequate funding, community conflict, and too many animals in need created a challenge for animal shelter work. However, flexibility and positive relationships in their workplace, feelings of duty to animals, and understanding animal needs were identified as positive factors. The perspectives of Indian animal shelter staff showed that certain issues are similar to those encountered in Western shelters; however, other issues are specific to social, political, and cultural influences. Context specific research in animal sheltering is needed to gain a broader world understanding of human–animal relationships.

Abstract: Animal shelters in India are at the forefront of efforts to improve free-ranging dog welfare and tackle animal overpopulation. In terms of cultural and political context, access to resources, and public health challenges, they operate in a very different environment than Western counterparts. Despite these distinctions, current sheltering literature is largely centered around countries such as the United States. The goal of this exploratory study was to examine the experiences of Indian animal shelter staff. Researchers conducted ten semi-structured interviews, in a mix of Hindi and English, with managers, veterinary nurses, and animal caretakers from three shelters. Using thematic analysis, shelter challenges as well as resiliency factors that enable staff to cope with these challenges were identified. Key challenges were inadequate funding, community conflict, and high intake numbers. Resiliency factors included flexibility, duty of care, co-worker relationships, and understanding animal needs. The results of this qualitative study revealed that the experiences of shelter staff are shaped by social, political, and cultural factors and that there is a need for further, context specific research on Indian sheltering rather than only relying on Western perspectives.

Keywords: animal shelter; companion animal; free-ranging dog; India; interviews; occupational health; qualitative research

Citation: Srinivasa, D.; Mondal, R.; Von Rentzell, K.A.; Protopopova, A. Interviews with Indian Animal Shelter Staff: Similarities and Differences in Challenges and Resiliency Factors Compared to Western Counterparts. *Animals* **2022**, *12*, 2562. https://doi.org/10.3390/ani12192562

Academic Editor: Betty McGuire

Received: 17 August 2022
Accepted: 23 September 2022
Published: 26 September 2022

1. Introduction

In the past 25 years, the goals and activities of shelters in Western countries have changed drastically [1]. Rowan and Kartal (2018) attribute these shifts, in part, to a drastic decrease in animal overpopulation [2]. In 1970, street dogs comprised 25% of the national dog population in the United States (US) and the budgets of humane societies were largely dedicated to population control efforts [3]. Over the last three decades, animal overpopulation in the country has declined significantly, with unowned dogs largely absent from

US communities today [2]. This shift has allowed shelters to direct greater energy and resources to other activities, including adoption and humane education.

Rowan and Kartal (2018) suggest that successful shelter practices in the US can act as a useful "template for other countries with large street dog populations", such as India. However, given that Indian and Western shelters operate in very different socio-cultural and political environments [4], there is a need to examine the extent to which we can 'export' sheltering research and practices from the West. With a focus on staff experiences, this study offers insight into sheltering in the Indian context and contributes to the discussions around cross-cultural collaboration, information sharing, and representation in the sheltering field. For the purposes of this paper, we will use the imperfect term "Global South" to refer to countries with low to upper-middle income located in Africa, Asia, Oceania, Latin America, and the Caribbean [5].

1.1. Animal Overpopulation in India

Like many countries in the Global South, India faces a significant dog overpopulation problem, with recent estimates projected at 59 million [6]. Free-ranging dogs in the country may experience welfare challenges, including malnutrition, poor skin condition, parasite infections, and human mistreatment, in the form of beating and poisoning [7]. Furthermore, India has the highest number of dog-related rabies cases, accounting for 35% of global fatalities and approximately 20,847 human deaths per year [8].

Until the late 1990s, Indian dog populations were controlled through mass culling, using methods such as poisoning and electrocution. From 1993 to 1999, this "catch-and-kill" approach was gradually replaced with high volume spay-neuter and vaccination programs, with Delhi becoming the first city to implement an Animal Birth Control—Anti Rabies Vaccination (ABC-ARV) facility in 1993 [9]. In 2001, these efforts were formalized under the new Animal Birth Control (Dogs) Rules. This legislation bans the culling, relocation, or removal of street dogs, and requires all state and municipal governments to budget for ABC-ARV work [10].

While the ABC Rules identify the government's responsibility for funding population control efforts, Animal Welfare Organizations (AWOs) are tasked with the actual implementation of these programs. Animal Welfare Organizations are shelters or SPCAs officially registered with the Animal Welfare Board of India [11]. The terms 'Animal Welfare Organization' and 'animal shelter' will be used synonymously in this paper.

While many run spay-neuter programs, Indian AWOs are not official regulators of population control. Sterilization practices are government regulated, and outlined in the 2009 Standard Operating Procedures for the Sterilization of Stray Dogs. Under this legislation, Indian AWOs are responsible for using state and municipal funding to run ABC programs "with a standard code of professional practice" [12]. This includes meeting standards for the capture and transportation of dogs, kennel management and ventilation, record keeping, and post-operative care. Thus, it appears that the responsibilities of Indian animal shelters are heavily centered around free-ranging dog care and population control. However, there is an overall lack of peer-reviewed research on the activities of Indian organizations and the experiences of staff in a shelter environment.

1.2. Occupational Health of Animal Shelter Staff

Despite the lack of India-specific research, a broad body of literature has looked at the occupational health of animal shelter staff in other countries. Studies conducted in the US, Canada, Australia, and the United Kingdom (UK) have identified the risk of poor mental health outcomes, such as compassion fatigue and moral injury, amongst animal shelter staff.

Arluke and Sanders (1996) define compassion fatigue as a form of secondary trauma, wherein staff in 'caring professions' are psychologically impacted by their distressed patients [13]. Surveying 2879 animal care workers in the US, Hill and colleagues (2020) identified that degree of exposure to cruelty cases was a strong indicator of vulnerability to compassion fatigue [14]. Rank and colleagues (2009) found relationships between the severity of compassion fatigue and the level of involvement in the euthanasia of shelter animals amongst US shelter workers [15]. After assisting with pet euthanasia, staff displayed an elevated heart-rate and lower heart-rate variability. Further, staff directly engaged in selecting which animal to euthanize experienced elevated secondary traumatic stress [15]. Shelter employees involved with euthanasia may also demonstrate higher levels of 'moral injury' from engaging in an activity that violates one's ethical beliefs [16].

At the same time, reduced exposure to euthanasia does not always correspond to improved employee well-being. Andrukonis and Protopopova (2020) examined shelter staff's experiences in the US. Here, although job satisfaction increased with less euthanasia within animal shelters, the levels of burnout, moral injury, and secondary traumatic stress were also higher [17]. Regardless of the quantity of animal death in a facility, shelter employees are at risk of poor mental health outcomes. Baran and colleagues (2012) studied 102 shelter employees across eight US states, identifying a 27% turnover rate within the first two months of employment [18]. This indicates that new staff may be unprepared for their working environment's high emotional and physical demands.

Previous research also demonstrates how staff exhibit 'resiliency factors' that help them with the stressors of their jobs. These may include positive shelter activities, social support, and personal attitude [19]. Positive actions, such as community outreach and companion animal adoption, may benefit staff by offsetting the impact of emotionally draining ones, such as euthanasia. Through qualitative interviews with shelter professionals in Florida, USA, Reeve and colleagues (2004) showed that staff benefited from engaging in proactive programs, like adoption drives, and had an increased sense of contributing directly to animal welfare [19].

Shelter staff with strong social support may also be more resilient in their jobs. Amongst 150 shelter staff from Melbourne, Australia, satisfaction with professional and personal relationships was the most significant predictor of euthanasia-related traumatic stress [20]. Employees may also rely on one another to cope with the loss of shelter animals and process challenging cases of animal abandonment or abuse [21]. At the same time, Baran and colleagues (2012) found that US shelter employees who engaged with euthanasia were less likely to discuss struggles outside the workplace, leading them to lose essential forms of social support in their personal lives [18].

Finally, personal attitudes may influence staff outcomes in a shelter environment. Reeve and colleagues (2004) found higher rates of job satisfaction amongst euthanasia technicians with an attitude of acceptance towards euthanasia [19]. Schabram and Maitlis (2017) conducted interviews with 50 animal shelter workers in the US and identified that a staff member's 'calling' (reason for pursuing shelter work) significantly impacted job satisfaction. Individuals who were 'practice oriented' (committed broadly to shelter work and animal care) displayed greater resilience than those that were 'contribution oriented' (motivated by individual skills and impact) [22]. Holy-Gerlach, Ojha and Arkow (2021) also identified the potential use of social work in animal shelters to reduce or mitigate occupational stress [23].

It appears that animal shelter staff are at risk of a range of negative mental health outcomes from the demands of their jobs. Further, findings on resilience suggest that interventions to mitigate compassion fatigue can focus on increasing positive job activities, social connections, and personal feelings of acceptance to one's job.

1.3. Research Gap and Study Rationale

Existing literature on sheltering and animal care is centered around Western countries, despite significant differences in animal welfare challenges and socio-cultural contexts in the Global North and South.

This may force Indian organizations to draw from research and resources that are not locally relevant. Cultural factors have been identified as a close predictor of the effectiveness of occupational health interventions [24]. Thus, exploring if and where Indian and Western counterparts diverge is particularly important in the context of staff experiences.

The current study provides initial insight into the experiences of Indian shelter staff through qualitative interviews with managers, animal caretakers, and veterinary nurses. This exploratory research will identify future opportunities for cross-cultural collaboration as well as the areas of Indian sheltering that require context-specific research and interventions.

1.4. Reflexivity Statement

Deyvika Srinivasa is an undergraduate Global Health student at the University of British Columbia. She identifies as being female and of South Indian heritage. Deyvika spent her early years in the US but grew up primarily in Bengaluru, India. Throughout her childhood, she was an active volunteer at animal rescues and community-based spay-neuter programs in her city. Kai von Alain Rentzell is a graduate student in the Animal Welfare Program at the University of British Columbia. He has experience with qualitative research and a specific interest in cross-country dog importation. Kai identifies as male and is of Japanese and German upbringing. He has engaged in companion animal care in multiple contexts through working at veterinary clinics in Japan and volunteering at Canadian dog rescues. Dr. Rubina Mondal (RM) is a postdoctoral scholar at the Indian Institute of Science. She is educated in behavioral ecology and has a particular interest in the welfare and adoption of free-ranging dogs. Rubina identifies as female, lives and works in Kolkata, India, and is actively engaged in dog rescue work in her community. Dr. Alexandra Protopopova, the study supervisor, is an assistant professor in the Animal Welfare Program at the University of British Columbia. Alexandra identifies as female, is of Russian and Swedish upbringing, and has lived and worked in the US, primarily with animal shelters in the south, and Canada. She is educated in behavior analysis and ethology and is particularly interested in animal sheltering and dog behavior and welfare.

All four authors offered distinct academic and cultural perspectives, which allowed us to adopt 'insider' and 'outsider' positions as researchers. An insider perspective was developed by having an investigator of Indian heritage (RM) conduct several interviews in Hindi. In this way, staff were able to engage in the research process and share their experiences in a language and with an interviewer with whom they were comfortable. At the same time, our team's background in American and Canadian animal sheltering placed us in an outsider position. This was beneficial during the analysis process, allowing us to draw cross-cultural comparisons between Western and Indian shelters. However, an outsider positioning may have also led to biases and the interpretation of participants' responses based on previous researchers' previous understandings of Western shelter staff. To address these shortcomings, researchers engaged in ongoing self-reflection and implemented a collaborative approach to interview coding and writing.

The Section 2.5 includes further details about measures taken to reduce bias and improve study rigor. While prioritizing reflexivity, we also recognized the value of a multidisciplinary and multicultural research team and used our different backgrounds to engage sensitively with participants, and eventually draw meaningful and hopefully accurate conclusions from the narratives of Indian shelter staff.

2. Materials and Methods

2.1. Participant Recruitment and Demographics

Participants were recruited between July and August 2021 through non-probabilistic, convenience sampling techniques (i.e., researchers contacted shelters that had openly available contact information on their websites or social media).

The study utilized convenience sampling, wherein participants were recruited from three shelters that researchers had already established relationships with. As a result, participants were only recruited from three states (Karnataka, Rajasthan, Himachal Pradesh), and results do not represent sheltering across the country. The impact of a restricted convenience sample is further examined in the Limitations section.

All three shelters had been registered as official Non-Governmental Organizations (NGOs) under the Indian Trust Act for at least eight years. Shelter names and cities have been removed to retain confidentiality, but basic descriptive data collected from shelter managers is presented in Table 1.

Table 1. Animal Shelter Characteristics.

Location [State]	Primary Animal Types	Total Number of Employees	Annual Animal Intake
Karnataka	Dogs, Cats, Rabbits	44	400
Himachal Pradesh	Dogs	#	1500
Rajasthan	Dogs, Cats, Cows, Bulls	100	11,182

Data unavailable.

During recruitment, shelter managers were emailed a study flyer that outlined the purpose and format of the research and asked to pass on details to eligible staff. Interested employees were then instructed to contact researchers directly by email or phone. The study flyer included a local mobile number (belonging to the co-author based in India) to ensure easy communication between participants in India and the research team. In the two weeks following recruitment, posters were sent to shelters managers, the authors received replies from ten participants, who reached out directly using the phone or email indicated on the poster. At this point, the shelter managers were informed that recruitment was complete, and we no longer required additional participants at this time.

To be eligible, participants had to be over the age of 18, living in India, currently employed at an Indian animal shelter, and working as a manager, veterinary nurse, or animal caretaker.

Participants were recruited from three shelters in three different states (Karnataka, Himachal Pradesh, and Rajasthan).

At the time of the interviews, all ten participants were actively employed and working in person (not remotely) at their respective shelters. The final sample consisted of 10 animal shelter workers (Table 2). Five individuals identified as male and five as female. Participants ranged from 22 to 40 years old, with a mean age of 29.3 years old. Participants held a range of shelter positions: three were employed as managers, three as veterinary nurses and four as animal caretakers. Half of the participants reported their current position was their first formal job working with animals; previous occupational histories for these participants included carpentry, taxi driving, post-secondary education, and family care. Of the individuals with prior animal experience, three out of four had a background in wildlife conservation and one person had worked with a small-scale dog rescue group.

Table 2. Participant Demographics and Employment Data.

Participant	Job Title	Gender	Age	State	Interview Language
P1	Manager	Woman	31	Rajasthan	English
P2	Manager	Woman	26	Rajasthan	English
P3	Manager	Man	27	Karnataka	English
P4	Vet nurse	Man	32	Himachal Pradesh	Hindi
P5	Vet nurse	Woman	27	Himachal Pradesh	Hindi
P6	Vet nurse	Man	28	Rajasthan	Hindi
P7	Caretaker	Man	22	Karnataka	Hindi
P8	Caretaker	Woman	#	Karnataka	English
P9	Caretaker	Man	31	Rajasthan	Hindi
P10	Caretaker	Woman	40	Rajasthan	Hindi

Data unavailable.

2.2. Ethics

The study received approval from the University of British Columbia Human Ethics Board (H21-01759).

2.3. Interviews and Data Collection

The interviews were conducted between July and August of 2021. During recruitment, participants selected an interview time from three to four proposed slots. These times were outside employees' working hours to reduce the likelihood of staff engaging in interviews at their place of work. If participants were unavailable at these proposed times or only had access to the internet or a phone at the shelter, interviews were scheduled during work hours. Participants were asked if they preferred to conduct their interview in English or Hindi. Six out of ten participants opted for Hindi interviews, all of which were conducted by the same member of the research team (RM). The remaining four interviews were conducted in English by the first author (DS). There was no prior relationship between researchers and participants prior to study commencement. Participants received a consent form by email or text message to fill out prior to the interview and return back. The written consent form informed participants of the specific goals of the research, including an examination of the attitudes and experiences of Indian animal shelter staff.

A semi-structured interview guide was used, divided into three lines of questions: Occupational Health, Dog Management Strategies, and Perceptions of Animal Welfare. The authors aimed to gain a broad understanding of shelter experiences by asking questions on individual, organizational, and societal and cultural levels. The guide contained open-ended questions, such as "Can you talk about your relationship with your co-workers?" as well as closed-ended ones like "Do you feed community dogs?". The preliminary guide was pilot tested amongst members of the research team and shortened to ensure all questions could be asked in a 60 min period to improve interview flow. The final interview guide consisted of eleven open-ended questions and four closed-ended questions (Table 3). All respondents were asked questions from the interview guide in the same sequence. However, researchers posed additional, unplanned follow-up questions to clarify meaning or inquire about interesting topics raised by participants.

All interviews were conducted virtually over Zoom or phone call and audio only was recorded for transcription. The interview length ranged from 23 min to 1 h 9 min, with an average time of 35 min. At the start of the interview, participants were re-informed of the purpose of the study, and the approximate duration. Participants were also reminded of consent protocols, including their right to leave the interview at any time or refuse to answer questions. Each participant received INR 500 ($\sim$8.55 CAD) as compensation for their involvement. No repeat interviews were conducted due to technological errors or lost data and none of the participants withdrew their data post-interview.

2.4. Data Processing

The second author (RM) and the first author (DS) transcribed, manually verbatim, their respective conversations to generate written transcripts. English transcripts were edited to remove non-standard speech patterns and grammatical errors as laid out by Chang and Spector (2011) [24]. Transcripts were anonymized, to remove participant and shelter identifiers. Interview audio and transcripts were stored on hard drives and computers accessible only to the four members of the research team. Audio files were destroyed after data analysis completion and anonymized transcripts are stored according to university ethics guidelines.

Transcripts were not returned to participants for correction or comment. However, anonymized interview data was shared with an independent contractor to assess the accuracy of Hindi-English translations. The contractor produced 'back translations' of all Hindi interviews by converting English transcripts back to Hindi as laid out by Blauner (1987) [25]. They then compared the back translations with the original Hindi transcripts to identify differences in meaning and flag potential translation errors. All errors were reviewed by the research team. If they were found to be significant (i.e., impacted interview interpretation and analysis), the independent contractor was instructed to listen to Hindi interview audio, re-translate the incorrect section to English and modify the original transcript with the corrected sentence. Overall, back translation analysis revealed there were no significant errors in Hindi to English translations and moreover, that participant responses from Hindi interviews were accurately represented in English transcripts.

Table 3. Interview Guide.

Occupational Health
Can you describe your role and main responsibilities within (stated) organization?
What does a typical day of work look like for you and what are your working hours?
Can you talk about your relationship with your co-workers or supervisors?
Do you generally find the workload manageable? **[C.E]** [a]
What are the biggest challenges of your job?
What are the most rewarding and exciting aspects of your job?
What is the reaction of your friends and family to your job?
Shelter goals and practices
What are your shelter's main goals?
Does your shelter conduct low cost spay-neuter for street dogs that come into the shelter? **[C.E]**
What are current challenges with the work your organization does?
In your opinion, how different are the goals of Western animal NGOs from Indian animal NGOs?
What changes would you like to see in your organization or Indian animal NGOs as a whole?
Perceptions of animal welfare
Do you feed community dogs/free-ranging dogs in your neighborhood? **[C.E]**
In a 'perfect world', what would the lives of these dogs be like?
Do you think we should attempt to get all dogs off the streets into homes or can street dogs have a good quality of life if numbers are controlled? **[C.E]**

[a] **[C.E]**: close ended question.

2.5. Theoretical Approach

The present study drew from bounded relativist ontology, which defines knowledge as the common ideas within a 'bounded group' (for example, a specific cultural or political orientation) and constructionist epistemology, which identifies that knowledge is generated through interactions between members of this group. Drawing from these foundations, the study implemented an interpretivist theoretical perspective, attempting to situate and interpret participants' responses in local context and culture [26]. Researchers closely engaged with interpretivism, specifically the concepts of life worlds, situated freedom, and co-constitutionality [27], to design the research focus, methodology and methods, and data analysis processes.

Given the limited past literature on Indian sheltering and staff experiences, it was not possible to replicate an existing framework before conducting the interviews. Instead, a broad-based interview guide was designed that addressed Occupational Health, Shelter Goals, and Perceptions of Animal Welfare (Table 3). After conducting interviews, all four researchers reflected on the study goal (comparing Indian and Western staff experiences) to select an appropriate research framework. To facilitate cross-cultural comparisons, the team looked to previous literature on Western sheltering and eventually selected a framework from a 2020 study by Levitt and Genzinski, which focused on "Compassion Fatigue and Resiliency Factors" in interviews with American shelter staff [28]. From reviewing interviewed transcripts, researchers identified that participants in the present study placed greater emphasis on broad shelter challenges as opposed to individual struggles, such as compassion fatigue. Thus, *shelter challenges and resiliency factors* was used as the final research framework and aligned with both research objectives as well as participant narratives.

The study used a phenomenological methodology, centering subjective human lived experiences in descriptive data. By applying an interpretive lens, researchers were able connect staff's lived experiences to a broader social and cultural context. An interpretative phenomenological approach promoted a focus on staff's overt responses in the interviews as well as their broader experiences in the Indian context.

The theoretical approach influenced data interpretation and analysis. Researchers specifically incorporated the idea of situated freedom into the analysis process. This concept identifies that while individuals are the 'experts' in their own experiences, they are also impacted by their social, cultural, and political environment [29]. When analyzing interviews and identifying themes, researchers centered on participants' situated freedom and considered how staff may have overcome or been shaped by external forces. This process of 'situating' participants' stated experiences in a broader social, cultural, and political context was crucial throughout the analysis.

The use of an interpretivist lens in all aspects of the study generated a deep, contextual understanding of Indian sheltering, centered around participants' lived experiences. At the same time, in line with the idea of co-constitutionality, the results do not represent the only 'true meaning' that can be drawn from the interviews, but rather a blend of ideas and interests from participants and investigators.

2.6. Data Analysis

After transcription and back translation, the interviews were analyzed using thematic analysis. Analysis was performed as laid out by Vaismoradi and colleagues (2016), wherein themes are generated through a process of categorizing and summarizing qualitative data [30].

The first author (DS) read all ten interviews multiple times and developed short summaries for each response to a question. The summaries included both basic information provided in participants' answers as well as interpretation of how this reflects the shelter staff's attitudes and experiences. For example, the following summary was created based on Participant 5's response to the question *"Can you describe your role and main responsibilities within your organization?"*:

Participant 5 is involved in a range of shelter operations (treatment, feeding, surgery preparation). She can prioritize tasks depending on the shelter's needs on a given day and work with a flexible schedule.

The transcripts were then re-read to identify corresponding quotes and evidence for each summary. To assess the level of support for the analysis and establish coding reliability, the summaries were reviewed by the entire research team. All the summarized interviews were reviewed at least once by another co-author. Those with insufficient evidence were either discarded or modified. During this phase, co-authors also identified instances of leading questions and misinterpretation of questions caused by language barriers. The researchers identified one leading question in interviews with Participant 8 and Participant 10. Responses to leading questions were removed from the analysis and not incorporated as evidence for any summary.

Finally, the summaries for individual questions were compared across participants and used to generate broad themes and sub-themes. Through the entire analysis phase, researchers met weekly to discuss recurring ideas, select important evidence, and refine themes. RM and DS did not make consistent written field notes during interviews but reflected on firsthand interactions with interviewees during group discussions.

During group discussions, researchers also discussed data saturation and the potential need for additional interviews. Guest, Bunce, and Johnson (2006) define data saturation as the point at which no new themes emerge from the data and no additional data collection is needed [31]. In the context of cross-cultural research, the notion of data saturation can be misleading as it implies 'outsider' researchers have understood participants' experiences to completeness [32]. Thus, researchers in the current study determined sample size based on 'conceptual depth'. Nelson (2017) defines the latter as the point at which there is sufficient data for researchers to theorize and lays out ten criteria to assess the "sufficiency" of conceptual depth [33]. Data from the ten interviews met "sufficiency" criteria, including clear connections between themes, a wide range of evidence to illustrate themes, and resonance with existing literature. Researchers collectively determined that no further interviews were necessary within the scope of the study. While participants did not provide feedback on the preliminary analysis, they were given the option to receive a copy of the final report.

3. Results and Discussion

From the interview responses, the authors identified themes relating to shelter challenges as well as resiliency factors that enable staff to cope with these barriers. Luthar, Cicchetti and Becker (2000) describe 'resilience' as the process of rebounding from significant adversity and resultant stress [34]. As stated by Feder and colleagues (2013) identified that this can refer to a range of coping mechanisms, including morals, religion, physical fitness and social support [35]. In the context of literature on occupational health, the term typically refers to protective factors that help staff to mitigate the potential mental and physical health impacts of their jobs. For example, Brintzinger and colleagues (2021) identify 'emotional openness' as a resiliency factor against burnout amongst male and female health professionals [36].

The original interview guide included questions on three topics: occupational health, shelter goals and practices, and perceptions of animal welfare. Interestingly, participants' responses did not divide distinctly across these lines of inquiry. Instead, themes often overlapped in different sections of the interviews. For example, many participants identified that conflict with community members impacted their occupational health as well as their shelter's practices. Because of this overlap, results were not reported as three separate lines of inquiry, but rather in terms of broad Challenges and Resiliency factors described by staff. For further details on the selection of this research framework, see Section 2.5.

In the present study, key shelter challenges were inadequate funding, community conflict, and high intake numbers. In the face of these barriers, resilience factors were the duty of care, co-worker relationships, and understanding of animal needs. The themes and sub-themes are in Table 4. High-level challenges and resiliency factors reflected existing literature on sheltering in Western countries. However, sub-themes, such as government policy, religious beliefs, and a focus on community-based care revealed that the staff's experiences and assets were also specific to the Indian cultural, societal, and political context. The relationships between themes are sub-themes are detailed in Figure 1.

3.1. *Shelter Challenges*

3.1.1. High Intake

All participants spoke about the difficulties with managing high animal intake with limited resources. Participants described intake challenges in relation to pet abandonment, animal death, animal overpopulation, and seasonal fluctuations.

A. Pet Abandonment

Participant 1 identified the high rates of abandonment for purebred dogs in India, stating, "The more frustrating part is when people buy breed dogs, pedigree dogs, and they abandon them". He also highlighted the low outflow of animals from the shelter, explaining that "Local adoptions are not so popular. Nobody wants abandoned dogs".

Table 4. Shelter challenges and resiliency factors.

Section 3.1	**Challenges**	
	Themes	**Sub-Themes**
Section 3.1.1	High intake	A. Pet abandonment B. Animal overpopulation C. Seasonal fluctuations D. Animal death
Section 3.1.2	Inadequate funding	A. Lack of government support B. Government policy C. Cultural and religious beliefs
Section 3.1.3	Community conflict	A. Rescuer pressure B. Resident pushback C. Incorrect community care
Section 3.2	**Resiliency Factors**	
	Themes	**Sub-Themes**
Section 3.2.1	Flexibility and prioritization	
Section 3.2.2	Co-worker support	A. Collaboration B. Equity and safe space
Section 3.2.3	Duty of care	
Section 3.2.4	Understanding animal needs	A. Unrestricted movement B. Autonomy C. Community care

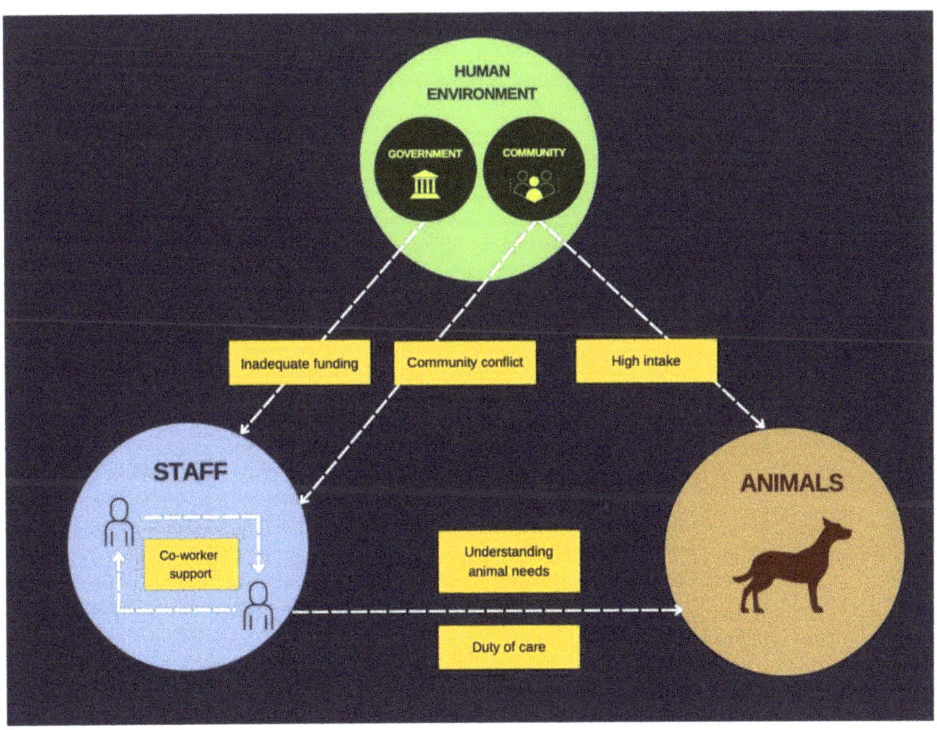

Figure 1. Connections between themes and sub-themes.

B. Animal Overpopulation

In addition to abandoned pets, staff also linked shelter numbers to large free-ranging dog populations. Participant 1 stated, "In India there is an uncontrolled population of stray dogs ... [It is] so hard for Indian organizations because there are too many cases, too many dogs that are on the street, that we have to help".

Participant 2 identifies high intake as a commonality between Indian and Western shelters, stating, "I do feel like we are similar in many ways. We [both] get a lot of animals in and are all overwhelmed ... So, I think that is one thing that really unifies us". Parallels in inflow patterns are also seen in existing literature in the US, where the intake of both stray dogs and abandoned pet animals to shelters is well documented. Data from the National Council on Pet Populations indicates that approximately 30% of dogs entered shelters as owner-surrenders from 1994–1995 [37], while stray intake is estimated to be 53–83% of shelter dog populations [38–41]

At the same time, there appear to be distinctions between Indian and Western shelter intake in terms of scale of inflow, the primary source of animals, and reasons behind pet relinquishment to shelters. When asked about the difference between Western and Indian shelters, Participant 2, stated:

> A lot of Western NGOs ... have quite different situations from what we have in India ... We are very [different] in terms of having stray animals as a part of Indian community.
>
> I know it's not physically possible right now. But I just wish we could reach that stage where all of our dogs are spayed and neutered.

Here, Participant 2 identifies that Indian organizations grapple with a significantly higher caseloads than those in Western countries and may find common objectives, such as population control, more challenging as a result.

Previous literature also indicates distinctions in terms of the primary source of animal inflow. Participants identified that most of their shelter's work involves the care of free-ranging dogs, through veterinary treatment or sterilization. Participant 2 explained that, when providing treatment for "dogs that are on the street ... and part of the Indian community", staff travel to the animal's street location as opposed to bringing them into the physical shelter environment. The Pet Care and Facilities Act identifies five primary categories for animal intake (stray, owner surrender, intrastate transfer, interstate transfer, and other) which are commonly used in the design of shelter tracking software [42]. It is evident that free-ranging dogs do not easily fit into these standard categories. For example, the 'stray' label includes animals which may have previously been pets (despite having no owner upon intake) and, thus, does not accurately describe an unowned, community animal. This may limit applicability of shelter software in the Indian context. Additionally, there is a question of what counts as 'animal intake' as well as 'outcome' for an Indian organization, when free-ranging dogs are treated inside and outside the shelter and returned to the community environment. Future investigations are needed into the modification of existing intake categories for Indian shelters and what additional metrics may be needed to track care for community animals. In fact, these metrics may additionally be useful in Western contexts as US animal shelters are moving to community-driven sheltering models [43].

Additionally, the inflow of abandoned pets may also be unique at Indian shelters. Participant 1 identifies that many owners "buy breed dogs, pedigree dogs, and they abandon them". This contrasts with the American context, where most owner-surrendered dogs at shelters are not purebred [44]. In a scoping review, Coe et al., (2014) identified housing barriers, aggressive companion animal behavior, and caretaker personal issues as the most investigated reasons for companion-animal relinquishment at US shelters [45]. The high rates of purebred dog abandonment, specifically, indicate that additional factors influence the purchase, and eventually relinquishment of dogs at Indian shelters. In their ethnographic analysis, Bhan and Bose (2020) describe purebred dogs as a symbol of middle- and upper-class Indian identity, with potential roots in colonial messaging that distinguished

British authorities and their "pedigree dogs" from Indian subjects and street dogs [46]. Volsche, Mohan, Gray and Rangawamy (2019) surveyed college students in Bangalore, India, with 89% of the total sample identifying as upper or upper-middle-class [47]. From the total respondents, 62.1% stated they preferred a purebred dog, while only 19.91% preferred Indian street dogs. Preferences for pedigree dogs appear to be connected to Indian class dynamics and potentially influenced by colonial rhetoric. Interventions to reduce pet abandonment at Indian shelters will require an understanding of this interaction between pet attitudes and historical and socioeconomic factors.

C. Seasonal Fluctuations

Participants also identified how intake numbers fluctuate throughout the year, increasing significantly during the monsoon (rainy) season, from June-September. Participant 1 stated, "Monsoons are pretty tough because there are lots and lots of cases of maggots and it's a mating season as well". Here, she identifies a seasonal spike in the intake of free-ranging dogs for emergency treatment and sterilization. Participant 5 also explains that animal recovery slows at this time:

> In the monsoon, healing takes so much time. In our animal birth control program, we have to release the dogs after like 10 days. Otherwise, in other seasons, we can release them after five days because healing processes [are] fast.

With the onset of dog mating, coupled with increased emergency cases, Indian shelters experience larger caseloads from June to September. Western shelters may see similar seasonal fluctuations: Janke and colleagues (2018), for example, report an increase in the admission rate of cats at the Guelph Humane society during spring "kitten season" [48]. Additionally, with global temperatures on the rise, seasonal intake patterns may be subject to change in the coming years. Protopopova, Ly, Eager and Brown (2021) identify that climate change outcomes, such as extreme weather events, are intrinsically linked with sheltering and companion-animal health [49]. The seasonal fluxes at Indian and Western shelters further indicate the sensitive relationship between animal intake and environmental conditions. In the Indian context, specifically, changes in the monsoon season could alter or exacerbate animal inflow. Modelling precipitation and runoff patterns, Clemens and colleagues (2021) indicate that such changes are possible with current greenhouse gas concentrations and project an increase in the quantity and variability of South Asian monsoon precipitation in the next decade [50]. Future studies on climate impacts and mitigation in India should account for their potential effects on both companion animals and sheltering systems.

D. Animal Death

Participant 3 describes a slightly different phenomenon; he focused on injured or ill free-ranging dogs, as opposed to abandoned pets, being 'dumped' at shelters:

> What happens is sometimes people feel that, you know, I don't want to see those animals dying in front of my house or inside my house ... [They say], 'if I can afford to pay 5000 rupees, I will send her to a shelter, but I will not see the animal dying in the shelter. What happens in the shelter is not my problem'.

Here, Participant 3 describes how many residents are inclined to take injured community dogs to shelters, perceiving them as a haven for animals. Participant 3 describes the contrast between these public perceptions of a shelter environment and the reality that staff experience:

> Our Indian ideology is that people think that we will pick up stray animals, put them in a cage and keep them there lifelong by giving them food. But that is not what we experience, no? That is not what we see. [We see] animals dying on us and, you know, it's very painful at times.

This response reveals the striking volume of animal death to which staff are exposed. It also appears that the public is largely unaware of the reality of overcrowded shelters. Participant 3 expanded on this, describing the typical struggles of new staff at her shelter:

> You're suddenly in a place with a hundred animals who are bleeding, who have wounds, who have maggots in their wounds, who are paralyzed, can't walk, and whatnot. You're suddenly in the middle of the room and you're like, okay, I have to take care of them. It's not that you are not alone, but you do feel alone.

These insights demonstrate how emergency cases contribute to the intensity of the Indian shelter environment, creating conditions in which staff must cope with an extremely high prevalence of death. The relationship between intake volume and staff wellbeing is well-documented. In a survey of 127 South Australian veterinary nurses, staff with longer work hours and higher contact with distressed clients or animals reported higher levels of work burnout [51]. Additionally, Reeve and colleagues (2004) found associations between seasonal influxes of puppies and kittens and downturns in staff wellbeing [19]. This research suggests that Indian shelter staff may experience greater vulnerability to emotional distress in the busy monsoon months and a reduced capacity to cope with challenging work situations. Leadership training at shelters should address these outcomes, focusing on developing managers' knowledge of common mental health challenges and their symptoms. This will allow shelter management to identify and extend additional support to vulnerable staff and prioritize positive workplace connections during busy seasons.

3.1.2. Inadequate Funding

All participants reported funding shortages as a significant barrier at their shelter. Participant 1, a shelter manager, stated, "Sometimes we are out of funds for the dogs, and we have to pay the staff less amount of money". Participant 3, a manager at a different shelter, expanded on this, describing how these shortages were exacerbated during the global pandemic of coronavirus disease 2019 (COVID-19):

> What happened is that the funding that we were supposed to get was all diverted to these COVID activities. A lot of big donors who said they are going to support us, at the last moment, they said, 'right now I think it's better to help people rather than animals.

Here, he identifies that, at the height of the pandemic, shelter donors shifted their focus from animal to human support. Participant 7, an animal caretaker from the same shelter, reaffirmed this, stating, "Because of the lockdown, there has been a decrease in the donations".

Inadequate funding has also been reported in a Western context: Turner and colleagues (2012) highlight the increasing importance of volunteers in the Canadian context, with many shelters unable to 'afford' enough paid staff to deliver animal care [52]. Financial challenges at shelters may have consequences for both animal well-being and the occupational health of staff. Lack of funds may result in the purchase of lower quality food and supplies, worsened facility hygiene, and inadequate animal husbandry due to staff shortages—all of which compromise animal care [53]. Additionally, research on Indian health care staff indicates that wage delays worsen occupational health. Kar and Suar (2014) surveyed nurses across 24 public hospitals in six Indian cities [54]. They found that participants, who reported frequent payment delays, wage cuts, and lack of compensation over time also experienced the highest levels of depersonalization and burnout.

In the current study, funding challenges at Indian shelters were influenced by three main factors: lack of government support, government policy, and cultural and religious beliefs.

A. Lack of Government Support

Participant 5, a veterinary nurse, "We don't get any support or help from the government. There are a few locals who help us. But we don't get any help from the government". Additionally, the public may lack an understanding of the inadequate support that shelters receive. Participant 3, for example, stated, "Most people think non-profit organizations are getting aid from the administration and the government, but that is not true".

B. Government Policy

In addition to a lack of government funding, government policy may create additional barriers. Participant 3 revealed a specific challenge with securing foreign grants. Here, he references the Foreign Contribution Regulation Act [55]—a set of laws that restrict the flow of foreign funding to the country through tighter registration protocols for both local non-profit organizations and international donors:

> The Indian government brought a law saying that, you know, you have to streamline your foreign contribution. That took us a really long time to get all the work done, opening your bank account. So that was again a little painful.

Di Russo (2011) describes the FCRA as the "primary source of the power of the Indian government over volunteer organizations" [56]. This policy lays out the protocol for a non-profit to become eligible for foreign funding and involves a complex evaluation to assess if 'welfare activities' benefit local communities. For Indian shelters, successful registration alone may be highly tedious, contingent on registration officers understanding the value of animal shelter work for local communities. Here, Participant 1 describes the 2020 FCRA Amendment Bill which introduced additional control on how foreign funds can be spent once received. These changes add further barriers for shelters and may prevent them from creating a budget that fits the needs of their organization and local communities.

This tight government control is quite distinct from the regulation of shelters in North America. Past research on the US charitable sector demonstrates that non-profits are under little federal control. With the assumption of their "good faith" intentions, the government relies on non-profits to "police themselves" [57]. This is understandable, given that non-profits in the US are less likely to be beneficiaries of international aid than Global South counterparts [58]. Without the inflow of foreign funds, governments may no longer see the need for regulatory policies, like the FCRA. However, current philanthropic law in the US may also leave room for unethical activity. Milofsky and Blades (1991) for example, describe the insufficient federal direction on recording financial transactions or flagging board member affiliations for health charities in the US and the consequences of unethical fundraising [59].

It is evident that Indian and Western non-profits face varying degrees of government regulation, and that the activity of Indian shelters is shaped by a unique political context. Further, there are broader differences in how shelters in the Global North and South sustain themselves: while the latter can rely on domestic financial resources, Global South countries, like India, appear to access both international and local funds.

C. Cultural and Religious Beliefs

Beyond government barriers, participants noted difficulties with gaining community donations. Dog rescues may struggle to gain local support because of the greater cultural and religious importance of large animals in Hindu communities. For example, Participant 1 stated, "We don't take large animal cases like cows and donkeys and all that stuff. People really don't want to donate for dogs. That's why we have very few donations". Participant 4, an animal caretaker at the same shelter, also spoke to the prioritization of large animal welfare and push back from locals when the shelter is unable to house cattle and goats:

> Many times, people accuse us [of not doing our jobs]. We tell them that yes, we are an animal rescue, but we don't have space to keep large animals. We can treat them, but we can't keep them. It feels bad to tell them that.

Participants 1 and 4, who lived and worked in the Northern states of Rajasthan and Himachal, respectively, both describe community members' frustration that shelters prioritize free-ranging dogs over the care of large animals, such as cows, buffalo, and donkeys. Large animals in India are part of a complex "cultural ecology": buffaloes and donkeys hold economic importance in the country's largely agricultural economy, while cows have great religious, social, and political significance [60]. Chigateri (2008) explores the intersection of religion and attitudes towards animals, highlighting the perceived sacredness of cows amongst dominant-caste Hindus [61]. Parikh and Miller (2019) explain how this narrative has been harnessed by political actors to subjugate minority Dalit and Muslim communities [62]. In this context, a shelter's ability to gain local support is not simply a function of the quality of care but rather whether the 'correct' animals are being cared for. Interventions to increase donations will require sensitivity given the religious and cultural standing of different species of urban animals, as well as the complex history of human race and ethnicity in India.

The relationship between religion, culture, and donations at Indian shelters illustrates another potential divergence from Western counterparts. Surveying communities in the US, Wang & Graddy (2008) found little relationship between religiosity and charitable behavior: though participants were more likely to donate to their specific religious group, they were no more likely to donate to secular causes [63]. Though little work has been done on identity and donation to animal causes in Western countries, religious dimensions do appear to be more significant in Indian sheltering. This indicates that any successful community engagement and funding strategies must account for these nuances and cannot merely replicate those used in Western organizations.

The current study was constrained to religion in the Indian context; Hindu communities in other countries may offer a different perspective. Past research has examined religion and animal attitudes in Bali, Indonesia, where over 80% of the province practices Hinduism [64]. Analyzing community perspectives towards dog meat consumption, Corrieri and colleagues (2018) identify how tenets of Balinese Hinduism have shaped and often promoted animal welfare in the country [65]. This includes the concept of "Pale Mahan" [harmony with one's natural environment] which encourages equal appreciation for all animals, including livestock, pets, and community dogs. Surveying residents across ten Balinese villages, Widyastuti and colleagues (2015), identified how Hindu beliefs might impact the treatment of free-ranging dogs [66]. When asked why they would not kill Balinese street dogs, residents cited the Hindu principle of ahimsa (non-violence).

Additionally, religious beliefs may not entirely dictate the treatment of animals. While Hindu principles prevented community involvement in dog culling, participants did not oppose the discarding of unwanted female puppies near garbage dumps or waterways [67]. In Balinese communities, positive religious attitudes towards dogs did not guarantee welfare-promoting behavior. Similarly, in the Indian context, while Hindu communities may promote cow welfare, individuals could act against these norms. When examining how religion affects Indian shelters and their funding, it is important to recognize how complex combinations of social factors, including, but not limited to, religion, impact human–animal interactions.

Participants in the current study demonstrated that, though financial instability is typical in animal welfare work, their experiences were shaped by a combination of religious, cultural, and political dimensions. A deep understanding of this complex local context is crucial to effectively tackling funding challenges at Indian shelters.

3.1.3. Community Conflict

Conflict with community members was another key challenge reflected on by participants. Participant 3, a shelter manager, explained the impact of clashes with the community: "See, handling animals is very easy, but handling people is very, very tough [laughs]. So that is something that really, you know, that takes a toll on you". These conflicts were of three main types: rescuer pressure, resident pushback, and incorrect community care.

A. Rescuer Pressure

All three shelter managers identified struggling with the large volume of calls from rescuers. Participant 1 stated: The biggest challenge for me is the helpline that I manage here. The people call up for rescues and many other things. I have to properly deal with them, make them understand. Participant 2 also explains her distress when sorting through these requests, identifying that "Every day we're getting around a hundred complaints to prioritize ... Whom do you need to reach first? Who will die?". There seemed to be a lack of understanding amongst the public about these experiences and the sheer volume of requests shelters receive. Participant 2 speaks to this issue, stating, "The other person just sees oh, the ambulance could not reach. Whereas we are actually sitting in between a hundred calls and going 'Oh god, what do we do?'". Beyond the volume of requests, staff also navigated difficult conversations with individual community members. When asked to recall a recent challenging case, Participant 3 described the following:

> There was this one scenario I still remember. There was a dog with a broken pelvic bone—the pelvic was broken into almost three to four pieces. So, there was no way to repair that dog ... But the rescuer said, 'No, I don't want to euthanize this animal.' She said she'd like to take it to some other place. So, she took the dog, did the surgery, and the dog died on the table.

Cases such as these, where an animal's life is at risk, emphasize the emotional burden placed on staff when speaking with community members. Participant 3 highlighted the intensity of their jobs', stating, "You have to deal with people who bring in those animals and sometimes it's a lot of emotions, you know? The working environment in a shelter is never not stressful". When asked about how community members impact her on a personal level, Participant 2 explained that "When somebody loses their animal, who is super, super attached to it. You get a load of people who are coming in and saying that you did not do enough".

B. Resident Pushback

In contrast to rescuers who blame staff for 'not doing enough', local residents may also oppose any shelter activity in their area. Participant 5 describes this challenge:

> There are some people who do not like shelters ... Sometimes, if we have to catch a dog, people will chase it away. They will not tell us where the dog is or if there is any problem or if they have to put in some effort.

Participant 3 also identified pushback to mandatory spay-neuter protocols at his shelter, stating, "A lot of time people are against it, but we tell them, this is mandatory". Despite receiving pushback, all shelter staff were willing to uphold shelter policies relating to animal birth control, even if it resulted in conflict with community members.

Shelter staff in Western countries appear to navigate very similar situations. Loyd and Miller (2010) surveyed Illinois homeowners and identified that most participants opposed their local shelter's TNR (trap-neuter-return) programs for controlling feral cat populations, favoring relocation of the animals instead [67]. Ashforth and Kreiner (2014) describes the stigmatization of shelter staff from wider communities and social construction of aspects of shelter work, such as euthanasia, as 'dirty work' [68]. Lopina and colleagues (2012) identified that such perceptions may heighten burnout and emotional strain amongst shelter staff. Interestingly, participants in the current study reported positive reactions from friends and peers to their jobs and a lack of community stigma. This is understandable given that the moral and physical 'taint' is most associated with the high-volume convenience animal euthanasia at Western shelters [69]. Additionally, Mendonca, D'Cruz and Noronha (2022) identify how 'dirty work' stigmatization for Indian cleaning workers may intersect with caste and class stigma; further investigations can consider whether the stigmatization of Indian shelter staff is similarly impacted by social position [70].

C. Incorrect Community Care

While there were challenges with residents who oppose shelter activities, Participant 3 also described issues created by individuals who provide incorrect care to community animals. He specifically identified the challenges with residents who feed free-ranging dogs:

> And there are a lot, a lot of unethical feeders. So, yeah, rather than solving any issues, it creates a lot of problems: These people are keeping them with milk and rice and non-veg. Milk and rice will give them loose motion. So, the dogs are going to be pooping near all these peoples' houses and no one will feel comfortable to clean it after feeding them. [People in the community] say: 'You know what? You take them to your house, look out for them in the house. Don't feed them here. We don't want these dogs here. So, there's a lot of conflict.

Here, Participant 3 identifies food, such as meat or eggs ("non-veg"), and milk and rice, that are commonly fed to free ranging, but potentially unsuitable for them. The manager sees a connection between unethical feeding, increased dog disruptions in the community, and heightened 'anti-dog' sentiment from residents who do not feed. Participant 3 also highlights the negative impact of incorrect feeding practices on free-ranging dog welfare:

> [And] you know, not like three meals a day, if you're feeding them, feed them every alternate day, because the animals shouldn't be dependent on one particular person. So, when you start feeding them on a daily basis, you are killing their survival instincts. You know, it becomes very difficult for the animals to survive.

According to Participant 3, the role of a feeder is to supplement community dogs' diets, without making them entirely dependent on human support. It seems that many residents struggle to strike this careful balance. Participant 3 further identifies that with a strong focus on daily feeding, other important activities may be neglected. He stated, "When you feed a stray dog, you need to take the responsibility to make sure that the animal is sterilized and vaccinated". While residents are eager to engage in low-barrier forms of care, such as feeding, they may be more reluctant to help coordinate sterilization and vaccinations programs that are crucial for long term animal welfare.

Here, it is apparent that Indian shelters must strike a careful balance between encouraging the feeding of community dogs while allowing them to retain independence and, in this way, performing an educational role similar to a wildlife rescue. Indian shelters must engage community members in nuanced discussions of the needs of free-ranging dogs. In contrast to Western shelters, who operate in areas with lower numbers of or no free-ranging dogs, Indian shelters may navigate a more complex set of responsibilities and community conflicts.

The parallels between challenges at Western and Indian shelters, in terms of funding, managing overcrowding, and navigating community conflicts indicate the potential for greater collaboration and information sharing on population control efforts and community engagement between the two countries. At the same time, context-specific factors, such as religious influence, government funding policy, and free-ranging dog feeding, highlight the unique barriers faced by Indian staff and need for context-specific interventions.

3.2. *Resiliency Factors*

In addition to identifying shelter challenges, interviews also revealed important resiliency factors that allowed staff to cope and succeed in their jobs. Researchers divided the resiliency factors into four main categories: flexibility and prioritization, co-worker support, duty of care, and understanding animal needs.

3.2.1. Flexibility and Prioritization

All participants displayed an immense amount of adaptability. Amongst animal caretakers, this was seen through their comfort with varied working hours and responsibilities. Participant 9, for example, stated, "It's not fixed working hours ... For three to four weeks, I worked in the 7-4 pm shift. Then 10-7 pm for four weeks. Now I am working the night shift". While Participant 10 described the fluctuations in her day-to-day tasks:

> If I am doing some work, like if there is some priority case, then we handle them first. If someone's clothes [i.e., bedding/bandages] are wet, we change them immediately. If they need hot water, we get it done ... If someone hasn't eaten, then we retry feeding them. If someone needs an extra egg, we give them to ensure that the feeding is complete. If someone's clothing is dirty, then we change those.

Rather than sticking to a rigid set of protocols, Participant 10 was capable of monitoring her environment and making decisions to optimize the comfort of shelter animals. From Participant 1's responses, we saw that adaptability is also important in a management role. Despite having a range of administrative duties, her priorities were very similar to that of animal caretakers - tasks directly related to animal care were placed at the top of her list:

> I am that sort of person who ends up taking more on her plate than she can manage, even if it's just going and checking up on somebody and spending 15 min there and I'm like 'Oh god I could do something else!', but that was important for me at that particular point in time.

Flexibility was also seen in staff's response to emergency cases. Participant 10, for example, described how her shelter contacts other frontline groups to facilitate large animal rescues: "Sometimes we contact the fire brigadiers. They send over a team. They are already trained for large animals. So, for large animals, they come when we ask them to". Participant 4, a veterinary nurse, described similar collaboration in a veterinary context:

> No ma'am, we don't have an X-ray machine. We go to a private clinic for those. There are some in [shelter city]. We have a CBC [Complete Blood Count] machine now. Any other biochemical tests are done in private clinics.

Participants' ability to adapt to fluctuating schedules, caseloads, and resource limitations, may indicate their high levels of 'psychological flexibility'. Kashdan & Rottenberg (2010) describe psychological flexibility as the human ability to adapt to situational demands, shift behavior, and remain open to new mindsets [71]. Previous literature has established the positive effects of this mindset on shelter staff. In a study on 170 non-profit service works, Biron and van Veldhoven (2012) found that personal psychological flexibility was associated with reduced emotional exhaustion as staff were inclined to accept, as opposed to repress, their emotions [72]. Psychological flexibility appears to be a powerful indicator of both well-being and work performance. This mindset may help staff to mitigate previously identified challenges, such as conflicts with rescuers and residents, and, in turn, prevent emotional exhaustion.

3.2.2. Co-Worker Support

A. Collaboration

The sentiment of collaboration was also seen amongst co-workers. Participants saw relationships as incredibly important for their wellbeing and described having an intimate, family-like environment at the shelter. Individuals that had moved to a city, from rural areas, to pursue employment had particularly deep connections, based on their shared backgrounds. Participant 4, for example, stated, "We are from the same village, four of us boys. We are from the same village, so it feels good to work together". Participant 4 also had positive interactions with higher level management and explained, "My boss is also good. Quite good. So, it's fun to work." Participant 9, another animal care worker, even noted an absence of rigid hierarchies in his organization: "Nobody thinks that he is a worker, he is a compounder or doctor, there is nothing like that. We speak to each other lovingly. They call me [omitted], my name". This significance of workplace friendship was also seen in management. Participant 2 described relying on her colleagues for social connection given the amount of time she spends at the shelter:

> So, nine hours of working plus like about two hours of traveling every day ... It almost consumes my entire life. So, it's like, my coworkers are the entire family and friends I have, my life is very sad [laughs].

These informal, family-like relationships seemed to have positive effects on the working environment. Participant 5, for example, explained that she trusts that her co-workers will support her and stated, "If there's any time that there's too much work and I am unable to handle it, I can go and tell someone that I am unable to handle or finish this job and ask them for help to do it. So, it gets managed". Participant 2 described her collaborative decision-making approach as a manager:

> If things are being changed then, I want them to understand that it's for the bigger animal welfare picture. I try to explain to them why a certain decision is being made. Or if I'm scheduling them somewhere, then why is it so important, why them and not somebody else.

Beyond including employees in decisions, Participant 2 also prioritized providing emotional support to staff:

> [I try to be] emotionally available for [new staff], because this [work] is so overwhelming . . . So, we try to gradually and slowly move them forward, and also be there and try and talk to them as to how they feel about it. I'm always trying to always find a balance where people can be able to express themselves and not get overwhelmed.

B. Equity and Safe Space

Managers also offered nuanced and individualized forms of support, with the goal of improving staff's equity and autonomy both in and outside the shelter. Participant 3, for example, explained how she created safe spaces for her female employees:

> A lot of women that we get from the local villages have so much responsibility. They need to go back home and cook for their husbands. And sometimes they are not in the best situations. So, I really want to make these women feel more comfortable, not just in their workspace. But also, that it's okay to say, 'I'm not in a good place at home'.
>
> If they're going through something at their home place and you see that someone is down, like their energies are not as they used to be, we try to talk to them sometimes and see if we can help them out sometimes. Because it's already too much to go through in the workplace—we are continuously stressed and you're working nine hours a day. And then you go back home, and you have another issue.

Participant 2 put her views on equity into action by implementing with tangible structural changes, such as promoting female staff to positions of authority:

> I would say about 35 to 40% women and then the rest of them are men. It's still predominantly men, but the shelter area is handled by women. [Name omitted] and [name omitted] two of our very strong women, they're like the best caregivers that we have. Any new staff who enters the shelter, irrespective of their gender, needs to know that both of them are their bosses.
>
> It's also important to make them feel empowered. You are working. It's you who is running the family. You are as independent as a man out there. So don't, in any area, feel like you don't do enough or feel like you are obliged to something.

Participant 2's approach to management, focused on promoting staff's professional and personal wellbeing, is quite distinct from shelter governance in North America. Yoffe-Sharp (2012) examined the culture in US humane societies, identifying rigid hierarchical operations that may heighten staff conflict, worsen communication and create perceptions of unfair treatment by management [73]. By contrast, Indian non-profits may avoid stringent professional norms when interacting with co-workers. Sharma and colleagues (2019) surveyed 100 non-profit employees in Jaipur, Rajasthan, identifying that job satisfaction increased with informal co-working gatherings and comfort with supervisors [74]. The same seems to apply to Indian shelters, with staff prioritizing intimate connections with one another and acknowledging the interconnectedness between personal and work challenges. In doing so, an environment of genuine care is created that helps staff cope with the chronic stressors of their jobs.

3.2.3. Duty of Care

In addition to a commitment to coworkers' wellbeing, participants also demonstrated a deep sense of duty towards animals. All staff saw their work for animals as an essential service, found ways to 'manage' high shelter intake, refusing to turn animals away, and extended care to community dogs outside their working hours.

This mindset was seen in Participant 1 who, after describing overcrowding at her shelter, stated, "We cannot neglect any rescue, [by] saying that we don't have space. We have to manage". Participant 4 reflected a similar sentiment. After explaining that there is "no space" for new rescues, he quickly emphasized, "We cannot refuse calls for large animal rescue either".

A duty of care mindset is also reflected in the high quality post-operative care protocols for sterilized animals. Participant 10, an animal caretaker, stated, "Our shelter has facilities for their stay, food and water, and good care. Some can stay ten days, some five days. Meaning, till the animal requires time to get better". Participant 5 reflects a similar practice at her shelter, stating, "If any animal is very weak, then we first nurse them back to health. Then we do the surgery and only then release it back". Despite dealing with a very high volume spay-neuter program, staff appear committed to the individual recovery and welfare of sterilized animals.

Non-negotiable care was also extended to community animals. For many participants, caring for free-ranging dogs outside of their work was part of their daily routines. When asked about this topic, Participant 2 stated, "I have nine dogs that I take care of every single day. They sleep in my house. I get beds for them". Participant 1 also reported caring for many local animals but focused on feeding dogs on the streets as opposed to sheltering them in her home. She stated, "I have 32 stray dogs with me that I have rescued myself. So, whenever I see a dog and I'm feeding and they come outside my gate, yes, I feed them. I love to feed dogs."

Some staff have family members who also care for community dogs. Participant 8 described feeding dogs along with her spouse:

> Me and my husband, daily we feed around 30 dogs. After we come back [from work], all the dogs are there. 'When they come, when they come!' They are waiting for their meal [laughs].

In contrast to staff who fed daily, some participants explained that they simply extended care as needed. This was the case for Participant 6, a veterinary nurse:

> Yes, I feed them sometimes. For example, if I come across some dogs on the road and they approach me, I give them something. And if I know some dog, especially the dogs suffering from mange, you see a lot of mange-infested dogs around, so for treating them I usually put the tablets in some food and give it.

While the exact type of care varied, all participants had strong emotional connection with their community animals. Participant 9, whose neighborhood dogs appeared to trust and have a strong bond with him:

> When I return, they get very happy. Sometimes they start fighting on seeing me or during feeding. They otherwise usually don't fight among themselves ... The moment they see me, they come to me running.

Further, staff did not view feeding as a burden on top of shelter duties. In fact, Participant 1 identified that her shelter role put her in a good position to care for community animals:

> I have made them different kennels, so they stay in their kennels. Every day I pick up their poop and all that stuff because I'm used to it, because I work in an NGO and it's my daily work.

It seems that many staff felt their jobs made them an asset to the community. Participant 4, a veterinary nurse, expressed his willingness to provide treatments outside the shelter, stating, "I have told them [hotel staff] that you are doing a very good job feeding them [dogs]. If there's any problem with any of them, call me, I will personally come to treat them".

High exposure to animal suffering, animal death, and large volumes of stray animals appear to increase shelter staff's vulnerability to compassion fatigue [15]. Despite being exposed to many of these risk factors, participants appeared to be highly resilient and did not indicate overt symptoms of compassion fatigue. This may stem from the 'duty of care' mindset: by viewing their jobs as an essential service, staff may feel an increased sense of pride and fulfillment, even under challenging circumstances. Additionally, having a sense of 'duty' towards animals places them in a position of autonomy, with the ability to take action and improve animals' outcomes. This is in contrast to experiences of euthanasia technicians in the US, who may experience feelings of helplessness from the requirements of their jobs [13]. With a greater sense of feelings of duty and control, Indian staff are perhaps more resilient when faced with similar stressors in their jobs.

Of all ten participants, only Participant 3 identified a hesitancy to care for community animals, citing fears of being 'harassed' by local residents:

> To be very, very honest, I don't feed any animals in my neighborhood. The reason is because, what happens is when I start feeding them people will start asking me or there have been cases where people will dump animals into my house. So, when they know that I'm associated with an association like this, they'll be like you know what, take away this dog. So, it becomes a huge problem for me and for my family members.
>
> People will ask you for medication, people will ask you for breed dogs, where do you get it, what do you do, how to get rid of this dog, cat. Answering all of these queries sometimes is really very stressful.

While a sense of duty towards animals may connect shelter staff to their work, Participant 3's responses also highlight the dangers of this mindset. When the protection of all animals in their shelter and home environment is seen as non-negotiable, staff may be unable to draw boundaries and combat feelings of overwhelm and stress.

3.2.4. Understanding Animal Needs

In addition to engaging in animal feeding and care, participants differentiated between free-ranging and pet dogs, and reflected on their unique needs. While staff identified the importance of human care for dogs' physical health, they stressed that animals' emotional wellbeing—that is, what they need 'to be happy'—was maximized with greater autonomy and reduced human intervention. Adopting a nuanced perspective, which acknowledges physical and emotional experiences, allowed staff to identify their specific responsibilities to community animals, while also acknowledging limits of their support. Participants specifically identified unrestricted movement, autonomy, and community care as the most important 'metrics' for free-ranging dog welfare.

A. Unrestricted Movement

All participants identified unrestricted movement as the most important aspect of good welfare for free-ranging dogs. Unrestricted movement referred to the ability of free-ranging dogs to move freely in their neighborhood, independent of human control. Participant 4 described his fears that free-ranging dogs would be uncomfortable if treated as traditional pets:

> Sometimes if we get them adopted, then they [the dog] starts wondering, why have I been restricted. For example, if we are suddenly asked to leave our house and start staying somewhere else, we will also feel odd and face issues.

Expanding on these ideas, Participant 8 recounted her personal experience with keeping free-ranging dogs in her home, stating, "One time I put them in my compound, they were very afraid. They felt uncomfortable. Now, they want food two times a day and they feel happy." By referring to their happiness and need for space, it is evident Participant 4 considered the animals' physical and emotional state as part of their overall welfare. Participant 5 expands on these ideas, identifying, "The street dogs that are there, they have a life. They like to stay open, unrestricted". She further draws a distinction between free-ranging and purebred abandoned dogs:

> Abandoned dogs cannot survive outside. They have no idea how to walk on the road, where to get food, water. They have no idea about anything. So, we should definitely try from our end to find them homes, good homes.

Distinguishing them from typical pet dogs, which require direct care and supervision from their owners, Participant 5 emphasizes that free-ranging dogs are highly robust and able to live independently, without human control.

B. Autonomy

The idea of autonomy was also connected to positive welfare. Participants described the community dogs as capable of making independent decisions. Participant 2 spoke to this idea:

> My [community] dogs have the best living situation as then they can go around and chase whoever they want. My home is forever open for them, so they can walk in whenever they want, and they can walk out wherever.

Here, Participant 2 emphasized the importance of the animal 'choosing' the human and making a conscious decision to return to their home. Participant 4 described a similar relationship with his community dogs, who are not kept in his house, but come back to him willingly:

> We don't tie up those dogs, so they roam around. We have a lot of open space here. They know they will get food in the evening. They come back at that time.

He identifies that community dogs can actively make decisions, informed by patterns in their environment, without directions by a human owner. Participant 6 extends the idea of independence to social behavior:

> I think these dogs can be kept at home, but they are street dogs. They should be allowed to roam out as well as allowed to stay inside the house. It shouldn't happen that the dog is kept inside the house 24 × 7 and only sees the humans of that house. They should mix with others too.

Here, we see the importance of varied social interactions for free-ranging dogs and their ability to forge relationships autonomously. Participant 6 identifies that community dogs thrive on interactions with both humans and conspecifics and thus, need to be able to move independently in their environment to form these varied social connections.

C. Community Care

Finally, many participants highlighted the importance of community-based care to ensure the welfare of free-ranging dogs. Participant 6 stressed that local residents can easily reduce injury and harm to community animals, stating, "They will have a better life on the streets ma'am. If people drive a little more carefully and if it's [the dog] taken care of, then they will be happier on the streets". Participant 6 further identified the potential involvement of community members in medical care and animal birth control programs. He suggested that a decentralized animal care system may ease the burden placed on animal shelters:

> Some dogs may be taken care of by the locals. If something is wrong with them, the medicines are handed over to their local caretakers. Then there is no need to send them to the shelter ... We need to make the local people aware of ABC and sterilization and that they can go to any shelter/NGO to get it done. Or if they are having trouble, then they can gather a few people for help and go to a government hospital and get that done.

This notion of community-based care was also seen in conversations with Participant 2. When asked if free-ranging dogs can have a 'good quality of life', she stated:

> I absolutely disagree to say that if they're living on the streets, then they don't have a good life if they have people in the community to take care of them. As long as these dogs on the street are community dogs, dogs that the entire community takes care of. I don't see an issue in it.

She further highlighted the power of collective actions, encouraging each community to take responsibility for a few animals in their locality and identified the importance of the first step of 'getting to know' your neighborhood dogs:

> If a community does decide to take care of these dogs, they don't have to take care of like a hundred dogs. They know that these nine dogs will stay in my lane. So, they will develop a relationship with these dogs because they stay there, and they know these dogs. They know, this one eats a lot. You know all those small details.

In relation to dogs' physical health, participants acknowledged the role of community members. However, when asked about the animals' emotional wellbeing, staff stressed the autonomy and unrestricted movement. This perspective, which promotes *reduced* human intervention, contrasts attitudes to animal welfare in the West. Tuan (2003) identifies that Western human-dog relationships are based on a combination of domination and affection: owned dogs are constrained physically within a home and restricted socially to a specific owner [75]. At the same time, they receive intense attention and love, often becoming integrated into the human family [76]. These perspectives have resulted in a rigid culture of 'responsible pet ownership' in the West, where to be a good owner is to be always in complete control over one's animal [77]. Additionally, Haraway (2003) identifies that pet-keeping standards inform dog welfare in other contexts, including shelters [78]. Thus, shelter staff may be under pressure to control and care for animals as pet owners do and feel disappointment if they fall short. Such experiences were documented in a 2017 study by Schabram and Maitlis, where emotions of shame, guilt, and personal disappointment were seen in narrative interviews with 50 shelter staff [79]. Indian shelter staff offer a different perspective, by valuing animal autonomy but also recognizing limits to which human caretakers can enhance welfare. Staff see themselves as a source of support, rather than control, for animals and, in this way, may relate more positively to their jobs and performance. A similar perspective, if applied in the Western context, may allow shelter workers to feel more successful and empowered in a demanding environment.

While animal attitudes vary between cultures, they may also fluctuate within them. In the present study, participants saw native free-ranging dogs as being independent but expressed that abandoned purebred dogs "cannot survive outside" and require "good homes". Their understanding of pet dogs resembles Western attitudes on animal ownership and control. It appears that the emphasis on autonomy was not applied ubiquitously to all dogs, but rather specifically to community dogs. Fluctuations in animal attitudes are also seen in the Western context: Jorgensen and Brown (2014) investigated leash-law on public beaches amongst pet owners in McConaughy, Nebraska and found that less than 25% of owners abided by regulations, despite expressing negative beliefs about unleashed dogs [79].

It is apparent that attitudes to animals are neither universal nor consistent within "cultural landscapes" [80]. Yet, there appears to be inherent value in understanding the different ways in which humans relate to animals and looking beyond dominant Western rhetoric [81]. The resiliency factors identified in the current study demonstrate the unique ways in which Indian shelter staff relate to animals and humans in their environment: staff prioritize connections with co-workers and shelter animals and appreciate the independent relationships amongst community animals. This relationship-based coping may inform effective support interventions for shelter staff beyond the Indian context. Despite extensive evidence of compassion fatigue amongst Western shelter staff, administrators are often unprepared to provide support due to a lack of knowledge about cost and time effective interventions [18]. Past research has focused on external interventions, such as skills-based training, counselling, and stress and coping seminars to address compassion fatigue. This study suggests that a community and relationship-based approach, as exemplified by Indian staff, may have long-lasting effects on emotional well-being.

4. Limitations and Future Research

There were several limitations in the study that must be acknowledged. Participants were selected from only three states (Himachal, Rajasthan, and Karnataka). As a result, our results do not reflect the diversity of cultural experiences and animal care across Indian states. In the present study, participant recruitment was restricted as the research team was only able to conduct interviews in Hindi or English. Future projects can prioritize collaborations with translators who are to facilitate interviews with shelter staff in their local language.

There may be general challenges with conducting reliable trans-linguistic interviews. In this study, the translation of Hindi interviews to English was performed by one member of the research team and an independent contractor was used to identify any translator errors. Lopez and colleagues (2008) propose a more rigorous protocol for conducting reliable cross-cultural research. The authors describe a seven-step methodology which includes the use of multiple independent translators, and group discussions about variations in meaning based on region and dialect [82]. Future studies, particularly if facilitating interviews in a larger number of Indian languages, should implement a similar process to ensure translations represent participants' experiences as accurately as possible.

In the present study, the use of audio recordings alone may have limited the depth of the results as researchers were not able to record the subtleties of non-verbal expression. Furthermore, conducting interviews over Zoom may have impacted participants' responses if they were uncomfortable with using an online platform. Repeating this study with in-person interviews would allow researchers to pick up on the participant's body language and remove any barriers created by technology.

While this qualitative study offers initial insight into Indian sheltering, large-scale quantitative research may be needed to see whether the discovered themes represent overarching challenges and resiliency factors in a representative sample of Indian animal shelter staff. This methodological triangulation would allow for a more comprehensive understanding of staff experiences and improve the credibility of the current findings. This may be important when considering the impact of government policies, such as the FCRA, that regulate foreign funding. Future surveys can identify how many Indian shelters rely on foreign funding and are, in turn, impacted by such legislation. Additional investigation into the psychological experiences of shelter staff is also needed. Many participants were reluctant to discuss mental health and the emotional challenges of their jobs and denied experiencing any burnout or compassion fatigue. While this may reflect staff's resilience, it may also be the result of cultural stigma around mental health and openly addressing one's struggles. In future research, implementing a mixed-methods approach (by posing questions about mental health in an anonymous survey format as opposed to an interview) may make Indian staff more comfortable, if they were not, in opening up about the emotional experiences in a shelter environment.

5. Conclusions

While past literature has largely focused on the Global North, this study demonstrates the importance of recognizing Indian sheltering as a distinct area of interest. This is demonstrated by challenges, such as government regulation and cultural preferences for large animal care, that are specific to the Indian context and impact a shelter's ability to sustain animal care operations. At the same time, the identified resiliency factors indicate that Indian staff also cope with job stressors in unique ways. Participants in this study may harness relationships with both animals and humans to increase resilience and maintain their mental health. This 'relationship-centered' perspective can be applied to the Western context to design preventative measures against compassion fatigue that focus on deepening staff's connection with one another and the animals for which staff care. Furthermore, others can use insights about the identified needs of dogs, such as a need for autonomy, to reduce any potential ethnocentric biases in the determination and improvement of animal welfare. While such cross-cultural knowledge exchange may be very powerful, this study also highlights the need for far more research focused specifically on the Indian context. Further studies should investigate the specific challenges and staff experiences at Indian shelters and how socio-cultural and political factors influence the capacity to support both human and animal welfare.

Author Contributions: Conceptualization, D.S., A.P. and R.M.; methodology, D.S., A.P., K.A.V.R. and R.M.; validation, D.S., A.P., K.A.V.R. and R.M.; formal analysis, D.S., A.P. and R.M.; investigation, D.S. and R.M.; resources, A.P.; data curation, D.S., A.P. and R.M.; writing—original draft preparation, D.S.; writing—review and editing, D.S., A.P., K.A.V.R. and R.M.; visualization, D.S.; supervision, A.P.; project administration, A.P. All authors have read and agreed to the published version of the manuscript.

Funding: This research received no external funding.

Institutional Review Board Statement: The study was conducted in accordance with the Declaration of Helsinki, and approved by the Institutional Review Board (or Ethics Committee) of The University of British Columbia (H21-01759; approved on 14 June 2021).

Informed Consent Statement: Informed consent was obtained from all subjects involved in the study.

Data Availability Statement: The data presented in this study are available on request from the corresponding author. The data are not publicly available due to the need to protect anonymity of the participants.

Acknowledgments: We thank members of the Animal Welfare Program at the University of British Columbia for their ongoing excitement and support for this project. We also thank David Fraser, Katie Koralesky, and Nina von Keyserlingk for their feedback on drafts and valuable insight into qualitative approaches. Finally, we are grateful to all ten participants for sharing their stories with us and for all that they do for animals.

Conflicts of Interest: The authors declare no conflict of interest.

References

1. McCrea, R.C. *The Humane Movement: A Descriptive Survey*; The Columbia University Press: New York, NY, USA, 1910.
2. Rowan, A.; Kartal, T. Dog population & dog sheltering trends in the United States of America. *Animals* **2018**, *8*, 68.
3. Rowan, A.N.; Williams, J. The success of companion animal management prrams: A Review. *Anthrozoös* **1987**, *1*, 110–122. [CrossRef]
4. Srinivasan, K. The biopolitics of animal being and welfare: Dog control and care in the UK and India. *Trans. Inst. Br. Geogr.* **2013**, *38*, 106–119. [CrossRef]
5. The World Bank. South Asia. In *Global Economic Prospects*; World Bank Publications: Washington, DC, USA, 2015; pp. 149–156. [CrossRef]
6. Pal, S.K. Population ecology of free-ranging urban dogs in West Bengal, India. *Acta Theriol.* **2010**, *46*, 69–78. [CrossRef]
7. Massei, G.; Fooks, A.R.; Horton, D.L.; Callaby, R.; Sharma, K.; Dhakal, I.P.; Dahal, U. Free-roaming dogs in Nepal: Demographics, health and public knowledge, attitudes and practices. *Zoonoses Public Health* **2017**, *64*, 29–40. [CrossRef]
8. Taylor, L.H.; Hampson, K.; Fahrion, A.; Abela-Ridder, B.; Nel, L.H. Difficulties in estimating the human burden of canine rabies. *Acta Trop.* **2017**, *165*, 133–140. [CrossRef]
9. Radhakrishnan, S.; Vanak, A.T.; Nouvellet, P.; Donnelly, C.A. Rabies as a public health concern in India: A historical perspective. *Trop. Med. Infect. Dis.* **2020**, *5*, 162. [CrossRef]
10. Animal Birth Control (Dogs) Rules, 2001 (India). Available online: https://bombayhighcourt.nic.in (accessed on 16 August 2022).
11. Agnihotram, R. An overview of occupational health research in India. *Indian J. Indig. Med.* **2005**, *9*, 10. [CrossRef]

12. Standard Operating Procedures for Sterilization of Stray Dogs under the Animal Birth Control Program. (Animal Welfare Board of India). 2009. Available online: http://www.awbi.in/ (accessed on 16 August 2022).
13. Arluke, A.; Sanders, C.R. The institutional self of shelter workers. In *Regarding Animals*; Temple University Press: Philadelphia, PA, USA, 1996; pp. 82–106.
14. Hill, E.M.; LaLonde, C.M.; Reese, L.A. Compassion fatigue in animal care workers. *Traumatology* **2020**, *26*, 96–108. [CrossRef]
15. Rank, M.G.; Zaparanick, T.L.; Gentry, J.E. Nonhuman animal care compassion fatigue: Training as treatment. *Best Pract. Ment. Health Int. J.* **2009**, *5*, 40–61.
16. Andrukonis, A.; Hall, N.J.; Protopopova, A. The impact of caring and killing on physiological and psychometric measures of stress in animal shelter employees: A pilot study. *Int. J. Environ. Res. Public Health* **2020**, *17*, 9196. [CrossRef] [PubMed]
17. Andrukonis, A.; Protopopova, A. Occupational health of animal shelter employees by live release rate, shelter type, and euthanasia-related decision. *Anthrozoös* **2020**, *33*, 119–131. [CrossRef]
18. Baran, B.E.; Rogelberg, S.G.; Carello Lopina, E.; Allen, J.A.; Spitzmüller, C.; Bergman, M. Shouldering a silent burden: The toll of dirty tasks. *Hum. Relat.* **2012**, *65*, 597–626. [CrossRef]
19. Reeve, C.L.; Spitzmuller, C.; Rogelberg, S.G.; Walker, A.; Schultz, L.; Clark, O. Employee reactions and adjustment to euthanasia-related work: Identifying turning-point events through retrospective narratives. *J. Appl. Anim. Welf. Sci.* **2004**, *7*, 1–25. [CrossRef]
20. Bennett, P.; Rohlf, V. Perpetration-induced traumatic stress in persons who euthanize nonhuman animals in surgeries, animal shelters, and laboratories. *Soc. Anim.* **2005**, *13*, 201–220. [CrossRef]
21. Taylor, N. Animal shelter emotion management: A case of in situ hegemonic resistance? *Sociology* **2010**, *44*, 85–101. [CrossRef]
22. Schabram, K.; Maitlis, S. Negotiating the challenges of a calling: Emotion and enacted sensemaking in animal shelter work. *AMJ* **2017**, *60*, 584–609. [CrossRef]
23. Hoy-Gerlach, J.; Ojha, M.; Arkow, P. Social Workers in Animal Shelters: A Strategy Toward Reducing Occupational Stress among Animal Shelter Workers. *Front. Vet. Sci.* **2021**, *8*, 734396. [CrossRef]
24. Chang, C.-H.; Spector, P.E. Cross-cultural occupational health psychology. In *Handbook of Occupational Health Psychology*, 2nd ed.; American Psychological Association: Washington, DC, USA, 2011; Volume 2, pp. 119–137.
25. Blauner, B. Problems of editing in sociology. *Qual. Sociol.* **1987**, *10*, 46–64. [CrossRef]
26. Chen, H.-Y.; Boore, J.R. Translation and back-translation in qualitative nursing research: Methodological review. *J. Clin. Nurs.* **2010**, *19*, 234–239. [CrossRef]
27. Moon, K.; Blackman, D. A guide to understanding social science research for natural scientists. *Conserv. Biol.* **2014**, *28*, 1167–1177. [CrossRef] [PubMed]
28. Levitt, A.L.; Gezinski, L.B. Compassion fatigue and resiliency factors in animal shelter workers. *Soc. Anim.* **2020**, *28*, 633–650. [CrossRef]
29. Lopez, K.A.; Willis, D.G. Descriptive versus interpretive phenomenology: Their contributions to nursing knowledge. *Qual. Health Res.* **2004**, *14*, 726–735. [CrossRef]
30. Vaismoradi, M.; Jones, J.; Turunen, H.; Snelgrove, S. Theme development in qualitative content analysis and thematic analysis. *J. Nurs. Educ. Pract.* **2016**, *6*, 100–110. [CrossRef]
31. Guest, G.; Bunce, A.; Johnson, L. How many interviews are enough? *Field Methods* **2006**, *18*, 59–82. [CrossRef]
32. Dey, I. *Grounding Ground Theory: Guidelines for Qualitative Inquiry*; Emerald Group Publishing Limited: San Diego, CA, USA, 1999; pp. 249–269. [CrossRef]
33. Nelson, J. Using conceptual depth criteria: Addressing the challenge of reaching saturation in qualitative research. *Qual. Res. QR* **2017**, *17*, 554–570. [CrossRef]
34. Luthar, S.S.; Cicchetti, D.; Becker, B. Research on resilience: Response to commentaries. *Child. Dev.* **2000**, *71*, 573–575. [CrossRef]
35. Feder, A.; Ahmad, S.; Lee, E.J.; Morgan, J.E.; Singh, R.; Smith, B.W.; Southwick, S.M.; Charney, D.S. Coping and PTSD symptoms in Pakistani earthquake survivors: Purpose in life, religious coping and social support. *J. Affect. Disord.* **2013**, *147*, 156–163. [CrossRef]
36. Brintzinger, M.; Tschacher, W.; Endtner, K.; Bachmann, K.; Reicherts, M.; Znoj, H.; Pfammatter, M. Patients' style of emotional processing moderates the impact of common factors in psychotherapy. *Psychotherapy* **2021**, *58*, 472–484. [CrossRef] [PubMed]
37. Zawistowski, S.; Morris, J.; Salman, M.D.; Ruch-Gallie, R. Population dynamics, overpopulation, and the welfare of companion animals: New insights on old and new data. *J. Appl. Anim. Welf. Sci.* **1998**, *1*, 193–206. [CrossRef] [PubMed]
38. Wenstrup, J.; Dowidchuk, A. Pet overpopulation: Data and measurement issues in shelters. *J. Appl. Anim. Welf. Sci.* **1999**, *2*, 303–319. [CrossRef]
39. Lepper, M.; Kass, P.H.; Hart, L.A. Prediction of adoption versus euthanasia among dogs and cats in a California animal shelter. *J. Appl. Anim. Welf. Sci.* **2002**, *5*, 29–42. [CrossRef]
40. Marston, L.C.; Bennett, P.C.; Coleman, G.J. What happens to shelter dogs? An analysis of data for one year from three Australian shelters. *J. Appl. Anim. Welf. Sci.* **2004**, *7*, 27–47. [CrossRef]
41. Protopopova, A.; Gilmour, A.J.; Weiss, R.H.; Shen, J.Y.; Wynne, C.D.L. The effects of social training and other factors on adoption success of shelter dogs. *Appl. Anim. Behav. Sci.* **2012**, *142*, 61–68. [CrossRef]
42. Hawes, S.M.; Hupe, T.M.; Gandenberger, J.; Morris, K.N. Temporal trends in intake category data for animal shelter and rescue organizations in Colorado from 2008 to 2018. *J. Am. Vet. Med. Assoc.* **2021**, *260*, 311–319. [CrossRef]

43. Hurley, K.F. The evolving role of triage and appointment-based admission to improve service, care and outcomes in animal shelters. *Front. Vet. Sci.* **2022**, *9*, 809340. [CrossRef]
44. New, J.C.; Salman, M.D.; King, M.; Scarlett, J.M.; Kass, P.H.; Hutchison, J.M. Characteristics of shelter-relinquished animals and their owners compared with animals and their owners in U.Spet-owning households. *J. Appl. Anim. Welf. Sci. JAAWS* **2000**, *3*, 179–201. [CrossRef]
45. Coe, J.B.; Young, I.; Lambert, K.; Dysart, L.; Nogueira Borden, L.; Rajić, A. A scoping review of published research on the relinquishment of companion animals. *J. Appl. Anim. Welf. Sci.* **2014**, *17*, 253–273. [CrossRef]
46. Bhan, M.; Bose, P. Canine counterinsurgency in Indian-occupied Kashmir. *Crit. Anthropol.* **2020**, *40*, 341–363. [CrossRef]
47. Volsche, S.; Mohan, M.; Gray, P.B.; Rangaswamy, M. An exploration of attitudes toward dogs among college students in Bangalore, India. *Animals* **2019**, *9*, 514. [CrossRef] [PubMed]
48. Janke, N.; Berke, O.; Klement, E.; Flockhart DT, T.; Coe, J.; Bateman, S. Effect of capacity for care on cat admission trends at the Guelph Humane Society, 2011-2015. *J. Appl. Anim. Welf. Sci.* **2018**, *21*, 283–294. [CrossRef] [PubMed]
49. Protopopova, A.; Ly, L.H.; Eagan, B.H.; Brown, K.M. Climate change and companion animals: Identifying links and opportunities for mitigation and adaptation strategies. *Integr. Comp. Biol.* **2021**, *61*, 166–181. [CrossRef] [PubMed]
50. Clemens, S.C.; Yamamoto, M.; Thirumalai, K.; Giosan, L.; Richey, J.N.; Nilsson-Kerr, K.; Rosenthal, Y.; Anand, P.; McGrath, S.M. Remote and local drivers of Pleistocene South Asian summer monsoon precipitation: A test for future predictions. *Sci Adv.* **2021**, *7*, eabg3848. [CrossRef] [PubMed]
51. Black, A.F.; Winefield, H.R.; Chur-Hansen, A. Occupational stress in veterinary nurses: Roles of the work environment and own companion animal. *Anthrozoös* **2011**, *24*, 191–202. [CrossRef]
52. Turner, P.; Berry, J.; Macdonald, S. Animal shelters and animal welfare: Raising the bar. *Can. Vet. J.* **2012**, *53*, 893–896.
53. Kim, J. Social finance funding model for animal shelter programs: Public–private partnerships using social impact bonds. *Soc. Anim. Soc. Sci. Stud. Hum. Exp. Other Anim.* **2018**, *26*, 259–276. [CrossRef]
54. Kar, S.; Suar, D. Role of burnout in the relationship between job demands and job outcomes among Indian nurses. *Vikalpa* **2014**, *39*, 23–38. [CrossRef]
55. Agarwal, S. *Accountable Handbook FCRA 2010: Theory and Practice: Foreign Contribution (Regulation) Act, 2010, Foreign Contribution (Regulation) Rules, 2011*; Account Aid India: New Delhi, India, 2012.
56. Di Russo, A.A. American nonprofit law in comparative perspective. *Wash. Univ. Glob. Stud. Law Rev.* **2011**, *10*, 39–86.
57. Leslie, M. The wisdom of crowds? groupthink and nonprofit governance. *SSRN Electron. J.* **2009**, *62*, 1179–1226. [CrossRef]
58. Marx, J.D.; Carter, V.B. Factors influencing U.S charitable giving during the great recession: Implications for nonprofit administration. *Adm. Sci.* **2014**, *4*, 350–372. [CrossRef]
59. Milofsky, C.; Blades, S.D. Issues of accountability in health charities: A case study of accountability problems among nonprofit organizations. *Nonprofit Volunt. Sect. Q.* **1991**, *20*, 371–393. [CrossRef]
60. Nanda, A.S.; Nakao, T. Role of buffalo in the socioeconomic development of rural Asia: Current status and future prospectus. *Anim. Sci. J.* **2003**, *74*, 443–455. [CrossRef]
61. Chigateri, S. 'Glory to the cow': Cultural difference and social justice in the food hierarchy in India. *South Asia J. South Asian Stud.* **2008**, *31*, 10–35. [CrossRef]
62. Parikh, A.; Miller, C. Holy cow! Beef ban, political technologies, and Brahmanical supremacy in Modi's India. *ACME Int. J. Crit. Geogr.* **2019**, *18*, 835–874.
63. Wang, L.; Graddy, E. Social Capital, volunteering, and charitable giving. *Volunt. Int. J. Volunt. Nonprofit Organ.* **2008**, *19*, 23–42. [CrossRef]
64. McCreanor, K.; McCreanor, S.; Utari, A. Connected and interconnected: Bali people and Bali dogs. In *Dog's Best Friend?: Rethinking Canid-Human Relations*; Sorenson, J., Matsuoka, A., Eds.; McGill-Queen's University Press: Montreal, QC, Canada, 2019.
65. Corrieri, L.; Adda, M.; Miklósi, Á.; Kubinyi, E. Companion and free-ranging Bali dogs: Environmental links with personality traits in an endemic dog population of South East Asia. *PLoS ONE* **2018**, *13*, e0197354. [CrossRef]
66. Widyastuti, M.D.W.; Bardosh, K.L.; Sunandar; Basri, C.; Basuno, E.; Jatikusumah, A.; Arief, R.A.; Putra, A.A.G.; Rukmantara, A.; Estoepangestie, A.T.S.; et al. On dogs, people, and a rabies epidemic: Results from a sociocultural study in Bali, Indonesia. *Infect. Dis. Poverty* **2015**, *4*, 30. [CrossRef]
67. Loyd, K.A.; Miller, C.A. Factors related to preferences for trap–neuter–release management of feral cats among Illinois homeowners. *Wildfire* **2010**, *74*, 160–165. [CrossRef]
68. Ashforth, B.E.; Kreiner, G.E. Dirty work and dirtier work: Differences in countering physical, social, and moral stigma. *Manag. Organ. Rev.* **2014**, *10*, 81–108. [CrossRef]
69. Lopina, E.C.; Rogelberg, S.G.; Howell, B. Turnover in dirty work occupations: A focus on pre-entry individual characteristics. *J. Occup. Organ. Psychol.* **2012**, *85*, 396–406. [CrossRef]
70. Mendonca, A.; D'Cruz, P.; Noronha, E. Identity work at the intersection of dirty work, caste, and precarity: How Indian cleaners negotiate stigma. *Organization* **2022**, 13505084221080540. [CrossRef]
71. Kashdan, T.B.; Rottenberg, J. Psychological flexibility as a fundamental aspect of health. *Clin. Psychol. Rev.* **2010**, *30*, 865–878. [CrossRef] [PubMed]
72. Biron, M.; van Veldhoven, M. Emotional labour in service work: Psychological flexibility and emotion regulation. *Hum. Relat.* **2012**, *65*, 1259–1282. [CrossRef]

73. Yoffe-Sharp, B. Administrative issues. In *Shelter Medicine for Veterinarians and Staff*; Wiley: Hoboken, NJ, USA, 2012; pp. 49–58.
74. Sharma, A.; Arora, P.; Bagga, S. Impact of job satisfaction on organizational performance among the employees working in a non-government organization (NGO) in Jaipur. *Anwesh Roorkee* **2019**, *4*, 36–45.
75. Tuan, Y.F. Dominance and affection. In *Dominance and Affection*; Yale University Press: New Haven, CT, USA, 2003.
76. Charles, N.; Davies, C.A. My family and other animals: Pets as kin. *Sociol. Res. Online* **2008**, *13*, 13–26. [CrossRef]
77. Borthwick, F. Governing pets and their humans. *Griffith Law Rev.* **2009**, *18*, 185–201. [CrossRef]
78. Haraway, D.J. *The Companion Species Manifesto: Dogs, People, and Significant Otherness*; Prickly Paradigm Press Chicago: Chicago, IL, USA, 2003; Volume 1.
79. Jorgensen, J.G.; Brown, M.B. Piping Plovers Charadrius melodus and dogs: Compliance with and attitudes toward a leash law on public beaches at Lake McConaughy, Nebraska, USA. *Wader Study Group Bull.* **2014**, *121*, 71–76.
80. Elder, G.; Wolch, J.; Emel, J. Race, place, and the bounds of humanity. *Soc. Anim.* **1998**, *6*, 183–202. [CrossRef]
81. Savvides, N. Living with dogs: Alternative animal practices in Bangkok, Thailand. *Anim. Stud. J.* **2013**, *2*, 28–50.
82. Lopez, G.I.; Figueroa, M.; Connor, S.E.; Maliski, S.L. Translation barriers in conducting qualitative research with Spanish speakers. *Qual. Health Res.* **2008**, *18*, 1729–1737. [CrossRef] [PubMed]

animals MDPI

Article

You Are *Not* Welcome! A Media Analysis of Risk Factors, Prevalence and Management of Free-Roaming Dogs in Iran

Farshad Amiraslani

Independent Researcher, 53A, Millburn Road, Coleraine BT52 1QT, UK; fami8991@alumni.sydney.edu.au

Simple Summary: An increase in the number of free-roaming dogs and the related consequences (e.g., dog biting) has been observed across Iranian cities. So far, no clear scientific reasons for this increase were demonstrated, and a few existing domestic studies have only investigated the behaviours and frequencies of these animals in the cities. Based on online news articles, the on-the-ground reasons, challenges and dog-controlling activities mentioned by key urban and health authorities are examined in this paper. A scientific link is established between increased urbanisation, its features and its implications for dogs and humans.

Abstract: History has witnessed a long-term relationship between humans and animals. Historical documents and modern findings prove that humans' needs to use animals for companions or services are commonplace in many parts of the world, leading to the domestication of certain animals. Yet, modern societies have degraded many natural habitats for wildlife, confining them to small patches of landscapes or urban areas. Whether a domesticated/free-roaming animal or a wild species, their close contact with humans can create cumbersome situations for both species. This paper explores a link between online media content and on-the-ground efforts to manage free-roaming dogs as a rare case study. As indicated by news articles, the municipal costs of managing free-roaming dogs in Iranian cities have increased, and this can potentially derail the control of such dogs in the long run. This paper lays out pivotal factors for recent increasing human–animal encounters, which have led to many challenges (e.g., rabies) across cities in Iran. We show that some urban features (e.g., topography) can influence the presence and behaviours of free-roaming animals in the cities. The findings of this paper can be related to other developing countries where the plague of rabies is rising.

Keywords: free roaming; free ranging; stray; dog; urban; Iran; online media; news articles; rabies; landscape; habitat

Citation: Amiraslani, F. You Are *Not* Welcome! A Media Analysis of Risk Factors, Prevalence and Management of Free-Roaming Dogs in Iran. *Animals* **2023**, *13*, 2347. https://doi.org/10.3390/ani13142347

Academic Editor: Betty McGuire

Received: 8 June 2023
Revised: 13 July 2023
Accepted: 17 July 2023
Published: 18 July 2023

1. Introduction

1.1. A Short History of the 'Human-Animal' Intimacy Linkage

Foley and Gamble [1] describe the ecology of humanity's social transitions. As one of these social transition eras, the so-called T5 evolutionary era (ecological intensification: 200–10 ka) portrayed an 'energy-rich ecology through aquatic resources, cereal harvesting, hunting and domestication of animals' (Table 3: [1]). It was a time when most of the population possibly lived in separate groups and landscapes without proper communication and social interactions, yet early settlers felt the need to initiate new forms of human communities and social interactions. The *H. sapiens* gradually made intergroup and regional social structures and networks, while the social system was dominated by resource ownership, defence and control (ibid.). The era was continued by a 'fully sedentary, agricultural and ethnically complex world in the last 15 ka' (ibid.).

Over the past millennia, humans have embraced the idea of domesticating animals for their benefit. The most notable cases are the domestication of goats (*Capra hircus*) that occurred 10,000 years ago [2]. A study reveals that dogs were independently domesticated several millennia ago in Eastern and Western Eurasia from distinct wolf populations [3].

'The earliest historical mention of domestic cats originates from Persia [now Iran] and dates to the sixth century CE, . . . women kept cats as pets, dying their fur, adorning them with jewellery, and letting them sleep in their beds' [4]. There is also evidence of domestic cats (*Felis catus* L., 1758) from Kazakhstan, dating to 775–940 cal. CE (ibid.).

1.2. Contemporary World: Dominance of Human Species

The contemporary world has changed dramatically. Over the past centuries, unprecedented population growth and resource utilisation have disrupted the equilibrium between humans and animals. Crowded urban areas and surrounding segmented forests and rangelands have created a chaotic world for millions of birds, mammals, insects, etc. A diverse range of wet markets, open/closed fish markets, local vendors of live animals, etc., have exacerbated this linkage to the point that a wave of health and hygienic risks in cities is imminent. Those earlier domesticated animals, known these days as pets, have approached our private life, but with associated risks. Common zoonotic diseases between humans and pets, especially dogs, have endangered our health and well-being (e.g., [5]). They can also be potential conservation threats to wildlife (e.g., [6,7]).

The dysconnectivity between humans and dogs as companions in urban areas has culminated into an urban challenge due to the rising number of free-roaming dogs. The issue has become a cross-disciplinary subject intersecting veterinary science, urban planning, anthropology, epidemiology and other health-related sciences. The positive psychological and mental effects of this human–dog intimacy (e.g., [8]) could lead to serious health issues posed by free-roaming dogs (e.g., [9]). A range of humanistic adverse side effects (e.g., rabies), pertinent curing costs for humans and dogs and exerted costs on urban management have enticed city planners to re-evaluate their encounter with this growing public health concern worldwide.

Contemporary growing urbanisation has made the relationship between humans and wildlife/domesticated animals much more special, but complicated [10–14]. We encounter and enjoy wildlife in our cities. We signify some animal species on our aircrafts (e.g., the oryx for Qatar Airways) and shelter them in our places for nesting, breeding and feeding [15]. Sometimes, though, we (un)intentionally injure or kill them (e.g., road accidents).

Urbanisation has caused natural landscapes to become fragmented, disturbed and degraded. Few natural habitats have remained untouched and healthy in urban areas, and their fragile condition will worsen due to land use changes. The legacy of our toxic materials introduced to these territories in order to develop our built-up areas (e.g., cemented pavements), for our daily usage (e.g., face masks), or consequences of our habits (e.g., vehicle smoke) remains forever, with dire impacts on animals' lives and health (e.g., Urban environment and cancer in wildlife; [16]). A strong link between urbanisation and the prevalence of pathogens in populations of free-roaming wildlife was established, mediated by an altered habitat structure and changes to the resource availability, leading to a low biodiversity and declining trends in species richness [13].

While at some point in the past, wildlife resided in large landscapes, currently, they are desperate to forage and explore resources outside of their original habitats, as we have changed these habitats to 'urban' areas (Figure 1). Even in the best situations, however, the unique settings of urban areas (e.g., buildings, highways) cause wildlife distraction due to lights or noises at night. These animals are thus being referred to as 'urban adapters', 'urban exploiters' or 'urban avoiders' [17].

Figure 1. (Clockwise, from top left): Birds desperate for human leftovers on the street, Leeds—UK; nesting on a house's boiler pipes, Tehran—Iran; foraging in a small pond within a university campus, Nanjing—China (all photos are from the author's archives).

Such an intricate and usually one-way human–animal relationship/impact has become a hot topic in many biodiversity and urban studies (e.g., [16–18]). Among these urban biodiversity topics is the consideration of free-roaming (free-ranging, stray) dogs or cats as a public health concern in many countries (e.g., [11,19]). Even the definitions and functions of each group within the same species differ. For instance, one researcher identifies the following three types of cats: 'Confined, Free-Roaming and Feral' [11]. While changed urban lifestyles and management have been critical factors in this trend, personal interests and circumstances could also be highlighted. The private ownership of pets, human treatment of these animals and their (un)intentional release afterwards are still debatable and controversial in animal welfare studies. Unowned pet cases are rising in societies. One research study found that the total urban unowned cat population in the UK could be 247,429 versus more than 10 million owned cats [20].

In urban realms, dogs, which are the most prominent and ubiquitous urban pets, are loved and hated [10]. Guard dogs and companion dogs have a relatively privileged position; in contrast, street and stray dogs are indications of a lower-status urban neighbourhood and are regarded as problematic urban subjects and virus transmitters, and the presence of 'nuisance' dogs can cause the eviction of residents, etc. [10]. Global reports prove that there are high public risks as a result of close contact or unprovoked attacks by free-roaming dogs [21]. Free-roaming animals are especially prevalent in the Global South. Reports indicate the correlation between these animals and the rising number of rabies cases in urban and suburban areas (e.g., Afghanistan: [5]; India: [6]).

1.3. Research Logic: The Problem Statement

Most studies on free-roaming dogs in the cities only address demographic, physiological, epidemiological and behavioural aspects. Less attention has been paid to fundamental urban reasons for creating suitable environments and conditions for such dogs to adapt to the cities.

This research intends to fill this gap by focusing on Iran as a case study. There has been a surge in the number of free-roaming dogs and the related consequences (e.g., rabies) in urban and suburban areas in Iran (e.g., [22–25]). According to the Ministry of Health in Iran, animal bite cases increased from 35 per 100,000 people in 1987 to 177 per 100,000 people in 2016 [22]. As such, more general public and official concerns rose to inquire about the increasing dog attack incidents, processes to tackle these growing encounters and possible solutions to eradicate them.

The goals of this research are three-fold as follows: first, to explore the spatial distribution and frequency of free-roaming dogs in various parts of Iran; second, to evaluate typical narrations of news articles to inform the public about free-roaming dogs (views on dog impacts); and third, to assess post-capturing operations publicised in online news platforms.

As rare research for Iran, this paper reviews the less-told challenges regarding existing domestic dog-controlling facilities and activities based on relevant news articles published in an online Iranian news platform. For instance, this paper explains the relationship between extensive land use changes in Iran that have consequently led to the removal of natural wildlife corridors between urban and suburban areas. This paper bases its discussion and explanations on qualitative information from news articles published in Persian. So far, no particular study or investigation has covered the possible spatial reasons for this surge in the number of free-roaming dogs in Iran, though information on the free-roaming dog population in Iran is very poor [26]. Existing domestic studies have only evaluated these rising concerns from epidemiological and medical points of view (e.g., [24,25]).

2. Data and Methodology

'The popular press (defined as print or online news articles meant for a general audience, as opposed to technical or trade publications) is an important source of information' ([27], p.144). Media (print or online) may influence people's understanding and views or inform people about topics they do not personally experience [27].

News article text or content analysis could be a practical approach to gathering on-the-ground and reliable data and information for many research topics and subjects. The content analysis could be regarded as a randomised sampling as no predetermined sampling methods, geographical areas or guidelines are utilised, though the original content could be biased. Also, they could cover much larger geographical regions and include a more diverse group of beneficiaries or stakeholders who are involved. Data gathering and analysis techniques are cost-effective, more straightforward and fast. A similar content analysis approach was used to comprehend the messages and outlines of wildlife news published in Iranian newspapers over a 7-year period [15]. Also, we found a global content analysis of media regarding free-roaming cats [27].

A popular Iranian online news broadcasting platform (ISNA: Iranian Students' News Agency) was examined here. It is an online news platform that attracts a diverse range of audiences. News articles from this platform, published in Persian, on 'free-roaming dogs' covering three months (March–May 2023) were collated. Each news article was individually screened, and key messages and other features (e.g., stakeholders) were extracted.

The dataset constituted 27 news articles covering several cities across 12 provinces (out of 31 provinces) as follows: East Azarbaijan (Tabriz), Mazandaran (several cities), Tehran (Tehran and suburban areas), South Khorasan (Birjand), Isfahan (Ardestan), Alborz, Qom (Dastjerd), Fars, Kerman (Narmashir), Khorasan Razavi, Khoramabad (Boroujerd) and Zanjan.

Although the selected news articles may not constitute a large dataset, the dataset is deemed a suitable proxy for revealing rising public concerns regarding free-roaming dogs across many provinces in the country. In this research, the dog-related challenges of the remaining provinces were also explained using other Iranian papers published on this subject. Discussions were compared with international findings and reports.

3. Results

3.1. Spatial Distribution and Frequency

The spatial distribution of free-roaming dogs reveals that most large provinces are affected by this issue. In our research, the case studies covered cities located in humid, semi-humid and dryland areas. Regardless of their geographical and climatic differences, some cities in each province reported similar trends in rising concerns about the high number of dogs in the streets. For instance, one news article refers to the recent annual 600% increase in street dog numbers. Some news articles indicate that 'street/free-roaming dogs' were among the top five municipality complaints raised by local people. Nevertheless, some news articles refer to this rise being due to foodstuff availability in streets and deliberate dog feeding by some citizens.

3.2. Views on Dog Impacts

Figure 2 illustrates the principal issues reflected in news articles in an Iranian online news platform (ISNA) during a 3-month timeframe in 2023. The language and tones of the narrators or reporters of all news articles regarding free-roaming dogs are negative. One news article points out a municipality in a small city and its problems that needed particular attention due to the high maintenance costs associated with clinics for caring for dogs. Larger cities also raised concerns about such costs being unpredicted in their annual budget. One news article refers to one child's death and several hundred casualties resulting from dog biting incidents.

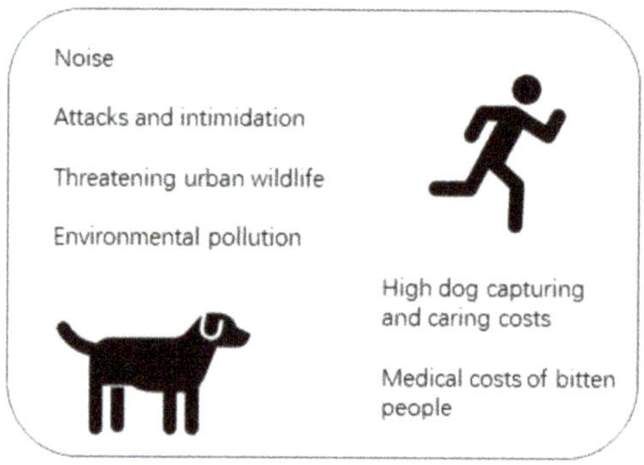

Figure 2. Infographic of causes of concerns reflected by news articles in an Iranian online news platform.

The impacts of free-roaming dogs were mentioned mainly by urban officials. A diverse range of health, financial and administrative concerns were reported. It was found that most of the news articles were reported by municipalities (41%), followed by veterinary institutes (23%) (Figure 3). In Iran, municipalities encompass specific operational 'Waste Management' departments that collect/recycle solid wastes on the street. Now, they have also been tasked to manage free-roaming animals.

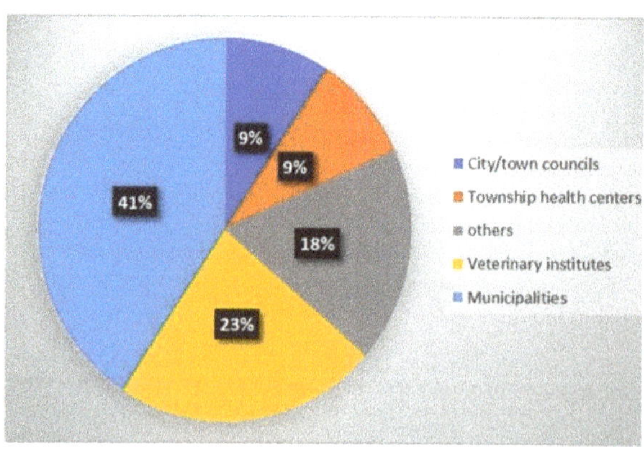

Figure 3. Share of organisations commenting on free-roaming dogs in online news articles.

The positive side is that these two entities (municipalities and veterinary institutes) are home to experts with relevant knowledge and qualifications regarding environmental public health and zoology. Nonetheless, no NGOs, policymakers or research organisations were involved in preparing, commenting or reporting free-roaming animals in online news.

3.3. Post-Capturing Operations/Clinics

The news articles did not mention how dogs are captured, though the procedure necessitates physical contact and trapping. Several news articles stated the presence of newly constructed caring houses (dog clinics) for such dogs in many cities, ranging from 0.3 to 8.5 ha in area, as a combination of buildings, open yards and veterinary clinics. These dog clinics have particular kitchens, separation spaces, treatment rooms, etc. The news articles also refer to stakeholders being involved, including those who directly capture dogs from the streets, veterinarians and clinics' cleaners. Several operations are implemented after capturing free-roaming dogs, outlined as follows:

- Segregation of male/female dogs.
- Early medical examination: Before being mixed with other group members, early medical interventions provide information about the body condition, weight, possible scars or other relevant health evidence about the new arrivals.
- Sterilisation: Predominant messages of news articles highlighted the need for current operations to sterilise dogs after capturing them. The process needs human expertise, medical facilities and other costly treatment procedures.
- Vaccination.
- Treating and caring for disabled/ill animals.

Given the limitation of space and finance for caring, dogs are usually released to nature after being vaccinated/sterilised, though many will return to cities for food and shelter afterwards (Figure 4).

Figure 4. Vaccinated free-roaming dogs spotted in a top Tehran park—Jamshidieh Park (photos taken by the author).

4. Discussion

4.1. Elaboration of Dog Clinics/Shelters in Iran

Our pool of news articles published on free-roaming dogs in Iranian cities reveals a growing concern among all urban and non-urban authorities. The issue has become spatially more diverse, from a limited number of provinces years ago to covering almost all provinces (Section 3.1).

The concept of publicly managed and funded dog clinics/shelters is new in Iran. Earlier attempts to capture and shelter free-roaming dogs in Iran were made based on people's interests and funds. People erected a handful of suburban shelters with no governmental support. Nevertheless, given the rising public health issues in many cities, mayors or other authorities started to accept that a large-scale movement is needed (see Section 3.2).

Based on reliable news articles, our research elaborates on some of the critical medical and health concerns (Section 3.3). For instance, a news article mentioned the segregation of male/female dogs. Although there is no national research on this issue, one study showed a significant male-to-female ratio (3.2/1) among the free-roaming dog population in Kerman City [26]. Therefore, such segregation assists in better controlling mating and keeping an average balance among dogs in the shelter.

Also, the news articles generally mentioned the sterilisation of dogs as a key activity in their centres. This procedure guarantees the control of the future propagation of these dogs. Elsewhere, research conducted in Kerman City suggests that more female neutering coverage can hugely decrease the population size of free-roaming dogs [26].

4.2. Rising Concerns in Iran

The news articles refer to major concerns described in Figure 2. Here, we elaborate on them in more detail as follows:

- Increasing incidences of dangerous and unprovoked dog attacks in the streets: Domestic research conducted in a city in central Iran revealed that over 92% of dog attacks were unprovoked [28]. Free-roaming dogs and their associated diseases (e.g., rabies) could be regarded as both rural and urban challenges. It was found that rabies has two epidemiological cycles, which are an urban cycle and a sylvatic cycle [29].
- Human health concerns of dog bites (e.g., rabies): It was shown that the prevalent rabies cases in Iran occur due to dog biting (over 95% of rabies cases) [30].
- Threatened outdoor activities of families: A recent 30-year global study shows that there is a correlation between the participation of people (especially men) in outdoor activities and the likelihood of biting incidences and infection with the rabies virus [31]. The physical exposures of humans to free-roaming dogs in outdoor environments increase the risks of dog attacks and biting even without any animal annoyance.
- Increasing municipalities' costs of controls: All news articles highlight the increasing costs of controlling, curing and keeping free-roaming dogs. The news articles refer to the costs of running dog clinics and their personnel, dog vaccines, part-time veterinarians' expenses, medicines, etc. These dog clinics are new entities, and many cities do not have the human, land and technical resources to establish them.
- Impacts on water and soil resources: The news articles did not explain the impacts, nor was any research found on this matter in Iran. Nevertheless, the urination and defecation of dogs could add hazardous materials to the environment, pollute soil and water resources and affect human health (e.g., [32,33]).
- Threats to wildlife health and survival: Although the news articles did not address this important point, there are numerous global reports on the impact of free-roaming animals on wildlife (e.g., [7]).
- Noise pollution: The excessive barking and soiling of community spaces are enumerated as usual free-roaming dog behaviours causing a public nuisance [34].

4.3. Influencing Factors

Urbanisation is regarded as one of the potential factors contributing to increasing rabies cases (e.g., [35]). The recent trends in biting incidences in Iran reveal that most animal bites occur in urban areas, contrasting with national reports [36]. The analysed news articles referred to some, but not *all*, urban features concerning rabies cases in Iran. In the following list, the author explains the most important urban characteristics causing rabies cases:

- Availability of foodstuff in streets and deliberate dog-feeding (mentioned by news articles): The rising inclination of people to feed animals has become a severe challenge for urban authorities in Iran. There is a sign stating 'no feeding to animals' in almost every corner of greenery spaces and parks. Nevertheless, some people still insist on following their passion for feeding animals. This behaviour could encourage free-roaming animals to stay, breed and expand their territories in the cities. The research conducted shows that the likelihood of sighting an ideal or overweight dog in the city was 14.9 times higher than in the suburbs in Kerman City, indicating a much higher food availability for dogs in the city [26].

- No facilities for surveillance and monitoring of free-roaming dogs (mentioned by news articles): This reason is pertinent to the lack of overall animal/pet keeping and management systems and guidelines, including in zoos in Iran [15]. Also, small cities nationwide encompass various types of local illegal wet/bird markets without the proper human health and animal hygiene systems.
- City topography (mentioned by news articles): Iranian research emphasises the regional and geographical heterogeneity of rabies cases in Iran [29]. The news articles mention this factor without any further description, though this is one of the critical parameters in urban areas favouring free-roaming animals, including dogs. For example, Tehran, which is known as a valley city, is extended on the hillslopes of adjacent mountainous regions. Such topographical features offer various options to animals for breeding, hiding, roaming and escaping. Also, many districts in the city possess polluted water channels, favouring free-roaming animals. This latter issue was found to be relevant in transmitting the rabies virus in Arequipa (Peru) [37].
- Lost transitional zones between the urbanised areas and the surrounding forested/mountainous areas (mentioned by news articles): These transitional zones that are adjacent to cities could provide open spaces (buffers) for animal/wildlife roaming and movement without being interfered by human activities. For instance, in one of the cities in Central Iran (Kerman), vacant lots located in the older parts of the city with ruins of abandoned old buildings had the largest number of free-roaming dogs [26]. Nevertheless, such a factor was not deemed relevant in a study conducted in Argentina [38].
- Urbanisation and unbalanced relationships between wildlife and the surrounding territories: The conditions changed after rapid urbanisation started in Iran in the 1930s when urban populations increased due to rural–urban migrations. The oil-funded urbanisation in Iran was initiated earlier than in many other neighbouring countries. A massive flux of rural–urban migration created a chaotic condition for the then-small city [39]. This unprecedented urban sprawl resulted in an unequal distribution of urban amenities, crimes, improper solid waste and sewage management, among others. These bottlenecks favour many free-roaming animals.
- Emerging new urban slums: This is particularly relevant to most Iranian cities, as new illegal settlements and urban slums were developed due to rising land and housing prices in cities. It takes time for these areas to become recognised by municipalities as a city (to receive urban facilities and services) or by other governmental organisations as legal entities for land registration and utilities. These areas lack proper hygienic arrangements, including bin collection. Research shows that the abundance of unowned cats is increased in more deprived urban areas with a higher human population density [20]. Therefore, such a chaotic situation makes these areas perfect places for encountering free-roaming dogs/cats. Global research establishes a link between certain infectious diseases (e.g., cholera) and slums due to poor infrastructure and a lack of access to safer water and better sanitation [14].
- Changed lifestyle and behaviours: Culturally and religiously, keeping pets, especially dogs, is not a common custom in Iran, though this case was also mentioned for Afghanistan [5]. Dogs were always key working partners for rural people, including farm owners, herders and farmers in Iran. Currently, more and more Iranians live alone, which may also be a factor in choosing to live with a pet. Also, as some unofficial reports suggest, younger Iranian generations prefer to keep pets rather than raise children at home. Nevertheless, this does not necessarily indicate that these people will keep their pets forever. Given the rising living expenses in cities, including the costs of pets' health care and food, there is a high chance that pet owners will leave dogs in surrounding areas.

Moreover, over half of the owned dogs in Iran have not been vaccinated annually [25], which can increase the health risks for animals and humans.

In addition to urban characteristics, the other factors affecting the spread and control of rabies are as follows:

- The COVID-19 pandemic: For most of the two years during the COVID-19 pandemic, like other countries, many activities of service-providing organisations, including municipalities, were halted or reduced in Iran. Such a gap in controlling urban pests and maintaining health orders were also affected, although there were rumours regarding the correlation between free-roaming animals and COVID-19 at that time. Nevertheless, a few cities in Iran recorded lower levels of rabies incidence during the COVID-19 period due to fewer outdoor activities or the avoidance of medical services for rabies treatment (e.g., [28]).
- Ecologically, free-roaming dogs exhibit broad temporal and spatial plasticity, broad distribution and population explosion and tolerance to environmentally different areas [40]. A report shows that some dogs in Kenya travel up to 24 km daily [34]. Such adaptabilities make controlling free-roaming dogs more difficult.
- The seasonality of rabies incidence must be highlighted. A meta-analysis of published research on rabies incidences in Iran shows that dog biting mainly occurred in the 'spring' seasons [29]. In China, however, a study revealed that 'August' was the peak month for rabies in 29 years [35]. Another Iranian research study also attributes some incidences to long-term drought and its associated lack of water and food in natural habitats in the Kerman province, located in the dryland areas of Iran [28]. High temperatures are positively correlated with the risk of rabies incidence, as dogs are more irritable and more likely to bite people in hot weather [41]. Moreover, proper national/provincial vaccination policies and educational programmes could effectively reduce rabies cases (e.g., China: [41]).

The above-mentioned dog-mediated issues raised by Iranian news articles could be observed in almost all cities in Iran, but Tehran, as a capital city, has drawn attention to public health. The city has the largest population in the country, which makes the city more vulnerable to public health challenges. The quality of life in Tehran is uneven and unjust, and each district suffers from urban inequality, inefficiency and/or deficiency [39]. In particular, Tehran is the most exposed location to free-roaming dogs for several reasons. Tehran is surrounded by expansive barren lands and non-built areas, which makes it the best living and hiding option for such dogs. The city has the densest entertaining and eating places that provide free food waste for free-roaming animals (e.g., rats and dogs). For decades, the rat plague has been common in Tehran. As such, the Tehran municipality has tried to identify the hotspots of groups (usually inside uncovered street water ditches or garbage sites) and eradicate them using poison baits. An Indian report shows the presence of free-roaming dogs close to garbage bins, predominantly within a 20 m radius in urban settings [42]. Finally, Tehran has one of the largest greenery spaces and parks in Iran, and these places provide food (by means of people feeding dogs) during the day and shelter at night for free-roaming dogs.

4.4. *Worrying Trends of Rabies Occurrences in Iran*

Worldwide, rabies still kills about 60,000 people a year, varying among countries and population age groups [43]. Dog vaccination is the most effective measure of rabies control [43].

In Iran, rabies is endemic, and old documents prepared by Iranian scholars such as Avicenna (Ibn Sina, 980–1037 AD) and others have described the transmission and treatment methods of rabies [44]. Modern scientific research on rabies and controlling measures have been followed in Iran since 1924, upon the establishment of the Pasteur Institute in Iran [25]. The rabies disease was fully controlled in Iran by 1977 [30].

Over the past decades, the frequency and distribution of rabies in Iran have changed, reflecting the social and economic conditions of people. The mortality rate due to rabies decreased from 0.9 per million people in the 1980s to 0.02–0.03 in recent years in the country [29]. Recent urban rabies cases are rising in Iran [28,36], and the new surge highlights the complexity of unpredictable risks in contemporary lifestyles and societies' desires.

4.5. Raising Awareness of Free-Roaming Dogs in Iran

As indicated in the previous sections, an online news platform regularly released news on this matter during our study period. Such dissemination of information is necessary for all people, especially vulnerable people, such as disabled or senior citizens, due to fewer mobility and self-defence opportunities. Equally important, children and pregnant women must be informed as they may show similar weaknesses when encountered by free-roaming animals. The problem is that only some of these affected groups can access such news platforms, and thus, urban authorities must convene workshops, TV programmes and school sessions for further effectiveness. Nevertheless, a previous study on publicising wildlife-related news in Iranian newspapers highlights the general low frequency and ineffective news dissemination [15].

5. Conclusions

Here, for the first time in Iran, we utilised a news articles analysis for covering free-roaming dogs in the streets. Our study encompasses broader geographical areas (12 out of 31 provinces) and a diverse pool of commenting/involved stakeholders. It reveals the growing concerns regarding the frequency and impacts of increasing free-roaming dogs and subsequent rabies in certain cities in Iran. The importance of the information used in this research is that these news articles cover the most recent reliable concerns of public and governmental organisations regarding the rising number of dog attacks, health-related challenges and rising costs of protective and treatment operations in urban/suburban areas.

It was beyond the scope of this paper to assess the communicative language or tone of Iranian news articles. Nevertheless, the Iranian news articles intend to present these animals' negative image, perhaps for public health.

We also highlighted the urban features that influence the presence and behaviours of free-roaming animals in the cities (e.g., topography, slums, etc.). These features have great potential to be revisited in other urban contexts, as these studies offer solutions to resolve diseases linked to free-roaming animals. The subject could also be revisited by others to understand the whole cycle of this challenge in urban and associated social concerns.

Current scant datasets regarding the free-roaming animals in Iran downgrade many conclusions and findings to speculations. Data gathered at finer spatial resolutions (e.g., citizen science techniques) and/or targeted ground-based data can improve our understanding of the dynamics and behaviours of free-roaming animals in our cities.

Finally, despite the negative image of dogs (free-roaming types) in this research and other similar papers, we must emphasise the positive roles of tame dogs who are real companions for supporting human mental health, guiding blind people or assisting disabled/deaf people or people who suffer from dementia. They remain our friends during long-term hardships or illnesses when many people leave or cannot support us.

Funding: This research received no external funding.

Institutional Review Board Statement: Not applicable.

Informed Consent Statement: Not applicable.

Data Availability Statement: No data is available.

Conflicts of Interest: The author declares no conflict of interest.

References

1. Foley, R.; Gamble, C. The ecology of social transitions in human evolution. *Philos. Trans. R. Soc. B Biol. Sci.* **2009**, *364*, 3267–3279. [CrossRef] [PubMed]
2. Zeder, M.A.; Hesse, B. The Initial Domestication of Goats (*Capra hircus*) in the Zagros Mountains 10,000 Years Ago. *Science* **2000**, *287*, 2254–2257. [CrossRef] [PubMed]
3. Frantz, L.A.F.; Mullin, V.E.; Pionnier-Capitan, M.; Lebrasseur, O.; Ollivier, M.; Perri, A.; Linderholm, A.; Mattiangeli, V.; Teasdale, M.D.; Dimopoulos, E.A.; et al. Genomic and archaeological evidence suggest a dual origin of domestic dogs. *Science* **2016**, *352*, 1228–1231. [CrossRef] [PubMed]
4. Haruda, A.F.; Miller, A.R.V.; Paijmans, J.L.A.; Barlow, A.; Tazhekeyev, A.; Bilalov, S.; Hesse, Y.; Preick, M.; King, T.; Thomas, R.; et al. The earliest domestic cat on the Silk Road. *Sci. Rep.* **2020**, *10*, 11241. [CrossRef] [PubMed]
5. Nasiry, Z.; Mazlan, M.; Noordin, M.M.; Lila, M.A.M. Evaluation of Dynamics, Demography and Estimation of Free-Roaming Dog Population in Herat City, Afghanistan. *Animals* **2023**, *13*, 1126. [CrossRef]
6. Punjabi, G.A.; Athreya, V.; Linnell, J.D.C. Using Natural Marks to Estimate Free-Ranging Dog *Canis Familiaris* Abundance in a Mark-Resight Framework in Suburban Mumbai, India. *Trop. Conserv. Sci.* **2012**, *5*, 510–520. [CrossRef]
7. Young, J.K.; Olson, K.A.; Reading, R.P.; Amgalanbaatar, S.; Berger, J. Is Wildlife Going to the Dogs? Impacts of Feral and Free-roaming Dogs on Wildlife Populations. *Bioscience* **2011**, *61*, 125–132. [CrossRef]
8. Bhattacharjee, D.; Bhadra, A. Humans Dominate the Social Interaction Networks of Urban Free-Ranging Dogs in India. *Front. Psychol.* **2020**, *11*, 2153. [CrossRef]
9. Lyons, M.A.; Malhotra, R.; Thompson, C.W. Investigating the free-roaming dog population and gastrointestinal parasite diversity in Tulúm, México. *PLoS ONE* **2022**, *17*, e0276880. [CrossRef]
10. Hubbard, P.; Brooks, A. Animals and urban gentrification: Displacement and injustice in the trans-species city. *Prog. Hum. Geogr.* **2021**, *45*, 1490–1511. [CrossRef]
11. Jaroš, F. The Cohabitation of Humans and Urban Cats in the Anthropocene: The Clash of Welfare Concepts. *Animals* **2021**, *11*, 705. [CrossRef] [PubMed]
12. Cox, D.T.; Shanahan, D.F.; Hudson, H.L.; Fuller, R.A.; Gaston, K.J. The impact of urbanisation on nature dose and the implications for human health. *Landsc. Urban Plan.* **2018**, *179*, 72–80. [CrossRef]
13. Hassell, J.M.; Begon, M.; Ward, M.J.; Fèvre, E.M. Urbanization and Disease Emergence: Dynamics at the Wildlife–Livestock–Human Interface. *Trends Ecol. Evol.* **2016**, *32*, 55–67. [CrossRef] [PubMed]
14. Neiderud, C.-J. How urbanization affects the epidemiology of emerging infectious diseases. *Infect. Ecol. Epidemiol.* **2015**, *5*, 27060. [CrossRef]
15. Amiraslani, F.; Dragovich, D. Wildlife and Newspaper Reporting in Iran: A Data Analysis Approach. *Animals* **2021**, *11*, 1487. [CrossRef]
16. Sepp, T.; Ujvari, B.; Ewald, P.W.; Thomas, F.; Giraudeau, M.; Tuul, S.; Beata, U.; Ewald, P.W.; Frédéric, T.; Mathieu, G.; et al. Urban environment and cancer in wildlife: Available evidence and future research avenues. *Proc. R. Soc. B Boil. Sci.* **2019**, *286*, 20182434. [CrossRef]
17. Bateman, P.W.; Fleming, P.A. Big city life: Carnivores in urban environments. *J. Zoo* **2012**, *287*, 1–23. [CrossRef]
18. Serenari, C. Reconsidering the role of the built environment in human–wildlife interactions. *People Nat.* **2020**, *3*, 104–114. [CrossRef]
19. Jackman, J.; Rowan, A. Free-roaming dogs in developing countries: The benefits of capture, neuter, and return programs. In *The State of the Animals*; Salem, D.J., Rowan, A.N., Eds.; Humane Society Press: Washington, DC, USA, 2007; pp. 55–78.
20. McDonald, J.L.; Skillings, E. Human influences shape the first spatially explicit national estimate of urban unowned cat abundance. *Sci. Rep.* **2021**, *11*, 20216. [CrossRef]
21. Hudson, E.G.; Brookes, V.J.; Ward, M.P.; Dürr, S. Using roaming behaviours of dogs to estimate contact rates: The predicted effect on rabies spread. *Epidemiol. Infect.* **2019**, *147*, e135. [CrossRef]
22. Bay, V.; Jafari, M.; Shirzadi, M.R.; Bagheri, A.; Asl, I.M. Trend and epidemiological patterns of animal bites in Golestan province (Northern Iran) between 2017 and 2020. *PLoS ONE* **2021**, *16*, e0252058. [CrossRef] [PubMed]
23. Leylabadlo, H.E.; Baghi, H.B. Rabies Elimination by 2030: What Challenges Does Iran Face? *Iran. J. Public Health* **2020**, *49*, 1397–1398. [CrossRef]
24. Nadalian, M.G.; Tadjbakhsh, H.; Mokhber Dezfuli, M.R.; Rezakhani, A.; Simani, S.B.M. Rabies and Its Present Situation in Iran. *Iran. J. Rumin. Health Res.* **2016**, *1*, 32–34. [CrossRef]
25. Gholami, A.; Fayaz, A.; Farahtaj, F. Rabies in Iran: Past, Present and Future. *J. Med. Microbiol. Infect. Dis.* **2014**, *2*, 1–10. Available online: http://jommid.pasteur.ac.ir/article-1-49-en.html (accessed on 20 May 2023).
26. Shamsaddini, S.; Gohari, M.A.; Kamyabi, H.; Nasibi, S.; Derakhshani, A.; Mohammadi, M.A.; Mousavi, S.M.; Baneshi, M.R.; Hiby, E.; Harandi, M.F. Dynamic modeling of female neutering interventions for free-roaming dog population management in an urban setting of southeastern Iran. *Sci. Rep.* **2022**, *12*, 4781. [CrossRef]
27. Gow, E.A.; Burant, J.B.; Sutton, A.O.; Freeman, N.E.; Grahame, E.R.M.; Fuirst, M.; Sorensen, M.C.; Knight, S.M.; Clyde, H.E.; Quarrell, N.J.; et al. Popular press portrayal of issues surrounding free-roaming domestic cats *Felis catus*. *People Nat.* **2021**, *4*, 143–154. [CrossRef]

28. Davarani, E.R.; Amiri Domari, A.; Haji Mahani, A.; Alian Samakkhah, S.; Raesi, R.; Daneshi, S. Epidemiological charac-teristics, injuries, and rabies post-exposure prophylaxis among children in Kerman county, Iran during 2019–2021. *Open Public Health J.* **2023**, *16*, e187494452303272. [CrossRef]
29. Abedi, M.; Doosti-Irani, A.; Jahanbakhsh, F.; Sahebkar, A. Epidemiology of animal bite in Iran during a 20-year period (1993–2013): A meta-analysis. *Trop. Med. Health* **2019**, *47*, 55. [CrossRef]
30. Fayaz, A.; Farahtaj, F.; Eslami, N.; Howaizi, N.; Biglari, P.; Janani, A.; Simani, S. A Survey of Human Rabies in Iran; 2000–2006. *Iran. J. Virol.* **2008**, *2*, 29–33. [CrossRef]
31. Gan, H.; Hou, X.; Wang, Y.; Xu, G.; Huang, Z.; Zhang, T.; Lin, R.; Xue, M.; Hu, H.; Liu, M.; et al. Global burden of rabies in 204 countries and territories, from 1990 to 2019: Results from the Global Burden of Disease Study 2019. *Int. J. Infect. Dis.* **2022**, *126*, 136–144. [CrossRef]
32. Allen, J.A.; Setälä, H.; Kotze, D.J. Dog Urine Has Acute Impacts on Soil Chemistry in Urban Greenspaces. *Front. Ecol. Evol.* **2020**, *8*, 615979. [CrossRef]
33. Penakalapati, G.; Swarthout, J.; Delahoy, M.J.; McAliley, L.; Wodnik, B.; Levy, K.; Freeman, M.C. Exposure to Animal Feces and Human Health: A Systematic Review and Proposed Research Priorities. *Environ. Sci. Technol.* **2017**, *51*, 11537–11552. [CrossRef] [PubMed]
34. Muinde, P.; Bettridge, J.M.; Sousa, F.M.; Dürr, S.; Dohoo, I.R.; Berezowski, J.; Mutwiri, T.; Odinga, C.O.; Fèvre, E.M.; Falzon, L.C. Who let the dogs out? Exploring the spatial ecology of free-roaming domestic dogs in western Kenya. *Ecol. Evol.* **2021**, *11*, 4218–4231. [CrossRef] [PubMed]
35. Zhou, H.; Vong, S.; Liu, K.; Li, Y.; Mu, D.; Wang, L.; Yin, W.; Yu, H. Human Rabies in China, 1960-2014: A Descriptive Epidemiological Study. *PLoS Neglected Trop. Dis.* **2016**, *10*, e0004874. [CrossRef]
36. Nejati, J.; Baygi, M.Z.; Salemroudi, S. Rabies: Repetition of an Old Tragic Story in Southeastern Iran. *Health Scope* **2022**, *11*, e121232. [CrossRef]
37. Raynor, B.; De Puente-león, M.; Johnson, A.; Díaz, E.W.; Michael, Z.; Recuenco, S.E.; Castillo-neyra, R. Movement patterns of free-roaming dogs on heterogeneous urban landscapes: Implications for rabies control. *Prev. Vet. Med.* **2020**, *178*, 104978. [CrossRef]
38. Arona, E.; Schiavini, A. Free-roaming dogs in Ushuaia City, Tierra del Fuego, Argentina. How many and why. *Urban Ecosyst.* **2023**, *26*, 559–574. [CrossRef]
39. Amiraslani, F. Analysis of quality of life across Tehran districts based on designated indicators and relational database management system. *Urban Gov.* **2021**, *1*, 107–114. [CrossRef]
40. Zanin, M.; Bergamaschi, C.L.; Ferreira, J.R.; Mendes, S.L.; Moreira, D.O. Dog days are just starting: The ecology invasion of free-ranging dogs (*Canis familiaris*) in a protected area of the Atlantic Forest. *Eur. J. Wildl. Res.* **2019**, *65*, 65. [CrossRef]
41. Li, H.; Li, Y.; Chen, Y.; Chen, B.; Su, Q.; Hu, Y.; Xiong, C. Mapping rabies distribution in China: A geospatial analysis of national surveillance data. *Int. J. Infect. Dis.* **2023**, *131*, 140–146. [CrossRef]
42. Tiwari, H.K.; Robertson, I.D.; O'Dea, M.; Vanak, A.T. Demographic characteristics of free-roaming dogs (FRD) in rural and urban India following a photographic sight-resight survey. *Sci. Rep.* **2019**, *9*, 16562, Correction in *Sci. Rep.* **2020**, *10*, 3757. [CrossRef] [PubMed]
43. World Health Organization; Food and Agriculture Organization of the United Nations; World Organisation for Animal Health. *Zero by 30: The Global Strategic Plan to End Human Deaths from Dog-Mediated Rabies by 2030: United against Rabies Collaboration: First Annual Progress Report: Global Strategic Plan to End Human Deaths from Dog-Mediated Rabies by 2030*; World Health Organization: Geneva, Switzerland, 2019; Available online: https://apps.who.int/iris/handle/10665/328053 (accessed on 20 May 2023).
44. Shabansalmani, N.; Pourhossein, B.; Khosravy, M.S.; Farahtaj, F.; Alavian, S.E. A Review of Rabies Treatment Methods in Ancient Iran. *J. Virol. Viral Dis.* **2022**, *2*, 1–7. [CrossRef]

animals

MDPI

Article

"*Do Your Homework as Your Heart Takes over When You Go Looking*": Factors Associated with Pre-Acquisition Information-Seeking among Prospective UK Dog Owners

Rebecca Mead *, Katrina E. Holland, Rachel A. Casey, Melissa M. Upjohn and Robert M. Christley

Dogs Trust, 17 Wakley Street, London EC1V 7RQ, UK
* Correspondence: rebecca.mead@dogstrust.org.uk

Simple Summary: Dogs are the most common pet animal species in the UK, with many households acquiring dogs every year. However, little is known about whether prospective owners undertake research prior to acquiring a dog and what factors may affect the likelihood of doing so. This paper reports findings from a mixed methods study into dog acquisition in the UK. We found that almost half of existing owners did not look for information or advice before getting their most recently acquired dog, and that those with previous dog ownership experience were less likely to do so than first-time owners. Younger prospective owners were more likely to have undertaken pre-acquisition research, as were those with formal education qualifications. Findings may be of interest to those who provide advice related to dog acquisition and dog ownership, in order to encourage more prospective owners towards effective pre-acquisition research.

Abstract: The factors influencing why and how people decide to acquire a dog are not well understood and little is known about the extent to which prospective owners undertake preparatory research. This study aimed to better understand what factors influence whether prospective dog owners in the UK conduct preparatory research. A 2019 online survey of current (n = 8050) and potential (n = 2884) dog owners collected quantitative and qualitative data. Additional qualitative data were collected through semi-structured interviews with current (n = 166) and potential (n = 10) dog owners. Of the current owners surveyed, 54% stated that they had looked for advice or information prior to acquiring their dog. Of potential owners, 68% reported already having looked for information, while a further 14% were planning to undertake research prior to acquiring a dog. Those with previous dog ownership experience were less likely to undertake pre-acquisition research, as were those who had worked with dogs. Demographic factors were also associated with the likelihood of conducting pre-acquisition research, with younger prospective owners being more likely to have undertaken research, as well as those with formal education qualifications. Among current owners, pre-acquisition research was more likely among those who acquired their dog through a breeder; a specific breed or a mix of two breeds; or as a puppy. Qualitative data were consistent with and added additional understanding and context to these findings. Almost half of current owners did not conduct pre-acquisition research, highlighting the need for increased awareness of its importance and the development of targeted interventions to encourage this activity. Understanding the different factors that influence whether dog owners undertake research may be of interest to animal welfare and veterinary organisations, in order to inform interventions to better prepare people for dog acquisition.

Keywords: dogs; dog acquisition; pre-acquisition research; pre-acquisition behaviours; preparatory research

Citation: Mead, R.; Holland, K.E.; Casey, R.A.; Upjohn, M.M.; Christley, R.M. "*Do Your Homework as Your Heart Takes over When You Go Looking*": Factors Associated with Pre-Acquisition Information-Seeking among Prospective UK Dog Owners. *Animals* **2023**, *13*, 1015. https://doi.org/10.3390/ani13061015

Academic Editor: Betty McGuire

Received: 2 February 2023
Revised: 1 March 2023
Accepted: 9 March 2023
Published: 10 March 2023

1. Introduction

Dogs (*Canis familiaris*) are the most popular companion animal species in the UK today, with an estimated 13 million dogs and 34% of households comprising one or more dogs [1].

There are therefore thousands of prospective owners looking to acquire a new dog each year, and the decisions they make during the acquisition process can have widespread implications for dog welfare. Puppies have been in high demand in the UK, a situation that was particularly notable during the COVID-19 pandemic [2]. Demand has grown for dogs of particularly "fashionable" breeds, including brachycephalic breeds such as French Bulldogs and Pugs [3], despite the health problems faced by these breeds [4–8]. The demand for dogs was such that it was not satisfied by legitimate and reputable UK breeders alone, leading to an increase of unscrupulous commercial breeders (often called "puppy farms") in the UK, and a rise in the importation of dogs, including illegal "puppy smuggling" from overseas [9,10]. The practices associated with the breeding and supply of dogs are closely tied to the health and welfare of these animals. Intensive breeding through puppy farming, including those illegally imported into the UK, has negative impacts on dog health and behaviours [10,11]. Increased prospective owner awareness of the implications of sourcing on dog welfare may improve the acquisition decisions made and benefit canine welfare. It has also been the case that while people continue to acquire dogs, many dogs have been relinquished every year [12,13]. There are numerous reasons for this, including owners not having access to information about dogs' needs or understanding the time, effort, and costs involved in dog ownership [14–17]. Access to this information, through pre-acquisition research, may improve owner expectations and reduce the risk of relinquishment [17]. In the UK, various online resources have been developed by charities and professional organisations that are designed to provide information for prospective owners about responsibly acquiring dogs—particularly puppies—and dog ownership, e.g., [18–21]. However, the extent to which these resources are accessed or influence subsequent buying behaviour is not known [22]. There is an urgent need to better understand acquisition decisions so that appropriate strategies can be developed to enable prospective owners to make informed decisions and help safeguard dog welfare.

There is currently limited knowledge about how prospective dog owners decide how and where to acquire their dogs. In particular, we know little about whether prospective owners undertake preparatory research. Previous research suggests that many prospective dog owners undertake some sort of research prior to acquiring a dog. The majority (84.3%) of dog owners who responded to a 2015 survey conducted by Packer et al. [23] stated that they had undertaken pre-acquisition research. Similarly, Kuhl et al.'s 2013 study [24] found that over three-quarters (78.9%) of owners reported looking for information prior to acquiring their dogs. Kuhl (2021) suggested that some owners may consider themselves to have adequate experience so as not to require further research [25]. A study into so-called "Pandemic Puppies" also investigated pre-purchase behaviours of UK owners who acquired their dog as a puppy (aged under 16 weeks) during 2019 or the 2020 COVID-19 pandemic [2]. In this population, almost half (46.7%) of owners who acquired their puppy in 2019 completed pre-purchase research but half (50.3%) did not because they considered themselves to be experienced dog owners. More owners who purchased a puppy in 2020 undertook research (58.1%); however, the difference between years was explained by ownership experience, as for the latter group, only 38.9% considered themselves experienced enough to have not needed to undertake pre-purchase research. Just 3% of owners across both year groups had both not completed any pre-purchase research and considered themselves to be inexperienced owners [2]. Kuhl et al. [24] found that owners of pedigree dogs were more likely than owners of non-pedigree dogs or a mix of pedigree and non-pedigree dogs to have sourced information prior to acquiring their dog (85.6% versus 71.9%). Burnett et al. (2022) found that pre-purchase research was more common amongst owners of designer crossbreeds (73.0%) compared to purebred puppies (48.6%); however, a higher proportion of purebred owners considered themselves to already be experienced dog owners [26].

This study aimed to better understand the preparatory research undertaken by current and potential dog owners in the UK. Specifically, we were interested in understanding, prior to acquiring a dog: (1) do prospective owners look for information or for advice; and

(2) what factors influence whether prospective owners undertake research? This study is the first that we are aware of to investigate a number of factors such as owner and dog demographics in relation to pre-acquisition information gathering. As such, this adds novel insights to the extant research into this area.

2. Materials and Methods

This study used a convergent mixed methods design, with data collected through a survey and interviews, as shown in Figure 1 after Ref. [27]. Data were collected largely in parallel, analysed independently, and interpreted together in a comparative and contrasting way [28]. Data were collected as part of a wider study investigating various aspects of dog acquisition [27,29].

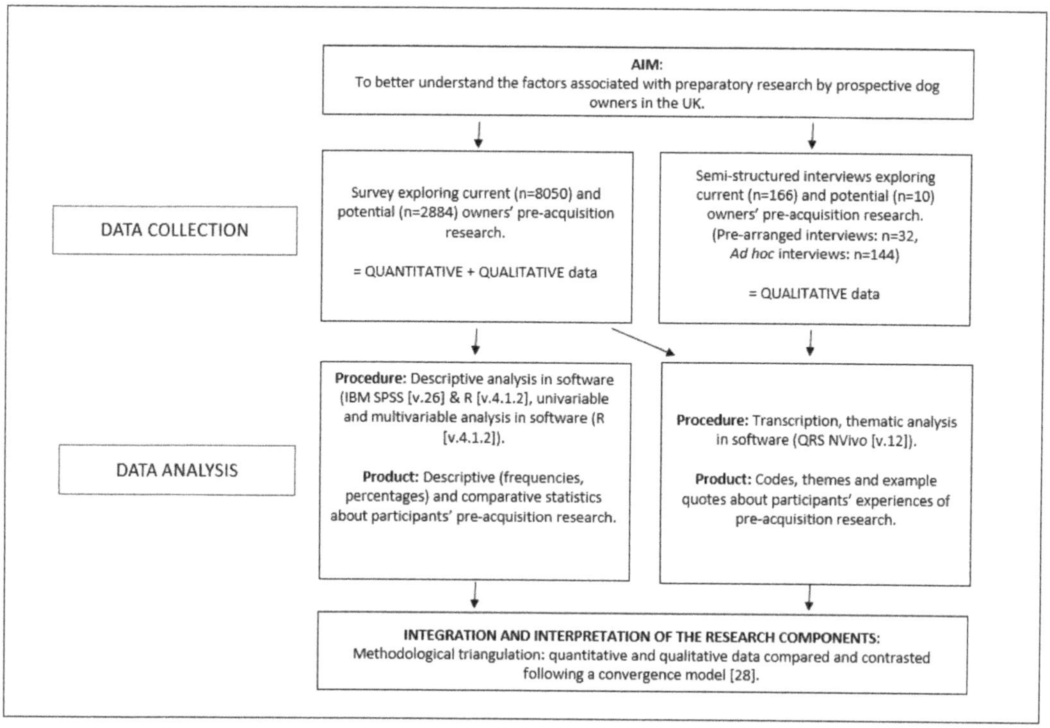

Figure 1. Procedural diagram of the mixed methods study design after Ref. [27].

2.1. Ethics Statement

Ethical approval for this study was granted by the Dogs Trust Ethical Review Board (reference numbers: ERB018 and ERB019). All participants were provided with an informed consent statement prior to participation in the study. This outlined the purpose of the study, described how data—especially any personal data—would be stored and used, explained that participation was voluntary, and provided instructions on how to withdraw from the study. Informed consent was obtained on the first page of the survey from survey participants, by signature for participants who were interviewed face-to-face, or verbally (audio recorded) for participants who were interviewed remotely. Participants were required to be aged 18 years or over and living in the UK. No payment or incentives were offered to participants.

2.2. *Data Collection*

2.2.1. Survey Design and Content

A self-completion online survey was designed to collect predominantly quantitative data about the experiences of current and potential dog owners. Qualitative data were also collected through free text responses. Questions were developed by the authors and were informed by a review of the current literature [30]. The survey was hosted on the online survey platform SmartSurvey^{TM} (https://www.smartsurvey.co.uk/ (accessed on 4 October 2021)). Prior to the launch, the survey was piloted twice: firstly, with 12 members of Dogs Trust staff who were not involved in developing the survey and, secondly, with 110 current or potential dog owners, who were recruited through two promotional posts on the Facebook page of "Generation Pup" (Generation Pup (https://generationpup.ac.uk/ (accessed on 2 May 2019)) is an ongoing longitudinal cohort study of dogs. Generation Pup has ethical approval from the University of Bristol Animal Welfare Ethical Research Board (UIN/18/052), the Social Science Ethical Review Board at the Royal Veterinary College (URN SR2017-1116), and the Dogs Trust Ethical Review Board (ERB009). Survey links were posted on the Generation Pup Facebook page (https://www.facebook.com/generationpup/) (accessed on 29 April 2019))). Following these pilots, minor changes were made to the logic of the survey to make respondents' progression easier.

The survey took approximately 20 min to complete and asked a variety of questions related to pre-acquisition research as part of a wider study on dog acquisition. It also included questions about the demographics of owners and (where applicable) their dogs and asked whether participants would be willing to be contacted about further research opportunities. Participants were asked if they currently owned at least one dog (thus were a "current" owner) and if they were considering acquiring a/another dog(s) in the future ("potential" owners). Owners who owned more than one dog were asked to provide responses for the dog they had acquired most recently. If more than one dog was acquired at the same time, respondents were asked to answer for the dog whose name was first alphabetically. For the purposes of this study, current owners who were considering getting another dog were only asked questions retrospectively about any research they undertook prior to acquiring their current dog: they were not asked questions about whether they were currently looking (or planning to look) for information or advice with regards to their future dog(s). All relevant survey questions and response options can be found in Supplementary Materials.

Current owners (n = 8050) were asked the question "Did you look for any information or ask anyone for advice before getting your dog?". Those respondents who selected "No" were not asked any further questions about pre-acquisition research. Those respondents who selected "Yes" were asked further about their pre-acquisition research, including "What information or advice did you look for before getting your dog?" (open-ended question). Although details of the nature of this information or advice are not reported in this article (but are the focus of a forthcoming paper), some respondents included information within this survey item about why they undertook research; thus, relevant responses have been included in the analysis for this study. All current owners (regardless of whether they had undertaken any research) were also asked the optional open-ended question "What advice would you give to potential dog owners about buying or rehoming a dog?".

Potential owners (n = 2884) were asked similar questions to those outlined above for current owners. The question "Have you looked for any information or asked anyone for advice about getting a dog?" could be answered in four ways: "Yes", "No but I plan to", "No and I don't plan to", or "I haven't thought about this yet".

2.2.2. Survey: Participant Recruitment

The survey was live for three months at the end of 2019 (25th September 2019–31st December 2019). It was promoted predominantly through Dogs Trust via social media posts, correspondence with supporters (e.g., e-newsletters and magazine), the Dogs Trust Contact Centre and rehoming centres. Thus, the sample was a convenience sample; however, some promotions were targeted towards males and those who were not supporters of Dogs Trust, through a paid Facebook advertisement. This was to improve reach and to increase representation of these groups, given that male participants are underrepresented in studies of human–animal relationships [31]. Further information on how participants were recruited to the survey can be found in Supplementary Materials, Table S1.

2.2.3. Interviews

Interviews were conducted with current and potential dog owners to gain a deeper understanding of aspects of the dog acquisition process. We conducted two types of interviews as part of the study: pre-arranged and ad hoc. Each interview was conducted by one of three authors (R.M., K.E.H., or R.M.C). All interviews explored owners' experiences of dog acquisition, including whether they had conducted any research prior to acquiring their current or prospective dog. Both types of interviews followed a semi-structured guide. This was piloted in a similar manner to the survey: firstly with 12 members of Dogs Trust staff; secondly with 5 respondents to the pilot survey. The interview guide was not changed following pilot interviews, thus data from pilot interviews were included in the analysis for the overall study. Interview guides can be found in Supplementary Materials.

Pre-arranged interviews were used to stimulate in-depth discussion about the dog acquisition process and were conducted between April 2019 and March 2020. They were conducted with current (n = 24) and potential (n = 8) dog owners. Of the 24 current owners, 3 were considering acquiring another dog in the near future. Interviewees were recruited through the survey (n = 15), pilot survey (n = 5), or were members of Dogs Trust staff (n = 12). Interviews were conducted remotely via telephone (n = 22) or face-to-face (n = 10). The majority of interviews were with individuals (n = 25), but 7 individuals were involved in group interviews with 2 or 3 participants in each (these were all with Dogs Trust staff and were trialled as part of the pilot: it was decided to focus on individual interviews following this, mainly for logistical purposes). All interviewees participated once, with the exception of one participant who also was involved in two additional follow-up interviews following acquisition of their dog. Interviews lasted between 17 and 60 min in length (mean = 33 min, median = 29 min). With participants' consent, interviews were audio recorded and later transcribed *intelligent verbatim*, (i.e., false starts, pauses, laughter, and filler words such as "um" and "err" were omitted).

Ad hoc interviews were undertaken to collect data from a broader range of dog owners than were likely to be reached through the pre-arranged interviews. These were conducted at 23 "Responsible Dog Ownership" events across the UK, organised by Dogs Trust Regional Campaigns staff, between May and December 2019. At these events, dog owners could gain advice on diet, exercise, and enrichment, and (excluding Northern Ireland events) veterinary nurses provided free basic health checks and microchipping. The locations for these events were determined using findings from Dogs Trust Stray Dog Survey data, e.g., [32], and discussions with community partners (e.g., dog wardens and housing association staff) about local hotspots for dog-related issues and areas of deprivation. Thus, these events were sometimes held in association with local authorities or housing associations and often took place in parks or community centres.

For ad hoc interviews, event attendees were asked whether they would be happy to be interviewed, either while they were waiting to be seen by a member of staff, or after they (and their dog(s)) had been attended to. If they were willing to participate, they were interviewed on-the-spot. In total, 142 current dog owners or carers (or sets of owners, where a dog was accompanied by more than one person) and 2 potential owners were interviewed. With consent, 44.4% of these interviews were audio recorded. Recorded interviews lasted between 2 and 26 min in length (mean = 11 min, median = 11 min) and were transcribed as per the pre-arranged interviews. Where interviews were not recorded (either due to participants not giving consent or where events were thought to be too noisy to enable clear audio recordings), the researcher made handwritten notes during and immediately after interviews, which were subsequently digitised by the researcher who conducted the interview.

2.3. Data Analysis

2.3.1. Quantitative Data Analysis

Initial data cleaning was completed in Microsoft Excel and IBM SPSS (v. 26). Responses to relevant closed-ended survey questions were summarised with descriptive statistics using IBM SPSS (v. 26) and R v. 4.1.2; [33]. Univariable and multivariable analyses were also completed using R (v. 4.1.2). These were used to compare responses given by different groups of respondents and to determine the relative contributions and relationships between variables.

2.3.2. Qualitative Data Analysis

The aims of the qualitative analysis were to identify further factors that may influence dog owners' decisions to seek information or advice before acquiring their dog and explore the range of dog owners' experiences of conducting such research. Interview transcripts and relevant free-text survey responses were imported into NVivo (v. 12). These data were then analysed using a thematic analysis approach [34] by two authors (R.M. and K.E.H.). The process of thematic analysis began with familiarisation of the data by reading through a selection of free-text survey responses and interview transcripts. Initial coding of data was inductive, meaning that the coding was driven by the content of the responses, rather than using pre-determined codes. Codes were then grouped into categories and then into potential themes that linked the concepts within categories and represented overall patterns in the data.

All interview data were initially coded and collated into potential themes by one author (K.E.H) as part of the wider acquisition study. Subsequently, interview data extracts relevant to the current study (i.e., data related to pre-acquisition information seeking) were independently coded and grouped into potential themes by R.M., prior to collaborative review of coding and theme development (by R.M. and K.E.H.).

Free-text survey data relevant to this study were collected across four open-ended questions. During initial coding, the researchers performed an ongoing appraisal of the "information power" [35] of the responses to each question. Information power refers to the amount of information contained within a sample, relevant to the study, and suggests that fewer participants are required in samples with more information power [35]. For two of the questions, the researchers determined that a subsample of responses had sufficient information relevant to answer the question and met the aim of this study's qualitative analysis, as outlined above. Once it was recognised that a subsample likely offered ample information power, and that incorporating additional responses was unlikely to elicit new codes or generate new understanding for codes [36], a quasi-random sampling approach was applied to the remaining data. This involved coding every 25th additional response. Further detail on how qualitative survey data were coded is presented in Supplementary Materials, Table S2.

Where direct quotes are included within this study, names have been omitted to protect participant confidentiality. Instead, a unique identifier and brief description of the participant (dog ownership status and mode of data collection) are given.

3. Results

3.1. Survey Results

The survey was started 15,350 times. Following data cleaning and deduplication, 11,265 of these responses were suitable for analysis. These comprised 8381 current owners and 2884 potential owners; however, of these current owners, 115 had bred their own dogs and 216 were not involved in the decision to acquire a dog. Therefore, the initial sample sizes reported here are for 8050 current and 2884 potential owners.

3.1.1. Participant Demographics

The majority of survey participants were female (88.3% of current and 79.9% of potential owners). Respondents represented age groupings from 18 to 85 years and above, with 45–54 years being the most common age category for current owners (23.8%) and 55–64 years being the most common for potential owners (20.8%). Respondents resided in all four UK nations, but the majority were based in England (84.2% of current and 82.4% of potential owners). Additional information on participant demographics can be found in Supplementary Materials, Table S3.

3.1.2. Dog Demographics

The majority of current owners (62.8%) had acquired their dog within 5 years prior to the completion of the survey. Over half of the dogs (54.4%) were acquired as puppies of 6 months or younger. Just over half (54.9%) were a specific breed (e.g., Labrador Retriever) and 22.9% were a mix of two specific breeds (e.g., Cockerpoo/Cocker Spaniel × Poodle cross). Most dogs were acquired from a charity or rehoming centre (43.6%) or from a breeder (39.6%). Further information on dog demographics can be found in Supplementary Materials, Table S4.

3.2. Do Prospective Owners Look for Information or Advice before Acquiring a Dog?

Of the current owners surveyed, just over half stated that they had looked for advice or information prior to acquiring their dog (54.4%, 95% CI [53.3%, 55.5]). Two-thirds of potential owners reported already having looked for information (67.8%, 95% CI [66.1%, 69.5%]) and a further 13.7% (95% CI [12.5%, 15.0%]) were planning to undertake research. Potential owners were significantly more likely to report having undertaken research than current owners, X^2 $(1, N = 10{,}934) = 155.1, p < 0.001$.

3.3. What Factors Influence Whether Prospective Owners Undertake Research Prior to Acquiring a Dog?

Multivariable analysis using survey responses suggested numerous factors that contribute to the likelihood of research being undertaken, for current (Table 1) and potential (Table 2) owners.

Previous dog ownership appeared to be an important factor: those who had lived with a dog or dogs as a child and an adult, and thus may be considered to have the most previous experience of ownership, were the least likely to undertake research (44.8% of current owners and 77.4% of potential owners). Compared to those who had owned dogs as an adult and a child, those who had previously lived with a dog or dogs as an adult (only) were 1.2 times more likely for current owners, and 1.4 times more likely for potential owners, to conduct research. Those who had (only) lived with a dog or dogs as a child were 2.5 times (current) and 3.6 times (prospective) more likely to conduct research. Those who had never lived with a dog previously were the most likely to have undertaken research (4.6 times and 11.1 times, respectively).

Among current owners, this pattern appeared to be mirrored to some extent based on ownership of the same breed or type of dog, with those current owners who had not previously owned the same breed or type of dog as their current one being significantly more likely to have undertaken research. However, this variable was not included simultaneously in the multivariable analysis with the variable "previous dog ownership" due to issues of collinearity. Further information can be found in Supplementary Materials, Table S5.

Table 1. Multivariable analysis of factors that affect whether research was undertaken prior to acquiring a dog for current owners (multivariable analysis n = 7279; univariable analyses n are given for each variable).

	Undertook Research				Univariable Analysis					Multivariable Analysis (n = 7279)				
Variable	Yes	Total	%	95% CI	Odds Ratio [1]	2.50%	97.50%	z Val.	p Val.	Odds Ratio	2.50%	97.50%	z Val.	p Val. [2]
Previous ownership (n = 8050)														
Previously lived with a dog/dogs as an adult and as a child	1588	3546	44.78%	43.15%, 46.42%	Ref				<0.0001	Ref				<0.0001
Previously lived with a dog/dogs as an adult	1236	2458	50.28%	48.31%, 52.26%	1.25	1.13	1.38	4.20	<0.0001	1.22	1.09	1.36	3.39	0.0007
Previously lived with a dog/dogs as a child	850	1188	71.55%	68.92%, 74.04%	3.10	2.69	3.58	15.58	<0.0001	2.53	2.16	2.96	11.60	<0.0001
First time lived with a dog	707	858	82.40%	79.71%, 84.81%	5.77	4.79	6.97	18.30	<0.0001	4.61	3.75	5.66	14.56	<0.0001
Age of owner (n = 7987)														
75 years or older	65	196	33.16%	26.94%, 40.03%	Ref				<0.0001	Ref				<0.0001
65–74 years	463	1126	41.12%	38.28%, 44.02%	1.41	1.02	1.94	2.09	0.0364	1.31	0.92	1.87	1.48	0.1388
55–64 years	904	1821	49.64%	47.35%, 51.94%	1.99	1.46	2.71	4.32	<0.0001	1.55	1.10	2.20	2.49	0.0128
45–54 years	1036	1916	54.07%	51.83%, 56.29%	2.37	1.74	3.24	5.45	<0.0001	1.59	1.12	2.25	2.60	0.0092
35–44 years	731	1206	60.61%	57.83%, 63.33%	3.10	2.25	4.27	6.95	<0.0001	1.76	1.23	2.52	3.08	0.0021
18–24 years	311	474	65.61%	61.22%, 69.75%	3.85	2.70	5.47	7.49	<0.0001	2.02	1.35	3.03	3.43	0.0006
25–34 years	839	1248	67.23%	64.57%, 69.78%	4.13	3.00	5.69	8.69	<0.0001	2.28	1.58	3.28	4.43	<0.0001
Work with dogs (n = 7850)														
Currently work with dogs	386	776	49.74%	46.23%, 53.25%	Ref				<0.0001	Ref				<0.0001
Previously worked with dogs	380	786	48.35%	44.87%, 51.84%	0.95	0.78	1.15	-0.55	0.5810	1.23	0.99	1.53	1.83	0.0667
Never worked with dogs	3409	6116	55.74%	54.49%, 56.98%	1.27	1.10	1.48	3.18	0.0015	1.48	1.25	1.75	4.52	<0.0001
Highest level of education (n = 7373)														
No formal qualifications	107	321	33.33%	28.40%, 38.66%	Ref				<0.0001	Ref				<0.0001
GCSE/National 5 or equivalent	656	1384	47.40%	44.78%, 50.03%	1.80	1.40	2.33	4.53	<0.0001	1.42	1.08	1.86	2.50	0.0123
A level/Scottish Higher or equivalent	525	965	54.40%	51.25%, 57.52%	2.39	1.83	3.11	6.45	<0.0001	1.51	1.13	2.01	2.81	0.0049
Foundation degree/Higher National Diploma (HND) or equivalent	570	1085	52.53%	49.56%, 55.49%	2.21	1.71	2.87	5.97	<0.0001	1.56	1.18	2.06	3.10	0.0019
University degree (e.g., BA, BSc) or equivalent	1362	2305	59.09%	57.07%, 61.08%	2.89	2.26	3.70	8.44	<0.0001	1.83	1.40	2.38	4.44	<0.0001
Postgraduate degree (e.g., MA, MBA, MSc, PhD) or equivalent	831	1313	63.29%	60.65%, 65.86%	3.45	2.66	4.46	9.41	<0.0001	2.27	1.72	3.00	5.76	<0.0001

Table 1. *Cont.*

Variable	Undertook Research				Univariable Analysis					Multivariable Analysis (n = 7279)				
	Yes	Total	%	95% CI	Odds Ratio [1]	2.50%	97.50%	z Val.	*p* Val.	Odds Ratio	2.50%	97.50%	z Val.	*p* Val. [2]
Source of dog (n = 8050)														
Friends or family/community	393	979	40.14%	37.12%, 43.25%	Ref				<0.0001	Ref				<0.0001
Private/third party seller	208	454	45.81%	41.29%, 50.41%	1.26	1.01	1.58	2.02	0.0431	1.17	0.91	1.50	1.20	0.2299
Charity/rehoming centre	1655	3427	48.29%	46.62%, 49.97%	1.39	1.21	1.61	4.50	<0.0001	1.52	1.28	1.80	4.74	<0.0001
A dog breeder	2125	3190	66.61%	64.96%, 68.23%	2.98	2.57	3.45	14.49	<0.0001	2.38	1.99	2.83	9.61	<0.0001
Breed or type of dog (n = 7596)														
Mix of breeds or types	844	979	45.82%	43.56%, 48.10%	Ref				<0.0001	Ref				<0.0001
Mix of two specific breeds	1012	3190	56.60%	54.29%, 58.88%	1.54	1.35	1.76	6.48	<0.0001	1.18	1.01	1.38	2.06	0.0395
Specific breed	2525	3427	57.13%	55.66%, 58.58%	1.58	1.41	1.76	8.15	<0.0001	1.37	1.19	1.57	4.43	<0.0001
Year acquired					1.05	1.04	1.07	8.53	<0.0001	1.06	1.04	1.07	7.83	<0.0001
Age of dog when acquired [3] (n = 8050)														
Senior adult (7 to <12 years)	195	465	41.94%	37.53%, 46.47%	Ref				<0.0001	Ref				0.0004
Juvenile (>6 months to <1 year)	30	65	46.15%	34.59%, 58.15%	1.13	0.88	1.46	0.98	0.3295	1.11	0.84	1.47	0.72	0.4713
Geriatric (12+ years)	334	755	44.24%	40.73%, 47.80%	1.19	0.70	2.00	0.64	0.5196	1.21	0.67	2.18	0.64	0.5224
Young adult (1 to <2 years)	394	888	44.37%	41.13%, 47.65%	1.07	0.85	1.34	0.55	0.5818	1.23	0.95	1.58	1.58	0.1137
Mature adult (2 to <7 years	755	1496	50.47%	47.94%, 53.00%	1.38	1.12	1.70	3.04	0.0024	1.41	1.12	1.77	2.94	0.0033
Puppy (0–6 months)	2673	4381	61.01%	59.56%, 62.45%	2.17	1.78	2.63	7.81	<0.0001	1.59	1.25	2.03	3.80	0.0001

[1] Ordering is based on multivariable model, hence some Odds Ratios < 1; [2] The *p* values given for each reference row are likelihood ratio test *p* values; [3] Age categorisations based on [37].

Table 2. Multivariable analysis of factors that affect whether research was undertaken prior to acquiring a dog for potential owners (multivariable analysis n = 2272; univariable analyses n are given for each variable).

Variable	Undertook Research				Univariable Analysis					Multivariable Analysis (n = 2272)				
	Yes	Total	%	95% CI	Odds Ratio [1]	2.50%	97.50%	z Val.	*p* Val.	Odds Ratio	2.50%	97.50%	z Val.	*p* Val. [2]
Previous ownership (n = 2861)														
Previously lived with a dog/dogs as an adult and as a child	1086	1403	77.41%	75.14%, 79.52%	Ref				<0.0001	Ref				<0.0001
Previously lived with a dog/dogs as an adult	689	880	78.30%	75.45%, 80.90%	0.72	0.50	1.02	−1.83	0.0679	1.36	1.06	1.74	2.43	0.0151
Previously lived with a dog/dogs as a child	284	303	93.73%	90.86%, 96.00%	3.15	1.26	7.87	2.45	0.0141	3.56	2.02	6.27	4.39	<0.0001
Never lived with a dog	269	275	97.82%	95.21%, 99.11%	5.09	1.59	16.28	2.74	0.0061	11.10	4.07	30.28	4.70	<0.0001
Age of owner (n = 2865)														
65–74 years	274	374	73.26%	68.55%, 77.50%	Ref				<0.0001	Ref				<0.0001
55–64 years	449	599	74.96%	71.33%, 78.27%	1.30	0.78	2.18	1.01	0.3120	1.05	0.74	1.49	0.26	0.7932
75 years or older	60	88	68.18%	57.84%, 77.01%	0.71	0.32	1.56	−0.85	0.3960	1.09	0.59	1.99	0.27	0.7870
45–54 years	457	573	79.76%	76.27%, 82.85%	1.17	0.70	1.95	0.61	0.5406	1.45	1.00	2.10	1.95	0.0515
35–44 years	388	453	85.65%	82.11%, 88.59%	3.58	1.72	7.48	3.40	0.0007	1.60	1.07	2.40	2.29	0.0218
25–34 years	507	562	90.21%	87.46%, 92.42%	2.76	1.47	5.18	3.17	0.0015	2.95	1.90	4.57	4.83	<0.0001
18–24 years	198	216	91.67%	87.14%, 94.73%	5.75	1.73	19.12	2.85	0.0044	3.34	1.76	6.36	3.68	0.0002
Work with dogs (n = 2716)														
Currently work with dogs	275	361	76.20%	71.5%, 80.3%	Ref				0.0113	Ref				0.0124
Previously worked with dogs	133	155	85.80%	79.4%, 90.5%	0.12	0.02	0.93	−2.03	0.0428	1.04	0.59	1.83	0.13	0.8948
Never worked with dogs	1813	2200	82.40%	80.8%, 83.9%	0.13	0.02	0.92	−2.04	0.0413	1.58	0.94	2.64	1.74	0.0817

[1] Ordering is based on multivariable model, hence some Odds Ratios < 1; [2] The *p* values given for each reference row are likelihood ratio test *p* values.

Whether someone currently or had previously worked with dogs was significantly associated with whether research was undertaken. Current and potential owners who had experience of working with dogs (at the time of, or previous to, survey completion) were less likely to have undertaken research than those who had never worked with dogs. Current owners who had never worked with dogs were 1.5 times more likely to have undertaken research than those who worked with dogs at the time of survey completion.

The age of owner was a significant factor, with younger prospective owners being more likely to undertake research prior to acquiring a dog: just a third of current owners aged 75 years or older had undertaken any research prior to acquiring their most recent dog compared to approximately two-thirds of 25–34 (67.2%) and 18–24 year olds (65.6%). Current owners aged 25–34 years old were 2.3 times more likely to have undertaken research prior to acquiring their most recent dog, compared to those aged 75 years or older, and similar patterns were seen among potential owners.

Among current owners, those with formal education qualifications were also significantly more likely to have undertaken research prior to acquiring a dog. A third of those who had no formal qualifications had undertaken research. The odds of undertaking research increased with increasing levels of formal education, with those with a postgraduate qualification having 2.3 times greater odds of having undertaken research (63.3%). These differences were not significant among potential owners when included in the multivariable analysis. However, the pattern of effect was similar when education was considered in isolation, with research more likely among people with formal education (Supplementary Materials, Table S6).

A number of factors related to the dog acquired also appeared important in terms of whether any pre-acquisition research was undertaken by current owners. Specifically, the odds of undertaking research were higher when prospective owners went on to acquire their dog from a breeder, when a dog was a specific breed or a mix of two specific breeds, when a dog was acquired as a puppy, and when a dog was acquired more recently. If a causal relationship exists with these variables, the direction of the effect is unclear, nor whether these associations could be due to the confounding effect of other, unmeasured, variables. Further information, including univariable results relevant to gender, having children under 18 living at home, employment status, and whether research was undertaken, can be found in Supplementary Materials, Tables S7–S10.

Qualitative data offer additional suggestions as to why prospective owners choose to undertake research prior to acquiring a dog. Some owners commented on how important it was for them to have looked for information prior to acquiring a dog, and many recommended that all potential owners should *"do your research"*. For some prospective owners, searching for information was an important part of deciding whether to acquire a dog:

> *"Whether I would have the time and resources needed to give a dog a good home."* (Current owner, survey ID 3136)

Others discussed how they were motivated to undertake research as they had their potential dogs' interests in mind. Several commented on how they sought information to ensure that they could offer a good home and appropriate lifestyle that would ensure a dog's wellbeing. Thus, undertaking research appeared important in preparing for the arrival of a dog:

> *"Everything I may need to know to look after him/her to the best of my ability. And make sure she has everything to fit her needs."* (Potential owner, survey ID P3287)

Furthermore, some prospective owners appeared to recognise the value of completing research in preparation for managing the emotional drivers associated with acquisition:

> *"Do your homework as your heart takes over when you go looking."* (Current owner, survey ID 1274)

As identified with quantitative analysis, previous ownership and experiences with dogs appeared to be an important factor when undertaking research. Those with little previous experience, including of a breed or life stage, sometimes commented on this as being a motivator when choosing to undertake research:

> *"As she's our first dog, we did a lot of research into different breeds and their personalities."* (Current owner, survey ID 1549)

> *"We haven't had a puppy between us before [...] so we had a lot of studying to do about puppies."* (Current owner, survey ID 1828)

A number of prospective owners who had previously owned dogs described how they conducted research to update their knowledge prior to looking for their next dog:

> *"I have previously owned 3 dogs and have researched each one. Over the last month, I've been giving myself a refresher course."* (Potential owner, survey ID 2841)

In contrast, some who had previously owned dogs noted that they only needed to find information related to specific areas that they were less familiar with, often highlighting that they thought they knew about other aspects of dog ownership and thus did not need any more information on these:

> *"Quite knowledgeable on dogs in general, so was more specific to specific dog and situation and breed that I've never had before."* (Current owner, survey ID 1900)

Sometimes the need to undertake research appeared to be linked to the amount of time since a prospective owner had last had a dog, with those without recent experience suggesting that they needed to update their knowledge or find current advice:

> *"As I had not owned a dog for 20 years."* (Current owner, survey ID 8340)

Some prospective owners noted that, regardless of previous experience of similar breeds or circumstances, they were always keen to learn more:

> *"We did have experience of this [rehoming a Greyhound], having rehomed two retired racers, but the more advice the better!"* (Current owner, survey ID 8292)

Despite the value many prospective owners placed on seeking information or advice, a considerable proportion of prospective owners did not conduct any research prior to acquiring a dog. In interviews, a number of barriers became apparent. Some suggested that due to their previous experiences as a dog owner they felt that they did not need any information or advice. For example, one interviewee described how they did not conduct any research before acquiring their current dog as they had owned dogs before and felt they knew everything they needed to:

> *"I didn't [do any research] because I'm one of these people that thinks they know everything; do you know what I mean? Because I had had animals and had dogs and done dog training and dog trials, I had a very high opinion of myself. What, you know dog temperaments and how to train them and all that kind of thing, so no I didn't get any advice whatsoever."* (Current owner, interview ID B1RM1201)

Previous experiences of a chosen breed seemed a particularly important barrier against undertaking any research:

> *"We didn't [do any research] as we'd had that breed before."* (Current owner, survey ID 2279)

4. Discussion

Undertaking research prior to acquiring a dog is thought to be important for a successful dog–owner bond and the dog's future wellbeing [14–17]. Despite this, few studies have investigated the links between pre-acquisition research, acquisition behaviours, and ownership. This study used mixed methods to understand factors influencing pre-acquisition research.

This study is in keeping with the findings of previous studies, e.g., [23,24,38] in that the majority of prospective dog owners undertook some research prior to acquiring a dog, although we found that this was less common than in these previous studies. Our study found that those aspiring to acquire a dog at the time of survey completion were more likely to have undertaken pre-acquisition research (or planned to complete research) than those who already owned a dog did prior to acquiring their current dog. These differences might be due to multiple factors, including recall bias among current owners, recent shifts in behaviour towards undertaking pre-purchase research, or the source that potential owners planned to use (a far higher proportion of potential than current owners found out about our survey through Dogs Trust when enquiring about rehoming a dog—see Supplementary Materials, Table 1). It is also possible that potential owners who had not yet completed any research, but planned to, were over-optimistic in their research aspirations.

The factors that influence pre-acquisition research are varied. Among those who did not undertake research, previous experience with dogs was an important factor: those with more dog ownership experience were less likely to undertake research, as in previous studies [2,25]. Those working with dogs at the time of the survey completion were also less likely to have undertaken any research. Although self-described "experience" appears important, we do not know what this experience entailed for our respondents. For example, the number of dogs, length of dog ownership, or extent of dog caring responsibility are unknown. The relative success of previous dog ownership experiences in terms of dog wellbeing and strength of human–animal bonds are also unknown [39]. Equally, we do not know any details about the nature or period of working with dogs, or—for current owners—whether they had worked with dogs at the time they acquired their most recent dog. Regardless, qualitative data confirm that perceived views of their own experience with dogs was an important factor for prospective owners and is worthy of consideration by those involved in designing resources and interventions to influence decisions related to dog acquisition.

The age of the owner was an important factor, with younger prospective owners being more likely to have sought information or advice. This might suggest a greater importance placed by younger people on research, or that younger people are more easily able to access resources due to greater internet use and digital literacy [40–42]. Younger age groups will also likely have less dog ownership experience. Education was also important, with those current owners who attained the highest level of formal study being the most likely to have undertaken research. This may be indicative of an awareness of the importance of research or the ability to access resources. Regardless, these findings highlight that interventions to increase research and preparation prior to dog acquisition should reach across age and demographic groups.

Among current owners, the likelihood of conducting research was associated with the source of their current dog. Those who acquired their dog from breeders were most likely to have undertaken research, followed by those who acquired from rehoming centres. It is not clear whether research leads prospective owners to a particular source or whether those who intend to use a particular source (i.e., a breeder) are more motivated to undertake research. This may be influenced by the degree of financial investment with different sources. Those who had acquired their dog from friends, family, or the community may be more likely to be unplanned or less planned acquisitions, potentially driven by emotions more than intention [29], and may not allow time for pre-acquisition research. The type of dog acquired was also a motivator to pre-acquisition research: similar to previous research by Kuhl et al. [24], those who acquired specific breeds or mixes of two specific breeds appeared more likely to undertake research than those who acquired a dog of mixed breeds or unknown type. This may be confounded with source and possibly represents similar motivations for undertaking research. People tended to be more likely to undertake research if they went on to acquire a puppy. Although the causality of this relationship is unknown, qualitative research suggested that life stage was a driver for research amongst prospective owners. Those who acquired their dog more recently were more likely to have undertaken research. This might be indicative of increased awareness of the value of conducting research and greater visibility of, or easier access to, resources. Alternatively, recall bias may have affected responses from those who acquired their dogs longer ago.

There are likely other motivators and barriers to pre-acquisition research, and it should be noted that no questions were asked about why participants did *not* search for information or advice. It is possible that those who chose not to undertake research were unaware of the benefits of research (as suggested by [24]). Equally, prospective owners may have been restricted by practicalities: they may have acquired a dog within a time frame that did not allow for preparatory research, for reasons including unplanned acquisitions [29].

This study provides insights into the motivators and barriers to undertaking pre-acquisition research that may be useful for developing advice related to dog acquisition. Some experienced owners may be less receptive to messaging as they are more likely to rely on their existing knowledge and believe further research to be unnecessary. Successful messaging directed towards those who have previously owned dogs will need to overcome this barrier, perhaps by focusing on the importance of updating knowledge or targeting particular gaps in knowledge. Less experienced owners are more likely to undertake research, so reaching them may involve providing readily available resources in a format that are accessible to them. Given that almost a fifth of first-time dog owners did not recall undertaking any research before acquiring their current dog, there is clearly an opportunity to better understand how to reach more prospective first-time owners. Our research also demonstrates how demographics may affect the likelihood of undertaking pre-acquisition research. Providing resources which are relevant to different demographic groups, through different media channels and styles, may increase accessibility.

4.1. Strengths and Limitations

This is the UK's largest study of pre-acquisition research among prospective dog owners that we are aware of and adds to the existing evidence base. The mixed methods approach allowed the integration of quantitative and qualitative data, enabling a comparison of findings and greater credibility. The collection of free text survey items allowed for a wide breadth of responses to be documented, whereas semi-structured interviews yielded deeper insights into pre-acquisition behaviours. Collecting data from both current and potential owners enabled understandings from retrospective and current perspectives to be observed. Follow-up surveys will add further insight into this area.

This study has several limitations. The methods involved using a convenience sample, with a bias towards Dogs Trust supporters. This was particularly prominent among potential dog owners, of whom a large proportion were invited to participate in the survey on enquiring to Dogs Trust about adopting a dog. Respondents were self-selecting and there was a bias towards female respondents. Although this is common in dog-related surveys [37], the underrepresentation of males and those who prefer to self-identify means that further research is needed. Caution is needed if attempting to generalise findings to other populations. All survey responses were self-reported; thus, it is not possible to validate findings. Current owners completed the survey retrospectively. Although the majority (62.8%) of current owners had acquired the dog that they completed the survey about within the 5 years prior to survey completion, there is the potential for recall bias. Indeed, a small number of owners had acquired their dogs as long as 19 years prior to survey completion. Certain demographic information was reported at the time of data collection and may not have been the same as at the point of dog acquisition (e.g., owner age category or highest level of education). A cautionary note should be applied to the analysis of data related to potential owners: while these data can be used for hypothesis generating, further research is warranted.

4.2. Future Work

This study reports primarily on prospective owners who undertook research. Although our analyses reveal interesting insights into the factors which may act as motivators or barriers to research, we did not ask questions about why people did not undertake research, either as specific survey items, or in interviews. Inclusion of this insight is worthy of consideration in future studies. This study has not considered how ownership of another dog at the time of acquiring a new dog influences whether research is undertaken. Although we found that ownership experience is an important barrier to undertaking research, we do not know if or how ownership of multiple dogs may influence this. We also have limited understanding of unplanned acquisition and how this relates to pre-acquisition research. Although we know that many prospective owners undertake research, we cannot comment on the quality of research nor whether the information or advice received was correct. Nor do we know how different facets of pre-acquisition research may impact dog acquisition, future ownership, relinquishment, development of the human–animal bond, and dog welfare. These are complex areas to consider but future research that attempts to account for these could be of considerable value. Additional data collection with a wider demographic reach may allow generalisations across different populations and be of use to those interested in targeted interventions related to pre-acquisition behaviours.

Data reported here are from a wider study into dog acquisition, which includes more detailed investigation of pre-acquisition research. A forthcoming publication will focus on where prospective owners look for information, what information prospective owners search for, how long prospective owners spend on research, and whether prospective owners find all the information they want. We have collected additional data from prospective owners through two follow-up surveys, subsequent to the survey reported here, and hope that these may offer additional insights as to the nature of the research which those potential owners planning to undertake research actually did.

5. Conclusions

In conclusion, undertaking research before dog acquisition is important for, and valued by, many prospective owners. However, we found that almost half of (current) owners reported undertaking no research prior to acquiring their most recent dog. Those with previous dog ownership experience are less likely to undertake research than first-time owners and those who are younger or who have achieved a higher level of education are more likely to undertake research. Prospective owners who go on to acquire puppies, dogs of a specific breed, and who source their dog from a breeder are more likely to complete pre-acquisition research. Our findings may be of interest to organisations involved in improving pet welfare, especially those who provide advice related to dog acquisition, and provide insights which could support improved messaging and interventions.

Supplementary Materials: The following supporting information can be downloaded at https://www.mdpi.com/article/10.3390/ani13061015/s1: File S1: survey questions relevant to pre-acquisition research; pre-arranged interview schedule; ad hoc interview schedule; participant recruitment; coding of survey free text responses; participant demographics; dog demographics; factors that influence whether people undertake research prior to acquiring a dog.

Author Contributions: Conceptualisation: R.A.C. and M.M.U.; methodology: R.M., K.E.H. and R.M.C.; survey development: R.M., K.E.H. and R.M.C.; formal analysis: R.M., K.E.H. and R.M.C.; data curation: R.M.; writing—original draft preparation, R.M.; writing—review and editing, R.M., K.E.H., R.A.C., M.M.U. and R.M.C. All authors have read and agreed to the published version of the manuscript.

Funding: This work was supported by Dogs Trust, the UK's largest canine welfare charity. All authors are salaried employees of Dogs Trust. The authors would like to thank Dogs Trust for providing funding for open access publication. Generation Pup is funded by Dogs Trust and the Dogs Trust Canine Welfare Grants Committee, grant number CWG012.

Institutional Review Board Statement: This study was conducted in accordance with the Declaration of Helsinki and approved by the Dogs Trust Ethical Review Board (reference numbers: ERB018 and ERB019, 8 July 2019). Generation Pup has ethical approval from the University of Bristol Animal Welfare Ethical Research Board (UIN/18/052, 20 June 2018), the Social Science Ethical Review Board at the Royal Veterinary College (URN SR2017-1116, 22 August 2017), and the Dogs Trust Ethical Review Board (ERB009, 14 October 2018).

Informed Consent Statement: Informed consent was obtained from all subjects involved in this study.

Data Availability Statement: The data presented in this study are available on request from the corresponding author. The data are not publicly available due to ethical approval of participant informed consent that included survey respondents being informed that we will remove all personally identifiable information before sharing data with universities and/or research institutions.

Acknowledgments: The authors would like to thank all current and potential dog owners and carers involved in this study, Dogs Trust Campaigns staff for supporting our research at their events, and Dogs Trust staff across rehoming centres, retail shops, Dog School, and the Contact Centre, who helped to promote the survey.

Conflicts of Interest: The authors declare no conflict of interest.

References

1. PFMA. UK Pet Population. Available online: https://www.ukpetfood.org/information-centre/statistics/uk-pet-population.html (accessed on 31 May 2022).
2. Packer, R.M.A.; Brand, C.L.; Belshaw, Z.; Pegram, C.L.; Stevens, K.B.; O'neill, D.G. Pandemic puppies: Characterising motivations and behaviours of UK owners who purchased puppies during the 2020 COVID-19 pandemic. *Animals* **2021**, *11*, 2500. [CrossRef] [PubMed]
3. Kennel Club. Breed Registration Statistics. 2020. Available online: https://www.thekennelclub.org.uk/media-centre/breed-registration-statistics/ (accessed on 13 May 2022).
4. Packer, R.M.A.; Hendricks, A.; Tivers, M.S.; Burn, C.C. Impact of facial conformation on canine health: Brachycephalic obstructive airway syndrome. *PLoS ONE* **2015**, *10*, e0137496. [CrossRef] [PubMed]
5. Honey, L. Future health and welfare crises predicted for the brachycephalic dog population. *Vet. Rec.* **2017**, *181*, 550. [CrossRef]
6. O'Neill, D.G.; Jackson, C.; Guy, J.H.; Church, D.B.; McGreevy, P.D.; Thomson, P.C.; Brodbelt, D.C. Epidemiological associations between brachycephaly and upper respiratory tract disorders in dogs attending veterinary practices in England. *Canine Genet. Epidemiol.* **2015**, *2*, 10. [CrossRef]
7. O'Neill, D.G.; Baral, L.; Church, D.B.; Brodbelt, D.C.; Packer, R.M.A. Demography and disorders of the French Bulldog population under primary veterinary care in the UK in 2013. *Canine Genet. Epidemiol.* **2018**, *5*, 3. [CrossRef]
8. Asher, L.; Diesel, G.; Summers, J.F.; McGreevy, P.D.; Collins, L.M. Inherited defects in pedigree dogs. Part 1: Disorders related to breed standards. *Vet. J.* **2009**, *182*, 402–411. [CrossRef]
9. Dogs Trust. Puppy Smuggling: Puppies Still Paying as Government Delays. 2020. Available online: https://www.dogstrust.org.uk/downloads/2020%20Puppy%20smuggling%20report.pdf (accessed on 13 May 2022).
10. Wyatt, T.; Maher, J.; Biddle, P. Scoping Research on the Sourcing of Pet Dogs From Illegal Importation and Puppy Farms 2016–2017, Scottish Government. 2017. Available online: https://www.gov.scot/publications/scoping-research-sourcing-pet-dogs-illegal-importation-puppy-farms-2016/documents/ (accessed on 13 May 2022).
11. Wauthier, L.M.; Williams, J.M. Using the Mini C-BARQ to Investigate the Effects of Puppy Farming on Dog Behaviour. *Appl. Anim. Behav. Sci.* **2018**, *206*, 75–86. [CrossRef]
12. Stavisky, J.; Brennan, M.L.; Downes, M.; Dean, R. Demographics and economic burden of un-owned cats and dogs in the UK: Results of a 2010 census. *BMC Vet. Res.* **2012**, *8*, 163. [CrossRef]
13. Clark, C.C.A.; Gruffydd-Jones, T.; Murray, J.K. Number of cats and dogs in UK welfare organisations. *Vet. Rec.* **2012**, *170*, 493. [CrossRef]
14. Jensen, J.B.H.; Sandøe, P.; Nielsen, S.S. Owner-related reasons matter more than behavioural problems—A study of why owners relinquished dogs and cats to a Danish animal shelter from 1996 to 2017. *Animals* **2020**, *10*, 1064. [CrossRef]
15. Lambert, K.; Coe, J.; Niel, L.; Dewey, C.; Sargeant, J.M. A systematic review and meta-analysis of the proportion of dogs surrendered for dog-related and owner-related reasons. *Prev. Vet. Med.* **2015**, *118*, 148–160. [CrossRef]
16. Eagan, B.H.; Gordon, E.; Protopopova, A. Reasons for Guardian-Relinquishment of Dogs to Shelters: Animal and Regional Predictors in British Columbia, Canada. *Front. Vet. Sci.* **2022**, *9*, 448. [CrossRef]
17. Diesel, G.; Brodbelt, D.; Pfeiffer, D.U. Characteristics of relinquished dogs and their owners at 14 rehoming centers in the united kingdom. *J. Appl. Anim. Welf. Sci.* **2010**, *13*, 15–30. [CrossRef]
18. Dogs Trust. Getting or Buying a Dog. Available online: https://www.dogstrust.org.uk/help-advice/getting-or-buying-a-dog/ (accessed on 13 May 2022).
19. Kennel Club. When Buying a Dog. Available online: https://www.thekennelclub.org.uk/getting-a-dog/buying-a-dog/ (accessed on 13 May 2022).
20. Royal Society for the Prevention of Cruelty to Animals. Buying a Puppy. Available online: https://www.rspca.org.uk/adviceandwelfare/pets/dogs/puppy (accessed on 13 May 2022).
21. People's Dispensary for Sick Animals (PDSA). Getting a Dog. Available online: https://www.pdsa.org.uk/pet-help-and-advice/looking-after-your-pet/puppies-dogs/getting-a-dog (accessed on 13 May 2022).
22. Upjohn, M. Going beyond "educating": Using behaviour change frameworks to help pet owners make evidence-based welfare decisions. *Vet Rec.* **2022**, *190*, 412–414. [CrossRef]

23. Packer, R.M.A.; Murphy, D.; Farnworth, M.J. Purchasing popular purebreds: Investigating the influence of breed-Type on the pre-purchase motivations and behaviour of dog owners. *Anim. Welf.* **2017**, *26*, 191–201. [CrossRef]
24. Kuhl, C.A.; Dean, R.; Quarmby, C.; Lea, R.G. Information sourcing by dog owners in the UK: Resource selection and perceptions of knowledge. *Vet. Rec.* **2021**, *190*, e1081. [CrossRef]
25. Kuhl, C. An Investigation of Pedigree Dog Breeding and Ownership in the UK: Experiences and Opinions of Veterinary Surgeons, Pedigree Dog Breeders and Dog Owners. Ph.D. Thesis, The University of Nottingham, Nottingham, UK, 2017.
26. Burnett, E.; Brand, C.L.; O'Neill, D.G.; Pegram, C.L.; Belshaw, Z.; Stevens, K.B.; Packer, R.M.A. How much is that doodle in the window? Exploring motivations and behaviours of UK owners acquiring designer crossbreed dogs (2019–2020). *Canine Med. Genet.* **2022**, *9*, 8. [CrossRef]
27. Holland, K.E.; Mead, R.; Casey, R.A.; Upjohn, M.M.; Christley, R.M. Why Do People Want Dogs? A Mixed-Methods Study of Motivations for Dog Acquisition in the United Kingdom. *Front. Vet. Sci.* **2022**, *9*, 877950. [CrossRef]
28. Creswell, J.W.; Clark, V.L.P. *Designing and Conducting Mixed Methods Research*; SAGE Publications, Ltd.: Los Angeles, CA, USA, 2011.
29. Holland, K.E.; Mead, R.; Casey, R.A.; Upjohn, M.M.; Christley, R.M. "Don't Bring Me a Dog ... I'll Just Keep It": Understanding Unplanned Dog Acquisitions Amongst a Sample of Dog Owners Attending Canine Health and Welfare Community Events in the United Kingdom. *Animals* **2021**, *11*, 605. [CrossRef]
30. Holland, K.E. Acquiring a pet dog: A review of factors affecting the decision-making of prospective dog owners. *Animals* **2019**, *9*, 124. [CrossRef]
31. Herzog, H. Women Dominate Research on Human-Animal Bond. Am. Psychol. J. 2021. Available online: https://www.wellbeingintlstudiesrepository.org/cgi/viewcontent.cgi?article=1010&context=sc_herzog_compiss (accessed on 18 May 2022).
32. Dogs Trust. Stray Dogs Survey Report 2017–18. 2018. Available online: https://www.dogstrust.org.uk/help-advice/research/research-papers/stray-dogs-survey (accessed on 18 May 2022).
33. R Core Team. *R: A Language and Environment for Statistical Computing*; R Foundation for Statistical Computing: Vienna, Austria, 2021. Available online: https://www.R-project.org/ (accessed on 18 May 2022).
34. Braun, V.; Clarke, V. Using thematic analysis in psychology. *Qual. Res. Psychol.* **2006**, *3*, 77–101. [CrossRef]
35. Malterud, K.; Siersma, V.D.; Guassora, A.D. Sample Size in Qualitative Interview Studies: Guided by Information Power. *Qual. Health Res.* **2016**, *26*, 1753–1760. [CrossRef] [PubMed]
36. Hennink, M.M.; Kaiser, B.N.; Marconi, V.C. Code Saturation Versus Meaning Saturation: How Many Interviews Are Enough? *Qual. Health Res.* **2017**, *27*, 591–608. [CrossRef] [PubMed]
37. Harvey, N.D. How Old Is My Dog? Identification of Rational Age Groupings in Pet Dogs Based Upon Normative Age-Linked Processes. *Front. Vet. Sci.* **2021**, *8*, 643085. [CrossRef]
38. People's Dispensary for Sick Animals (PDSA). The PDSA Animal Wellbeing (PAW) Report 2020. 2020. Available online: https://www.pdsa.org.uk/media/10540/pdsa-paw-report-2020.pdf (accessed on 16 May 2022).
39. Rault, J.L.; Waiblinger, S.; Boivin, X.; Hemsworth, P. The Power of a Positive Human–Animal Relationship for Animal Welfare. *Front. Vet. Sci.* **2020**, *7*, 590867. [CrossRef]
40. Friemel, T.N. The digital divide has grown old: Determinants of a digital divide among seniors. *New Media Soc.* **2016**, *18*, 313–331. [CrossRef]
41. Ofcom. Adults' Media Use & Attitudes: Report 2020. 2020. Available online: https://www.ofcom.org.uk/__data/assets/pdf_file/0031/196375/adults-media-use-and-attitudes-2020-report.pdf (accessed on 23 June 2022).
42. Office for National Statistics. Internet Users, UK. 2021. Available online: https://www.ons.gov.uk/businessindustryandtrade/itandinternetindustry/bulletins/internetusers/2020 (accessed on 23 June 2022).

animals

MDPI

Article

Selection Factors Influencing Eventual Owner Satisfaction about Pet Dog Adoption

Ian R. Dinwoodie [1,2,†], Vivian Zottola [2,3,†], Karla Kubitz [2,4] and Nicholas H. Dodman [1,2,*]

1 Cummings School of Veterinary Medicine, Tufts University, North Grafton, MA 01536, USA
2 Center for Canine Behavior Studies, Salisbury, CT 06068, USA
3 Boston K9 Concierge LLC, Boston, MA 02127, USA
4 Department of Kinesiology, Towson University, Towson, MD 21204, USA
* Correspondence: nicholas.dodman@tufts.edu
† These authors contributed equally to this work.

Simple Summary: Pet dog adopters are influenced by a variety of complex factors some of which are ethical, emotional, and humanitarian, including wanting companionship for themselves or other pets, the dog's breed, age, appearance, temperament, or behavior, or to provide a home for a homeless dog. However, not all adoptions are successful and managing owner expectations preadoption is difficult to navigate. Using a self-reporting questionnaire, we found that consideration of a dog's personality and behavior had a positive effect on eventual owner satisfaction. Owners who adopted a dog for companionship were more likely to be satisfied than owners compelled by any other motive. In addition, less forethought prior to adoption, ideally less than one week, was found to increase the likelihood of eventual owner satisfaction. We suggest that consideration of these factors prior to adoption may lead to more successful pet dog adoption outcomes.

Abstract: Personal likes, experience, and deep-rooted interests to satisfy emotional needs such as companionship, affection, empathy, and security are some of the underlying human motivations for acquiring a pet companion. In this study, we asked how long the owner took to decide whether to adopt a dog, who their dog was adopted from, their primary motivation for adoption, a ranking of characteristics considered during the adoption process, and how satisfied they were with the eventual outcome. Participants (n = 933) to this Center for Canine Behavior Studies survey completed an online questionnaire with responses representing 1537 dog/owner pairs. A majority of participants reported satisfaction with at least one of their adopted dogs. Odds of eventual satisfaction are higher for participants who spent less than a week considering an adoption or were seeking a pet to provide companionship and affection. Participants that prioritized personality as an adoption criteria were more likely to be satisfied with their adopted dogs. A mast majority (91%) of participants reported they would consider adopting another dog in the future. Selection criteria rankings that participants indicated they would employ for future adoptions tended to shift away from physical to behavior characteristics when compared to selection criteria priorities of prior adoptions.

Keywords: dogs; adoption; survey; behavior; expectations

Citation: Dinwoodie, I.R.; Zottola, V.; Kubitz, K.; Dodman, N.H. Selection Factors Influencing Eventual Owner Satisfaction about Pet Dog Adoption. *Animals* **2022**, *13*, 2264. https://doi.org/10.3390/ani12172264

Academic Editor: Betty McGuire

Received: 27 July 2022
Accepted: 30 August 2022
Published: 1 September 2022

1. Introduction

In the United States, most people acquire pet dogs from various distribution channels including commercial and backyard breeders, pet stores, rescues, and shelters where dogs may be selected in person or online [1,2]. Whatever the source, when making decisions adopters are influenced by a variety of complex factors some of which are ethical, emotional, and humanitarian, including wanting companionship for themselves or other pets, the dog's breed, age, appearance, temperament, or behavior, or to provide a home for a homeless dog [1,3]. Other motivations for acquiring dogs satisfy more functional needs such as wanting an exercise and adventure partner, enhanced social interactions with other people, needing protection, or seeking an assistance or therapy dog [4]. If the qualities a prospective owner is looking for, such as a particular temperament, are not met the failure of expectations can affect adopted dogs' later welfare and even survival [5]. Veterinarians in private practice deal with clients closely, answering questions, discussing training methods, and helping with problems and difficulties before or after they arise [6]. Kidd et al. [6] found that owners of adopted dogs who avail themselves of veterinary advice are less likely to relinquish their adopted charge. In another study, Kidd et al. [7] found that greater adoption success was attained with owners who had previously owned pets, that men rejected a significantly higher percentage of pets than women, parents rejected more pets than non-parents, and that specific expectations differed considerably between men and women, parents and nonparents, and retainers and rejecters of adopted pets.

Dogs acquired from breeders or pet shops are generally chosen because the new owner wants a particular breed, though other factors, such as appearance, age, size, and temperament or behavior, also factor into the decision [8]. Humanitarian motivation features strongly in the adoption of rescue dogs. Rescue dogs are dogs of any breed or age that have been abandoned or relinquished and placed in a temporary boarding system ("shelter"). These dogs may have been found as strays or simply been surrendered by their previous human caretakers for a plethora of reasons. Rescue dogs can be successfully adopted from shelters, though some are subsequently brought back to the shelter if the new owner is disappointed or dissatisfied [9–11].

Not all adoptions are successful and managing owner expectations preadoption is difficult to navigate. If owner expectations are high and skewed based on limited knowledge, they may be difficult to achieve and therefore returns are more likely [12]. Well respected scientists have attempted to understand this phenomenon and investigate underlying motivating factors influencing long term success in pet dog adoptions. For example, personality—both dog and human—alone cannot determine successful outcome of adoptions [13]. Ensuring compatibility with all household members prior to adoption was looked at and was found to improve rehoming success rates [14], as was providing training post adoption [15]. In assessing the personality of 88 dogs and their owners from Oklahoma, Curb et al. [16] found four correlations between owner satisfaction scores and dog–owner personality match; these were (1) the tendency to share possessions, (2) love of running outside, (3) likeliness of being destructive, and (4) ability to have a good relationship with others. The authors suggest that prospective dog owners may want to consider adopting dogs that match their personality on these characteristics. In a similar Danish study of 421 dog-owner pairs, it was found that owner characteristics appeared to influence the dog-owner relationship more than dog personality traits [17]. The results of these two latter studies, while informational, require that owners assess their own personality and match it to the prospective adoptee dog to increase the odds of eventual owner satisfaction. Neither study found dogs' physical appearance, breed, signalment, personality/behavior, or trainability to be a helpful predictor of owner satisfaction, though the study by Curb et al. [16] alluded to compatibility with other pets, destructiveness, and a love of exercise as potentially helpful factors to consider when adopting a dog. Dog personality/behavior was loosely represented by destructiveness (presumably the lack thereof) and the love of exercise in the latter study. In another study of owned dogs, van Herwijnen at al. [18] found that increasing levels of aggressive behaviour significantly lowered the chance of owners being very satisfied with their dog. This study also found that disobedience was associated inversely with ownership satisfaction to a similar degree, possibly because unwanted behaviour by the dog, including aggression, was found to coincide with high perceived costs, including re-training expenses.

Discerning baseline behavior of shelter dogs is difficult as depending on the individual dog and their learned history, these may be suppressed while living in the shelter system and only presenting post adoption when living in a new home environment. In many cases post adoption may be too late [19]. A one-year cohort study including 14 centers conducted in the United Kingdom found 14% of adopted dogs were returned to shelters within six months of adoption [20]. Other studies found return periods as short as hours to days [21]. Whether an adopter returns a dog to a shelter has been shown to depend on the severity of the behaviors presented as well as the training provided post adoption. Untreated behaviors have been shown to harm human pet dog relations and, in some cases, quickly. This is likely due to an underestimated commitment of time, money and emotional investment required without guarantees [20,22].

To understand consumer interests and improve matching efforts when adopting dogs, the Center for Canine Behavior Studies, Inc 501(c)(3) (CCBS) members and general public were asked to provide feedback by an online qualitative survey. In this study, involving 933 participants, we examined what factors consumers prioritized when acquiring a pet dog and their subsequent level of satisfaction with the owner/dog relationship, while other studies have been conducted to examine factors influencing potential adopters' preferences [23–25], none were designed to examine the result of selection criteria on the eventual outcome success of the adoption process. Correlation of selection factors on the eventual outcome was the purpose of the present study.

2. Materials and Methods

2.1. Data Collection

The questionnaire for this study was developed and hosted on Typeform, an online survey service platform. A link to the public questionnaire was posted on social media platforms (Twitter, Instagram, and Facebook) and distributed to CCBS members via email. Data collection spanned one year starting from 13 February 2020. Participants were willing dog owners who voluntarily completed the online questionnaire. All data in this study was self-reported by participants (e.g., primary motivation is owner self-reported primary motivation).

Logically, the distributed questionnaire could be considered in three parts: (1) demographic information about the owner, (2) information about a single dog that was adopted (i.e., taken into a relationship by choice), and (3) information about future dog adoption. Participants with more than one dog were prompted to fill out the second part of the questionnaire for each of their dogs. The owner demographics component of the questionnaire was used to gather the age and gender of the owner. The individual dog component of the questionnaire was used to gather the length of time spent thinking about an adoption (choices: less than one week, one week to six months, greater than six months to less than six years, or greater than six years), the owner's primary motivation for the adoption (choices: companionship/affection, social interaction, exercise/adventure partner, protection, someone else in the house wanted a dog, or companion for another pet), the adoption source (choices: breeder, online pet shop, local pet shop, shelter/rescue, family member/friend, found, or another country/island), the ranking of characteristics considered for the adoption, if the adopted dog was currently living with the owner, and whether the owner's expectations for that dog had been met (choices: yes, partially, or no). The characteristics that owners were asked to rank were: age, appearance, breed, compatibility with other pets, personality, size, and trainability. Two additional questions were presented to owners when they indicated they were not living with the adopted dog: how long the owner had lived with the dog (choices: less than one week, one week to six months, greater than six months to less than six years, or greater than six years) and the living situation of the dog (choices: rehomed, surrendered, euthanized, lost, or passed). The future adoption component of the questionnaire was used to gather whether the owner would consider adopting a dog again in the future and, if so, their revised ranking of characteristics for a future adoption. The characteristics presented to users for ranking were identical to the characteristics from the previous component.

Email addresses provided by users were recorded as randomly generated MD5 checksums (i.e., 128-bit hashes) to avoid retaining any personally identifying information. Responses across the three parts of the questionnaire were linked using these hashes. As a result, participants remained anonymous to all personnel involved and no ethical approval was required for undertaking this study.

An annotated copy of the questionnaire used for study is available in Appendix A.

2.2. Data Preprocessing

The study data set was tidied using the R programming language (version 4.2.0) provided by the R Foundation for Statistical Computing [26] and packages from the `tidyverse` library [27], namely the `dplyr` [28], `tidyr` [29], `forcats` [30], and `purrr` [31] packages. Repeat responses were those that had non-unique owner/dog pairs. Of the repeat responses, only the most recent response was retained for the study. For each submitted questionnaire the ranked list of characteristics was exploded such that each characteristic was given its own column and assigned a value corresponding to its position in the list (i.e., the first ranked characteristic was assigned a value of 1 and the last ranked characteristic was assigned a value of 7). The local pet store and online pet store responses categories were collapsed down to a single local/online pet store category to minimize unnecessary complexity.

2.3. Inclusion Criteria

Only responses from adults, defined as participants with a reported age of 18 years or older, were retained for analysis. Complete responses were those that included data considered necessary to build relevant statistical models: owner demographics information, the length of time spent thinking about an adoption, the owner's primary motivation for the adoption, the adoption source, and the ranking of characteristics considered for the adoption. Incomplete responses were excluded from the study sample.

2.4. Descriptive Analysis

The study data set was exported from Typeform as a comma-separated values (CSV) file. All analyses were performed using the R programming language (version 4.2.0) provided by the R Foundation for Statistical Computing [26]. Descriptive statistics were calculated. Ranges were provided for all medians. Nonreduced proportions were provided for all percentages for clarity. Graphics were created using methods provided by the `ggplot2` package [32].

2.5. Analytical Modeling

The significance level was set to $\alpha \leq 0.05$ for all regression models in this manuscript. Corrections for multiple testing were performed using the Benjamini and Hochberg adjustment [33].

2.5.1. Change in Rankings

Mann–Whitney tests were performed using methods from the `stats` package [26] to evaluate the difference between past and future adoption characteristic rankings. The null hypothesis was that there was no difference.

2.5.2. Impact of Adoption Criteria

A hierarchy of binary logistic regression models was built to assess the impact of factors influencing adoption on adoption satisfaction. An owner was considered satisfied with an adoption if they had indicated that the dog had fully met their expectations. A null model (i.e., intercept only) was built to serve as a frame-of-reference. The first, second, and third steps of the hierarchy added terms to account for background variables, adoption source, and primary adoption motivation, respectively. Background variables consisted of the owner's age, gender, and time spent thinking about the adoption. Each model resulting from this process was compared with its preceding model.

Due to the compositional nature of ranked data (i.e., the constant-sum constraint across a complete set of rankings), models were limited to the inclusion of a single characteristic rank term. To evaluate all characteristics, the fourth and final step of the hierarchy was broken into seven parallel models; each model added a single characteristic rank term to the model from the third step of the hierarchy.

Binary logistic regression models were built using the generalized linear model (GLM) fitter provided by the `stats` package [26]. Two-way contingency tables were constructed for categorical predictors to verify a minimum of ten responses for the least frequent outcome (i.e., the "rule of ten") [34]. Odds ratios (ORs) were calculated as a measure of effect size. Confidence intervals were calculated using the bootstrapped samples with replacement (N = 10,000) using methods provided by the `rsample` package [35]. Multicollinearity was assessed by variance inflation factor (VIF) using methods provided by the `car` package [36]. Overall evaluation was performed for each model via likelihood ratio test. Fit of the models against actual outcomes was assessed using the Hosmer–Lemeshow goodness-of-fit test (H–L) [37] implemented by the `performance` package [38]. Pseudo-R^2s were calculated using methods proposed by Nagelkerke [39] and implemented by the `performance` package [38] to provide an additional measure of goodness-of-fit.

3. Results

3.1. Demographics

A total of 1595 responses were provided by 971 participating dog owners. Four percent (n = 58/1595) of the responses were dropped due to their failure to meet the inclusion criteria. The resulting sample study consisted of complete responses for 1537 owner/dog pairs across 933 participants.

Ninety-two percent (n = 859/933) of the participating dog owners were female; the remainder (n = 74/933) were male. Participants averaged an age of 51 years old (range: 18 to 85 years old). Female participants averaged 52 years old (range: 18 to 85 years old) while male participants averaged 58 years old (range: 20 to 75 years old). The median number of dogs per household was one (range: 1 to 10 dogs per household). More than half (n = 495/933) of the participants represented single dog households. Ninety-one percent (n = 851/933) of owners indicated that they would consider another dog adoption in the future, seven percent (n = 61/933) indicated that they would not consider dog adoption again in the future, and two percent (n = 21/933) elected not to provide a response.

Eighty-eight percent (n = 1347/1537) of the dogs lived with their participating owner. Of the minority subset of dogs that were not living with the participating owner, 75% (n = 142/190) had passed away, 18% (n = 34/190) were euthanized, four percent (n = 7/190) were not living with their owner for reasons other than those asked on the questionnaire, three percent (n = 5/190) had been rehomed, less than one percent (n = 1/190) had been lost, and less than one percent (n = 1/190) had been surrendered. Seventy-eight percent (n = 149/190) of the dogs not living with their participating owner had lived with their owner for more than six years, 19% (n = 36/190) for greater than six months but less than six years, two percent (n = 4/190) for one week to six months, and less than one percent (n = 1/190) for less than one week.

3.2. Time Spent Thinking

Forty-nine percent (n = 752/1537) of the dogs were acquired with one week to six months of forethought, 34% (n = 528/1537) with between six months and six years of forethought, 13% (n = 196/1537) with less than one week of forethought, and four percent (n = 61/1537) with greater than six years of forethought.

3.3. Adoption Source

Fifty-five percent (n = 847/1537) of the dogs were adopted from a rescue/shelter, 31% (n = 480/1537) from a breeder, eight percent (n = 130/1537) from friends or family members, four percent (n = 68/1537) from local or online pet shops, three percent (n = 51/1537) were found, and less than one percent (n = 12/1537) were adopted from a foreign country. Owners had an average age of 51 (range: 18 to 85 years old) for rescue/shelter adoptions, 56 (range: 19 to 84 years old) for breeder adoptions, 51 (range: 20 to 75 years old) for adoptions from friends or family members, 48 (range: 20 to 73 years old) for adoptions from a local or online pet shop, 49 (range: 22 to 74 years old) for owners of dogs that were found, and 53 (range: 36 to 65 years old) for adoptions from foreign countries.

3.4. Primary Motivation for Adoption

Fifty-five percent (n = 846/1537) of the dogs were acquired for companionship and affection, 15% (n = 224/1537) as a companion for another pet in the household, 12% (n = 187/1537) for reasons other than those explicitly asked for on the questionnaire, seven percent (n = 100/1537) for working or sporting, five percent (n = 77/1537) as an exercise and adventure partner, five percent (n = 70/1537) due to other members of the family wanting a dog, one percent (n = 21/1537) for social interaction, and less than one percent (n = 12/1537) were adopted for protection.

3.5. Characteristic Ranks

All characteristics, for both past and future adoptions, received rank positions that ranged from 1.00 to 7.00 (i.e., all characteristics had at least one response per minimum and maximum rank positions). Full distributions of ranks for each characteristic are displayed in Figure 1.

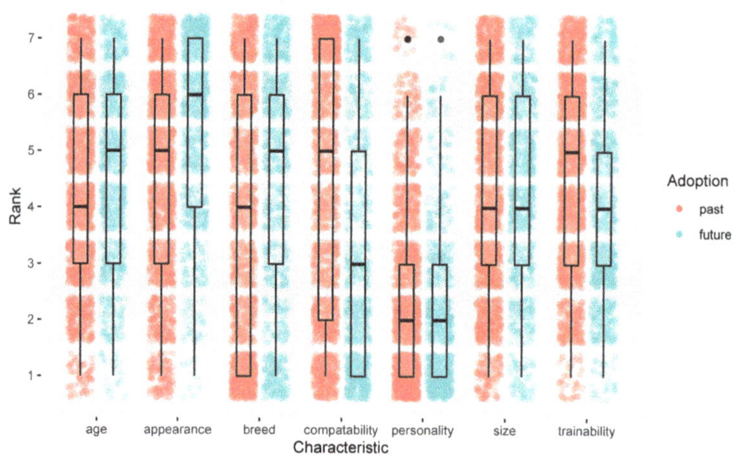

Figure 1. Jitter plot of characteristic rank positions for past and future adoptions. Box plots are overlaid to indicate quartiles and mean.

Characteristic rank means for past and future adoptions along with the difference between past and future rank means by characteristic are provided in Table 1. The change in rank position for each characteristic except for size was statistically significant.

Table 1. Average characteristic rank positions for past and future adoptions. A lower rank position indicates a higher priority characteristic.

Characteristic	Average Rank Position ($\bar{x}$)		$\Delta\bar{x}$	p
	Past Adoptions	Future Adoptions		
Age	4.31	4.59	0.28	<0.001
Appearance	4.54	5.47	0.93	<0.001
Breed	3.63	4.38	0.75	<0.001
Compatibility	4.49	3.33	−1.16	<0.001
Personality	2.5	2.16	−0.34	<0.001
Size	4.28	4.3	0.02	0.787
Trainability	4.65	4.01	-0.64	<0.001

3.6. Owner Satisfaction

Eighty-three percent (n = 1282/1537) of the adopted dogs fully met their owner's expectations, 16% (n = 248/1537) partially met their owner's expectations, and less than one percent (n = 7/1537) failed to meet their owner's expectations. Ninety percent (n = 843/933) of owners indicated satisfaction with at least one dog and 76% (n = 707/933) of owners indicated satisfaction with all their reported dogs. Ninety percent (n = 777/859) of females and 89% (n = 66/74) of males indicated satisfaction with at least one dog. Seventy-six percent (n = 649/859) of females and 78% (n = 58/74) of males indicated satisfaction with all their reported dogs.

Ninety-one percent (n = 851/933) of participating owners indicated they would consider another dog in the future. Ninety-one percent (n = 644/707) of owners who were satisfied with all their dogs indicated they would consider another dog in the future; three percent (n = 44/707) indicated they would not. Ninety-six percent (n = 131/136) of owners that indicated satisfaction with some, but not all, of their dogs indicated they would consider another dog in the future; three percent (n = 4/136) indicated they would not. Eighty-four percent (n = 76/90) of owners who were not satisfied with any of their dogs indicated they would consider another dog in the future; 14% (n = 13/90) indicated they would not.

Owners indicated that their expectations had been met for 100% (n = 1/1) of the dogs that had been lost, 94% (n = 32/34) of the dogs that had been euthanized, 86% (n = 6/7) of the dogs not living with their owner for reasons other than those asked on the questionnaire, 83% (n = 118/142) of the dogs that had died, 20% (n = 1/5) of the dogs that had been rehomed, and none (n = 0/1) of the dogs that had been surrendered.

3.7. Impact of Adoption Criteria on Owner Satisfaction

3.7.1. Additional Preprocessing

Two primary motivation categories (protection and social interaction) and two adoption source categories (found and imported from a foreign country or island) were dropped from all models due to violations of the "rule of ten" [34].

3.7.2. Models Accounting for Characteristic Ranks

According to the model fit to the data using personality rank in addition to primary motivation, adoption source, background variables, and intercept, a one-unit deprioritization in rank for personality (i.e., a move from 1st place to 2nd place) resulted in 0.86 (95% CI: 0.79–0.94; p = 0.001) the odds of indicating satisfaction with an adoption. Owners who indicated primary motivations of an exercise partner (OR: 0.45; 95% CI: 0.25–0.82; p = 0.014), requests from another family member (OR: 0.29; 95% CI: 0.16–0.53; $p < 0.001$), for working/sporting (OR: 0.47; 95% CI: 0.26–0.87; p = 0.022), or some unspecified reason (OR: 0.40; 95% CI: 0.26–0.62; $p < 0.001$) had decreased odds of indicating satisfaction with an adoption compared to those who indicated companionship as their primary motivation. In addition, owners with greater than six years of forethought (OR: 0.25; 95% CI: 0.13–0.56; p = 0.001) or greater than six months but less than six years of forethought (OR: 0.51; 95% CI: 0.31–0.84; p = 0.023) had decreased odds of indicating satisfaction with an adoption compared to those with less than one week of forethought. No other independent variable was found to have a significant correlation with the outcome due to either exceeding the study significance level or confidence interval disqualification.

No other model accounting for characteristic rank was found to provide a better fit to the data than the model accounting for primary motivation. A hierarchical overview of the models used in this study is provided in Table 2.

Table 2. Hierarchical regression analysis of adoption criteria on owner satisfaction.

Predictor	Logistic Regression Coefficient (β)									
	Model 1	Model 2	Model 3	Model 4a	Model 4b	Model 4c	Model 4d	Model 4e	Model 4f	Model 4g
Step 1. Background										
Owner is male	0.01	0.19	0.18	0.18	0.20	0.18	0.17	0.23	0.17	−0.18
Owner age (in years)	0.00	0.00	0.00	0.00	0.00	0.00	0.00	0.00	0.00	0.00
Forethought (ref: <1 week)										
1 week to 6 months	−0.19	−0.27	−0.47	−0.46	−0.49	−0.46	−0.47	−0.49	−0.47	−0.47
>6 months to 6 years	−0.12	−0.32	−0.59 *	−0.59 *	−0.63 *	−0.59 *	−0.60 *	−0.67 *	−0.60 *	−0.59 *
>6 years	−0.75	−1.00 *	−1.35 **	−1.34 **	−1.38 **	−1.34 **	−1.38 **	−1.38 **	−1.37 **	−1.34 **
Step 2. Adoption Source (ref: pet store)										
Breeder		0.42	0.42	0.41	0.37	0.45	0.42	0.46	0.44	0.42
Family/friend		−0.37	−0.32	−0.32	−0.35	−0.32	−0.30	−0.31	−0.31	−0.32
Shelter/rescue		−0.25	−0.33	−0.33	−0.37	−0.34	−0.31	−0.40	−0.33	−0.33
Step 3. Motivation (ref: companionship)										
Companion for another pet			−0.51 *	−0.52 *	−0.55 *	−0.52 *	−0.44	−0.44	−0.51 *	−0.51 *
Exercise			−0.82 *	−0.82 *	−0.85 **	−0.82 *	−0.82 *	−0.80 *	−0.81 *	−0.81 *
Requested by Family			−1.27 ***	−1.27 ***	−1.28 ***	−1.28 ***	−1.26 ***	−1.25 ***	−1.28 ***	−1.27 ***
Working/sporting			−0.74 *	−0.74 *	−0.79 *	−0.74 *	−0.73 *	−0.75 *	−0.72 *	−0.74 *
Other			−0.90 ***	−0.90 ***	−0.94 ***	−0.91 ***	−0.85 ***	−0.91 ***	−0.89 ***	−0.90 ***
Step 4. Characteristics Rankings										
a. Age				0.02						
b. Appearance					0.07					
c. Breed						0.02				
d. Compatability							0.03			
e. Personality								−0.15 **		
f. Size									−0.02	
g. Trainability										0.00
Metric										
Δdf	5	3	5	1	1	1	1	1	1	1
L^2 (χ^2)	4.81	17.14 ***	35.42 ***	0.21	3.18	0.20	0.87	12.08 ***	0.38	0.01
R^2_{nk}	0.01	0.02	0.06	0.06	0.07	0.06	0.06	0.08	0.06	0.06
H–L (χ^2)	0.01	0.02	0.06	0.06	0.07	0.06	0.06	0.08	0.06	0.06

* = $p \leq 0.05$; ** = $p \leq 0.01$; *** = $p \leq 0.001$.

4. Discussion

Over half of the dogs in this study were acquired from a shelter. Participants who acquired a dog from a rescue or shelter had an average age of 51 years, which is interesting since other studies exploring social and ethical methods of acquisition found a high percentage of adopters in the millennial age group [8,40]. A study that employed probability-based panel to generate a nationally representative sample of adults in the United States yielded a similar average participant age of 48 years (range: 18 to 94 years old) [41]. The majority of dog owners (92%) responding to the survey were female, which is very typical of these types of surveys [42–45].

A high percentage (83%) of participants were fully satisfied with their choice of dog with less than one percent being dissatisfied. A possible reason for this favorable outcome may have been population bias; participants in this study were a group of dedicated dog owners who chose to engage in our online platform in their spare time. Additionally, we cannot rule out the possibility that dog owners may be less inclined to share their adoption experiences when none of their adopted dogs have met their expectations. An even higher percentage of owners (95%) whose dogs had been euthanized indicated that their expectations had been met. One likely explanation for this slightly different finding is that these dogs were euthanized for medical reasons. Logically, owner satisfaction was much lower for dogs that had been rehomed or surrendered (20% and 0%, respectively).

Various selection criteria that we presented as possibly relevant to prospective dog owners, the dog's personality was the only one found to have a significant positive effect on eventual owner satisfaction. This is surprising since determining personality is not easy even for pet professionals [46] and, in addition, shelters and rescues are typically loud and visually over-stimulating for dogs making selection based on a dog's personality even more difficult [47]. A dog in this situation will likely be over-exuberant or displaying fearful behavior [48]. None of the other selection criteria we suggested, including age, appearance, breed, compatibility with other pets, size, or trainability, was found to have a significant effect on eventual owner satisfaction. Neither did we find breed to be an adoption criterion that led to greater or less eventual owner satisfaction. This is in distinction from the results of a study by Posage et al. [49] who found that certain breeds, including toys and terriers, were more successfully adopted.

Although Garrison and Weiss [50] found that no particular adoption criterion drove the dog adoption process, we found the dog's personality was, on average, the most highly prioritized factor in adoption. As far as future adoptions were concerned, our participants were more likely to prioritize behavioral elements such as the dog's personality, compatibility with other pets, and trainability. Conversely, participants were more likely to deprioritize the ranking of physical attributes such as age, appearance, and breed for future adoptions.

Compared to owners who adopted dogs primarily for companionship, those who selected their dog as an exercise partner, at the request of another family member, for working/sporting purposes, or some unspecified reason were less likely to indicate eventual satisfaction with their adoption.

Owners with six or more months forethought had decreased odds of satisfaction with an adoption than those with less than one week of forethought. In other words, a lengthy thought process about desired adoption criteria has a detrimental effect on eventual owner satisfaction, possibly because such owners have fastidious requirements for their future charge. On the other hand, owners who already know what they are looking for in a dog to be adopted would be expected to make a more rapid decision and to fare much better because of their definitive and clear-cut selection criteria. It is also possible that youthful and therefore behaviorally malleable dogs were the ones adopted more quickly and integrated more successfully. In support of this contention, Normando et al. [51] found that "young age" was the most important factor leading to quick pet dog adoption. Another possible explanation is that older dogs, a known and relatively immutable commodity, were more rapidly and successfully adopted. Unfortunately, we did not collect data regarding dog age at the time of adoption in this study; therefore, we are unable to formally verify the age group of dogs at the time of adoption as an explanation for the success of rapid adoptions.

The results of this study indicate that potential adopters should seriously consider the personality and behavior of a dog rather than breed and size. Shelters would be well advised to take this to heart and advise potential adopters accordingly.

5. Conclusions

While our primary interest was to identify key characteristics that adopters use to select dogs and how satisfied they were with the result of their selection, we found the dog's personality/behavior as the only characteristic with a positive effect on eventual owner satisfaction. Less forethought, ideally less than one week, was found to have a positive effect on eventual owner satisfaction. Adopting a dog for companionship and affection was more likely to lead to eventual owner satisfaction than any other motive. In our study, the greatest degree of owner satisfaction was assessed in dogs that were eventually euthanized, possibly because euthanasia was performed at the end of the dog's life and/or for medical reasons. In comparison to the selection criteria priorities of prior adoptions, the selection criteria rankings that participants had indicated they would employ for future adoptions tended to shift away from physical to behavior characteristics.

Author Contributions: Conceptualization, N.H.D.; methodology, V.Z. and I.R.D.; data curation, I.R.D.; formal analysis, I.R.D. and K.K.; writing, N.H.D., V.Z. and I.R.D. All authors have read and agreed to the published version of the manuscript.

Funding: A generous gift from the Jan Corning and Mac Emory via the American Foundation, amplified by additional financial contributions from board members of the Center for Canine Behavior Studies and its membership at large.

Institutional Review Board Statement: Not applicable.

Informed Consent Statement: Informed consent was obtained from all subjects involved in the study.

Data Availability Statement: The data presented in this study are openly available in Zenodo at 10.5281/zenodo.6910622, reference number [52].

Acknowledgments: We would like to acknowledge the online service "How I Met My Dog" and Donna Gleason for their valuable input drafting the initial survey.

Conflicts of Interest: The authors declare no conflict of interest.

Abbreviations

The following abbreviations are used in this manuscript:

MDPI	Multidisciplinary Digital Publishing Institute
CCBS	Center for Canine Behavior Studies
CSV	comma-separated values
GLM	generalized linear model
OR	odds ratio
VIF	variance inflation factor
H–L	Hosmer–Lemeshow goodness-of-fit test

Appendix A

This section of the appendix contains annotated questionnaire components that respondents completed to participate in the study. An asterisk was used to indicate that a response to the question is required for submission. Angle brackets were used to indicate text that the survey platform would substitute with data from a previous entry. Text that appears in square brackets are annotations meant to clarify survey platform constraints/logic that were employed.

Appendix A.1

Owners were asked to complete the following questions once.

1. What is your email? * [email]
2. What is your postal code (zip code)? [string]
3. What is your gender? [choose 1]
   (a) Female
   (b) Male
4. What is your age? [integer]

Appendix A.2

Owners were asked to complete the following questions once per dog. Questions 1 through 15 were presented each iteration. Questions 16 through 23 were presented only when the owner indicated they had no remaining dogs to complete the questionnaire for.

1. What is the name of the dog you would like to submit a response for? * [string]
2. How long were you thinking about adding a new dog to the family before you brought <dog> home? * [choose 1]
   (a) Less than 1 week
   (b) 1 week–6 months
   (c) 6+ months–6 years
   (d) 6+ years
3. What was your primary motivation for getting <dog>? * [choose 1]
   (a) Companionship and affection
   (b) Social interaction
   (c) Exercise, adventure partner
   (d) Protection
   (e) Working or sporting
   (f) Someone else in the home wanted a dog (e.g., kids, spouse)

   (g) Companion for another dog or pet
4. Where did you acquire <dog> from? * [choose 1]
   (a) Breeder
   (b) Online
   (c) Local pet shop
   (d) Shelter/Rescue
   (e) Family member or friend
   (f) Found roaming
   (g) Another country or island
5. Please select the characteristic you found most important when choosing <dog>? * [choose 1]
   (a) Appearance
   (b) Breed
   (c) Size
   (d) Age
   (e) Individual personality/behavior
   (f) Trainability
   (g) Compatibility with other pets in the home
6. What is the next most important characteristic? * [choose 1]
   (a) Appearance
   (b) Breed
   (c) Size
   (d) Age
   (e) Individual personality/behavior
   (f) Trainability
   (g) Compatibility with other pets in the home
7. What is the next most important characteristic? * [choose 1]
   (a) Appearance
   (b) Breed
   (c) Size
   (d) Age
   (e) Individual personality/behavior
   (f) Trainability
   (g) Compatibility with other pets in the home
8. What is the next most important characteristic? * [choose 1]
   (a) Appearance
   (b) Breed
   (c) Size
   (d) Age
   (e) Individual personality/behavior
   (f) Trainability
   (g) Compatibility with other pets in the home
9. What is the next most important characteristic? * [choose 1]
   (a) Appearance
   (b) Breed
   (c) Size
   (d) Age
   (e) Individual personality/behavior
   (f) Trainability
   (g) Compatibility with other pets in the home
10. What is the next most important characteristic? * [choose 1]

(a) Appearance
(b) Breed
(c) Size
(d) Age
(e) Individual personality/behavior
(f) Trainability
(g) Compatibility with other pets in the home

11. Were your primary relationship expectations met? * [choose 1]
    (a) Yes
    (b) Partially, I grew to love the dog over time
    (c) Partially, I accepted the dog and made it work
    (d) No, the dog was not a good match
12. Do you still live with <dog>? * [choose 1]
    (a) Yes [skip to 15]
    (b) No
13. How long did you and <dog> live together? * [choose 1]
    (a) Less than 1 week
    (b) 1 week–6 months
    (c) 6+ months–6 years
    (d) 6+ years
14. Where is <dog> now? * [choose 1]
    (a) Rehomed
    (b) Surrendered to shelter/rescue
    (c) Euthanized
    (d) Ran away/got lost
    (e) Passed
15. Would you like to enter information for another dog you own(ed)? [choose 1]
    (a) Yes [submit and restart at 1]
    (b) No
16. Would you ever consider bringing another dog into your life? * [choose 1]
    (a) Yes
    (b) No [submit and exit]
17. Which characteristic would you find most important when considering another dog to bring into your life? [choose 1]
    (a) Appearance
    (b) Breed
    (c) Size
    (d) Age
    (e) Individual personality/behavior
    (f) Trainability
    (g) Compatibility with other pets in the home
18. What is the next most important characteristic? * [choose 1]
    (a) Appearance
    (b) Breed
    (c) Size
    (d) Age
    (e) Individual personality/behavior
    (f) Trainability
    (g) Compatibility with other pets in the home
19. What is the next most important characteristic? * [choose 1]

(a) Appearance
(b) Breed
(c) Size
(d) Age
(e) Individual personality/behavior
(f) Trainability
(g) Compatibility with other pets in the home

20. What is the next most important characteristic? * [choose 1]
    (a) Appearance
    (b) Breed
    (c) Size
    (d) Age
    (e) Individual personality/behavior
    (f) Trainability
    (g) Compatibility with other pets in the home
21. What is the next most important characteristic? * [choose 1]
    (a) Appearance
    (b) Breed
    (c) Size
    (d) Age
    (e) Individual personality/behavior
    (f) Trainability
    (g) Compatibility with other pets in the home
22. What is the next most important characteristic? * [choose 1]
    (a) Appearance
    (b) Breed
    (c) Size
    (d) Age
    (e) Individual personality/behavior
    (f) Trainability
    (g) Compatibility with other pets in the home
23. If you were to bring another dog into your life, where are you likely to get that dog? * [choose 1]
    (a) Breeder
    (b) Online
    (c) Local pet shop
    (d) Shelter/Rescue
    (e) Family member or friend
    (f) Roaming
    (g) Another country or island

References

1. Bir, C.; Widmar, N.; Croney, C. Stated Preferences for Dog Characteristics and Sources of Acquisition. *Animals* **2017**, *7*, 59. [CrossRef] [PubMed]
2. Lampe, R.; Witte, T.H. Speed of Dog Adoption: Impact of Online Photo Traits. *J. Appl. Anim. Welf. Sci.* **2015**, *18*, 343–354. [CrossRef] [PubMed]
3. Norman, C.; Stavisky, J.; Westgarth, C. Importing rescue dogs into the UK: Reasons, methods and welfare considerations. *Vet. Rec.* **2020**, *186*, 248. [CrossRef] [PubMed]
4. Carlisle, G.K. Pet Dog Ownership Decisions for Parents of Children with Autism Spectrum Disorder. *J. Pediatr. Nurs.* **2014**, *29*, 114–123. [CrossRef]
5. Taylor, K.D.; Mills, D.S. The development and assessment of temperament tests for adult companion dogs. *J. Vet. Behav.* **2006**, *1*, 94–108. [CrossRef]
6. Kidd, A.H.; Kidd, R.M.; George, C.C. Veterinarians and Successful Pet Adoptions. *Psychol. Rep.* **1992**, *71*, 551–557. [CrossRef]

7. Kidd, A.H.; Kidd, R.M.; George, C.C. Successful and Unsuccessful PET Adoptions. *Psychol. Rep.* **1992**, *70*, 547–561. [CrossRef]
8. Holland, K.E. Acquiring a Pet Dog: A Review of Factors Affecting the Decision-Making of Prospective Dog Owners. *Animals* **2019**, *9*, 124. [CrossRef]
9. McMillan, F.D.; Duffy, D.L.; Serpell, J.A. Mental health of dogs formerly used as 'breeding stock' in commercial breeding establishments. *Appl. Anim. Behav. Sci.* **2011**, *135*, 86–94. [CrossRef]
10. McMillan, F.D. Behavioral and psychological outcomes for dogs sold as puppies through pet stores and/or born in commercial breeding establishments: Current knowledge and putative causes. *J. Vet. Behav.* **2017**, *19*, 14–26. [CrossRef]
11. Miller, D.D.; Staats, S.R.; Partlo, C.; Rada, K. Factors associated with the decision to surrender a pet to an animal shelter. *J. Am. Vet. Med. Assoc.* **1996**, *209*, 738–742. [PubMed]
12. Powell, L.; Lee, B.; Reinhard, C.L.; Morris, M.; Satriale, D.; Serpell, J.; Watson, B. Returning a Shelter Dog: The Role of Owner Expectations and Dog Behavior. *Animals* **2022**, *12*, 1053. [CrossRef] [PubMed]
13. Walker, S.L.; Weng, H.Y.; Ogata, N.; Anderson, K.G. Do Canine and Human Personality Assessments Predict Successful Adoptions? *Anthrozoös* **2022**, 1–17. [CrossRef]
14. Shore, E.R. Returning a Recently Adopted Companion Animal: Adopters' Reasons for and Reactions to the Failed Adoption Experience. *J. Appl. Anim. Welf. Sci.* **2005**, *8*, 187–198. [CrossRef] [PubMed]
15. Marston, L.C.; Bennett, P.C.; Coleman, G.J. Adopting shelter dogs: Owner experiences of the first month post-adoption. *Anthrozoös* **2005**, *18*, 358–378. [CrossRef]
16. Curb, L.A.; Abramson, C.I.; Grice, J.W.; Kennison, S.M. The Relationship between Personality Match and Pet Satisfaction among Dog Owners. *Anthrozoös* **2013**, *26*, 395–404. [CrossRef]
17. Meyer, I.; Forkman, B. Dog and owner characteristics affecting the dog–owner relationship. *J. Vet. Behav.* **2014**, *9*, 143–150. [CrossRef]
18. Herwijnen, I.R.v.; van der Borg, J.A.M.; Naguib, M.; Beerda, B. The existence of parenting styles in the owner-dog relationship. *PLoS ONE* **2018**, *13*, e0193471. [CrossRef]
19. Clay, L.; Paterson, M.B.A.; Bennett, P.; Perry, G.; Phillips, C.C.J. Do Behaviour Assessments in a Shelter Predict the Behaviour of Dogs Post-Adoption? *Animals* **2020**, *10*, 1225. [CrossRef]
20. Diesel, G.; Pfeiffer, D.; Brodbelt, D. Factors affecting the success of rehoming dogs in the UK during 2005. *Prev. Vet. Med.* **2008**, *84*, 228–241. [CrossRef]
21. Mondelli, F.; Prato Previde, E.; Verga, M.; Levi, D.; Magistrelli, S.; Valsecchi, P. The Bond That Never Developed: Adoption and Relinquishment of Dogs in a Rescue Shelter. *J. Appl. Anim. Welf. Sci.* **2004**, *7*, 253–266. [CrossRef] [PubMed]
22. Powell, L.; Duffy, D.L.; Kruger, K.A.; Watson, B.; Serpell, J.A. Relinquishing Owners Underestimate Their Dog's Behavioral Problems: Deception or Lack of Knowledge? *Front. Vet. Sci.* **2021**, *8*, 734973. [CrossRef] [PubMed]
23. Hecht, J.; Horowitz, A. Seeing Dogs: Human Preferences for Dog Physical Attributes. *Anthrozoös* **2015**, *28*, 153–163. [CrossRef]
24. Siettou, C.; Fraser, I.M.; Fraser, R.W. Investigating Some of the Factors That Influence "Consumer" Choice When Adopting a Shelter Dog in the United Kingdom. *J. Appl. Anim. Welf. Sci.* **2014**, *17*, 136–147. [CrossRef]
25. Weiss, E.; Miller, K.; Mohan-Gibbons, H.; Vela, C. Why Did You Choose This Pet?: Adopters and Pet Selection Preferences in Five Animal Shelters in the United States. *Animals* **2012**, *2*, 144–159. [CrossRef] [PubMed]
26. R Core Team. *R: A Language and Environment for Statistical Computing*; R Foundation for Statistical Computing: Vienna, Austria, 2022.
27. Wickham, H.; Averick, M.; Bryan, J.; Chang, W.; McGowan, L.D.; François, R.; Grolemund, G.; Hayes, A.; Henry, L.; Hester, J.; et al. Welcome to the tidyverse. *J. Open Source Softw.* **2019**, *4*, 1686. [CrossRef]
28. Wickham, H.; François, R.; Henry, L.; Müller, K. *dplyr: A Grammar of Data Manipulation*. 2022. Available online: https://dplyr.tidyverse.org (accessed on 26 July 2022).
29. Wickham, H.; Girlich, M. *tidyr: Tidy Messy Data*. 2022. Available online: https://tidyr.tidyverse.org (accessed on 26 July 2022).
30. Wickham, H. *forcats: Tools for Working with Categorical Variables (Factors)*. 2022. Available online: https://forcats.tidyverse.org (accessed on 26 July 2022).
31. Henry, L.; Wickham, H. *purrr: Functional Programming Tools*. 2022. Available online: https://purrr.tidyverse.org (accessed on 26 July 2022).
32. Wickham, H. *ggplot2: Elegant Graphics for Data Analysis*; Springer: New York, NY, USA, 2016.
33. Benjamini, Y.; Hochberg, Y. Controlling the False Discovery Rate: A Practical and Powerful Approach to Multiple Testing. *J. R. Stat. Soc. Ser. B (Methodol.)* **1995**, *57*, 289–300. [CrossRef]
34. Peduzzi, P.; Concato, J.; Kemper, E.; Holford, T.R.; Feinstein, A.R. A simulation study of the number of events per variable in logistic regression analysis. *J. Clin. Epidemiol.* **1996**, *49*, 1373–1379. [CrossRef]
35. Silge, J.; Chow, F.; Kuhn, M.; Wickham, H. *rsample: General Resampling Infrastructure*. 2022. Available online: https://rsample.tidymodels.org (accessed on 26 July 2022).
36. Fox, J.; Weisberg, S. *An R Companion to Applied Regression*, 3rd ed.; Sage: Thousand Oaks, CA, USA, 2019.
37. Lemeshow, S.; Hosmer, D.W. A Review of Goodness of Fit Statistics for Use in the Development of Logistic Regression Models. *Am. J. Epidemiol.* **1982**, *115*, 92–106. [CrossRef]
38. Lüdecke, D.; Ben-Shachar, M.S.; Patil, I.; Waggoner, P.; Makowski, D. performance: An R Package for Assessment, Comparison and Testing of Statistical Models. *J. Open Source Softw.* **2021**, *6*, 3139. [CrossRef]
39. Nagelkerke, N.J.D. A note on a general definition of the coefficient of determination. *Biometrika* **1991**, *78*, 691–692. [CrossRef]

40. Bir, C.; Olynk Widmar, N.; Croney, C. Exploring Social Desirability Bias in Perceptions of Dog Adoption: All's Well that Ends Well? Or Does the Method of Adoption Matter? *Animals* **2018**, *8*, 154. [CrossRef] [PubMed]
41. Mueller, M.K.; King, E.K.; Callina, K.; Dowling-Guyer, S.; McCobb, E. Demographic and contextual factors as moderators of the relationship between pet ownership and health. *Health Psychol. Behav. Med.* **2021**, *9*, 701–723. [CrossRef] [PubMed]
42. Kubinyi, E.; Turcsán, B.; Miklósi, A. Dog and owner demographic characteristics and dog personality trait associations. *Behav. Process.* **2009**, *81*, 392–401. [CrossRef]
43. Oyama, M.A.; Citron, L.; Shults, J.; Cimino Brown, D.; Serpell, J.A.; Farrar, J.T. Measuring Quality of Life in Owners of Companion Dogs: Development and Validation of a Dog Owner-specific Quality of Life Questionnaire. *Anthrozoös* **2017**, *30*, 61–75. [CrossRef]
44. Rayment, D.J.; Peters, R.A.; Marston, L.C.; Groef, B.D. Investigating canine personality structure using owner questionnaires measuring pet dog behaviour and personality. *Appl. Anim. Behav. Sci.* **2016**, *180*, 100–106. [CrossRef]
45. Doane, M.; Sarenbo, S. A modified combined C-BARQ and QoL for both the companion dog and its owner. An embryo to a companion dog welfare assessment? *Appl. Anim. Behav. Sci.* **2019**, *213*, 91–106. [CrossRef]
46. Patronek, G.J.; Bradley, J. No better than flipping a coin: Reconsidering canine behavior evaluations in animal shelters. *J. Vet. Behav.* **2016**, *15*, 66–77. [CrossRef]
47. Coppola, C.L.; Enns, R.M.; Grandin, T. Noise in the Animal Shelter Environment: Building Design and the Effects of Daily Noise Exposure. *J. Appl. Anim. Welf. Sci.* **2006**, *9*, 1–7. [CrossRef]
48. Kogan, L.R.; Schoenfeld-Tacher, R.; Simon, A.A. Behavioral effects of auditory stimulation on kenneled dogs. *J. Vet. Behav.* **2012**, *7*, 268–275. [CrossRef]
49. Posage, J.M.; Bartlett, P.C.; Thomas, D.K. Determining factors for successful adoption of dogs from an animal shelter. *J. Am. Vet. Med. Assoc.* **1998**, *213*, 478–482. [PubMed]
50. Garrison, L.; Weiss, E. What Do People Want? Factors People Consider When Acquiring Dogs, the Complexity of the Choices They Make, and Implications for Nonhuman Animal Relocation Programs. *J. Appl. Anim. Welf. Sci.* **2015**, *18*, 57–73. [CrossRef] [PubMed]
51. Normando, S.; Stefanini, C.; Meers, L.; Adamelli, S.; Coultis, D.; Bono, G. Some factors influencing adoption of sheltered dogs. *Anthrozoös* **2006**, *19*, 211–224. [CrossRef]
52. Dinwoodie, I.R.; Zotolla, V.; Dodman, N.H. Replication Data for: Factors Influencing Adoption of Pet Dogs and Eventual Owner Satisfaction. 2022. Available online: https://zenodo.org/record/6910691#.Yw3ME3ZBxPY (accessed on 26 July 2022).

animals

MDPI

Article

Creation and Validation of a Tool for Evaluating Caregiver Burnout Syndrome in Owners of Dogs (*Canis lupus familiaris*) Diagnosed with Behavior Disorders

Carmen Luz Barrios [1,*], Vanessa Gornall [1], Carlos Bustos-López [2], Rosa Cirac [3] and Paula Calvo [4]

1 Escuela de Medicina Veterinaria, Facultad de Ciencias, Ingeniería y Tecnología, Universidad Mayor, Camino La Pirámide 5750, Huechuraba 8580745, Región Metropolitana, Chile; vgornall@veterinaria.uchile.cl
2 Departamento de Ciencias Básicas, Facultad de Ciencias, Universidad Santo Tomás, Av. Ejército Libertador 146, Santiago 8320000, Región Metropolitana, Chile; cbustoslopez@santotomas.cl
3 Investigador Independiente, Sant Andreu del Palomar, 08030 Barcelona, Spain; rcirac2010@gmail.com
4 Cátedra Animales y Sociedad, Facultad de Ciencias de la Salud, Universidad Rey Juan Carlos, 28933 Madrid, Spain; paula.calvo@antrozoologia.com
* Correspondence: carmen.barrios@umayor.cl; Tel.: +56-02-328-1000

Simple Summary: Globally, over half of the human population has at least one companion animal, with the domestic dog (*Canis lupus familiaris*) standing out above other species. While this bond has mutual benefits, research shows that it can break due to canine behavioral disorders, leading to consequences including abandonment and/or euthanasia. The wear and tear implied by the physical, psychological and emotional demands when facing the care of a sick animal can lead to a continuous and prolonged level of stress, which in human medicine is referred to as caregiver burnout syndrome. Parallels can be drawn with dog owners handling animals with behavioral disorders, which makes it necessary to have a validated measurement instrument for this problem. The exhaustion of the caregiver of dependent people is evaluated through the Zarit Scale. The present study, through the Delphi method technique, modified and validated this scale to measure this overload in people with dogs with behavioral disorders. Three levels of overload were obtained (Low, Medium-Low and Medium-High Overload). Having an instrument that allows assessing the level of exhaustion of caregivers of dogs with behavior problems will provide information to help these people, and consequently their dogs, avoid the negative consequences of bond degradation.

Abstract: Currently, domestic dogs (*Canis lupus familiaris*) are the most common species among companion animals. The close bond that can grow between owners and their dogs could be worn out and finally broken due to various causes. One main cause is canine behavioral problems, leading to dogs being abandoned or euthanized due to the costs faced by the owner when caring for the animal. Tools have been developed to evaluate the mental and emotional cost of caring for humans, but there is currently no validated tool for evaluating this particular problem. The objective of this study was to develop a questionnaire to evaluate caregiver burnout syndrome for owners of dogs with behavioral disorders. The methodology used consisted of drafting the tool, peer validation using the Delphi methodology and internal validation via Cronbach's alpha. Non-linear snowball sampling was used (n = 156 participants). A questionnaire with 35 questions was obtained which referred to various aspects of caregivers' lives. Regarding the description of the sample used, 50% had Low Burnout, 41% had Medium-Low Burnout and 9% had Medium-High Burnout. Furthermore, regarding the internal validation of the questionnaire, the general Cronbach's alpha coefficient was α = 0.9468. We can thus conclude that the questionnaire is valid for measuring caregiver burnout syndrome in owners of dogs with behavioral disorders.

Keywords: caregiver burnout; behavioral disorders; human–animal bond; instrument validation; psychometric scales

Citation: Barrios, C.L.; Gornall, V.; Bustos-López, C.; Cirac, R.; Calvo, P. Creation and Validation of a Tool for Evaluating Caregiver Burnout Syndrome in Owners of Dogs (*Canis lupus familiaris*) Diagnosed with Behavior Disorders. *Animals* **2022**, *12*, 1185. https://doi.org/10.3390/ani12091185

Academic Editor: Betty McGuire

Received: 10 March 2022
Accepted: 27 April 2022
Published: 5 May 2022

1. Introduction

Worldwide, over half the human population has at least one animal companion in their care [1]. Latin America is a world leader in pet owner percentages, with people mainly preferring the domestic dog (*Canis lupus familiaris*) as a companion [1]. For example, in Chile, 52% of homes have a dog as a companion animal [2]. It is conjectured that the canine–human bond is stronger than that with other animals, due to the years of interaction, which due to various motives (competition, cooperation, co-evolution) have generated a high level of codependence, attachment and ease of interspecies communication, to the point that they have established a niche in our society like no other [3–5]. It should be mentioned that, within the definitions of a bond, there is the kind that is indicated as an affectionate and enduring interaction with a unique individual, non-interchangeable with another and established according to four principles: security, intimacy, affinity and constancy [5]. In turn, a companion animal, according to the concept defined in the study by [4], is one that is "under human control, linked to a home, sharing intimacy and proximity with its caregivers and receiving affection, care and attention, guaranteeing its health", i.e., one that becomes part of a family [4]. Studies indicated that animal caregivers perform emotional and financial efforts to enrich the human–animal bond, driven by the various benefits it provides: physiologically, psychologically, socially and therapeutically [4,6,7].

In spite of these points, research has shown that the human–dog link can be broken due to behavioral disorders in the latter party, which can lead to negative consequences for dogs, such as abandonment or even euthanasia [6,8–13]. A behavior disorder is a behavior pattern that is dangerous or annoying for humans, creating a communication dysfunction between both species and compromising their mutual well-being [13]. Authors including [11], among others, describe how some of the reasons for dogs being abandoned in shelters include behaviors considered dangerous or annoying including aggression and biting, although they can also include separation anxiety as a frequent cause [9,11]. Studies in the USA and UK show that over 80% of companion animal owners have identified some type of behavioral alteration in them, with aggressiveness standing out [8]. This suggests the importance of investigating the topic to avoid animal abandonment and euthanasia. Behavioral disorders, as well as diseases in animals, also carry physical, psychological and emotional demands on their caregivers [6,7,14,15], due to the burnout involved with informal care. Informal care is when people provide altruistic free care and attention to dependent individuals, motivated by ties of affection [16,17]. The consequence is caregiver burnout syndrome; continuous, prolonged stress caused by care for a dependent individual, causing physical, psychological and emotional exhaustion along with a rupture in the bond between the two parties [6,16–19]. It is important to consider that the concept mentioned above differs from compassion fatigue, which is understood as the ability of caregivers of other living beings such as non-human animals to notice the pain of the individual they care for [20]. The latter is very common in animal shelter staff, as a result of the link between the keepers and the animals that are housed in this type of facility. These animals frequently arrive in poor physical condition, have been subjected to previous mistreatment or finally, as a result of various reasons, they must be euthanized, which emotionally ends the people who live with them daily [21].

It should be mentioned that a review of studies on caregiver burnout syndrome in human–animal relations has produced studies by [6,14,15], which principally refer to cases of animals with chronic illnesses rather than any particular behavioral disorders, thus leaving room for interest in developing this topic [6,14,15]. Thus, considering that a dog with behavioral disorders and a person with caregiver burnout syndrome can lead to broken bonds and severe negative consequences, mainly for the animal, this study aimed to develop and validate a specific tool for the topic in question.

In a complementary way, the experience of guardians of companion animals with behavioral problems has been studied qualitatively (Buller and Ballantyne, 2020) [22]. Here, it is evident how difficult it is to be able to care for an animal with these characteristics, which can affect both the psychological and physical wear and tear of the caregiver.

Based on the information gathered, one cause that can be inferred for breaches in human–dog bonds is prolonged care for dogs with behavior disorders by overburdened owners, since people generate bonds with dogs similar to how they would with other people. This makes it possible to consider the existence of caregiver burnout syndrome for cases of human–dog bonds involving behavioral disorders. It should be mentioned that this syndrome in human bonds is highly prevalent and severe [16]. A validated measurement instrument makes it possible to confirm the existence of this syndrome in owners of dogs that have behavioral disorders. For this, the specific objectives were to develop the evaluation tool and then to validate it.

2. Materials and Methods

2.1. Participants

An in-person self-applied survey was done online (Google Forms®) using a Likert scale. The inclusion criteria were people over 18 years old who owned dogs diagnosed with some type of behavioral disorder in the previous six months. Exclusion criteria were people over 18 who had psychiatric and/or physical pathologies and/or were in the care of dependent people, whether this was for disease, old age and/or special care.

Non-linear snowball sampling was used to recruit the sample for this study (n = 156). This type of sampling consists of recruiting study subjects based on contacts of the first participants, who manage the incorporation of other people to contribute to the study. This process, which can be repeated over and over, gives the possibility for the researchers to find people that they would not otherwise have had access [23].

The participants were contacted by ethologists, who had previously cared for their animals and also recruited by acquaintances of said guardians.

This study was approved by the Scientific Ethics Committee of the Universidad Mayor de Chile on 20 March 2019 with folio 0098.

2.2. Methodology

Three work phases were considered: Question design, Question validation via the Delphi Method and Internal validation of the questionnaire via the Cronbach's Alpha Coefficient. These are described herein:

2.2.1. Question Modification (Step I)

The original version Zarit instrument was modified [24]. This evaluated caregiver syndrome among humans and was mainly altered regarding the species of the subject being cared for (changed from humans to dogs with behavioral disorders) and with the contextualization of each question. New questions were also made based on available scientific literature for animals [14], along with expertise from each researcher. This questionnaire classified the questions into 7 pillars: Perceived overload, abandonment of self-care (health and image), discomfort with the presence or behavior of the dog, irritability, fear for the dog's health or future, loss of family and socioeconomic role, and guilt for not doing enough (Table 1) and consisted of 35 questions, each with five response alternatives (Likert style) (Table 2).

2.2.2. Question Validation via the Delphi Method (Step II)

Following modification and the creation of new questions, the instrument was validated with a panel of experts (seven specialized professionals; three in clinical psychology and four in clinical ethology), whose observations were gathered and analyzed via the Delphi method [25]. This process began with the analysis of the Curriculum Vitae of each professional invited to participate in the study as an expert. For this, the years of experience performing in their area of expertise, current employment and professional recognition by their peers were considered. They each received an invitation letter by email detailing the different aspects of their anonymous and confidential participation, along with a brief introduction to the topic and the instructions to provide the corresponding observations. Expert feedback was done via email on two occasions, which were done with three weeks' separation between them. The researcher group finally made a joint decision on the final structure for the tool via a qualitative concordance analysis. In other words, answer frequencies were analyzed to categorize and eventually correct the questionnaire items. Experts' replies were categorized as answers with observations regarding Form (F), according to question format, Content (C) referring to modifications which had to do with the outlook of each specialist consulted, AQ (Add Question) if they wanted to add any questions relevant to the study and E (Eliminate) if questions were considered not germane to the survey.

Table 1. Question classification by Tool evaluation pillar.

Name of Question Distribution Pillar	Distribution Pillar Detail	# of Questions Per Pillar
Abandonment of self-care	This pillar evaluates attrition in dog owners' care for their image and health, relating the time that they have to dedicate to their dog.	1, 8, 12, 20, 23, 25, 28, 29, 31, 34
Perception of wear and tear	This pillar evaluates how the owner perceives stress, tiredness, angst and the resulting desire to leave the care of their dog to another person and/or stop caring for the animal.	2, 7, 14, 18, 21, 30, 32, 35
Discomfort over dog presence or behavior	This pillar evaluates when the owner feels permanently on edge and avoids exposure to any third party along with their dog.	3, 10, 24
Irritability	This evaluates how worn out the owner feels based on how much they feel bothered in contexts where they care for their animal.	4
Loss of family and social role	This evaluates owners' wear and tear based on they perceive negative changes in their daily routines due to caring for their dog, especially decay in their social bonds, tendency to feel alone, trapped, isolated and unsupported.	5, 9, 13, 19, 27
Fear for the health or future of the dog	These questions evaluate owners' wear and tear regarding how they perceive their own worry about the well-being of their dog.	6, 15
Economic	Evaluation of owners' attrition based on their perception of expenses incurred by caring for the dog.	11
Guilt over not doing enough	These questions focus on evaluating caregivers' wear and tear based on how they perceive nonconformity and non-fulfillment of their expectations as a caregiver and the lack of professional support to advise them.	16, 17, 22, 26, 33

Table 2. Measurement instrument for caregiver burnout syndrome among owners of dogs with behavioral disorders.

		Never	Almost Never	Sometime	Almost Always	Always
#	**Question**	**1**	**2**	**3**	**4**	**5**
1	Do you feel that because of the time you spend on your dog you no longer have enough time for yourself?					
2	Do you feel exhausted when you have to look after your dog and also handle other responsibilities?					
3	Do you feel uncomfortable with how your dog behaves?					
4	Do you feel angry when you think about your dog and everything involved in its care?					
5	Do you think that having to care for your dog negatively impacts your relationship with friends and other members of your family, even your other pets?					
6	Are you afraid for the future awaiting your dog?					
7	Do you feel exhausted when you have to be with your dog?					
8	Do you feel like your health has deteriorated due to caring for your dog?					
9	Do you believe your social life has been affected by having to care for your dog?					
10	Do you feel uncomfortable inviting friends to your house because of your dog?					
11	Do you feel like you lack enough money to care for your dog apart from your other expenses?					
12	Do you feel demotivated since your dog started to show undesired behavior?					
13	Do you feel a loss of control over your life since your dog started to show undesired behavior?					
14	Would you like to have other people take care of your dog?					
15	Do you feel insecure about how to handle your dog's behavior?					
16	Do you feel like you should be doing more than you currently do for your dog?					
17	Do you believe you could care for your dog better than you currently do?					
18	Do you generally feel very overburdened due to caring for your dog?					
19	Do you feel unsupported or lonely because you have to care for your dog?					
20	Do you feel that due to the time you spend on your dog you no longer have enough time to care for your physical appearance?					
21	Do you often consider giving your dog up for adoption?					
22	Do you feel like you've lacked professional support from veterinarians or other similar professionals to face this situation?					
23	Since your dog began showing unwanted behavioral problems, have you had trouble sleeping?					
24	Do you often feel like you have to remain alert and vigilant to avoid any incidents caused by your dog's behavior?					
25	Do you feel tired from caring for your dog?					
26	Do you feel responsible for your dog's behavior problems?					
27	Do you believe you modify your lifestyle by caring for your dog?					
28	Do you believe your quality of life has declined due to caring for your dog?					
29	Do you feel stressed or nervous when facing your dog's care?					
30	Do you often avoid interacting with your dog?					
31	Do you feel anxiety when you think about having to go home and care for your dog?					
32	Do you feel like you won't be able to take care of your dog for much longer?					
33	Do you feel incompetent to care for your dog?					
34	Do you feel distraction or lack of concentration in other activities since your dog began to have behavior problems?					
35	Do you have recurring desires to get rid of your dog?					
	Total Score					
	Level Obtained					

2.2.3. Tool Application (Step III)

Dog owners were selected who fulfilled the inclusion criteria (n = 156) detailed in the Participants item. Each one of them received the questionnaire via Google Forms®, to be answered in a self-explanatory and anonymous fashion within 45 min, following acceptance of informed consent. Subsequently, all data was gathered in Excel for statistical analysis.

2.3. Statistical Analysis

Descriptive statistics were applied through the construction of frequency tables and determination of percentages for the responses of each variable. The internal statistical stability of the tool was evaluated using Cronbach's Alpha coefficient. Minitab® software was used both to perform descriptive statistics and for Cronbach's Alpha analysis. A statistical significance level of 5% was considered.

3. Results

3.1. Question Validation with the Delphi Method (Step I and II)

Regarding validation via the Delphi Method, in the first evaluation, out of a total of 35 questions, 28.57% (10 questions) were observed. In turn, 60% of the questions observed were from the point of view of Content and 30% were from Form (Table 3). After the second expert evaluation, full approval for the survey were obtained from four professionals (57%). The other 43% made new observations regarding six questions from a total of 35 (17.14%), principally about Content (50%) (Table 4).

Table 3. Observation frequencies by category (%). First round of observations.

	# of Questions (Item 2)									
Observation Type	**10**	**12**	**14**	**15**	**21**	**22**	**25**	**32**	**33**	**35**
Form	14.28%	28.57%		14.28%						
Content	14.28%		14.28%				14.28%	14.28%	14.28%	14.28%
Add question						14.28%				
Eliminate question					14.28%					

Total of experts 7 (100%).

Table 4. Observation frequency by category (%). Second round of observations.

	# of Questions (Item 2)					
Observation Type	**4**	**19**	**21**	**22**	**25**	**35**
Form	14.28%				14.28%	
Content			14.28%	14.28%		14.28%
Add question		14.28%				
Eliminate question						

Total of experts 7 (100%).

As previously said, the criteria to keep, correct or cut questions were managed via analysis and discussion among the research team. In the first analysis round, the team decided to take the form and content observations, without adding or eliminating any questions. In the second analysis round, only the form observations were taken into account. Thus, the final questionnaire had a total of 35 questions (Table 2).

3.2. Internal Validation of the Questionnaire via the Cronbach's Alpha Coefficient (Step III)

Regarding the internal validity of the evaluation tool, the general Cronbach's alpha coefficient were $\alpha = 0.9468$ while the Cronbach's alpha coefficient per item also presented results above $\alpha = 0.9$ (Table 5).

Table 5. Cronbach's alpha values obtained by questions.

Questions (#)	Cronbach's Alpha
1.	0.9457
2.	0.9444
3.	0.9450
4.	0.9441
5.	0.9447
6.	0.9450
7.	0.9448
8.	0.9450
9.	0.9448
10.	0.9456
11.	0.9463
12.	0.9442
13.	0.9443
14.	0.9461
15.	0.9455
16.	0.9468
17.	0.9470
18.	0.9432
19.	0.9453
20.	0.9469
21.	0.9461
22.	0.9471
23.	0.9455
24.	0.9462
25.	0.9434
26.	0.9480
27.	0.9455
28.	0.9443
29.	0.9437
30.	0.9466
31.	0.9447
32.	0.9456
33.	0.9445
34.	0.9453
35.	0.9456

In total, 156 people were consulted in the study and classified in different levels of burnout (Low Burnout 35–69 points; Medium Low burnout 70–104 points; Medium High Burnout 105–139 points and High Burnout 140–175 points). Overall, 50% presented Low Burnout, 41% showed Medium Low Burnout and 9% had Medium High Burnout. The median was a burnout score of 73.16, which is within the range of Medium Low burnout.

4. Discussion

There is still no consensus regarding the construction, adaptation and validation of psychometric studies in humans [26], so these results have to be taken as part of a preliminary study, waiting for its reproducibility in latter applications to allow it to be perfected.

On the other hand, developing measurement tools for problems related with human and animal health must have the opinion of area experts—the more the better. This is why using the Delphi methodology and its collective intelligence principle is highly useful for generating consensus about these opinions regarding question formulation for a psychometric measurement tool as in our case, where both clinical ethologists and psychologists take part. However, while Delphi is a frequently used and recommended method, it is still more intuitive than rational by nature, which carries undeniable biases, principally due to varying interpretations of each question apart from those referring to their formulation and method of application [25]. It should also be noted that the Delphi method has another characteristic complicating its application, consisting in the time that experts must have to be able to repeatedly analyze the questions from the instrument being developed which can be generally scarce among these professionals.

The Cronbach's alpha coefficient value of the present study ($\alpha = 0.9468$) is considered satisfactory for both the items level and the instrument reliability level, according to the literature [6,14,24]. The usual preference is for values between 0.8 and 0.9. However, according to Luján and Cardona (2015), values above $\alpha = 0.9$ indicate that there could be duplication in the questions [26]. Thus, the variation of this coefficient should be analyzed with more information according to question modification, or else redundant items should be eliminated. While questions considered redundant were eliminated, eventually this pattern may still be present due to how individual perception of wear and tear in the face of this problem carries a common emotional burden for most of the questions, however diverse they may be. This, along with other individual factors, may influence the answers received.

Finally, it is worth highlighting the importance of having tools to evaluate these items, which are validated like the ones developed in the present study. These can be useful for identifying factors which can, for example, degrade the human–animal bond in order to prevent the consequences arising from it, or else to manage treatment from both a human and an animal perspective.

The main limitation of the present study refers to the bias of incorporating participants with a low burnout level, who could be more inclined and interested in participating in a research process of this type and justify the high percentage of participants with this result. This could be due to the convenience sampling used in this study. However, it should be mentioned that this bias had no direct influence on validating the previously developed instrument, but rather would affect the burnout levels recorded by participants.

Additional and no less important limitations correspond to those difficulties already mentioned above, such as the variety of factors that affect preliminary studies in mental-emotional health, individual burnout perception, intuitive methodologies, and specific conditions demonstrated. Finally, do not forget those limitations due to the resources associated with the feasibility of carrying out this research.

5. Conclusions

According to the results obtained, the questionnaire developed in the present study is a valid measurement instrument to evaluate caregiver burnout syndrome among owners of dogs with behavior disorders. This is fundamental, since before now, there were no instruments with these characteristics to address a problem as notable as burnout among owners in these cases, which can directly impact the wellbeing of both the people caring for the animal and the dogs themselves.

Author Contributions: Conceptualization: C.L.B., V.G., P.C., C.B.-L. and R.C.; methodology: C.L.B., V.G., P.C., C.B.-L. and R.C.; validation: C.L.B., V.G., P.C., C.B.-L. and R.C.; formal analysis: C.L.B., V.G., P.C., C.B.-L. and R.C.; investigation: C.L.B., V.G., P.C., C.B.-L. and R.C.; data curation: C.L.B., V.G. and C.B.-L.; initial drafting-preparation: C.L.B., V.G., P.C., C.B.-L. and R.C.; drafting—review and editing: C.L.B., V.G., P.C., C.B.-L. and R.C. All authors have read and agreed to the published version of the manuscript.

Funding: This research received no external funding.

Institutional Review Board Statement: This study was approved with folio #0098 by the Ethics Committee at Universidad Mayor, Temuco Campus on 20 March 2019, subject to the extant rules in Chile regarding personal data protection (Laws#19628, #20120, #20584).

Informed Consent Statement: Informed consent was obtained from all subjects involved in the study.

Acknowledgments: We would like to give thanks to the experts involved in the validation process for the tool developed in the present study: Soledad Torres, Constanza Stuart, Erick Lucero, Gaspar Romo, Karen Baeza, Paz González y Eduardo Tapia.

Conflicts of Interest: The authors declare that they have no conflict of interest.

References

1. PrNewswire. La Mayoría de Quienes Tienen Mascota radican en América Latina, Rusia y los EE.UU. 2016. Available online: https://www.prnewswire.com/news-releases/la-mayoria-de-quienes-tienen-mascota-radican-en-america-latina-rusia-y-los-eeuu-300273313.html (accessed on 24 November 2020).
2. Growth from Knowledge. Microestudio GFK: Los Chilenos y Sus Mascotas. 2018. Available online: https://www.gfk.com/fileadmin/user_upload/country_one_pager/CL/GfK_Los_chilenos_y_sus_mascotas.pdf (accessed on 30 November 2020).
3. Díaz Videla, M.; López, P. La oxitocina en el perro: Revisión bibliográfica y análisis de futuras áreas de investigación. *Interdisciplinaria* **2017**, *34*, 73–90. Available online: http://www.redalyc.org/articulo.oa?id=18052925005 (accessed on 30 November 2019).
4. Díaz, M.; Olarte, M. Animales de compañía, personalidad humana y los beneficios percibidos por los custodios. *PSIENCIA Rev. Latinoam. Cienc. Psicol.* **2016**, *8*, 1–19.
5. Gutiérrez, G.; Granados, D.; Piar, N. Interacciones humano-animal: Características e implicaciones para el bienestar de los humanos. *Rev. Colomb. Psicol.* **2007**, *16*, 163–183.
6. Britton, K.; Galioto, R.; Tremont, G.; Chapman, K.; Hogue, O.; Carlson, M.; Spitznagel, M. Caregiving for a Companion Animal Compared to a Family Member: Burden and Positive Experiences in Caregivers. *Front. Vet. Sci.* **2018**, *5*, 10. [CrossRef] [PubMed]
7. Meléndez-Samó, L. El vínculo humano-animal y sus implicaciones para la psicología en Puerto Rico. *Rev. Puertorriqueña Psicol* **2014**, *25*, 160–182.
8. Carreño, C. Diagnóstico Situacional de Problemas Comportamentales que Presentan los Perros que Asisten a la Clínica Veterinaria "Snoopy" Ubicada en la Concordia, Mediante Encuestas. Tesis para Optar al Título de Médico Veterinario Zootecnista. Universidad de las Américas, Quito. 2018. Available online: http://dspace.udla.edu.ec/handle/33000/8707 (accessed on 30 December 2020).
9. Chávez, G.; Bustos, C.; Ubilla, M. Frecuencia de Trastornos Conductuales en Perros. *Hosp. Vet.* **2012**, *4*, 119–124. Available online: https://issuu.com/revistahospitalesveterinarios/docs/revista_vol4_n_4_diciembre (accessed on 25 January 2019).
10. Fundación Affinity. Infografía Estudio de Abandono y Adopción. 2017. Available online: http://www.fundacion-affinity.org/observatorio/infografia-estudio-de-abandono-y-adopcion-2017 (accessed on 17 February 2021).
11. Salman, M.; Hutchison, J.; Ruch-Gallie, R.; Kogan, L.; New, J.; Kass, P.; Scarlett, J. Behavioral Reasons for Relinquishment of Dogs and Cats to 12 Shelters. *J. Appl. Anim. Welf. Sci.* **2010**, *3*, 93–106. [CrossRef]
12. Elgier, Á.M.; Jakovcevic, A.; Mustaca, A.E.; Bentosela, M.; Barrera, G. Problemas de comportamiento en los perros domésticos (*Canis familiaris*): Aportes de la psicología del aprendizaje. *Rev. Psicol.* **2009**, *18*, 123–146. [CrossRef]
13. Mugford, R.A. Behavioural disorders of dogs. In *The Behavioural Biology of Dogs*; Jensen, P., Ed.; Cromwell Press: Trowbridge, UK, 2007; pp. 225–242.
14. Spitznagel, M.; Jacobson, D.; Cox, M.; Carlson, M. Caregiver burden in owners of a sick companion animal: A cross-sectional observational study. *Vet. Rec.* **2017**, *181*, 321. [CrossRef] [PubMed]
15. Christiansen, S.; Kristensen, A.; Sandoe, P.; Lassen, J. Looking After Chronically III Dogs: Impacts on the Caregiver's Life. *Anthrozoös* **2013**, *26*, 519–533. [CrossRef]
16. Peña, F.; Álvarez, M.; Melero, J. Sobrecarga del cuidador informal de pacientes inmovilizados en una zona de salud urbana. *Enferm. Glob.* **2016**, *43*, 100–111. [CrossRef]
17. Díaz, D.; Arango, J.; Espinosa, I.; Rivera, A. Necesidades y sobrecarga en cuidadores primarios informales de pacientes con traumatismo craneoencefálico. *Psicol. Caribe* **2014**, *31*, 393–415. Available online: http://www.redalyc.org/articulo.oa?id=21332837002 (accessed on 19 February 2021).
18. Navarro, Y.; López, M.; Climent, J.; Gómez, J. Sobrecarga, empatía y resiliencia en cuidadores de personas dependientes. *Gac. Sanit* **2017**, *33*, 268–271. [CrossRef] [PubMed]
19. Herrera, P.; Fernández, B.; Araya, A.; Caro, S.; Centro Estudios de Vejez y Envejecimiento UC, Santiago. Estudio Sistematización y Descripción de los Perfiles de las Cuidadoras de Personas Dependientes, las Demandas de Apoyo que las Cuidadoras Presenten y los Programas Existentes para Aliviar el Trabajo de Cuidado. UC 2015. Licitación N° 711841-LE15. Santiago de Chile. Available online: http://www.ministeriodesarrollosocial.gob.cl/btca/txtcompleto/Final_Perfil_de_Cuidadoras.pdf (accessed on 10 August 2021).
20. Monaghan, H.; Rohlf, V.; Scotney, R.; Bennett, P. Compassion fatigue in people who care for animals: An investigation of risk and protective factors. *Traumatology* **2020**. [CrossRef]
21. Rohlf, V.I. Interventions for occupational stress and compassion fatigue in animal care professionals—A systematic review. *Traumatology* **2018**, *24*, 186–192. [CrossRef]
22. Buller, K.; Ballantyne, K.C. Living with and loving a pet with behavioral problems: Pet owners' experiences. *J. Vet. Behav.* **2020**, *37*, 41–47. [CrossRef]
23. Espinosa, P.; Hernández, H.; López, R.; Lozano, S. Muestreo de Bola de Nieve. México: UNAM. Departamento de Probabilidad y Estadística. 2018. Available online: https://es.scribd.com/document/379661920/Proyectofinal-Bola-de-Nieve (accessed on 18 June 2021).
24. Breinbauer, H.; Vásquez, H.; Mayanz, S.; Guerra, C.; Millán, T. Validación en chile de la escala de sobrecarga del cuidador de Zarit en sus versiones original y abreviada. *Rev. Méd. Chile* **2009**, *137*, 657–665. [CrossRef]

25. Riaño, C.; Palomino, M. Diseño y elaboración de un cuestionario acorde con el método Delphi para seleccionar laboratorios virtuales (LV). *Sophi* **2015**, *11*, 129–141.
26. Luján, J.; Cardona, J. Construcción y validación de escalas de medición en salud: Revisión de propiedades psicométricas. *iMedPubJournalsA* **2015**, *11*, 1–10.

animals

MDPI

Article

Impact of Dog's Age and Breed on Dog Owner's Physical Activity: A German Longitudinal Study

Benedikt Hielscher-Zdzieblik [1,2,*], Ingo Froboese [1], James Serpell [3] and Udo Ganslößer [2]

1 Institute of Movement Therapy and Movement-Oriented Prevention and Rehabilitation, German Sport University Cologne, 50933 Cologne, Germany; i.froboese@dshs-koeln.de
2 Institute of Zoology and Evolutionary Research with Phyletic Museum, Friedrich Schiller University Jena, 07743 Jena, Germany; udo@ganslosser.de
3 Department of Clinical Studies VHUP, School of Veterinary Medicine, University of Pennsylvania, Philadelphia, PA 19104, USA; serpell@vet.upenn.edu
* Correspondence: benedikt.hielscher@gmx.de

Simple Summary: Dog ownership has been linked to physical activity of the owners in several countries. Physical activity is also affected by age, size and energy level, as perceived by the owners, of the dogs. Earlier studies were mostly cross-sectional, which does not allow causal conclusions. This study aimed to find differences and changes in the physical activity behavior of owners of ten different dog breeds that were selected based on their size and energy level. Nine dog breed groups were used and owners filled out an online physical activity questionnaire once per year for three years. The results show that dog owners' total and dog-related physical activity as well as their leisure time and dog walking decreased over time. Owners of the dog breed groups differed in all physical activity variables. If only participants who completed the study were analyzed, no changes in any physical activity variable were found. At baseline, owners of different dog breeds differed in the types of reported dog-related activities. Overall, the results indicate that physical activity behavior in dog owners is stable over time. However, no clear pattern could be found based on the age, size and energy level of the dogs.

Citation: Hielscher-Zdzieblik, B.; Froboese, I.; Serpell, J.; Ganslößer, U. Impact of Dog's Age and Breed on Dog Owner's Physical Activity: A German Longitudinal Study. *Animals* **2022**, *12*, 1314. https://doi.org/10.3390/ani12101314

Academic Editor: Betty McGuire

Received: 16 March 2022
Accepted: 19 May 2022
Published: 20 May 2022

Abstract: Dog ownership contributes positively to physical activity (PA). The impact of different dog breeds and age on PA is less investigated in longitudinal studies. This study aimed to evaluate PA changes in dog owners as their dogs' ages increased and to explore whether there are differences in PA between owners of different breeds over a three-year period. Owners of different dog breeds were categorized into nine groups according to the perceived energy level and size of the breed. PA was monitored using an online questionnaire for three consecutive years. Linear mixed models (LMM) showed a small, but significant decrease in total PA, leisure time walking, dog-related PA and dog walking over three years. No decreases were found if only participants who attended at all time points were included. In all LMM analyses, a significant relationship between the dog breed and the outcomes of PA were shown. At baseline, dog owners performed different types of activities depending on their dog breed. In conclusion, owners of different dog breeds differ in their types of PA. The study emphasizes that age, size and energy level of the dog does not per se have an impact on dog owners PA.

Keywords: dog-related physical activity; dog walking; longitudinal; agility; obedience; activity types

1. Introduction

Recent studies have shown that dog ownership is associated with increased physical activity (PA) in Australia [1], Canada [2], the Czech Republic [3,4], Finland [5], Germany [6], Japan [7], the United Kingdom [8–14], South Korea [15] and the USA [16]. Dog owners have also a higher number of steps per day on average than non-dog owners [4,8]. Much of the PA of dog owners consists of dog walking [6,17,18]. Most previous studies of the relationship between dog ownership and PA were cross-sectional, and only a few longitudinal studies give information about the causal relationship between dog ownership and owners' PA. An early study from the UK found significant increases in PA after acquiring a dog [14]. An Australian study showed that dog acquisition leads to an increase in dog walking but not total PA [19]. Two other investigations detected an initial increase in daily steps after dog acquisition that was diminished at a second follow-up period [20,21]. Therefore, it could be concluded that dog acquisition might increase PA in prospective dog owners.

Physical inactivity is associated with poor health and an increased mortality risk [22–26]. It has been documented that dog ownership also correlates inversely with the risk of several diseases and mortality [27] and that dog owners have a better cardiovascular condition than people who do not own any pets [3]. However, the authors of a recent meta-analysis identified only a non-significant reduction in the mortality risk for dog owners and, therefore, advocate treating these previous results with caution [28].

It has been shown that the size of a dog [29–35] and their owner-perceived energy level [31,36] are positively associated with dog walking. Moreover, it has been shown that the average energy levels of dogs, as perceived by their owners, vary between breeds and breed groups [37,38].

The age and the health status of the dog have also been shown to be associated with dog walking [35,36,39–42]. The probability of being walked is smaller for older dogs and dogs with a poorer health status as compared to younger and healthier dogs [35,36]. Furthermore, dog walking behavior changes as dogs develop health problems [41]. However, since the aging process of dogs of different sizes varies significantly [43], dogs of different sizes might influence dog owners PA differently with increasing age.

The aim of this study was, therefore, to investigate differences at baseline and changes over time in the PA of dog owners of different dog breeds. It was hypothesized that owners of smaller dogs with a lower energy level would be less physically active overall. It was expected that the PA of dog owners would decline over the years as the dogs aged. Furthermore, if a dog died it was expected that the PA behavior of the owner would decline, if there was no dog left in the household.

2. Materials and Methods

2.1. Participants

Study participants were required to be at least 18 years old and to have at least one dog of ten specified dog breeds at the time of recruitment. Participants that owned more than one of the selected breeds were categorized as a separate group. In order to achieve a sufficiently large sample, an average number of 500 puppies registered in the German subsection of the FCI (VDH) per year in the period from 2010 to 2014 [44] was used as a criterion. The dog breeds were then selected based on two criteria:

First, the dog breeds were divided into categories based on the height at the withers as specified in the Féderation Cynologique International (FCI) standard (small $\leq$ 40 cm; medium = 40–59 cm; large $\geq$ 60 cm). If the breed standard recorded a size that exceeded the defined size limits, the breed was placed in the larger category. Then, the energy level was evaluated, as measured by the Canine Behavioral Assessment and Research Questionnaire (C-BARQ). We used data from the C-BARQ project at the University of Pennsylvania (https://vetapps.vet.upenn.edu/cbarq/, assessed on 15 March 2022) to estimate the energy levels of the different breeds. Participants in the current study did not complete the C-BARQ questionnaire.

The C-BARQ is an online survey that allows owners to evaluate the behavior of individual dogs [45]. Energy level is one of 14 behavioral dimensions evaluated by the C-BARQ and it consists of two questionnaire items: "playful, puppyish, boisterous" and "active, energetic, always on the go" (Serpell and Duffy, 2014, p. 48) [37]. Both items are scored on five-point scales from 0 to 4, with a higher score indicating that the behavior is exhibited more frequently [37,46]. The score for energy level represents the average of the scores for these two items.

The selected breeds are:

1. Cavalier King Charles Spaniel (CKCS) [47] (small size, low energy)
2. West Highland White Terrier (WHWT) [48] (small size, medium energy)
3. Jack Russell Terrier (JRT) [49] (small size, high energy)
4. Parson Russell Terrier (PRT) [50] (small size, high energy)
5. Whippet (WHIP) [51] (medium size, low energy)
6. Labrador Retriever (LAB) [52] (medium size, medium energy)
7. Border Collie (BC) [53] (medium size, high energy)
8. Bernese Mountain Dog (BMD) [54] (large size, low energy)
9. Rottweiler (ROTT) [55] (large size, medium energy)
10. Belgian Shepherd Dog (BSD) [56] (large size, high energy)

To take part in the study, participants needed to be able to understand, read and write in German. They could own a maximum of 5 dogs. Participants were excluded if they reported not owning a purebred dog of the selected breeds or if they reported that they did not complete the questionnaire accurately. Furthermore, they were excluded if they had help from another person, because this might bias the results, e.g., because of social desirability [57].

Participants were recruited using groups that focused on the selected dog breeds on social media. A description of the study was posted alongside a link that led to the online questionnaire. Permission was obtained from the group administrators before the link was posted. For each dog breed two to four groups agreed to the posting. The number of group members per group varied between 203 and 7778. Further, several sub-organizations of the German Kennel Club (VDH) that care for the welfare of the dog breeds were contacted. Some of these associations published an appeal in their club newspapers or contacted their members directly.

2.2. Measurements

A 15-min online questionnaire was used. By answering question 1 participants gave informed consent actively (see Supplementary Materials). Participants self-reported sociodemographic and anthropometric data. Body mass index (BMI) was calculated from self-reported height and weight as kg/m^2. Information on the age, sex, sizes (measured standing at the withers in cm), weight in kg, neuter status and breed of each dog was also provided. Participants completed the questionnaire only once per time point.

Participants also completed the Physical Activity, Exercise and Sport Questionnaire (Bewegungs- and Sportaktivitätsfragebogen [BSA-F], Version 1.0) by Fuchs et al. [58]. An English translation of the BSA-F is available for download at the University of Freiburg (https://www.sport.uni-freiburg.de/de/institut/psychologie/messinstrumente/Messung_der_Sport_und_Bewegungsaktivitaet, assessed on 15 March 2022) [59]. It measures PA in minutes per week over the previous four weeks. The BSA-F was validated by Fuchs et al. and correlates with physical fitness [58].

In addition to the BSA-F, questions were included that specifically asked about PA performed together with the dog (dog-related PA). The questions were based on the BSA-F. Participants were asked about the frequency and duration they walked their dog or rode a bicycle with their dog. Finally, they could report five other dog-related PAs in a semi-open question design. All PA related outcomes were calculated as hours per week (h/week). This approach has been used in earlier studies [6,17].

2.3. Procedure

Participants were recruited from 1 August 2017 until 31 July 2018. At baseline (T0) they completed the questionnaire. Participants were asked to create an individual code from their initials and their date of birth in order to enable the data to be assigned to the different points in time.

The questionnaires were made available on the data survey tool https://www.soscisurvey.de (assessed on 15 March 2022). Soscisurvey is a German company that complies with the German and European data protection guidelines [60].

At the first (T1), second (T3) and third (T4) year of follow-up participants received three e-mails within 20 days that reminded them to participate in the study. In addition to the BSA-F, they were asked to report any changes in dog ownership status. In particular, they were asked whether any of the dogs had died. Not completing the questionnaire at follow-up dates was interpreted as withdrawal from the study. The data collection ended two weeks after the last participant received the last reminder at T3.

2.4. Statistical Analysis

Unless otherwise specified, descriptive values are reported as mean (M) $\pm$ standard deviation (SD). Outliers were identified using the mean values $\pm$ 3 SDs. Outliers outside this range were winsorized and changed to the calculated maximum or minimum value.

Baseline values of all parameters were compared between the study groups to show accordance for demographic and anamnestic parameters. For all tests of the descriptive analysis: In case of normally distributed continuous data (examined using a Shapiro Wilk test) t-tests were used for group comparison. Non-normally distributed continuous data, and ordinal data were tested via Kruskal Wallis tests. Categorical data were tested by χ^2-tests.

Changes over time were analyzed using a linear mixed model (LMM). A maximum likelihood approach was used. Linear, quadratic and cubic time trends were tested as described by Shek and Ma [61]. Breed groups were used as predictors with owners of CKCS being the reference group, since owners of the smallest breed with the lowest energy levels were hypothesized to be least active. A random intercept was used for subjects. All other variables were defined as fixed. The time points were nested within individuals. The model was built using a step-by-step approach, adding one predictor at a time. First, the time trends were added one by one. If an added variable (e.g., quadratic trend) did not improve the model, the next stage was discarded (e.g., cubic trend). The best models were identified using the -2 log-likelihoods of the separate models and χ^2-tests as recommended by Field and Tabachnik and Fidell [62,63].

Additionally, a completer analysis was performed. For this purpose, only participants who had completed the whole study were examined using a repeated measures ANOVA. Due to the small sample sizes only within group analysis were performed.

If participants stated at one point in time that they owned several of the selected breeds and at another point in time only one of the selected breeds, only the latter was retained in the LMM. Participants that owned more than one of the specified breeds at one point in time were excluded from the ANOVA because only the effects of the individual dog breeds should be examined.

The level of statistical significance was set at $\alpha = 0.05$ in all tests. All analyses were performed using IBM SPSS Statistics Version 27.

3. Results

3.1. Classification of Dog Breeds

JRT and PRT were merged together to form a single group in order to ensure a sufficiently large sample size. They do not differ in their energy level as evaluated by the C-BARQ ($t = 1.18$, $df = 413$, $p = 0.239$).

Overall, significant differences in energy level were found within the groups of small (CKCS, WHWT, JRT/PRT) (F (2, 665) = 19.03, $p < 0.001$), medium (WHIP, LAB, BC) (F (2, 2824) = 29.11, $p < 0.001$) and large (BERN, ROTT, BSD) (F (2, 934) = 54.45, $p < 0.001$) dog breeds. Linear trends were analyzed and shown to be significant (F_{small} (1, 665) = 34.39, $p_{small} < 0.001$; F_{medium} (1, 2824) = 51.12, $p_{medium} < 0.001$; F_{large} (1, 934) = 92.45, $p_{large} < 0.001$). Compared to WHWT (2.01 $\pm$ 0.97), CKCS (1.78 $\pm$ 1.01) had a lower and JRT/PRT (2.37 $\pm$ 1.07) had a higher energy level. In contrast to LAB (2.02 $\pm$ 1.08), WHIP (1.63 $\pm$ 0.98) exhibited a lower energy level, while BC exhibited higher energy levels (2.25 ± 1.05). Taking the energy levels of ROTT (1.97 ± 1.06) as reference, the energy levels of BERN (1.78 ± 0.96) were lower and of BSD (2.80 ± 0.99) were higher.

3.2. Baseline Characteristics of the Study Population and Dog Breeds

At T0 435 dog owner participated, of which 84 completed the study. Thus, the dropout rate was 80.7%. For a more detailed description see Figure 1. The number of participants per breed at T0 varied from 20 (WHWT) to 98 (BC).

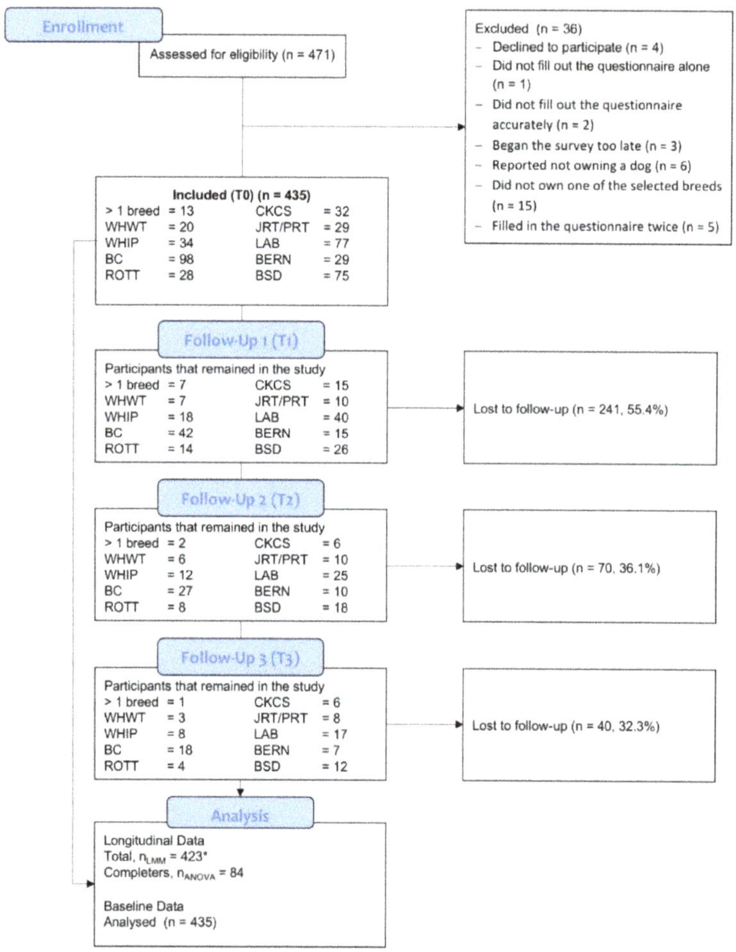

Figure 1. Flow chart of participants over the course of the study. * 12 participants reported owning more than one of the specified dog breeds at all times of the study and were, therefore, excluded from the LMM. BC, Border Collie; BERN, Bernese Mountain Dog; BSD, Belgian Shepherd Dog; CKCS, Cavalier King Charles Spaniel; JRT/PRT, Jack and Parson Russell Terrier; LAB, Labrador Retriever; PA, physical activity; ROTT, Rottweiler; WHIP, Whippet; WHWT, West Highland White Terrier.

An overview of the statistically significant differences of the sociodemographic and anthropometric variables at baseline is summarized in Table 1. There were differences in the distribution in smoking status (χ^2 = 20.47, df = 9, p = 0.015), educational attainment (H = 43.41, df = 9, p < 0.001), income in € (H = 25.40, df = 9, p = 0.003), children under 18 years of age living in the household (χ^2 = 16.92, df = 9, p = 0.050), age of the participant

($H = 27.71$, $df = 9$, $p < 0.001$) and the number of dogs living in the household ($H = 37.36$, $df = 9$, $p < 0.001$). No differences in the distribution between the dog owner groups were detected in terms of gender ($\chi^2 = 10.92$, $df = 9$, $p = 0.282$), completer status ($\chi^2 = 4.76$, $df = 9$, $p = 0.855$), relationship status ($\chi^2 = 7.89$, $df = 9$, $p = 0.545$), employment status ($\chi^2 = 23.04$, $df = 9$, $p = 0.189$), size of hometown ($H = 6.49$, $df = 9$, $p = 0.690$), people over the age of 59 years living in the household ($\chi^2 = 4.32$, $df = 9$, $p = 0.889$), garden ownership ($\chi^2 = 14.76$, $df = 9$, $p = 0.098$), chronic diseases of participants ($\chi^2 = 8.94$, $df = 9$, $p = 0.443$) or the BMI ($H = 16.55$, $df = 9$, $p = 0.056$). On average, participants engaged in 26.6 ± 15.8 h/week total PA, 14.7 ± 8.5 h/week dog-related PA, 12.0 ± 7.2 h/week leisure time walking and 11.3 ± 6.9 h/week dog walking at baseline.

Differences between the dogs of different breeds are displayed in Table 2.

3.3. Baseline Comparison of PA

At baseline, the outcomes of PA were analyzed with all participants who owned one of the specified dog breeds, but without the owners that reported having more than one of the specified breeds. Statistically significant differences in dog-related PA (H (8) = 26.99, $p < 0.001$, Figure 2) and dog walking (H (8) = 16.49, $p = 0.036$, Figure 3) were found between the owners of the specified dog breeds. Differences in total PA (H (8) = 13.78, $p = 0.088$, Figure 4) and leisure time walking (H (8) = 15.46, $p = 0.051$, Figure 5) did not reach statistical significance. The following dog-related activities or activities that could be performed with a dog were reported most frequently by the participants: bicycle riding (n = 108), ball work (activities that were indicated as using a ball with the dog, like "ball play" or "fetching the ball") (n = 75), jogging (n = 72), (rally)obedience (activities in which the dog's obedience is practiced on a course) (n = 72) and agility (n = 67). Statistically significant differences between the owners of the selected dog breeds were identified in ball work, (rally)obedience and agility (Table 3).

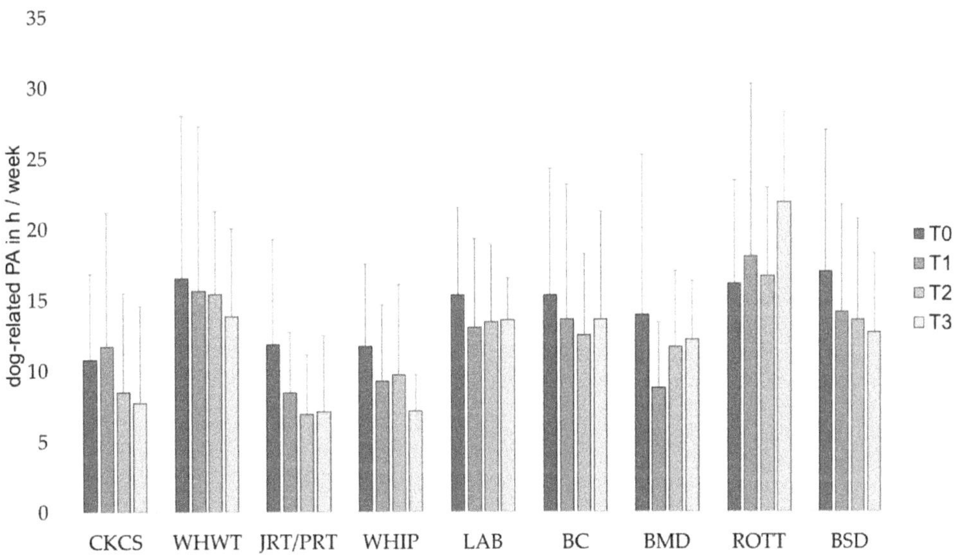

Figure 2. Dog-related PA in h/week (mean ± *SD*) of owners of different dog breeds at baseline (T0), after 1 (T1), 2 (T2) and 3 (T3) years. BC, Border Collie; BERN, Bernese Mountain Dog; BSD, Belgian Shepherd Dog; CKCS, Cavalier King Charles Spaniel; JRT/PRT, Jack and Parson Russell Terrier; LAB, Labrador Retriever; PA, physical activity; ROTT, Rottweiler; WHIP, Whippet; WHWT, West Highland White Terrier.

Table 1. Sociodemographics of the participants split by owned breed at T0.

Variable	Manifestation	More than 1 Breed	CKCS	WHWT	JRT/PRT	WHIP	LAB	BC	BERN	ROTT	BSD
Smoking status	Yes n (%)	5 (38.5)	8 (25.0)	10 (50.0)	7 (24.1)	7 (20.6)	15 (19.5)	30 (30.6)	7 (24.1)	14 (51.9)	30 (40.0)
	No n (%)	8 (61.5)	24 (75.0)	10 (50.0)	22 (75.9)	27 (79.4)	62 (80.5)	68 (69.4)	22 (75.9)	13 (48.1)	45 (60.0)
Education	No degree n (%)	1 (7.7)	0	0	0	0	0	0	0	1 (3.7)	1 (1.3)
	Secondary modern school qualification n (%)	1 (7.7)	2 (6.3)	3 (15.0)	3 (10.3)	0	1 (1.3)	6 (6.1)	1 (3.4)	5 (18.5)	1 (1.3)
	Intermediate high school certificate n (%)	7 (53.8)	12 (37.5)	10 (50.0)	9 (31.0)	9 (26.5)	22 (28.6)	36 (36.7)	12 (41.4)	16 (59.3)	37 (49.3)
	University of applied science qualification or high school diploma n (%)	2 (15.4)	10 (31.3)	5 (25.0)	8 (27.6)	11 (32.4)	21 (27.3)	29 (29.6)	10 (34.5)	3 (11.1)	16 (21.3)
	College or university degree n (%)	2 (15.4)	7 (21.9)	2 (10.0)	6 (20.7)	12 (35.3)	24 (31.2)	26 (26.5)	6 (20.7)	2 (7.4)	20 (26.7)
	Dissertation n (%)	0	1 (3.1)	0	3 (10.3)	2 (5.9)	9 (11.7)	1 (1.0)	0	0	0
Income in €	<1000 n (%)	3 (30.0)	5 (20.0)	2 (10.5)	2 (8.7)	3 (12.5)	5 (8.8)	8 (9.3)	1 (4.5)	3 (12.5)	12 (16.9)
	1000–1999 n (%)	4 (40.0)	12 (48.0)	5 (26.3)	5 (21.7)	2 (8.3)	14 (24.6)	38 (44.2)	7 (31.8)	10 (41.7)	23 (32.4)
	2000–2999 n (%)	2 (20.0)	4 (16.0)	6 (31.6)	7 (30.4)	6 (25.0)	13 (22.8)	22 (25.6)	4 (18.2)	6 (25.0)	13 (18.3)
	3000–3999 n (%)	0	3 (12.0)	3 (15.8)	8 (34.8)	8 (33.3)	9 (15.8)	12 (14.0)	5 (22.7)	1 (4.2)	12 (16.9)
	4000–5999 n (%)	1 (10.0)	1 (4.0)	3 (15.8)	1 (4.3)	5 (20.8)	8 (14.0)	5 (5.8)	4 (18.2)	1 (4.2)	10 (14.1)
	6000–7999 n (%)	0	0	0	0	0	5 (8.8)	1 (1.2)	0	1 (4.2)	0
	8000–9999 n (%)	0	0	0	0	0	2 (3.5)	0	0	1 (4.2)	0
	≥10,000 n (%)	0	0	0	0	0	1 (1.8)	0	1 (4.5)	1 (4.2)	1 (1.4)
Children under 18 years living in the household	Yes n (%)	6 (46.2)	11 (35.5)	1 (5.0)	4 (13.8)	4 (11.8)	16 (21.1)	20 (20.6)	4 (13.8)	8 (29.6)	18 (24.0)
	No n (%)	7 (53.8)	20 (64.5)	19 (95.0)	25 (86.2)	30 (88.2)	60 (78.9)	77 (79.4)	25 (86.2)	8 (70.4)	57 (76.0)
Age of the participant	*M (SD)*	37.5 (12.8)	40.8 (11.9)	50.2 (10.9)	44.7 (12.5)	41.2 (12.1)	46.3 (12.2)	39.3 (11.1)	44.1 (12.2)	42.4 (14.5)	43.3 (10.7)
Number of dogs owned	*M (SD)*	2.6 (1.0)	2.1 (1.2)	1.6 (1.1)	1.8 (0.9)	2.1 (0.8)	2.0 (1.3)	2.2 (1.1)	1.8 (1.1)	1.5 (.8)	2.4 (1.1)

Notes: Only statistically significant differences are depicted. BC, Border Collie; BERN, Bernese Mountain Dog; BSD, Belgian Shepherd Dog; CKCS, Cavalier King Charles Spaniel; JRT/PRT, Jack and Parson Russell Terrier; LAB, Labrador Retriever; PA, physical activity; ROTT, Rottweiler; WHIP, Whippet; WHWT, West Highland White Terrier.

Table 2. Characteristics of the different dog breeds at baseline.

Variable	Manifestation	Other Breed/Mix	CKCS	WHWT	JRT/PRT	WHIP	LAB	BC	BMD	ROTT	BSD	Statistics (*df*)	*p*
Sex	Male n (%)	84 (48)	23 (41.8)	18 (58.1)	27 (52.9)	43 (69.4)	60 (42.6)	87 (47.5)	21 (48.8)	13 (39.4)	50 (41.7)	18.51 [a] (9)	0.03
	Female n (%)	91 (52)	32 (58.2)	13 (41.9)	24 (47.1)	19 (30.6)	81 (57.4)	96 (52.5)	22 (51.2)	20 (60.6)	70 (58.3)		
Neutering status	Neutered n (%)	97 (55.4)	12 (21.8)	15 (48.4)	31 (59.6)	10 (16.1)	29 (20.6)	50 (27.3)	8 (18.6)	17 (51.5)	29 (24.2)	96.32 [a] (9)	<0.001
	Intact n (%)	78 (44.6)	43 (78.2)	16 (51.6)	21 (40.4)	52 (83.9)	112 (79.4)	133 (72.7)	35 (81.4)	16 (48.5)	91 (75.8)		
Chronic diseases of dogs	Yes n (%)	36 (20.7)	11 (20)	4 (12.9)	5 (9.6)	6 (9.8)	20 (14.2)	18 (9.8)	4 (9.3)	5 (15.2)	13 (10.8)	14.53 [a] (9)	0.105
	No n (%)	138 (79.3)	44 (80)	27 (87.1)	47 (90.4)	55 (90.2)	121 (85.8)	165 (90.2)	39 (90.7)	28 (84.8)	107 (89.2)		
Age in years	*M* (*SD*)	6.6 (4)	4.3 (3)	5.3 (3.1)	6.5 (4.2)	4.5 (3.2)	5.2 (3.7)	4.4 (3.7)	3.6 (2.7)	4.4 (2.9)	4.8 (3.5)	6.36 [b] (9, 878)	<0.001

Notes: [a], χ^2-statistics; [b], *F*-value for ANOVA; BC, Border Collie; BERN, Bernese Mountain Dog; BSD, Belgian Shepherd Dog; CKCS, Cavalier King Charles Spaniel; JRT/PRT, Jack and Parson Russell Terrier; LAB, Labrador Retriever; PA, physical activity; ROTT, Rottweiler; WHIP, Whippet; WHWT, West Highland White Terrier.

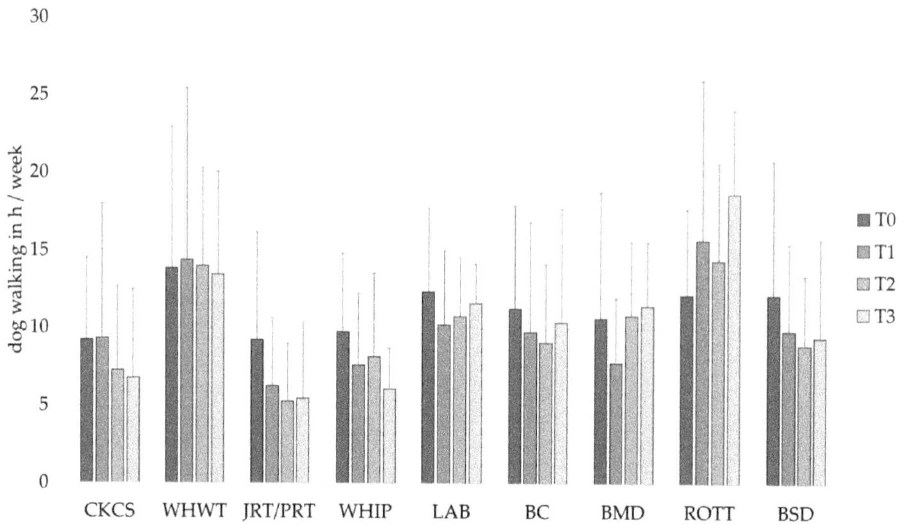

Figure 3. Dog walking in h/week (mean ± *SD*) of owners of different dog breeds at baseline (T0), after 1 (T1), 2 (T2) and 3 (T3) years. BC, Border Collie; BERN, Bernese Mountain Dog; BSD, Belgian Shepherd Dog; CKCS, Cavalier King Charles Spaniel; JRT/PRT, Jack and Parson Russell Terrier; LAB, Labrador Retriever; ROTT, Rottweiler; WHIP, Whippet; WHWT, West Highland White Terrier.

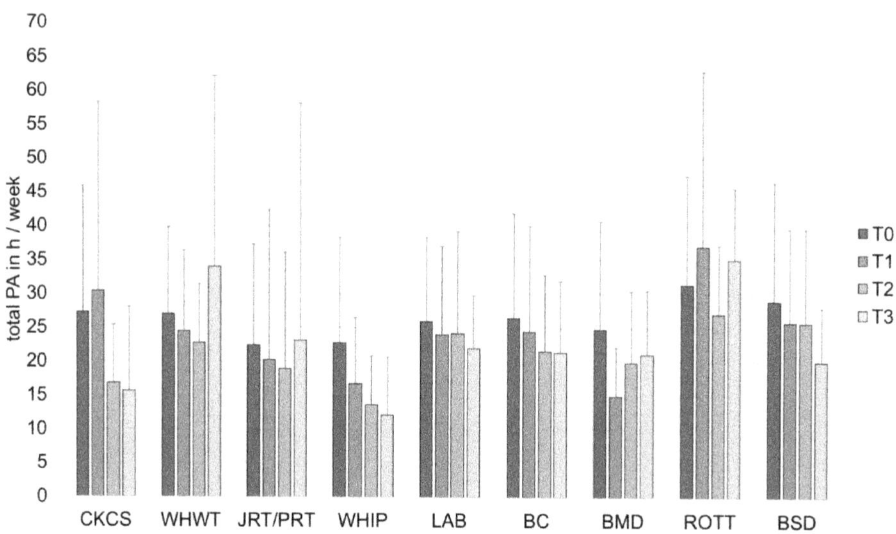

Figure 4. Total PA in h/week (mean ± *SD*) of owners of different dog breeds at baseline (T0), after 1 (T1), 2 (T2) and 3 (T3) years. BC, Border Collie; BERN, Bernese Mountain Dog; BSD, Belgian Shepherd Dog; CKCS, Cavalier King Charles Spaniel; JRT/PRT, Jack and Parson Russell Terrier; LAB, Labrador Retriever; PA, physical activity; ROTT, Rottweiler; WHIP, Whippet; WHWT, West Highland White Terrier.

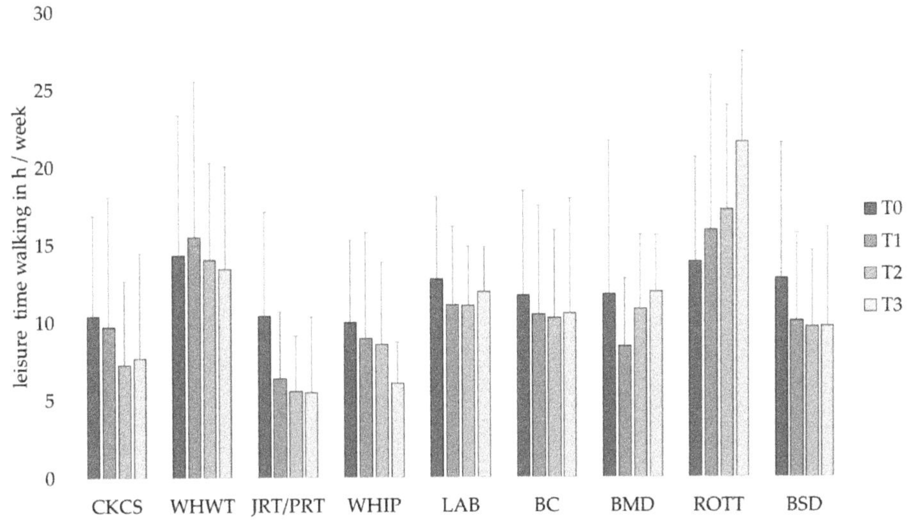

Figure 5. Leisure time walking in h/week (mean ± *SD*) of owners of different dog breeds at baseline (T0), after 1 (T1), 2 (T2) and 3 (T3) years. BC, Border Collie; BERN, Bernese Mountain Dog; BSD, Belgian Shepherd Dog; CKCS, Cavalier King Charles Spaniel; JRT/PRT, Jack and Parson Russell Terrier; LAB, Labrador Retriever; ROTT, Rottweiler; WHIP, Whippet; WHWT, West Highland White Terrier.

Table 3. Participation of owners of different dog breeds in different types of activities.

Type of Exercise	CKCS n (%)	WHWT n (%)	JRT/PRT n (%)	WHIP n (%)	LAB n (%)	BC n (%)	BMD n (%)	ROTT n (%)	BSD n (%)	χ^2-Value (*df*)	*p*
Riding the bicycle with the dog	6 (18.8)	3 (15)	7 (24.1)	14 (41.2)	17 (22.1)	26 (26.5)	5 (17.2)	4 (14.3)	26 (34.7)	13.06 (8)	0.11
Ball work	3 (9.4)	8 (40)	9 (31)	5 (14.7)	9 (11.7)	13 (13.3)	4 (13.8)	5 (17.9)	19 (25.3)	18.58 (8)	0.017
Jogging	3 (9.4)	0	4 (13.8)	3 (8.8)	11 (14.3)	21 (21.4)	6 (20.7)	4 (14.3)	20 (26.7)	14.35 (8)	0.073
(Rally)Obedience	0	0	3 (10.3)	0	4 (5.2)	24 (24.5)	4 (13.8)	9 (32.1)	28 (37.3)	56.6 (8)	<0.001
Agility	2 (6.3)	0	4 (13.8)	6 (17.6)	3 (3.9)	40 (40.8)	2 (6.9)	1 (3.6)	9 (12)	65.85 (8)	<0.001

Notes: Only participants who participated in the mentioned activity are displayed. Only activities that were mentioned at least 20 times are displayed; BC, Border Collie; BERN, Bernese Mountain Dog; BSD, Belgian Shepherd Dog; CKCS, Cavalier King Charles Spaniel; JRT/PRT, Jack and Parson Russell Terrier; LAB, Labrador Retriever; ROTT, Rottweiler; WHIP, Whippet; WHWT, West Highland White Terrier.

3.4. Changes of PA over Time

The results of total PA show statistically significant variability across participants, $Var(u_{0j}) = 149.07$, standard error (SE) = 15.29, $Wald\ Z = 9.75$, $p < 0.001$. There is a statistically significant linear decrease of total PA over time, $F\,(1, 498.63) = 6.85$, $p = 0.009$. The breed groups were found to differ from one another, $F\,(8, 406.28) = 2.21$, $p = 0.026$. However, no significant differences were found if the individual estimates of dog breeds were compared to CKCS (Figure 4, Table 4).

Dog-related PA shows statistically significant variability among individuals, $Var(u_{0j}) = 49.37$, $SE = 4.58$, $Wald\ Z = 10.79$, $p < 0.001$. The results demonstrate a statistically significant linear decrease over time, $F\,(1, 440.56) = 12.58$, $p < 0.001$. Additionally, the owners of the different dog breeds were found to differ statistically significantly from each other, $F\,(8, 400.20) = 3.46$, $p = 0.001$. It was found that owners of WHWT, LAB, BC, ROTT and BSD engage in significantly more dog-related PA than owner of CKCS (Figure 2, Table 4).

Table 4. Linear mixed models of total PA, dog-related PA, leisure time walking and dog walking over time in hours per week.

Outcome	Predictor	Estimate	*SE*	95% CI	*p*
Total PA in h/week	Intercept	26.39	2.53	21.42, 31.37	<0.001
	Linear trend over time per month	−0.09	0.03	−0.15, −0.02	0.009
	CKCS	Ref.			
	WHWT	0.14	4.13	−7.99, 8.27	0.973
	JRT/PRT	−3.58	3.7	−10.85, 3.70	0.334
	WHIP	−5.82	3.52	−12.74, 1.11	0.099
	LAB	−0.63	3.01	−6.54, 5.28	0.834
	BC	−0.12	2.92	−5.86, 5.62	0.966
	BMD	−2.24	3.55	−9.21, 4.74	0.529
	ROTT	6.99	3.69	−0.26, 14.24	0.059
	BSD	2.76	3.04	−3.21, 8.74	0.364
Dog-related PA in h/week	Intercept	11.05	1.36	8.37, 13.73	<0.001
	Linear trend over time per month	−0.05	0.02	−0.08, −0.02	<0.001
	CKCS	Ref.			
	WHWT	4.85	2.24	0.45, 9.24	0.031
	JRT/PRT	0.56	2	−3.37, 4.50	0.778
	WHIP	0.16	1.91	−3.60, 3.91	0.935
	LAB	3.99	1.63	0.80, 7.19	0.015
	BC	4.34	1.58	1.24, 7.44	0.006
	BMD	2.19	1.89	−1.53, 5.91	0.248
	ROTT	6.00	2.00	2.07, 9.92	0.003
	BSD	5.63	1.64	2.40, 8.85	0.001
Leisure time walking in h/week	Intercept	10.56	1.19	8.22, 12.91	<0.001
	Linear trend over time per month	−0.1	0.05	−0.20, 0.00	0.048
	CKCS	Ref.			
	WHWT	3.81	1.95	−0.02, 7.64	0.051
	JRT/PRT	−0.38	1.75	−3.81, 3.05	0.828
	WHIP	−0.51	1.67	−3.79, 2.78	0.762
	LAB	2.06	1.43	−0.74, 4.86	0.15
	BC	1.14	1.38	−1.57, 3.85	0.41
	BMD	0.53	1.69	−2.79, 3.85	0.755
	ROTT	3.12	1.75	−0.32, 6.56	0.076
	BSD	1.98	1.43	−0.84, 4.80	0.168
	CKCS*Linear trend over time per month	Ref.			
	WHWT*Linear trend over time per month	0.01	0.08	−0.15, 0.17	0.898
	JRT/PRT*Linear trend over time per month	0.04	0.07	−0.09, 0.17	0.532
	WHIP*Linear trend over time per month	0.03	0.06	−0.10, 0.15	0.673
	Lab*Linear trend over time per month	0.04	0.06	−0.07, 0.15	0.502
	BC*Linear trend over time per month	0.09	0.06	−0.02, 0.20	0.124
	BMD*Linear trend over time per month	0.10	0.07	−0.04, 0.23	0.153
	ROTT*Linear trend over time per month	0.27	0.07	0.12, 0.41	<0.001
	BSD*Linear trend over time per month	0.05	0.06	−0.07, 0.17	0.407
Dog walking in h/week	Intercept	9.35	1.11	7.17, 11.53	<0.001
	Linear trend over time per month	−0.11	0.04	−0.18, −0.03	0.006
	Quadratic trend over time per month	0.002	0.001	0.000, 0.004	0.036
	CKCS	Ref.			
	WHWT	4.28	1.82	0.71, 7.84	0.019
	JRT/PRT	−0.12	1.63	−3.32, 3.07	0.94
	WHIP	0	1.55	−3.05, 3.04	0.998
	LAB	2.74	1.32	0.14, 5.34	0.039
	BC	2.01	1.28	−0.51, 4.53	0.117
	BMD	1.16	1.55	−1.87, 4.20	0.452
	ROTT	4.22	1.62	1.04, 7.41	0.01
	BSD	2.52	1.33	−0.10, 5.14	0.059

Notes: BC, Border Collie; BERN, Bernese Mountain Dog; BSD, Belgian Shepherd Dog; CKCS, Cavalier King Charles Spaniel; JRT/PRT, Jack and Parson Russell Terrier; LAB, Labrador Retriever; PA, physical activity; ROTT, Rottweiler; WHIP, Whippet; WHWT, West Highland White Terrier.

Leisure time walking was identified to differ statistically significantly between individuals, $Var(u_{0j}) = 34.62$, $SE = 3.27$, $Wald\ Z = 10.59$, $p < 0.001$. A statistically significant decrease of leisure time walking was identified over time, $F\ (1, 446.88) = 3.87$, $p = 0.050$. Additionally, the interaction term between the linear time trend and the breed groups was found to be statistically significant, $F\ (8, 444.16) = 2.36$, $p = 0.017$, indicating that the leisure time walking changes over time, depending on the dog breed. However, no effect was found for the breed groups itself, $F\ (8, 461.78) = 1.50$, $p = 0.153$. The interaction term demonstrates that the owners of ROTT increase their leisure time walking in comparison to owners of CKCS (Figure 5, Table 4).

Statistically significant individual differences were found in the participants in dog-walking, $Var(u_{0j}) = 31.46$, $SE = 3.03$, $Wald\ Z = 10.39$, $p < 0.001$. A negative linear trend was identified over time, F (1, 425.62) = 7.77, p = 0.006. However, a quadratic increase over time was also found to be statistically significant, F (1, 396.47) = 4.41, p = 0.036. This suggests that there is a steeper decrease in the beginning of the study. Further, the breed groups were found to differ significantly, F (8, 396.09) = 2.23, p = 0.025. It was found that owners of WHWT, LAB and ROTT engaged in significantly more dog walking than owners of CKCS (Figure 3, Table 4).

3.5. Changes of PA in the Completers Population

To identify if those who completed all questionnaires did not differ from those who completed questionnaires only at some time points, a completers analysis was carried out. The significant differences of completers and non-completers in the sociodemographic variables are shown in Table 5. There were differences between completers and non-completers in gender (χ^2 = 5.15, df = 1, p = 0.024), smoking status (χ^2 = 6.59, df = 1, p = 0.012), educational attainment (U = 11,062, z = (3.70, p < 0.001), employment status (χ^2 = 9.40, df = 2, p = 0.009), income in € (U = 13,167, z = 3.59, p < 0.001) and chronic diseases of the participants (χ^2 = 8.93, df = 1, p = 0.002). No differences were detected in breed that the participant owns (χ^2 = 4.76, df = 9, p = 0.855), relationship status (χ^2 = 0.07, df = 1, p = 0.787), size of hometown (U = 13,957, z = −0.79, p = 0.428), garden ownership (χ^2 = 1.91, df = 1, p = 0.205), children under the age of 18 years (χ^2 = 0.01, df = 1, p = 1.000) or adults over the age of 59 years (χ^2 = 0.10, df = 1, p = 0.867) living in the household, age of the participant (t = (1.14, df = 139, p = 0.257), BMI (t = 0.02, df = 423, p = 983), number of dogs in the household (t = 1.18, df = 140.2, p = 0.242), age of the dog (t = −0.42, df = 886, p = 0.676), size of the dog (t = −0.05, df = 891, p = 0.961) or weight of the dog (t = 0.36, df = 893, p = 0.719).

Table 5. Sociodemographic status of completers vs. non-completers.

Variable	Manifestation	Completer	Non-Completer
Gender	Male n (%)	4 (7.7)	48 (92.3)
	Female n (%)	80 (20.9)	302 (79.1)
Smoking	Yes n (%)	16 (12.0)	117 (88.0)
	No n (%)	68 (22.6)	233 (77.4)
Educational attainment	No degree n (%)	0	3 (100.0)
	Secondary modern school qualification n (%)	2 (8.7)	21 (91.3)
	Intermediate high school certificate n (%)	23 (13.5)	147 (86.5)
	University of applied science qualification or high school diploma n (%)	25 (21.7)	90 (78.3)
	College or university degree n (%)	25 (23.4)	82 (76.6)
	Dissertation n (%)	9 (56.3)	7 (43.8)
Employment status	Full time n (%)	42 (19.5)	173 (80.5)
	Part time n (%)	34 (26.0)	97 (74.0)
	Not employed n (%)	8 (9.2)	79 (90.8)
Income in €	<1000 n (%)	3 (6.8)	41 (93.2)
	1000–1999 n (%)	16 (13.3)	104 (86.7)
	2000–2999 n (%)	22 (26.5)	61 (73.5)
	3000–3999 n (%)	16 (26.2)	45 (73.8)
	4000–5999 n (%)	9 (23.1)	30 (76.9)
	6000–7999 n (%)	4 (57.1)	3 (42.9)
	8000–9999 n (%)	0	3 (100.0)
	≥10,000 n (%)	2 (50.0)	2 (50.0)
Chronic diseases of participants	Yes n (%)	12 (10.3)	104 (89.7)
	No n (%)	72 (23.3)	237 (76.7)

Notes: Only statistically significant differences are depicted.

Significant differences in total PA and dog-related PA were found between completers and non-completers at T0 and in total PA at T1. No other significant differences appeared in total PA and dog-related PA. In leisure time walking and dog walking no significant differences were found between completers and non-completers at any time point (Table 6). However, non-completers scored higher in leisure time and dog walking at T0 and T1, but lower at T2 than completers.

Table 6. Differences in PA between completers and non-completers.

Time	Variable	Completer		Non-Completer		*t* (*df*)	*p*	Cohens *d*
		M (*SD*)	n	*M* (*SD*)	n			
T0	Total PA in h/week	22.9 (12.3)	84	27.5 (16.4)	351	2.84 (161.5)	0.005	0.29
	drPA in h/week	13.3 (6.3)	84	15 (8.9)	351	2.04 (172.6)	0.043	0.2
	Leisure time walking in h/week	11 (5.4)	84	12.3 (7.5)	351	1.8 (166.8)	0.074	0.18
	Dog walking in h/week	10.4 (5.3)	84	11.5 (7.2)	351	1.62 (162.8)	0.107	0.16
T1	Total PA in h/week	21.3 (11.7)	84	26.4 (19.8)	110	2.23 (181.7)	0.027	0.3
	drPA in h/week	12.5 (6.9)	84	13 (9.3)	110	0.43 (192)	0.665	0.06
	Leisure time walking in h/week	10.3 (5.8)	84	10.8 (7.6)	110	0.47 (192)	0.642	0.07
	Dog walking in h/week	9.7 (5.8)	84	10.1 (7.5)	110	0.44 (192)	0.658	0.06
T2	Total PA in h/week	21.6 (11.7)	84	23.8 (16.5)	40	0.85 (122)	0.397	0.16
	drPA in h/week	13 (6.4)	84	11.7 (7)	40	−1.04 (122)	0.302	−0.2
	Leisure time walking in h/week	10.9 (5.3)	84	9.6 (6.2)	40	−1.17 (122)	0.243	−0.23
	Dog walking in h/week	10.2 (4.9)	84	9 (5.9)	40	−1.15 (122)	0.251	−0.24

Notes: PA, physical activity; drPA, dog-related PA.

If only completers were analyzed using a repeated measures ANOVA, no changes in any of the PA outcomes were detected (Table 7). Due to the small sample size for some of the breed groups, only changes over time were examined.

Table 7. Changes in PA and walking behavior of participants who completed the study.

		T0	T1	T2	T3			
Variable	*n*	*M* (*SD*)	*M* (*SD*)	*M* (*SD*)	*M* (*SD*)	F (*df*)	*p*	Partial h^2
Total PA in h/week	84	22.9 (12.3)	21.3 (11.7)	21.6 (11.7)	21.4 (14.8)	0.86 (3, 249)	0.464	0.01
drPA in h/week	84	13.3 (6.3)	12.5 (6.9)	13 (6.4)	12.2 (6.4)	1.45 (3, 249)	0.229	0.02
Leisure time walking in h/week	84	11 (5.4)	10.3 (5.8)	10.9 (5.3)	10.5 (6.3)	0.61 (3, 249)	0.611	0.01
Dog walking in h/week	84	10.4 (5.3)	9.7 (5.8)	10.2 (4.9)	10.1 (6)	0.67 (3, 249)	0.572	0.01

Notes: PA, physical activity; drPA, dog related PA.

3.6. Changes in PA after a Dog Died

At T1 24, T2 16 and T3 17 participants reported that at least one dog had died, respectively. As a consequence, 57 dogs died during the course of the study. However, in the subgroup of participants whose dog died, no significant changes in total

PA ($\Delta = 1.08 \pm 12.3$, $t = (0.66$, $df = 56$, $p = 0.510$) or leisure time walking ($\Delta = (0.12 \pm 4.7$, $t = 0.20$, $df = 56$, $p = 0.843$) were detected, after the dog had died.

Only one of the participants reported not owning another dog after the dog's death. Before the dog died this participant reported 10.5 h/week of leisure time walking and 16.7 h/week of total PA per week. All leisure time walking was performed as dog walking and dog-related PA accounted for 11.7 h/week (70.1%) of total PA. After the dog died, the participant reported 3 h/week of total PA. This corresponds to a decrease of 82% in total PA. No leisure time walking was reported after the death of the dog.

4. Discussion

The main purpose of the present study was to examine the influence of dog size, energy level and dog age on their owners' PA behavior. At baseline, no statistical group differences were identified for total PA and leisure time walking. In contrast, groups differed significantly in dog-related PA and dog walking as well as in the types of chosen activities.

According to the LMM analysis, total PA, dog-related PA, leisure time walking and dog walking decreased significantly over 3 years in all groups except for leisure time walking in owners of ROTT. In this group, leisure time walking increased over time. In the group of dog owners that completed the trial, no changes in total PA, dog-related PA, leisure time walking and dog walking could be observed.

These findings suggest that the type of PA—and as a potential consequence the intensity of PA—might have a greater impact on physical health of the owners of different dog breeds than just the duration of PA.

The duration and variability in the data of total and dog-related PA is slightly higher than in other German cross sectional studies that used the same questionnaire [6,17]. The reasons for this finding remain unclear. However, high individual variability is a well-known phenomenon in this field of study (e.g., [1,13,15,64,65]).

Old age of dogs is negatively correlated with PA of their owners [35,36,39,40,42]. Thus, it was hypothesized that PA levels of dog owners decrease over time. However, the results only partly support this hypothesis. Although a negative trend was found in the overall population, this trend was not supported if only completers were analyzed. The lack of PA decline could be explained by the fact that many of the dogs might not have been sufficiently old to display an age-related decline in PA. Given the results of Patronek et al. (1997), the mean physiological age of all dog breeds would have been at the younger end of the middle-aged spectrum (28 to 39 human years). Therefore, three years later, the dogs' mean physiological age would not exceed 55 human years. At T3, the dogs were probably not old enough and the dogs were still too healthy to cause a decrease in their owners' PA. Another explanation could be that non-completers reported higher amounts of all PA outcomes at T0 and T1 and in total PA at T2. Although not all of the comparisons were significant on a statistical level, the decreases in PA that were detected in the LMMs might be derived from participants who either overreported their PA or were more physically active, but did not complete the study and thus might bias the results. Thorpe et al. reported that in their population, dog walkers' PA levels decreased at the same rate as in all other groups after three years. However, the participants of Thorpe et al. were between 70 and 79 years old at baseline and, thus, not comparable to the population of the current investigation [66].

Earlier studies demonstrated that having multiple dogs deters dog owners from engaging in dog-related PA [36,67,68]. However, having multiple dogs might also help owners to remain active when one of the dogs gets old or sick. Only one participant reported that her dog died and that no other dog remained in the household afterwards. The level of PA dropped dramatically after the dogs' death. However, this is only a single case and cannot be extrapolated to a larger group of dog owners. Degeling and Rock report a similar case. They state that one of their participants was less motivated to exercise after the dog's death, but another participant reported the opposite [40]. In the present study, except for the one named case, there was always another dog living in the household when another dog had died. In these cases, no changes emerged in total PA and leisure time walking when a dog died. This indicates that, if at least one dog remains in the household, the death of one dog does not impact the PA behavior of the dog owner. Future studies are warranted to investigate the relationship of owning several dogs, dog death and PA of the owners.

The results do not show a clear pattern that owners of larger or more energetic dogs were more active than owners of dog breeds that are smaller or less energetic. This contradicts earlier findings [29,31,32,34,36]. It indicates that just the size and energy level of a dog breed are insufficient to predict how much PA the owner will engage in with and without their dog. It suggests that other factors need to be taken into account. However, the cited studies asked the owners for their perception of their dogs. The current study categorized the dog breeds based on their energy level a priori. Consequently, the energy level attributed to each dog by the owner may not match the category based on the breed-average C-BARQ scores. It is possible that the owner's perception of an individual dog's energy level may be more reliable than a level derived from averaging multiple assessments of dogs of the same breed.

Further, a cultural element may complicate the interpretation of the influence of a dog's energy level on the owners' PA. Nagasawa et al. found that dogs in Japan are perceived as more energetic and restless than dogs in the USA [69]. Therefore, the influence of the energy level of a dog on the PA behavior of their owner might differ between people from different cultural backgrounds. The current investigation used data from the C-BARQ study that takes place in the USA and is mainly performed in English [45]. Thus, it remains unclear whether the average perception of German and US-American dog owners of their dog's energy level match or whether there are slightly different.

Several differences in the selection of PA types were found between the owners of the different dog breeds. However, due to the limitations of study design, these differences cannot be explained. It could be assumed that some dog breeds are better suited for certain activities than others. Some dogs might not be able to engage in PA at an intensity that is beneficial to the owner. This could explain the lower volume of dog-related PA in CKCS compared to WHWT, LAB, BC, ROTT and BSD. Since no statistically significant differences between the owners of CKCS and owners of the other dog breeds were detected in total PA, this suggests that owners of CKCS engage in other non-dog-related activities more than other owners. This could in turn lead to greater health benefits for the owners of CKCS, due to increased intensities. This may be especially true since an earlier study indicated that dog-related PA are mostly not of a moderate intensity [17]. However, this study did not investigate the types of PA. Therefore, it is not possible to conclusively assess the quality of the non-dog-related PAs.

Some activities might also be performed with certain dog types more often. For example, (rally)obedience was mostly performed by owners of medium to large breed dogs, especially owners of BC, ROTT and BSD. Arhant et al. report that owners of larger dogs are more likely to be engaged in this activity [34]. Especially, owners of ROTT and BSD might perform these activities because they might be afraid that their dogs are strong enough to harm other people and need to be "under the control" of the owner. On the other hand, ROTT do not show increased stranger-directed, dog-directed and owner-directed aggression or dog rivalry as compared to other dog breeds [37]. This could indicate that ROTT owners either successfully take part in activities like (rally)obedience or dog school training. However, the reasons why certain dog owners engage in certain activities remain not fully understood.

There were great differences between the dog breeds in regards to neuter status. Especially JRT/PRT and ROTT were often neutered, while BERN, CKCS and WHIP were more often non-neutered. At the outset, this was not anticipated and the authors have no explanation for this finding. However, it is conceivable that there are owner beliefs about dogs of the selected breeds that have not been surveyed and might influence whether or not owners decide to neuter their dogs.

During the COVID-19 pandemic no serious decrease of PA was detected in this study. Earlier studies that focused on PA during the COVID-19 pandemic identified dramatic declines in moderate to vigorous PA [70,71] with potentially serious health effects [72]. Similar declines in dog walking and PA have been found in some [73,74] but not all [75] studies that focus on dog owners. Thus, the current study indicates that dog ownership could be a protective factor against the decline of PA during the pandemic and that dog owners might benefit greatly from their dogs in terms of PA during the COVID-19 pandemic. Still, it has to be emphasized that the legal framework varied greatly between different countries in regard to the lockdowns. The opportunities owners had to walk their dogs during lockdown varied greatly between different countries. For example, the lockdown in Spain and Serbia included dog walking [73,74], while leisure time walking was allowed during the lockdown in Germany [76]. Overall, it must be considered that the COVID-19 pandemic is an exceptional event that impacts the lives of people worldwide. Therefore, the study results are probably not generalizable, or only with limitations, to a time outside the pandemic.

Overall, this study has some limitations. As with most studies in this field, it relies exclusively on self-reported PA. Several studies in different populations show that over-reporting is a common problem in self-reported PA, especially moderate to vigorous PA [77–82]. This may also be true for this study. However, the results are similar to earlier studies that also used the BSA-F [6,17]. Thus, it is likely that the results are reliable.

Further, the BSA-F does not include an assessment of the intensities of PAs. Therefore, it remains unclear whether the intensity of the reported PAs is sufficient to produce health enhancing effects. Overall, results on the intensity of dog-related PA remain controversial. Hielscher et al. considered it likely that most of the dog-related PA failed to achieve moderate intensity [17], which would be necessary to reach the PA guidelines as specified by the World Health Organization (WHO) [83]. However, Richards et al. state that a considerable amount of dog-related PA is of at least moderate intensity [18]. Thus, dog-related PA could be considered to be health enhancing. Furthermore, recent studies highlight the positive impact of light intensity PA on health and mortality [23,84–87], even though moderate to vigorous PA is considered to be more effective [23,86,87]. Thus, the high levels of PA in this study show that dog owners are likely to benefit from their dogs due to increased levels of PA, regardless of the breed.

The recruitment design of the study was based solely on self-selection in a convenience sample. This might have biased the results because only the most motivated dog owners participated in the study. It is possible that the PA behavior of these owners differs systematically from owners who did not participate in the study. However, self-selection bias is a phenomenon that is not limited to online research, as the results of Oswald et al. show [88]. Nevertheless, interpreting and generalizing the data has to be treated with caution.

Participants were mostly recruited online. It is possible that dog owners who use dog-centric online media are more active with their dog than dog owners who are not organized in dog-related social media groups. This could be related to the fact that dog owners in dog-related online groups identify more strongly with their dog and the ownership of a dog and therefore have different attitudes than dog owners who are not organized in this way, which, in turn, might be reflected in their dog ownership behavior. This, together with self-selection bias, may limit the extent to which the findings can be generalized to the whole dog owner population.

The dropout rate in this study was high. It has been shown that a higher dropout rate is associated with a greater bias in statistical models [89,90]. The results of the completer analysis show that participants with a lower educational status dropped out of the current study more often. This is congruent with the results of Gustavson et al. [90]. This suggests caution when generalizing the current findings. Because the reasons for dropping out of the study could not be investigated, it remains unclear how this could bias the results. However, the fact that completers and non-completers differed in several ways suggests that the results may be biased in some way. Most participants who dropped out terminated their participation in the second year, thus, before the COVID-19 pandemic. Therefore, the authors do not believe that the pandemic influenced the decision to terminate participation to a great extent.

5. Conclusions

Overall, the study shows that the PA behavior of owners of the selected breeds is stable over time in this population. The aging of the dog was only found to have a minor influence on the PA of the owners. Anecdotal evidence suggests that losing one's dog might have a significant, negative impact on dog owners' PA.

The results also provide evidence that owners of different dog breeds differ in their choice of PA types, as in the duration of total PA, total dog-related PA, leisure time walking and dog walking. The extent to which this influences the health of the dog owner remains unclear and must be examined in future studies.

Supplementary Materials: The following supporting information can be downloaded at: https://www.mdpi.com/article/10.3390/ani12101314/s1, the questionnaire translated into English and used in this study.

Author Contributions: Conceptualization, B.H.-Z., I.F. and U.G.; methodology, B.H.-Z., I.F., U.G. and J.S.; formal analysis, B.H.-Z.; investigation, B.H.-Z.; data curation, B.H.-Z.; writing—original draft preparation, B.H.-Z.; writing—review and editing, B.H.-Z., I.F., J.S. and U.G.; visualization, B.H.-Z.; supervision, U.G. and I.F.; project administration, B.H.-Z. All authors have read and agreed to the published version of the manuscript.

Funding: This research received no external funding.

Institutional Review Board Statement: The study was in accordance with the Declaration of Helsinki and approved by the Ethics Committee of the German Sport University Cologne (No.: 108/2017).

Informed Consent Statement: Informed consent was obtained from all subjects involved in the study.

Data Availability Statement: The data on sociodemographics and physical activity are available upon request from Benedikt Hielscher-Zdzieblik, while the data on the C-BARQ are available upon request from James Serpell.

Acknowledgments: The authors would like to thank the dog owners for participating in the studies. Furthermore, the authors want to thank the above-mentioned organizations and social media groups that supported the recruitment of study participants.

Conflicts of Interest: The authors declare no conflict of interest.

References

1. Cutt, H.E.; Giles-Corti, B.; Knuiman, M.; Timperio, A.; Bull, F. Understanding Dog Owners' Increased Levels of Physical Activity: Results From RESIDE. *Am. J. Public Health* **2008**, *98*, 66–69. [CrossRef] [PubMed]
2. Brown, S.G.; Rhodes, R.E. Relationships Among Dog Ownership and Leisure-Time Walking in Western Canadian Adults. *Am. J. Prev. Med.* **2006**, *30*, 131–136. [CrossRef] [PubMed]
3. Maugeri, A.; Medina-Inojosa, J.; Kunzova, S.; Barchitta, M.; Agodi, A.; Vinciguerra, M.; Lopez-Jimenez, F. Dog Ownership and Cardiovascular Health: Results From the Kardiovize 2030 Project. *Mayo Clin. Proceed. Innov. Qual. Outcomes* **2019**, *3*, 268–275. [CrossRef] [PubMed]
4. Mičková, E.; Machová, K.; Dadóvá, K.; Svobodová, I. Does Dog Ownership Affect Physical Activity, Sleep, and Self-Reported Health in Older Adults? *Int. J. Environ. Res. Public Health* **2019**, *16*, 3355. [CrossRef]
5. Wasenius, N.S.; Laine, M.K.; Savola, S.; Simonen, M.; Tiira, K.; Lohi, H.; Eriksson, J.G.; Salonen, M.K. Dog Ownership From a Life Course Perspective and Leisure-Time Physical Activity in Late Adulthood: The Helsinki Birth Cohort Study. *Am. J. Health Behav.* **2018**, *42*, 11–18. [CrossRef]
6. Hielscher, B.; Gansloßer, U.; Froboese, I. Impacts of Dog Ownership and Attachment on Total and Dog-related Physical Activity in Germany. *Hum. Anim. Interact. Bull.* **2021**, *10*, 22–43.
7. Oka, K.; Shibata, A. Dog Ownership and Health-Related Physical Activity Among Japanese Adults. *J. Phys. Act. Health* **2009**, *6*, 412–418. [CrossRef]

8. Dall, P.M.; Ellis, S.L.H.; Ellis, B.M.; Grant, P.M.; Colyer, A.; Gee, N.R.; Granat, M.H.; Mills, D.S. The Influence of Dog Ownership on Objective Measures of Free-Living Physical Activity and Sedentary Behaviour in Community-Dwelling Older Adults: A Longitudinal Case-Controlled Study. *BMC Public Health* **2017**, *17*, 496. [CrossRef]
9. Ding, D.; Bauman, A.E.; Sherrington, C.; McGreevy, P.D.; Edwards, K.M.; Stamatakis, E. Dog Ownership and Mortality in England: A Pooled Analysis of Six Population-Based Cohorts. *Am. J. Prev. Med.* **2018**, *54*, 289–293. [CrossRef]
10. Feng, Z.; Dibben, C.; Witham, M.D.; Donnan, P.T.; Vadiveloo, T.; Sniehotta, F.; Crombie, I.K.; McMurdo, M.E.T. Dog Ownership and Physical Activity in Later Life: A Cross-Sectional Observational Study. *Prev. Med.* **2014**, *66*, 101–106. [CrossRef]
11. Mein, G.; Grant, R. A Cross-Sectional Exploratory Analysis Between Pet Ownership, Sleep, Exercise, Health and Neighbourhood Perceptions: The Whitehall II Cohort Study. *BMC Geriatr.* **2018**, *18*, 176. [CrossRef]
12. Solomon, E.; Rees, T.; Ukoumunne, O.C.; Metcalf, B.; Hillsdon, M. Personal, Social, and Environmental Correlates of Physical Activity in Adults Living in Rural South-West England: A Cross-Sectional Analysis. *Int. J. Behav. Nutr. Phys. Act.* **2013**, *10*, 129. [CrossRef] [PubMed]
13. Westgarth, C.; Christley, R.M.; Jewell, C.; Boddy, L.M.; Christian, H.E. Dog Owners are More Likely to Meet Physical Activity Guidelines Than People Without a Dog: An Investigation of the Association Between Dog Ownership and Physical Activity Levels in a UK Community. *Nat. Sci. Rep.* **2019**, *9*, 5704. [CrossRef] [PubMed]
14. Serpell, J. Beneficial Effects of Pet Ownership on Some Aspects of Human Health and Behaviour. *J. R. Soc. Med.* **1991**, *84*, 717–720. [CrossRef] [PubMed]
15. Park, M.; Park, H.K.; Hwang, H.S.; Park, K.Y.; Yim, H.H. The Relationship Between Dog Ownership and Physical Activity in Korean Adults. *Korean J. Fam. Med.* **2020**, *42*, 42–65. [CrossRef]
16. Gretebeck, K.A.; Radius, K.; Black, D.R.; Gretebeck, R.J.; Ziemba, R.; Glickman, L.T. Dog Ownership, Funcional Ability, and Walking in Community-Dwelling Older Adults. *J. Phys. Act. Health* **2013**, *10*, 646–655. [CrossRef]
17. Hielscher, B.; Gansloßer, U.; Froboese, I. More Than "Just" Walking: An Observational Study of Dog-Related Physical Activities. *People Anim. Int. J. Res. Pract.* **2020**, *3*, 7.
18. Richards, E.A.; Troped, P.J.; Lim, E. Assessing the Intensity of Dog Walking and Impact on Overall Physical Activity: A Pilot Study Using Accelerometry. *Open J. Prev. Med.* **2014**, *4*, 523–528. [CrossRef]
19. Cutt, H.E.; Knuiman, M.W.; Giles-Corti, B. Does Getting a Dog Increase Recreational Walking? *Int. J. Behav. Nutr. Phys. Act.* **2008**, *5*, 17. [CrossRef]
20. Potter, K.; Teng, J.E.; Masteller, B.; Rajala, C.; Balzer, L.B. Examining How Dog 'Acquisition' Affects Physical Activity and Psychosocial Well-Being: Findings from the BuddyStudy Pilot Trial. *Animals* **2019**, *9*, 666. [CrossRef]
21. Powell, L.; Edwards, K.M.; Bauman, A.; McGreevy, P.; Podberscek, A.; Neilly, B.; Sherrington, C.; Stamatakis, E. Does Dog Acquisition Improve Physical Activity, Sedentary Behaviour and Biological Markers of Cardiometabolic Health? Results From a Three-Arm Controlled Study. *BMJ Open Sport Exerc. Med.* **2020**, *6*, e000703. [CrossRef] [PubMed]
22. Matthews, C.E.; Keadle, S.K.; Troiano, R.P.; Kahle, L.; Koster, A.; Brychta, R.; Van Domelen, D.; Caserotti, P.; Chen, K.Y.; Harris, T.B.; et al. Accelerometer-Measured Dose-Response for Physical Activity, Sedentary Time, and Mortality in US Adults. *Am. J. Clin. Nutr.* **2016**, *104*, 1424–1432. [CrossRef] [PubMed]
23. Ekelund, U.; Tarp, J.; Steene-Johannessen, J.; Hansen, B.H.; Jefferis, B.; Fagerland, M.W.; Whincup, P.; Diaz, K.M.; Hooker, S.P.; Chernofsky, A.; et al. Dose-Response Associations Between Accelerometry Measured Physical Activity and Sedentary Time and all Cause Mortality: Systematic Review and Harmonised Meta-Analysis. *BMJ* **2019**, *366*, l4570. [CrossRef] [PubMed]
24. Garber, C.E.; Blissmer, B.; Deschenes, M.R.; Franklin, B.A.; Lamonte, M.J.; Lee, I.M.; Nieman, D.C.; Swain, D.P. American College of Sports Medicine Position Stand. Quantity and Quality of Exercise for Developing and Maintaining Cardiorespiratory, Musculoskeletal, and Neuromotor Fitness in Apparently Healthy Adults: Guidance for Prescribing Exercise. *Med. Sci. Sports Exerc.* **2011**, *43*, 1334–1359. [CrossRef]
25. Kyu, H.H.; Bachman, V.F.; Alexander, L.T.; Mumford, J.E.; Afshin, A.; Estep, K.; Veerman, J.L.; Delwiche, K.; Iannarone, M.L.; Moyer, M.L.; et al. Physical Activity and Risk of Breast Cancer, Colon Cancer, Diabetes, Ischemic Heart Disease, and Ischemic Stroke Events: Systematic Review and Dose-Response Meta-Analysis for the Global Burden of Disease Study. 2013. Available online: http://www.bmj.com/content/354/bmj.i3857 (accessed on 2 April 2017).
26. Paluch, A.E.; Bajpai, S.; Bassett, D.R.; Carnethon, M.R.; Ekelund, U.; Evenson, K.R.; Galuska, D.A.; Je, B.J.; Kraus, W.E.; Lee, I.-M.; et al. Daily Steps and All-Cause Mortality: A Meta-Analysis of 15 International Cohorts. *Lancet Public Health* **2022**, *7*, e219–e228. [CrossRef]
27. Mubanga, M.; Byberg, L.; Nowak, C.; Egenvall, A.; Magnusson, P.K.; Ingelsson, E.; Fall, T. Dog Ownership and the Risk of Cardiovascular Disease and Death—A Nationwide Cohort Study. *Sci. Rep.* **2017**, *7*, 15821. [CrossRef]
28. Bauman, A.; Owen, K.B.; Torske, M.O.; Ding, D.; Krokstad, S.; Stamatakis, E. Does Dog Ownership Really Prolong Survival?: A Revised Meta-Analysis and Reappraisal of the Evidence. *Circ. Cardiovasc. Qual. Outcomes* **2020**, *13*, e006907. [CrossRef]
29. Cutt, H.E.; Giles-Corti, B.; Knuiman, M. Encouraging Physical Activity Through Dog Walking: Why Don't Some Owners Walk With Their Dog? *Prev. Med.* **2008**, *46*, 120–126. [CrossRef]
30. Pickup, E.; German, A.J.; Blackwell, E.; Evans, M.; Westgarth, C. Variation in Activity Levels Amongst Dogs of Different Breeds: Results of a Large Online Survey of Dog Owners From the United Kingdom. *J. Nutr. Sci.* **2017**, *17*, e10. [CrossRef]
31. Lim, C.; Rhodes, R.E. Sizing up Physical Activity: The Relationships Between Dog Characteristics, Dog Owners' Motivations, and Dog Walking. *Psychol. Sport Exerc.* **2016**, *24*, 65–71. [CrossRef]

32. Schofield, G.; Mummery, K.; Steele, R. Dog Ownership and Human Health-Related Physical Activity: An Epidemiological Study. *Health Promot. J. Aust.* **2005**, *16*, 15–19. [CrossRef] [PubMed]
33. Oka, K.; Shibata, A. Prevalence and Correlates of Dog Walking Among Japanese Dog Owners. *J. Phys. Act. Health* **2012**, *9*, 786–793. [CrossRef] [PubMed]
34. Arhant, C.; Bubna-Littitz, H.; Bartels, A.; Futschik, A.; Troxler, J. Behaviour of Smaller and Larger Dogs: Effects of Training Methods, Inconsistency of Owner Behaviour and Level of Engagement in Activities With the Dog. *Appl. Anim. Behav. Sci.* **2010**, *123*, 131–142. [CrossRef]
35. Reeves, M.J.; Rafferty, A.P.; Miller, C.E.; Lyon-Callo, S.K. The Impact of Dog Walking on Leisure-Time Physical Activity: Results From a Population-Based Survey of Michigan Adults. *J. Phys. Act. Health* **2011**, *8*, 436–444. [CrossRef]
36. Richards, E.A.; McDonough, M.H.; Edwards, N.E.; Lyle, R.M.; Troped, P.J. Psychosocial and Environmental Factors Associated With Dog-Walking. *Int. J. Health Promot. Educ.* **2013**, *51*, 198–211. [CrossRef]
37. Serpell, J.; Duffy, D.L. Dog Breeds and Their Behavior. In *Domestic Dog Cognition and Behavior: The Scientific Study of Canis Familiaris*; Horowitz, A., Ed.; Springer: Berlin/Heidelberg, Germany, 2014; pp. 31–57, ISBN 978-3-642-53993-0.
38. Tonoike, A.; Nagasawa, M.; Mogi, K.; Serpell, J.; Ohtsuki, H.; Kikusui, T. Comparison of Owner-Reported Behavioral Characteristics Among Genetically Clustered Breeds of Dog (Canis familiaris). *Sci. Rep.* **2015**, *5*, 17710. [CrossRef]
39. Cutt, H.E.; Giles-Corti, B.; Wood, L.J.; Knuiman, M.W.; Burke, V. Barriers and Motivators for Owners Walking Their Dog: Results From Qualitative Research. *Health Promot. J. Aust.* **2008**, *19*, 118–124. [CrossRef]
40. Degeling, C.; Rock, M.J. "It was not Just a Walking Experience": Reflections on the Role of Care in Dog-Walking. *Health Promot. Int.* **2013**, *28*, 397–406. [CrossRef]
41. Belshaw, Z.; Dean, R.; Asher, L. Slower, Shorter, Sadder: A Qualitative Study Exploring How Dog Walks Change When the Canine Participant Develops Osteoarthritis. *BMC Vet. Res.* **2020**, *16*, 85. [CrossRef]
42. Downes, M.J.; Devitt, C.; Downes, M.T.; More, S.J. Understanding the Context for Pet Cat and Dog Feeding and Exercising Behaviour Among Pet Owners in Ireland: A Qualitative Study. *Ir. Vet. J.* **2017**, *70*, 29. [CrossRef]
43. Patronek, G.J.; Waters, D.J.; Glickman, L.T. Comparative Longevity of Pet Dogs and Humans: Implications for Gerontology Research. *Prev. Sch. Fail.* **1997**, *52A*, B171–B178. [CrossRef] [PubMed]
44. Verband für das Deutsche Hundewesen Welpenstatistik (Presse-Informationen). Available online: http://www.vdh.de/presse/welpenstatistik/ (accessed on 12 December 2016).
45. Serpell, J. University of Pennsylvania About the C-BARQ. Available online: https://vetapps.vet.upenn.edu/cbarq/about.cfm (accessed on 15 March 2022).
46. Hsu, Y.; Serpell, J.A. Development and Validation of a Questionnaire for Measuring Behavior and Temperament Traits in Pet Dogs. *J. Am. Vet. Med. Assoc.* **2003**, *223*, 1293–1300. [CrossRef] [PubMed]
47. FCI FCI—Standard N° 136. Cavalier King Charles Spaniel. Available online: http://www.fci.be/Nomenclature/Standards/136g09-en.pdf (accessed on 25 March 2017).
48. FCI FCI—Standard N° 85. West Highland White Terrier. Available online: http://www.fci.be/Nomenclature/Standards/085g03-en.pdf (accessed on 25 March 2017).
49. FCI FCI—Standard N° 345. Jack Russell Terrier. Available online: http://www.fci.be/Nomenclature/Standards/345g03-de.pdf (accessed on 28 December 2020).
50. FCI FCI—Standard N° 339. Parson Russel Terrier. Available online: http://www.fci.be/Nomenclature/Standards/339g03-en.pdf (accessed on 25 March 2017).
51. FCI FCI—Standard N° 162. Whippet. Available online: http://www.fci.be/Nomenclature/Standards/162g10-en.pdf (accessed on 25 March 2017).
52. FCI FCI—Standard N° 122. Labrador Retriever. Available online: http://www.fci.be/Nomenclature/Standards/122g08-en.pdf (accessed on 25 March 2017).
53. FCI FCI—Standard N° 297. Border Collie. Available online: http://www.fci.be/Nomenclature/Standards/297g01-en.pdf (accessed on 21 August 2019).
54. FCI FCI—Standard N° 45. Bernese Mountain Dog. Available online: http://www.fci.be/Nomenclature/Standards/045g02-en.pdf (accessed on 25 March 2017).
55. FCI FCI—Standard N° 147. Rottweiler. Available online: http://www.fci.be/Nomenclature/Standards/147g02-en.pdf (accessed on 25 March 2017).
56. FCI FCI—Standard N° 15. Chien de Berger Belge (Belgian Shepherd Dog). Available online: http://www.fci.be/Nomenclature/Standards/015g01-en.pdf (accessed on 25 March 2017).
57. Adams, S.A.; Matthews, C.E.; Ebbeling, C.B.; Moore, C.G.; Joan, E.; Fulton, J.; Hebert, J.R. The Effect of Social Desirability and Social Approval on Self-Reports of Physical Activity. *Am. J. Epidemiol.* **2005**, *161*, 389–398. [CrossRef] [PubMed]
58. Fuchs, R.; Klaperski, S.; Gerber, M.; Seelig, H. Messung der Bewegungs-und Sportaktivität mit dem BSA-Fragebogen. Eine methodische Zwischenbilanz. [Measuring Physical and Exercise Activity Using the BSA-Questionnaire. A Methodological Interim Balance]. *Z. Für. Gesundh.* **2015**, *23*, 60–76. [CrossRef]
59. University of Freiburg Measuring Exercise and Physical Activity [Messung der Sport-und Bewegungsaktivität]. Available online: https://www.sport.uni-freiburg.de/de/institut/psychologie/messinstrumente/Messung_der_Sport_und_Bewegungsaktivitaet (accessed on 15 March 2022).

60. SoSci Survey GmbH Privacy and Data Protection in Online Surveys. Available online: https://www.soscisurvey.de/en/privacy (accessed on 15 March 2022).
61. Shek, D.T.L.; Ma, C.M.S. Longitudinal Data Analyses Using Linear Mixed Models in SPSS: Concepts, Procedures and Illustrations. *Sci. World J.* **2011**, *11*, 42–76. [CrossRef]
62. Field, A. *Discovering Statistics Using IBM SPSS Statistics*, 4th ed.; Sage Publications Ltd.: Thousand Oaks, CA, USA, 2014; ISBN 978-1-4462-4917-8.
63. Tabachnik, B.G.; Fidell, L.S. *Using Multivariate Statistics*, 6th ed.; Pearson Education Limited: Harlow, UK, 2014; ISBN 1-292-02131-4.
64. Liao, Y.; Huang, P.-H.; Chen, Y.-L.; Hsueh, M.-C.; Chang, S.-H. Dog Ownership, Dog Walking, and Leisure-time Walking Among Taiwanese Metropolitan and Nonmetropolitan Older Adults. *BMC Geriatr.* **2018**, *18*, 85. [CrossRef]
65. Christian, H.; Giles-Corti, B.; Knuiman, M. "I'm Just a'-Walking the Dog" Correlates of Regular Dog Walking. *Fam. Community Health* **2010**, *33*, 44–52. [CrossRef]
66. Thorpe, R.J.; Simonsick, E.M.; Brach, J.S.; Ayonayon, H.; Satterfield, S.; Harris, T.B.; Garcia, M.; Kritchevsky, S.B. Dog Ownership, Walking Behavior, and Maintained Mobility in Late Life. *J. Am. Geriatr. Soc.* **2006**, *54*, 1419–1424. [CrossRef]
67. Westgarth, C.; Christian, H.E.; Christley, R.M. Factors Associated With Daily Walking of Dogs. *BMC Vet. Res.* **2015**, *11*, 116. [CrossRef]
68. Masters, A.M.; McGreevy, P.D. Dogkeeping Practices as Reported by Readers of an Australian Dog Enthusiast Magazine. *Aust. Vet. J.* **2008**, *86*, 18–25. [CrossRef]
69. Nagasawa, M.; Kanbayashi, S.; Mogi, K.; Serpell, J.; Kikusui, T. Comparison of Behavioral Characteristics of Dogs in the United States and Japan. *J. Vet. Med. Sci.* **2016**, *78*, 231–238. [CrossRef] [PubMed]
70. Wilke, J.; Mohr, L.; Tenforde, A.S.; Edouard, P.; Fossati, C.; González-Gross, M.; Ramírez, C.S.; Laiño, F.; Tan, B.; Pillay, J.D.; et al. A Pandemic Within the Pandemic? Physical Activity Levels Substantially Decreased in Countries Affected by COVID-19. *Int. J. Environ. Res. Public Health* **2021**, *18*, 2235. [CrossRef] [PubMed]
71. Ammar, A.; Brach, M.; Trabelsi, K.; Chtourou, H.; Boukhris, O.; Masmoudi, L.; Bouaziz, B.; Bentlage, E.; How, D.; Ahmed, M.; et al. Effects of COVID-19 Home Confinement on Eating Behaviour and Physical Activity: Results of the ECLB-COVID19 International Online Survey. *Nutrients* **2020**, *12*, 1583. [CrossRef] [PubMed]
72. Ammar, A.; Boukhris, O.; Halfpaap, N.; Labott, B.K.; Langhans, C.; Herold, F.; Grässler, B.; Müller, P.; Trabelsi, K.; Chtourou, H.; et al. Four Weeks of Detraining Induced by COVID-19 Reverse Cardiac Improvements From Eight Weeks of Fitness-Dance Training in Older Adults With Mild Cognitive Impairment. *Int. J. Environ. Res. Public Health* **2021**, *18*, 5930. [CrossRef] [PubMed]
73. Bowen, J.; García, E.; Darder, P.; Argüelles, J.; Fatjó, J. The Effects of the Spanish COVID-19 Lockdown on People, Their Pets, and the Human-Animal Bond. *J. Vet. Behav.* **2020**, *40*, 75–91. [CrossRef]
74. Vučinić, M.; Vučićević, M.; Nenadović, K. The COVID-19 Pandemic Affects Owners Walking With Their Dogs. *J. Vet. Behav.* **2020**, *48*, 1–10. [CrossRef]
75. Christley, R.M.; Murray, J.K.; Anderson, K.L.; Buckland, E.L.; Casey, R.A.; Harvey, N.D.; Harris, L.; Holland, K.E.; McMillan, K.M.; Mead, R.; et al. Impact of the First COVID-19 Lockdown on Management of Pet Dogs in the UK. *Animals* **2021**, *11*, 5. [CrossRef]
76. Die Bundeskanzlerin und die Regierungschefinnen und Regierungschefs der Länder Erweiterung der Beschlossenen Leitlinien zur Beschränkung Sozialer Kontakte. Besprechung der Bundeskanzlerin mit den Regierungschefinnen und Regierungschefs der Länder vom 22.03.2020. Available online: https://www.bundesregierung.de/breg-de/themen/coronavirus/besprechung-der-bundeskanzlerin-mit-den-regierungschefinnen-und-regierungschefs-der-laender-vom-22-03-2020-1733248 (accessed on 11 March 2022).
77. Hartley, S.; Garland, S.; Bennell, K.L.; Tay, I.; Gorelik, A. A Comparison of Self-Reported and Objective Physical Activity Measures in Young Australian Women. *JMIR Public Health Surveill.* **2014**, *1*, e14. [CrossRef]
78. Nicolaou, M.; Gademan, M.G.J.; Snijder, M.B.; Engelbert, R.H.H.; Dijkshoorn, H.; Terwee, C.B.; Stronks, K. Validation of the SQUASH Physical Activity Questionnaire in a Multi-Ethnic Population: The HELIUS Study. *PLoS ONE* **2016**, *11*, e0161066. [CrossRef]
79. Wagoner, C.W.; Choi, S.K.; Deal, A.M.; Lee, J.T.; Wood, W.A.; Muss, H.B.; Nyrop, K.A. Establishing Physical Activity in Breast Cancer: Self-Report Versus Activity Tracker. *Breast Cancer Res. Treat.* **2019**, *176*, 395–400. [CrossRef]
80. Pedersen, S.J.; Kitic, C.M.; Bird, M.L.; Mainsbridge, C.P.; Cooley, P.D. Is Self-Reporting Workplace Activity Worthwhile? Validity and Reliability of Occupational Sitting and Physical Activity Questionnaire in Desk-Based Workers. *BMC Public Health* **2016**, *16*, 836. [CrossRef] [PubMed]
81. Pedersen, E.S.L.; Mortensen, L.H.; Brage, S.; Bjerregaard, A.L.; Aadahl, M. Criterion Validity of the Physical Activity Scale (PAS2) in Danish Adults. *Scand. J. Public Health* **2018**, *46*, 726–734. [CrossRef] [PubMed]
82. Peters, T.M.; Shu, X.O.; Moore, S.C.; Xiang, Y.B.; Yang, G.; Ekelund, U.; Liu, D.K.; Tan, Y.T.; Ji, B.T.; Schatzkin, A.S.; et al. Validity of a Physical Activity Questionnaire in Shanghai. *Med. Sci. Sports Exerc.* **2010**, *42*, 2222–2230. [CrossRef] [PubMed]
83. Al-Ansari, S.; Biddle, S.; Borodulin, K.; Buman, M.; Cardon, G.; Carty, C.; Chaput, J.-P.; Chastin, S.; Chou, R.; Dempsey, P.; et al. WHO Guidelines on Physical Activity and Sedentary Behaviour. Available online: https://www.who.int/publications/i/item/9789240015128 (accessed on 28 April 2021).
84. Füzéki, E.; Engeroff, T.; Banzer, W. Health Benefits of Light-Intensity Physical Activity: A Systematic Review of Accelerometer Data of the National Health and Nutrition Examination Survey (NHANES). *Sport Med.* **2017**, *47*, 1769–1793. [CrossRef] [PubMed]

85. Ku, P.-W.; Hamer, M.; Liao, Y.; Hsueh, M.C.; Chen, L.J. Device-Measured Light-Intensity Physical Activity and Mortality: A Meta-Analysis. *Scand. J. Med. Sci. Sport.* **2020**, *30*, 13–24. [CrossRef] [PubMed]
86. Saint-Maurice, P.F.; Troiano, R.P.; Berrigan, D.; Kraus, W.E.; Matthews, C.E. Volume of Light Versus Moderate-to-Vigorous Physical Activity: Similar Benefits for All-Cause Mortality? *J. Am. Heart Assoc.* **2018**, *7*, e008815. [CrossRef]
87. Chastin, S.F.M.; De Craemer, M.; De Cocker, K.; Powell, L.; Van Cauwenberg, J.; Dall, P.; Hamer, M.; Stamatakis, E. How Does Light-Intensity Physical Activity Associate With Adult Cardiometabolic Health and Mortality? Systematic Review With Meta-Analysis of Experimental and Observational Studies. *Br. J. Sports Med.* **2019**, *53*, 370–376. [CrossRef]
88. Oswald, L.M.; Wand, G.S.; Zhu, S.; Selby, V. Volunteerism and Self-Selection Bias in Human Positron Emission Tomography Neuroimaging Research. *Brain Imaging Behav.* **2013**, *7*, 163–176. [CrossRef]
89. Pan, Y.; Zhan, P. The Impact of Sample Attrition on Longitudinal Learning Diagnosis: A Prolog. *Front. Psychol.* **2020**, *11*, 1051. [CrossRef]
90. Gustavson, K.; Von Soest, T.; Karevold, E.; Roysamb, E. Attrition and Generalizability in Longitudinal Studies: Findings From a 15-Year Population-Based Study and a Monte Carlo Simulation Study. *BMC Public Health* **2012**, *12*, 918. [CrossRef]

animals

MDPI

Article

The Role of Dogs in the Relationship between Telework and Performance via Affect: A Moderated Moderated Mediation Analysis

Ana Junça-Silva [1,2,*], Margarida Almeida [2] and Catarina Gomes [3,4,5,6]

1 ISCTE–IUL, Business Research Unit, Lisbon University Institute, Avenida das Forças Armadas, 1649-026 Lisboa, Portugal
2 Instituto Politécnico de Tomar, Escola Superior de Gestão de Tomar, Quinta do Contador, Estrada da Serra, 2300-313 Tomar, Portugal; margarida.almeida@ipt.pt
3 CIPES, Centro de Investigação em Política, Economia e Sociedade, Universidade Lusófona de Humanidades e Tecnologias, Campo Grande, 376, 1749-024 Lisboa, Portugal; catarinagomes04@gmail.com
4 Escola de Ciências Económicas e das Organizações, Universidade Lusófona de Humanidades e Tecnologias, 1749-024 Lisboa, Portugal
5 TRIE: Centro de Investigação Transdisciplinar Para o Empreendedorismo e Inovação Ecossistémica, Universidade Lusófona de Humanidades e Tecnologias, 1749-024 Lisboa, Portugal
6 CICPSI, Faculdade de Psicologia, Universidade de Lisboa, Alameda da Universidade, 1649-013 Lisboa, Portugal
* Correspondence: ana_luisa_silva@iscte-iul.pt or analjsilva@gmail.com

Simple Summary: In this research, we conducted a study with an overall sample of 401 individuals to test a mediating model between telework, positive affect, and self-reported job performance. Additionally, we analyzed whether dogs' physical closeness and emotional attachment would moderate this mediating path. The results showed that telework was significantly and positively related to positive affect, which in turn, increased self-reported job performance. Moreover, the mediation model was moderated by the dog's physical closeness while working and emotional attachment to them, in such a way that the relationship between telework on self-reported job performance, via positive affect was strengthened when the owner's physical and emotional closeness to their dogs was higher. In sum, telework might be an efficient strategy to improve performance among employees who have dogs at home, because working with them nearby, when emotionally attached to them, are factors that enhance the individual's self-perceived performance in telework.

Abstract: Although there is evidence that pets may help individuals facing significant daily stressors, and that they may enhance the well-being of their owners, little is known about the benefits of pets for job performance. Since the COVID-19 pandemic crisis, teleworking was a strategy implemented in many countries to reduce the virus widespread and to assure organizational productivity. Those who work from home and who own pets may work close to them. Based on the conservation of resources theory, this study aimed to analyze whether positive affect mediated the relationship between telecommuting and self-reported job performance and if psychological and physical closeness to the pet would moderate this relationship in such a way that it would be stronger for those who worked closer to their pet, and who were more emotionally attached to them. For this study, we collected data from 81 teleworkers who did not own pets, and from 320 teleworkers who owned pets. Both answered an online questionnaire. Findings: Results from the study showed the existence of significant differences between those who owned and who did not own pets regarding positive affect and performance, in which those who owned pets reported higher levels of positive affect and self-reported performance and perceived telework more positively. Moreover, positive affect mediated the relationship between telework and self-reported job performance. Furthermore, emotional and physical closeness moderated the mediating effect. This study contributes to a better understanding of the human-animal interaction and how pets can be a personal resource able to change their owners' affective experiences and job performance while they are working from home. The findings demonstrate that telework may be a suitable organizational strategy for pet-owners.

Citation: Junça-Silva, A.; Almeida, M.; Gomes, C. The Role of Dogs in the Relationship between Telework and Performance via Affect: A Moderated Moderated Mediation Analysis. *Animals* **2022**, *12*, 1727. https://doi.org/10.3390/ani12131727

Academic Editor: Betty McGuire

Received: 30 May 2022
Accepted: 29 June 2022
Published: 4 July 2022

Keywords: pet attachment; telework; positive affect; self-reported job performance; pet closeness; dog ownership

1. Introduction

"Dogs do not need space; they just need to be nearby you!" (Jesse Koz & Shurastey).

The quote above represents that dogs and the bonds they have with their owners are a personal resource for them, inclusive while working [1]. The conservation of resources theory (COR) [2] argues that individuals who possess greater resources are less vulnerable to resource loss and more capable of resource gain [3]. Thus, individuals higher in personal resources (e.g., pet closeness) are in a better position to invest resources that, in turn, may result in positive outcomes such as performance [4].

Telework is a work arrangement that allows employees to work from a remote location [5]. Since the pandemic crisis of COVID-19, organizations were forced to implement it to survive and reduce the chance of spreading the virus. Telecommuting has been found to be positively associated with autonomy and flexibility [6] that in turn appears to enhance job satisfaction [7] and positive affect [8].

Despite the correlational evidence of the benefits of dogs for positive outcomes, such as well-being or stress reduction [9], studies exploring their benefits for workplace outcomes are scarce. As such, this study attempts to determine why attitudes toward telecommuting predict the quality of self-reported job performance and when it will occur. Thus, drawing on the COR theory, we propose that pets, conceptualized as a resource, can influence the effects of attitudes towards telecommuting on affect and self-perceived performance, thus moderating the mediating effect.

2. The Relationship between Telework, Positive Affect, and Performance

Telework, originally proposed by Nilles in the 1970s, is a work arrangement also recognized as telecommuting or remote work and was originally defined as working from a remote location away from a standard office or work site [5]. It was defined as "working outside the conventional workplace and communicating by way of telecommunications or computer-based technology" [5] (p. 384). Similarly, Fitzer [10] defined telecommuting as a "work arrangement in which employees perform their regular work at a site other than the ordinary workplace, supported by technological connections" [11] (p. 336).

Telework has progressively spread over the last 40 years and has been strongly encouraged by the measures to limit the COVID-19 pandemic [12]. The Global Workplace Analytics (2019 cited in [13]) predicted that over 70% of the workforce will be working remotely (at least five days per month) in the next five years.

Gajendran and Harrison [14] evidenced that "telecommuting is mainly a good thing" (p. 1535) and showed that it was associated with increased perceptions of autonomy at work. Telecommuting also provides freedom and flexibility and offers many benefits such as positive affective experiences [6,7,15].

Positive affect includes brief and multidimensional affective responses to events or changes in the environment and is based on the individual's interpretation of these events or contexts [16,17]. Accordingly, positive affect leads individuals to engage in novel and larger behavioral repertoires; and is related to positive behaviors that are important for workers' performance, such as giving more attention to the tasks at hand [18]. In addition, positive affect serves to build personal resources that help workers to deal effectively with their daily life at work. These resources may encompass physical (e.g., energy), intellectual (e.g., knowledge), social (e.g., empathy), and psychological (e.g., engagement) aspects that buffer against the deleterious effects of daily hassles [19]. Past research suggested that resources are fundamental for individuals dealing with work-related hassles [20] and job demands [21], and for energizing performance [22]. In this regard, resources can include positive behaviors, such as perceived task performance.

There are differences between task and contextual performance [23]. While task performance is related to the core tasks of the individual and therefore is related to the organizational goals (e.g., goal attainment, judgment, and decision-making), the contextual performance is referred to all the activities and behaviours that contribute to the work's psychological climate and include, for example, helping colleagues engage in learning.

Working from home might lead to different subjective experiences from individual to individual. For instance, it is likely that many workers worked remotely for the first time, and many of them had no choice. Despite this and based on the social exchange theory and on the broad-and-brand theory, we expect that telecommuting, by promoting flexibility, autonomy and freedom will increase positive affect and this, in turn, will enhance job performance. Thus, we defined the following hypothesis:

H1. *The attitude toward telework will be positively related to self-perceived performance.*

H2. *Positive affect will mediate the positive relationship between the attitude toward telework and self-reported job performance.*

3. The Moderating Role of Pet Attachment and Physical Closeness

Human-pet interaction and bonding is an interspecies relationship that is historically old [24]. With respect to this Brickel [25], suggested that animals provide "one highly reliable association in a person's life ... more consistent and reliable than human-human" (p. 310). Moreover, Bradshaw [26], reinforced that: pets hold a "relationship of mutualism" with their owners; that is, pet owners believe they not only give but receive love and affection from their animals. Likewise, Cusak [27], argued that pets are human confidants with no risk of betrayal. Indeed, pets can create connections through their vivacity and ability to interact with humans. Plus, they are sensitive to the feelings of their owners and change their behavior in certain situations [28,29].

The importance of pets for human life has received some attention. Indeed, pets can provide individuals with many benefits, such as stress reduction and increased well-being [30]. However, these findings are controversial because some studies demonstrated significant positive effects whereas others showed non-significant ones [31,32]. While some authors argued that positive effects on health were non-significant, others demonstrated the opposite. For instance, Dotson and Hyatt [24], showed that pet ownership lowers blood pressure, helps to prevent heart disease, to fight depression, and therefore improves one's health. Moreover, pets can increase the mental health of their human owners [33]. Perhaps, this may justify the increased importance attributed to pets.

The number of individuals with pets is increasingly higher than some decades ago [31]; moreover, pets are considered, for many owners, as family members, best friends, companions, or "furry babies" [9]. As the number of pets increases, and as pets take a more central role in the lives of individuals, there is an increased need to consider how having a pet at home might affect an individual's work-related outcomes, such as job performance [9]. Although often overlooked, pets cross with organizations in relevant ways. More recently, organizations and managers have acknowledged it and some of them are becoming pet-friendly. For instance, there are organizations that let their workers take their pets to work, for instance, as Amazon or Google. By accommodating pets, organizations promote positive effects for workers, since many of them consider their pets as family members [34]. Many studies demonstrated that having a pet around increases positive affect and the number of prosocial behaviors [35]. Similarly, Wagner and Pina e Cunha [1], found that the presence of pets at work reduced stress, improved communication, and enhanced social cohesion. Moreover, Pina e Cunha et al. [36], stated that in companies where employees may bring their pets to work, problems tend to be addressed openly, and employees have more autonomy, with flexibility for breaks and greater tolerance for failure and errors. Hall and Mills [37] reported that workers who often took their pets to work reported higher work engagement, work-based friendship, and fewer turnover intentions, compared to those who never took their pets to work. Hall and Mills [37] also showed that those who frequently took their pet to work evidenced higher work-related quality of life, general wellbeing, home-work interface, job-career satisfaction, more control at work, and better-perceived working conditions compared to those who never took their pet to work.

One reason why people might benefit from working closer to their pets is that they represent an important source of social support. Hence, several studies showed that higher social support improved psychological and physiological health [38,39], greater self-esteem [38], and higher performance rates [40].

Following the recommendations of the World Health Organization, in March 2020, many governments swiftly enacted states of emergency, involving, for example, mandatory telecommuting, stay-at-home orders, physical distancing requirements, and quarantine measures for exposed individuals [41,42]. As COVID-19 pushes employees to work from home, many employees are working alongside their pets for the first time. Also, due to social distancing measures of the COVID-19 crisis, it is likely that pets become more important for individuals, particularly those who live alone. Pets may have the potential to help individuals cope with the loneliness and anxiety that may come from social distance and isolation that comes from telecommuting, as well as the uncertainness and worry that comes with thinking about COVID-19. This presents a unique context to better understand how working closer to pets may affect employee work behavior and attitudes and to better understand the pros and cons of working alongside pets.

The conservation of resources theory [2] is appropriate in further explaining employees' personal gains from working closer to their pets. Accordingly, the theory states that individuals who possess greater resources are less vulnerable to resource loss and more capable of resource gain [3]. Thus, when individuals are higher in personal resources (e.g., working close to their pets), they become less vulnerable to resource loss and are in a better position to invest resources into the engagement process and that, in turn, may result in positive outcomes, such as job performance [4].

Drawing on COR theory, we propose that pets, conceptualized as a resource, can assist in understanding how people face the effects of working from home, due to the coronavirus, on affect and performance, thus moderating this mediating effect. Specifically, we expect that individuals with higher levels of pet attachment and who work close to their pets will be able to focus on the tasks at hand, improving their self-reported job performance, while working from home (see Figure 1).

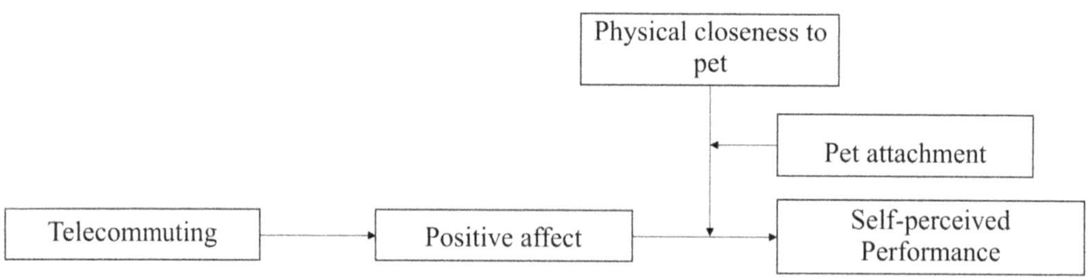

Figure 1. Conceptual model (moderated moderated mediation).

H3. *The relationship between the attitude toward telework and self-reported job performance, through positive affect, will be moderated by the physical closeness to pets while working, and by the emotional attachment to them, such that the indirect effect becomes stronger when individuals work closer to their pets (versus not working nearby them) and when they show higher levels of emotional attachment to them (versus lower levels) (moderated moderated mediation).*

4. Method

4.1. Participants and Procedure

Before conducting the study, this was approved by the ethics committee of the university, thereby we could proceed with the study. We used a non-probabilistic convenience sample as we resorted to participants from our professional networks. We collected data from two groups of teleworkers. One group (n = 81) did not own dogs, and the other group owned dogs (n = 320). All the participants, from the two groups, were Portuguese individuals transitioning from working at a traditional work location (e.g., an office) to working at home because of the coronavirus and the resultant mandatory confinement.

We collected data through a questionnaire-based survey on the second mandatory confinement due to the COVID-19 pandemic crisis (during February and March of 2021 which was one of the peak periods of COVID-19 in Portugal). Participants from our professional networks were emailed to participate in a study about attitudes at work and toward animals. In order to meet the ethical requirements, we ensured them the anonymity and confidentiality of their responses and we asked them to reply if they agreed to participate. Those who answered the email received another one with the link for the survey. Overall, 450 general questionnaires were distributed among teleworkers, from which 401 agreed to voluntarily participate in the study (response rate: 89.11%).

4.1.1. The Group without Pets

We collected data from 81 teleworkers who did not own pets. The mean age was 32.09 years old (SD = 9.48), of which 51% were male and 58% reported being single. The mean organizational tenure was 6.20 years (SD = 8.60), and the mean hours worked per week was 40.31 (SD = 12.10). On average, the household consisted of 2.5 individuals (SD = 1.30) and most of the participants did not have kids at home (73.6%). Participants worked in diverse occupational areas, being the majority in administrative functions (46%), marketing (32%), and teaching (22%).

4.1.2. The Group with Pets

Of the 320 participants with pets, most of them were female (62%), the mean age was 31.87 years old (SD = 9.50), and the mean organizational tenure was 5.13 years (SD = 7.78). Participants worked in several occupations, being the majority in administrative functions (58%), teaching (32%), and insurance salesman (10%). On average, participants worked about 41 h per week (SD = 11.12) and had 1.31 animals (SD = 1.31), of which 1.04 (SD = 1.22) lived indoor, and 0.30 (SD = 0.82) lived outside the house. All participants reported being owners of dogs (100%). On average, the household consisted of 2.82 individuals (SD = 1.59) and most of the participants did not have kids at home (68.4%).

4.2. Measures

The attitude toward telework was measured with the 17-item E-Work Life Scale [43]. This scale measures four aspects related to the perceived quality of telework experience: effectiveness/productivity (four items; e.g., "When e-working I can concentrate better on my work tasks"), organizational trust (three items; e.g., "My organisation provides training in e-working skills and behaviours."), the interference between personal and work-life (seven items; e.g., "My social life is not poor when e-working remotely"), and flexibility (three items; e.g., "My work is so flexible I could easily take time off e-working remotely, if and when I want to."). Participants answered on a 5-point scale (1—totally disagree; 5—totally agree). The Cronbach's alpha for this scale was 0.73.

Positive affect was measured with eight items from Multi-Affect Indicator [44] to assess the frequency of daily positive affect experienced at work on that day (e.g., "enthusiastic"). Participants answered on a 5-point scale (1—never; 5—always). The Cronbach's alpha was 0.90.

Self-reported job performance was assessed through the 6-item In-Role Performance Scale [45]: "Today, I achieved my job goals". Items were rated on a 5-point scale ranging from 1 (very little) to 5 (a great deal) ($\alpha = 0.84$).

The emotional attachment was assessed through the Lexington Attachment to Pets Scale (LAPS) [46]. This scale assesses the perception of individuals towards their pets, and is divided into three dimensions: (1) general attachment (11 items; e.g., "6. I play with my pet quite often."); (2) people substituting (seven items; e.g., "My pet means more to me than some friends"); (3) animals rights/animal welfare (five items; e.g., "I consider my pet to be part of the family"). Answers were given on a 5-point Likert scale ranging from (1) completely disagree to (5) completely agree. In this study, we only used the first dimension of the scale ($\alpha = 0.97$).

Physical closeness to pets was measured with four items focused on physical and interaction moments with pets while working. Responses were given on a five-point Likert scale (1—never; 5—always) (e.g., "In telework, I usually take breaks to interact with my pet"; "While you work from home, my pet is close to me when I am working", "In telework, my pet is not close to me while I work"; "While I am working, I use to interact with my pet"). The Cronbach's alpha was 0.87.

Control variables. We used sex and age as control variables. Sex may account for differences in experienced affect [44] and age may account for differences in the perceived experience of telework [43].

Each of the five surveys described above is included in Appendix A.

4.3. Data Analysis

Relationships were tested using PROCESS macro 3.1 [47] (in SPSS v. 25), specifically model 4 (mediation) and 18 (moderated moderated mediation). Control variables (age, sex) were entered in Step 1 with telework entered as the independent variable, positive affect as the mediator, and performance as the dependent one (mediating model). Then, physical closeness and emotional attachment to pets were entered as the moderator variables, the products were mean-centered, and bootstrapping (5000 times) was used to provide confidence intervals (moderated moderated mediation).

As both the predictor and the criterion variables were measured at the same time, we took some measures to avoid the issue of common method variance [48]. First, we shuffled the questions of various measures and then used various dummy questions (e.g., I like pets). Second, Harman's single factor test was used to assess the common method variance, and it was observed that the single factor accounted for only 22.95% variance, which was much below the standard value of 50% proposed by Podsakoff et al. [49], thus the common method variance issue was not severe for this study.

5. Results

5.1. Descriptive Statistics and Correlations

Table 1 shows the descriptive statistics and correlations between the variables.

5.2. Means Comparison between Groups

Before testing our hypotheses, we analyzed whether there were differences among the variables under study between the two groups of participants (pet owners and non-pet owners). Results showed statistically significant differences for positive perceived effects of telework ($F_{(398)} = 4.80$, $p < 0.001$), positive affect ($F_{(398)} = 4.27$, $p < 0.01$), and performance ($F_{(398)} = 3.39$, $p < 0.05$), suggesting that pet-owners had a more positive perception of telework ($M = 3.30$, $SD = 0.46$ versus $M = 3.20$, $SD = 0.45$) experienced more positive affect ($M = 3.21$, $SD = 0.54$ versus $M = 3.11$, $SD = 0.73$) and showed higher levels of self-reported performance than non-pet owners ($M = 4.10$, $SD = 0.55$; $M = 3.98$, $SD = 0.45$), respectively (see Table 2).

Table 1. Correlations and descriptive statistics of the variables under study.

Variables	*M*	*SD*	1	2	3	4	5	6	7
1. Telework	3.20 [1]	0.51	-						
2. Positive affect	3.11 [1]	0.69	0.50 **	-					
3. Self-reported job performance	4.05 [1]	0.54	0.31 **	0.37 **	-				
4. Pet closeness	2.63 [1]	1.29	0.00	0.01	0.07	-			
5. Pet attachment	3.78 [1]	0.99	−0.06	−0.06	0.17 **	0.69 **	-		
6. Sex	-	-	0.01	0.13 **	−0.01	−0.20 **	−0.18 **	-	
7. Age	31.87	9.50	0.02	0.07	0.14**	−0.07	−0.12 *	0.12 *	-

$n = 401$; * $p < 0.05$ ** $p < 0.001$; [1] 5-point Likert scale: attitudes toward telework and pet attachment: 1 = totally disagree; 5 = totally agree; positive affect and pet closeness: 1 = never, 5 = always; self-reported job-performance: 1 = very little; 5 = a great deal).

Table 2. Means comparisons between the two groups of participants (pet owners versus non-pet owners).

Groups	Pet-Owners	Non-Pet Owners	*F*
Variables	*M (SD)*	*M (SD)*	
Perceived telework effects	3.30 (0.46)	3.20 (0.45)	4.80 ***
Positive affect	3.21 (0.54)	3.11 (0.73)	4.27 **
Performance	4.10 (0.55)	3.98 (0.45)	3.39 *

Note. Groups: Pet-owners ($n = 320$); non-pet owners ($n = 81$). * $p < 0.05$, ** $p < 0.01$, *** $p < 0.001$.

5.3. Hypotheses Testing

Hypothesis 1 predicted that the attitude toward telework would be positively related to perceived job performance. The results evidenced that the attitude toward telecommuting was significantly related to perceived job performance ($B = 0.33$, $p < 0.001$). Therefore, the first hypothesis was supported by the data.

Hypothesis 2 expected that the attitude toward telework would positively influence self-reported job performance through positive affect. The results showed a significant indirect effect of positive affect (0.15 with a 95% CI [0.08, 0.22]). Moreover, the relationship between telework and positive affect ($B = 0.67$, $p < 0.01$) and between positive affect and self-reported job performance ($B = 0.22$, $p < 0.01$) were significant. The total effect ($B = 0.33$, $p < 0.01$) between the attitude toward telework and self-reported job performance was also significant. After entering positive affect, the effect of the attitude toward telework on self-reported job performance remained significant ($B = 0.18$, $p < 0.01$) suggesting a partial mediating relationship, and thus lending support to hypothesis 2 (see Table 3).

Table 3. Summary regression table of the mediation model (Hypothesis 2).

Model	Positive Affect (M)			Self-Reported Performance (Y)		
	B	*SE*	*t*	*B*	*SE*	*T*
Telework (X)	0.67 **	0.07	10.07	0.18 **	0.06	2.92
PA (M)	-	-	-	0.22 **	0.05	4.75
Age	0.00	0.00	0.89	0.01 *	0.00	2.30
Sex	0.17 *	0.07	2.48	−0.07	0.06	−1.21
Indirect Effect	Effect (γ)	BootSE			LLCI-ULCI	
PA	0.15	0.04			[0.08, 0.22]	

Note. $n = 320$; * $p < 0.05$ ** $p < 0.001$. *B* = Unstandardized coefficients; PA = Positive affect.

Hypothesis 3 expected that the indirect effect of perceived effects of telework on self-reported performance via positive affect would be moderated by pet physical closeness and emotional attachment, in such a way that the relationship would become stronger for those who were closer (versus distant) and more attached to their pets (versus lower attachment). To test this hypothesis, we followed the recommendations from Hayes [48] to perform the moderated moderated mediation. The results showed that the moderated moderated mediation index was 0.26 (CI 95% [0.02, 0.52]). This means that the indirect effect of telework on self-reported job performance (through positive affect) differs between individuals who work closer to their pets and with different pet attachment levels (see Table 4).

Table 4. Summary regression table of the moderated-moderated-mediation model (Hypothesis 3).

Model	Positive Affect (M)			Self-Reported Performance (Y)		
	B	*SE*	*T*	*B*	*SE*	*t*
Telework (X)	0.66 **	0.07	10.07	0.24 **	0.06	3.31
PA (M)	-	-	-	0.13 *	0.06	2.16
Pet closeness (Mod)	-	-	-	−0.19 *	0.08	−2.48
Pet attachment (Mod)	-	-	-	0.19 **	0.04	4.57
PA * Pet attachment * Pet closeness	-	-	-	0.39 **	0.12	2.46
Age	0.00	0.00	0.89	0.01 *	0.00	2.77
Sex	0.17 *	0.07	2.48	−0.02	0.05	−0.43
Index of mod-mod-med effect	Effect (γ)	BootSE			LLCI-ULCI	
PA	0.26	0.14			[0.02, 0.52]	
	$R^2 = 0.26$ $F_{(11,308)} = 10.45$, $p = 0.00$, $\Delta R^2 = 0.02$, $p = 0.01$					

Note. $n = 320$; * $p < 0.05$ ** $p < 0.001$. *B* = Unstandardized coefficients; PA = Positive affect.

Then, we followed the suggestion of Hayes [48] to probe the conditional indirect effect. Specifically, we examined the magnitude and significance of the indirect effect of telework on self-reported job performance through positive affect, conditional on physical closeness to pets (at −1SD, mean, and +1SD) for pet attachment levels. The slope analysis showed that the indirect effect was significant for (1) individuals that worked closer to their pets, and had higher attachment levels to them ($B = 0.20$, $p < 0.01$, with CI 95% [0.10, 0.31]), and for (2) individuals who did not work closer to their pets, but whose pet attachment was lower ($B = 0.07$, $p < 0.01$, with CI 95% [0.06, 0.34]) (Figure 2). Thus, the third hypothesis was supported (Figure 3).

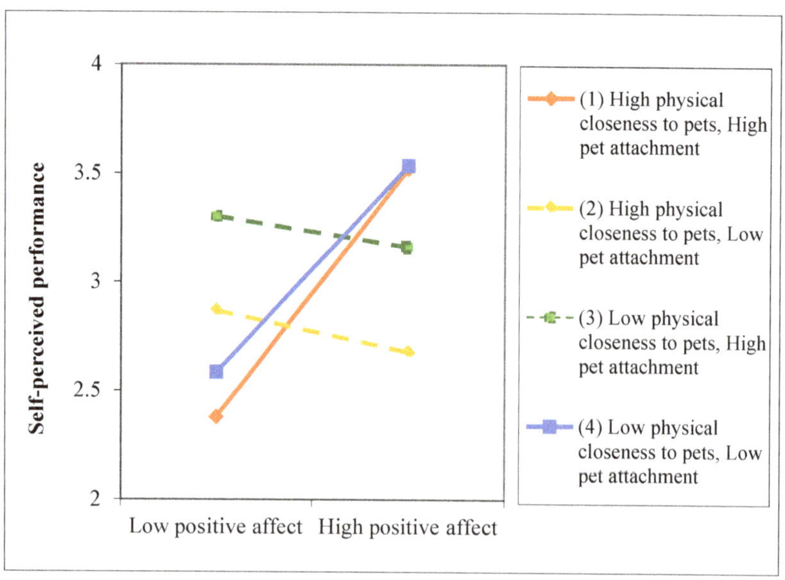

Figure 2. Indirect effect of telecommuting on self-perceived performance through positive affect conditional on pet closeness and attachment.

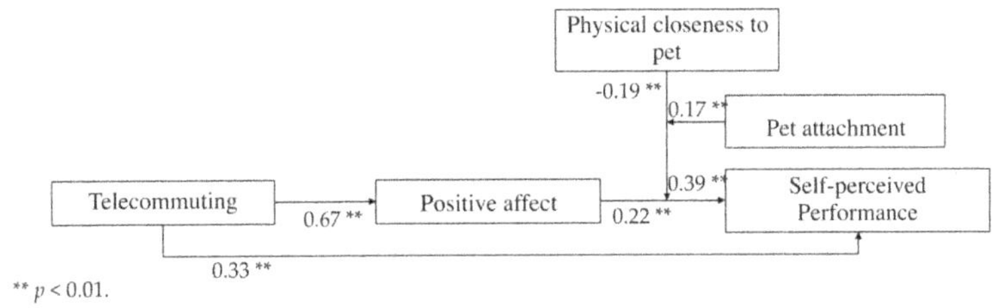

Figure 3. The moderated moderated mediation model with the results.

6. Discussion

The present study examined the role of positive affect on the self-reported job performance of teleworkers highlighting the importance of pet physical and emotional closeness on job performance of them. This study answers the call of studies from [9] to explore the benefits of pets for performance outcomes. Specifically, this study aimed to contribute to understanding the process and the conditions through which telecommuting improves self-reported job performance.

First, the results show that the attitude toward telecommuting is positively associated with self-perceived job performance, that is, while telecommuting workers appear to consider having a positive performance. This result is in line with other studies that have shown a positive effect of working from home on job performance [43].

Second, the results show that the attitude toward telecommuting improves positive affect which in turn enhances self-reported job performance. That is when individuals show higher levels of perceived life quality while teleworking, this tends to positively influence positive affect while working which in turn promotes self-reported job performance. This result is supported from a social exchange perspective. Accordingly, individuals tend to behave by weighing the costs and benefits that they expect to receive (e.g., flexibility) [7]. Thus, individuals who are telecommuters have more flexibility and autonomy at work, raising their feelings of obligation towards their organization, which in turn may enhance their positive affect while working promoting self-reported job performance. Moreover, telework is a model of work characterized by increased levels of flexibility, autonomy, and a sense of control over work [43]. These positive work characteristics trigger more frequently positive affect among workers [8,50], despite the pandemic times being lived. This is evidenced by the job characteristics model [51]. The model has been widely used to determine whether certain core characteristics of jobs (e.g., autonomy) do evoke some affective reactions by workers. Accordingly, autonomy has been consistently related to positive affect and other affective and motivational indicators [52]. Therefore, it is not surprising that telecommuting enhances positive affect at work. Additionally, the broaden-and-build theory suggests that positive affect broadens positive behaviors [15]. Accordingly, positive affect leads individuals to engage in novel and larger behavioral repertoires; and is related to positive behaviors that are important for workers' performance, such as giving more attention to the tasks at hand [18]. Plus, positive affect builds personal resources that help workers to energize performance [22]. Empirically, there is also evidence of the positive link between telework, positive affect, and self-reported job performance. For instance, Anderson et al. [53], showed that workers showed more positive emotions when they were teleworking when compared to days at the office. Similarly, Abdel Hadi et al. [54], in a diary study developed during the pandemic crisis of COVID-19, showed that individuals in telecommuting experienced fewer negative emotions and better performance rates.

Notwithstanding, this mediating effect seems to be conditional to the physical closeness to pets while working, and to the emotional attachment to them. Specifically, the results show that the mediation is stronger when telecommuters are working closer to their pets, and when they hold an emotional bond with them. That is, when telecommuting, positive affect is more frequent, leading to increases in self-reported job performance, for employees who work closer to their pets, and who demonstrate a high level of attachment to them. The mediation is also significant for individuals who do not work close to their pet, but whose emotional attachment is low. Although it is not a significant decrease, when emotional attachment to pets is high, and work is not being carried out closer to them, self-reported job performance tends to decrease, even after experiencing positive affect. There is evidence that the interaction between humans and their pets arouses oxytocin—a hormone responsible for well-being and love [36], making the individual feel happier and, as such, improving task performance. Barker et al. [55] showed that pets influence their owners through basic interactions such as observing and caring, which helps them to deal effectively with their daily tasks. Gee et al. [56], in an experimental study showed that performance on a memory task was better in the presence of a dog (compared to the absence, or the presence of a person). Therefore, the presence and interaction with pets, during work, when individuals are emotionally attached to them, can improve self-reported job performance. Attachment theory and the COR theory help explain these beneficial effects to workers. Attachment theory [57–59] suggests that a close emotional attachment between a pet and an individual provides psychological security, a source of social support, and advanced performance, for the individual, especially during this time of pandemic [60,61]. In addition, the COR theory [2] helps explain employees' personal gains from working with animals. Accordingly, pets may be viewed as a resource for the individual, thereby promoting attention to the tasks and improving their perceived performance, even though the pandemic times are being lived.

In this study, we focus on affect as a mechanism to explain how teleworking impacts self-reported job performance and we show that this is more beneficial for individuals who work closer to their pets, and when they hold an emotional bond with them. Thus, a day is not only better with a pet, but a pet also makes it a productive day.

6.1. Limitations and Future Research

Despite the positive features of this study, such as being a preliminary study in a relevant field, and with two working samples, it has some limitations. First, we must consider the differential sizes between each group (pet-owners and non-pet owners). The non-pet owner group was smaller than the other group; hence, the interpretation of the means comparison results should be regarded with some caution. Second, we used self-reported measures, which might account for common method variance [49–62], however, as referred before, we took some strategies to minimize it. Second, there are studies demonstrating that individual differences (e.g., personality traits) may influence how individuals perceive themselves, for instance regarding performance [63]. For instance, positive affectivity might positively influence self-perceived performance. Thus, future studies should examine whether positive affectivity or other personality traits (e.g., optimism) might influence perceived performance. Additionally, future studies could use other sources of information (e.g., colleagues, supervisors) regarding performance. Third, the fact that data was collected cross-sectional is a limitation. Therefore, future studies could replicate this study through a longitudinal or daily study. These designs would also safeguard type 1 errors—a safeguard to the internal validity of the study. Fourth, we only measured self-reported task performance because we were interested in this specific type of behavior. However, future studies might consider exploring contextual performance or creativity. Fifth, given the period of data collection—in mandatory confinement—there might have been some affect bias as people have experienced dramatic changes in affect, well-being, and mental health [64,65]. Thus, future studies should retest the model. Moreover, we did not measure some pet-related variables, such as the age of pets, and duration of pet ownership because we did not want a too long survey; however, we acknowledge that these variables might have some effect on diverse criterion variables such as performance or well-being. Thus, future studies would consider including such information.

These results open the way for future studies. First, the finding that pets are a condition through which telecommuting impacts affect and self-reported job performance, is relevant, as most studies have disregarded the importance of pets for human and organizational life [9]. Second, it would be interesting to test the model with other criterion variables, for instance, overall health. To do this, future studies could use objective measures of health, such as heart rate or blood pressure. Third, future studies should explore the role of different pet species (e.g., dogs, cats) because there are studies that demonstrated that different species had different effects. For instance, a study developed in a dentist's office showed that an aquarium full of fishes provided a relaxing climate and made the space calmer [54]. A study at Ferrari revealed that a cat interacting with people provided little distraction and, at the same time, lowered stress levels [66]. Similarly, diverse studies developed in technology companies showed that dogs interacted more, and needed other types of caring and attention, but could make the environment more dynamic, creative, and warming [30]. Fourth, future studies could retest the model and compare it between participants who own pets, and who do not. Moreover, because many workers were in mandatory remote work, future research should replicate this model, once the pandemic is over, to see if the results are the same. At last, in this study we only explored the benefits of being closer to pets regarding performance; however, it is possible that other pet-side benefits arise during their owners' teleworking, which strengthens the pet-human attachment. Thus, future studies might consider exploring other benefits for individuals (e.g., health). Plus, it would be relevant to analyze specific behavioral characteristics of pets that can impact human-animal interaction. For example, too much closeness and attachment to pets might be a distracting factor, decreasing one's working performance. As such, designing an experience sampling method would be relevant to analyzing daily fluctuations in human-animal interactions and subsequent distractions.

6.2. Practical Contributions

In sum, telework and affect are important variables for the prediction of performance. This study also emphasizes that this relationship is stronger when individuals work closer to their pets and to whom emotional attachment is higher. Thus, the relevance of pets at work has important implications for organizational theories and applied purposes, such as performance management, and employee development.

The results show that telecommuting is a way to assist workers' affective well-being and their self-reported job performance. Thus, adopting this model of work can be a strategy not only for pandemic times but also for the future. This strategy might be particularly important for workers with pets and with high levels of attachment to them.

However, it will not always be possible to have workers working from home (e.g., a hairstylist). In those cases, the presence of pets, as well as other practices related to them, seem to be relevant in organizations, to satisfy the needs of employees and their customers, and at the same time deliver benefits to organizations. It has been argued that the implementation of pet-friendly practices has reduced organizational costs, especially when compared to the benefits it has [46,67]. These benefits assert themselves even in the face of challenges related to health, safety, cultural issues, fears, phobias, and interruptions in the work environment [68,69]. However, practices must be implemented as baby steps. For instance, it should be a starting point to create a "pet day", which is an open day, in which workers and customers could take their pets to work. Another measure could include a license of bereavement following a pet death or allowing the owner to take their pet's birthday off. Other measures could include the extension of family-friendly practices to include pets. For instance, many organizations have aids for their workers' children's education. However, it could be extended to pet caring or to pets' daycare.

Given the importance associated with positive affect, managers can benefit from acknowledging its relevance for performance. Thus, they should create conditions for their workers to experience more frequently positive affect, for example, giving specific times to workers to make their task breaks, creating specific ways to regularly give feedback to them, and also creating a time and space for them to share it with each other.

7. Conclusions

Overall, this study shows that perceived quality of life while telecommuting is positively related to positive affect and performance and seems to be moderated by the pet's closeness and attachment. Specifically, this study sheds light on the power that pets play in this path, evidencing the positive interaction between pets' closeness and emotional attachment in the mediating path.

Author Contributions: A.J.-S. was responsible for conceptualization, methodology, software, formal analysis, resources, writing—original draft preparation, writing—review and editing funding acquisition. M.A. was responsible for investigation, data curation, visualization. C.G. was responsible for validation, supervision. All authors have read and agreed to the published version of the manuscript.

Funding: This work was supported by Fundação para a Ciência e a Tecnologia, grant UIDB/00315/2020.

Institutional Review Board Statement: The study was conducted according to the guidelines of the Declaration of Helsinki and approved by the Institutional Review Board (or Ethics Committee) of ISCTE—Lisbon University Institute.

Informed Consent Statement: Informed consent was obtained from all subjects involved in the study.

Data Availability Statement: Data will be made available upon reasonable request.

Conflicts of Interest: The authors declare no conflict of interest.

Appendix A.

Appendix A.1. Survey

Appendix A.1.1. The Attitude toward Telework

Please indicate your agreement or otherwise to the statements below

Appendix A.1.2. Effectiveness/Productivity

1. When e-working I can concentrate better on my work tasks.
2. E-working makes me more effective to deliver against my key objectives and deliverables.
3. If I am interrupted by family/other responsibilities whilst e-working from home, I still meet my line manager's quality expectations.
4. My overall job productivity has increased by my ability to e-work remotely/from home.

Appendix A.1.3. Organizational Trust

1. My organisation provides training in e-working skills and behaviours.
2. I trust my organisation to provide good e-working facilities to allow me to e-work effectively.
3. My organisation trusts me to be effective in my role when I e-work remotely.

Appendix A.1.4. Interference between Personal and Work-Life

1. My social life is not poor when e-working remotely.
2. My e-working does not take up time that I would like to spend with my family/friends or on other non-work activities.
3. When e-working remotely I do not often think about work related problems outside of my normal working hours.
4. I am happy with my work-life balance when e-working remotely.
5. Constant access to work through e-working is not very tiring.
6. I do not feel that work demands are much higher when I am e-working remotely.
7. When e-working from home I do not know when to switch off/put work down so that I can rest (R).

Appendix A.1.5. Flexibility

1. My work is so flexible I could easily take time off e-working remotely, if and when I want to.
2. My line manager allows me to flex my hours to meet my needs, providing all the work is completed.
3. My supervisor gives me total control over when and how I get my work completed when e-working.
4. Scale (1—totally disagree; 5—totally agree)

Appendix A.1.6. Positive Affect

Today, please indicate below approximately how often you have felt the following while you were working in your job. Everyone has a lot of overlapping feelings, so you'll have a total for all the items that are much greater than 100% of the time.

1. Enthusiastic.
2. Calm.
3. Joyful.
4. Relaxed.
5. Inspired.
6. Laid-back.
7. Excited.
8. At ease.

Scale (1) never to (5) always

Appendix A.1.7. Self-Reported Job Performance

On the last day (you worked), how well were you

1. Today, I achieved my job goals.
2. Today I made the right decisions.
3. Today I permed without mistakes.
4. Today I handled the responsibilities and daily demands of my work.
5. Today I get the things done.
6. Today I get along with others at work.

Scale (1) very little (5) a great deal

Appendix A.1.8. Emotional Attachment to Pets

Please consider the animal you have lived with the longest, and answer using the following criteria. (1) completely disagree to (5) completely agree.

1. My pet knows when I'm feeling bad.
2. I often talk to other people about my pet.
3. My pet understands me.
4. I believe that loving my pet helps me stay healthy.
5. My pet and I have a very close relationship.
6. I play with my pet quite often.
7. I consider my pet to be a great companion.
8. My pet makes me feel happy.
9. I am not very attached to my pet (R).
10. Owning a pet adds to my happiness.
11. I consider my pet to be a friend.

Appendix A.1.9. Physical Closeness to Pets

Please, think about the last day you were working remotely. Indicate the frequency through which the following situations with your pet occurred during the working day.

1. In telework, I usually take breaks to interact with my pet.
2. While you work from home, my pet is close to me when I am working.
3. During telework, my pet is not close to me while I work (R).
4. While I am working, I use to interact with my pet.

Scale (1) never to (5) always.

References

1. Wagner, E.; Pina e Cunha, M. Dogs at the Workplace: A Multiple Case Study. *Animals* **2021**, *11*, 89. [CrossRef] [PubMed]
2. Hobfoll, S.E. The influence of culture, community, and the nested-self in the stress process: Advancing conservation of resources theory. *Appl. Psychol.* **2001**, *50*, 337–421. [CrossRef]
3. Hobfoll, S.E.; Halbesleben, J.; Neveu, J.P.; Westman, M. Conservation of resources in the organizational context: The reality of resources and their consequences. *Annu. Rev. Organ. Psychol. Organ. Behav.* **2018**, *5*, 103–128. [CrossRef]
4. Hobfoll, S.E. Conservation of resources theory: Its implication for stress, health, and resilience. In *The Oxford Handbook of Stress, Health, and Coping*; Folkman, S., Ed.; Oxford University Press: Oxford, UK, 2011; pp. 127–147.
5. Kurland, N.B.; Bailey, D.E. When workers are here, there, and everywhere: A discussion of the advantages and challenges of telework. *Organ. Dyn.* **1999**, *28*, 53–58. [CrossRef]
6. Vander Elst, T.; Verhoogen, R.; Sercu, M.; Van den Broeck, A.; Baillien, E.; Godderis, L. Not extent of telecommuting, but job characteristics as proximal predictors of work-related well-being. *J. Occup. Environ. Med.* **2017**, *59*, e180–e186. [CrossRef] [PubMed]
7. Blau, P. *Exchange and Power in Social Life*; Wiley: New York, NY, USA, 1964.
8. Lemos, A.; Wulf, G.; Lewthwaite, R.; Chiviacowsky, S. Autonomy support enhances performance expectancies, positive affect, and motor learning. *Psychol. Sport Exerc.* **2017**, *31*, 28–34. [CrossRef]
9. Kelemen, T.K.; Matthews, S.H.; Wan, M.; Zhang, Y. The secret life of pets: The intersection of animals and organizational life. *J. Organ. Behav.* **2020**, *41*, 694–697. [CrossRef]
10. Fitzer, M.M. Managing from afar: Performance and rewards in a telecommuting environment. *Compens. Benefits Rev.* **1997**, *29*, 65–73. [CrossRef]
11. Fonner, K.L.; Roloff, M.E. Why teleworkers are more satisfied with their jobs than are office-based workers: When less contact is beneficial. *J. Appl. Commun. Res.* **2010**, *38*, 336–361. [CrossRef]
12. Magnavita, N.; Tripepi, G.; Chiorri, C. Telecommuting, off-time work, and intrusive leadership in workers' well-being. *Int Journ Environ. Res. Public Health* **2021**, *18*, 3330. [CrossRef] [PubMed]
13. Molla, R. How Remote Work is Quietly Remaking Our Lives. 9 October 2019. Available online: https://www.vox.com/recode/2019/10/9/20885699/remote-work-from-anywhere-change-coworking-office-real-estate (accessed on 29 May 2022).
14. Gajendran, R.S.; Harrison, D.A. The good, the bad, and the unknown about telecommuting: Meta-analysis of psychological mediators and individual consequences. *J. Appl. Psychol.* **2007**, *92*, 1524. [CrossRef]
15. Fredrickson, B.L. What good are positive emotions? *Rev. Gen. Psychol.* **1998**, *2*, 300–319. [CrossRef] [PubMed]
16. Cropanzano, R.; Mitchell, M.S. Social exchange theory: An interdisciplinary review. *J. Manag.* **2005**, *31*, 874–900. [CrossRef]
17. Fredrickson, B.L. Positive emotions broaden and build. In *Advances in Experimental Social Psychology*; Academic Press: Cambridge, MA, USA, 2013; Volume 47, pp. 1–53. [CrossRef]
18. Baron, R.A. Environmentally induced positive affect: Its impact on self-efficacy, task performance, negotiation, and conflict. *J. Appl. Soc. Psychol.* **1990**, *20*, 368–384. [CrossRef]
19. Bono, J.E.; Glomb, T.M.; Shen, W.; Kim, E.; Koch, A.J. Building positive resources: Effects of positive events and positive reflection on work stress and health. *Acad. Manag. J.* **2013**, *56*, 1601–1627. [CrossRef]
20. Hobfoll, S.E. Conservation of resources: A new attempt at conceptualizing stress. *Am. Psychol.* **1989**, *44*, 513. [CrossRef] [PubMed]
21. Bakker, A.B.; Demerouti, E. The job demands-resources model: State of the art. *J. Manag. Psychol.* **2007**, *22*, 309–328. [CrossRef]
22. Avey, J.B.; Reichard, R.J.; Luthans, F.; Mhatre, K.H. Meta-analysis of the impact of positive psychological capital on employee attitudes, behaviors, and performance. *Hum. Resour. Dev. Q.* **2011**, *22*, 127–152. [CrossRef]
23. Motowildo, S.J.; Borman, W.C.; Schmit, M.J. A theory of individual differences in task and contextual performance. *Hum. Perform.* **1997**, *10*, 71–83. [CrossRef]
24. Dotson, M.J.; Hyatt, E.M. Understanding dog–human companionship. *J. Bus. Res.* **2008**, *61*, 457–466. [CrossRef]
25. Brickel, C.M. 18/Pet Facilitated Therapies: A Review of the Literature and Clinical Implementation Considerations. *Clin. Gerontol.* **1986**, *5*, 309–332. [CrossRef]
26. Bradshaw, J.W.S. *Social Interactions Between Animals and People—A New Biological Framework*; Anthrozoology Institute, University of Southampton: Southampton, UK, 1995.
27. Cusack, O. *Pets and Mental Health*; Haworth: New York, NY, USA, 1988.

28. Levinson, B.M.; Úbeda, M.R. *Psicoterapia Infantil Asistida por Animales*; Fundación Affinity: Barcelona, Spain, 2006.
29. Nebbe, L. Nature Therapy. In *Handbook on Animal Assisted and Therapy: Theoretical Foundations and Guidelines for Practice*; Fine, A., Ed.; Academic Press: San Diego, CA, USA, 2000; pp. 385–414.
30. Graham, D.W.; Bergeron, G.; Bourassa, M.W.; Dickson, J.; Gomes, F.; Howe, A.; Kahn, L.H.; Morley, P.S.; Scott, H.M.; Wittum, T.E. Complexities in understanding antimicrobial resistance across domesticated animal, human, and environmental systems. *Ann. N. Y. Acad. Sci.* **2019**, *1441*, 17–30. [CrossRef] [PubMed]
31. Bradshaw, J. *The Animals Among Us: How Pets Make Us Human*; Basic Books: New York, NY, USA, 2017.
32. Herzog, H. *Some We Love, Some We Hate, Some We Eat*; Tantor Audio: Old Saybrook, CT, USA; HarperCollins: New York, NY, USA, 2010.
33. O'Haire, M. Companion animals and human health: Benefits, challenges, and the road ahead. *J. Vet. Behav.* **2010**, *5*, 226–234. [CrossRef]
34. Cohen, S.P. Can pets function as family members? *West. J. Nurs. Res.* **2002**, *24*, 621–638. [CrossRef] [PubMed]
35. Colarelli, S.M.; McDonald, A.M.; Christensen, M.S.; Honts, C. A companion dog increases prosocial behavior in work groups. *Anthrozoös* **2017**, *30*, 77–89. [CrossRef]
36. Pina e Cunha, M.; Rego, A.; Munro, I. Dogs in organizations. *Hum. Relat.* **2019**, *72*, 778–800. [CrossRef]
37. Hall, S.S.; Mills, D.S. Taking dogs into the office: A novel strategy for promoting work engagement, commitment and quality of life. *Front. Vet. Sci.* **2019**, *6*, 138. [CrossRef] [PubMed]
38. Harter, S. The development of self-representations during childhood and adolescence. In *Handbook of Self and Identity*; Leary, M.R., Tangey, J.P., Eds.; Guilford Press: New York, NY, USA, 2003; pp. 610–642.
39. McConnell, A.R.; Strain, L.M.; Brown, C.M.; Rydell, R.J. The simple life: On the benefits of low self-complexity. *Pers. Soc. Psychol. Bull.* **2009**, *35*, 823–835. [CrossRef] [PubMed]
40. Junça-Silva, A.; Lopes, C. Cognitive and affective predictors of occupational stress and job performance: The role of perceived organizational support and work engagement. *J. Econ. Adm. Sci.* 2021, *ahead of print*. [CrossRef]
41. Imai, N.; Dorigatti, I.; Cori, A.; Riley, S.; Ferguson, N.M. Estimating the Potential Total Number of Novel Coronavirus Cases in Wuhan City, China. 2020. Available online: https://www.imperial.ac.uk/media/imperial-college/medicine/mrc-gida/2020-01-17-COVID19-Report-1.pdf (accessed on 2 April 2022).
42. Sohrabi, C.; Alsafi, Z.; O'neill, N.; Khan, M.; Kerwan, A.; Al-Jabir, A.; Iosifidis, C.; Agha, R. World Health Organization declares global emergency: A review of the 2019 novel coronavirus (COVID-19). *Int. J. Surg.* **2020**, *76*, 71–76. [CrossRef] [PubMed]
43. Grant, C.A.; Wallace, L.M.; Spurgeon, P.C.; Tramontano, C.; Charalampous, M. Construction and initial validation of the E-Work Life Scale to measure remote e-working. *Empl. Relat.* **2019**, *41*, 16–33. [CrossRef]
44. Warr, P.; Bindl, U.K.; Parker, S.K.; Inceoglu, I. Four-quadrant investigation of job-related affects and behaviours. *Eur. J. Work Organ. Psychol.* **2014**, *23*, 342–363. [CrossRef]
45. Abramis, D.J. Relationship of job stressors to job performance: Linear or an inverted-U? *Psychol. Rep.* **1994**, *75*, 547–558. [CrossRef] [PubMed]
46. Johnson, T.P.; Garrity, T.F.; Stallones, L. Psychometric evaluation of the Lexington attachment to pets scale (LAPS). *Anthrozoös* **1992**, *5*, 160–175. [CrossRef]
47. Hayes, A.F. *Introduction to Mediation, Moderation, and Conditional Process Analysis: A Regression-Based Approach*; Guilford publications: New York, NY, USA, 2017.
48. Hayes, A.F. Partial, conditional, and moderated moderated mediation: Quantification, inference, and interpretation. *Commun. Monogr.* **2018**, *85*, 4–40. [CrossRef]
49. Podsakoff, N.P. A tutorial on the causes, consequences, and remedies of common method biases. *MIS Q.* **2017**, *35*, 293–334.
50. Martela, F.; Ryan, R.M.; Steger, M.F. Meaningfulness as satisfaction of autonomy, competence, relatedness, and beneficence: Comparing the four satisfactions and positive affect as predictors of meaning in life. *J. Happiness Stud.* **2018**, *19*, 1261–1282. [CrossRef]
51. Hackman, J.R.; Oldham, G.R. Development of the job diagnostic survey. *J. Appl. Psychol.* **1975**, *60*, 159. [CrossRef]
52. Siruri, M.M.; Cheche, S. Revisiting the Hackman and Oldham Job Characteristics Model and Herzberg's Two Factor Theory: Propositions on How to Make Job Enrichment Effective in Today's Organizations. *Eur. J. Bus. Manag. Res.* **2021**, *6*, 162–167. [CrossRef]
53. Anderson, A.J.; Kaplan, S.A.; Vega, R.P. The impact of telework on emotional experience: When, and for whom, does telework improve daily affective well-being? *Eur. J. Work Organ. Psychol.* **2015**, *24*, 882–897. [CrossRef]
54. Abdel Hadi, S.; Bakker, A.B.; Häusser, J.A. The role of leisure crafting for emotional exhaustion in telework during the COVID-19 pandemic. *Anxiety, Stress Coping* **2021**, *34*, 530–544. [CrossRef] [PubMed]
55. Barker, S.B.; Rasmussen, K.G.; Best, A.M. Effect of aquariums on electroconvulsive therapy patients. *Anthrozoös* **2003**, *16*, 229–240. [CrossRef]
56. Gee, N.R.; Fine, A.H.; Schuck, S. Animals in educational settings: Research and practice. In *Handbook on Animal-Assisted Therapy*; Academic Press: Cambridge, MA, USA, 2015; pp. 195–210. [CrossRef]
57. Bowlby, J. *Attachment and Loss: Volume 1. Attachment*; Basic Books: New York, NY, USA, 1969.
58. Bowlby, J. *Attachment and Loss: Volume 2. Separation: Anxiety and Anger*; Basic Books: New York, NY, USA, 1973.
59. Bowlby, J. *Attachment and Loss: Volume 3. Loss: Sadness and Depression*; Basic Books: New York, NY, USA, 1980.

60. Antonacopoulos, N.M.; Pychyl, T.A. An examination of the potential role of pet ownership, human social support and pet attachment in the psychological health of individuals living alone. *Anthrozoös* **2010**, *23*, 37–54. [CrossRef]
61. Podsakoff, P.M.; MacKenzie, S.B.; Podsakoff, N.P. Sources of method bias in social science research and recommendations on how to control it. *Annu. Rev. Psychol.* **2012**, *63*, 539–569. [CrossRef]
62. Stoermer, S.; Lauring, J.; Selmer, J. The effects of positive affectivity on expatriate creativity and perceived performance: What is the role of perceived cultural novelty? *Int. J. Intercult. Relat.* **2020**, *79*, 155–164. [CrossRef]
63. Lades, L.K.; Laffan, K.; Daly, M.; Delaney, L. Daily emotional well-being during the COVID-19 pandemic. *Br. J. Health Psychol.* **2020**, *25*, 902–911. [CrossRef]
64. Taylor, S.; Landry, C.A.; Paluszek, M.M.; Asmundson, G.J. Reactions to COVID-19: Differential predictors of distress, avoidance, and disregard for social distancing. *J. Affect. Disord.* **2020**, *277*, 94–98. [CrossRef]
65. Morgeson, F.P.; Humphrey, S.E. The Work Design Questionnaire (WDQ): Developing and validating a comprehensive measure for assessing job design and the nature of work. *J. Appl. Psychol.* **2006**, *91*, 1321. [CrossRef]
66. JunÃ§a-Silva, A.; Pombeira, C.; Caetano, A. Testing the affective events theory: The mediating role of affect and the moderating role of mindfulness. *Appl. Cogn. Psychol.* **2021**. [CrossRef]
67. AFP-JIJI. To Reduce Work Stress, Japan Firms Turn to Office Cats, Dogs and Goats. *The Japanese Times.* 2017. Available online: http://www.japantimes.co.jp/news/2017/05/25/national/reduce-work-stresse-japan-firms-turn-office-cats-dogs-goats/#.WTQbyuvyvIU (accessed on 2 April 2022).
68. Johnson, Y. A Pet-Friendly Workplace Policy to Enhance the Outcomes of an Employee Assistance Programme (EAP). Ph.D. Thesis, University of Pretoria, Pretoria, South Africa, 2006.
69. Hunter, C.; Verreynne, M.L.; Pachana, N.; Harpur, P. The impact of disability-assistance animals on the psychological health of workplaces: A systematic review. *Hum. Resour. Manag. Rev.* **2019**, *29*, 400–417. [CrossRef]

animals MDPI

Article

How Dog Behavior Influences Pet Owner's Perceptions of Dog Preference for Dental Chews

Anamarie C. Johnson [1,*], Holly C. Miller [2] and Clive D. L. Wynne [1]

1 Department of Psychology, Arizona State University, Tempe, AZ 85287, USA; cwynne1@asu.edu
2 General Mills, 1 General Mills Blvd, Golden Valley, MN 55426, USA; holly.miller@genmills.com
* Correspondence: acjohn47@asu.edu

Simple Summary: Most studies on dog food and treat preferences focus on owner reports about the product and how much the dog consumes. The aim of this study was to examine dog behavior and engagement in a home-environment with eight different dental chews. Owners submitted a video of their dogs which was analyzed to investigate any relationship between coded dog behavior and owner survey responses for preference among the chew types. Owner-reported dog preference related more to the video coded behavior than their own preference providing some preliminary guidance on what factors might relate to product preference and purchase and how analysis of in-home behavior may better guide pet product research.

Abstract: American pet owners spend billions of dollars on food and treats so it is important to understand what products they want and what they think their dog would enjoy. This study analyzed video recordings of dogs engaging in dental chews in their home environment and compared the observed appetitive behaviors to owner preference and owner-reported dog preference. Overall, appetitive behavior differed significantly between some dental chews. Owner preference for the chews correlated significantly with dog appetitive behavior, but the effect was small (r (702) = 0.22, $p = 0.001$), whereas owner-reported dog preference correlated significantly with dog appetitive behavior and showed a moderate effect size (r (702) = 0.43, $p = 0.001$)—similar in magnitude to findings when parents are asked to report on their children's behavior. By merging objective behavioral observation of owner-recorded videos with their survey responses, we were able to preliminarily parse out what factors owners may use to assess preference and encourage the future use of in-home video recordings to better understand dog and owner engagement and interaction with pet products.

Keywords: dog; pet food industry; human–dog interaction

Citation: Johnson, A.C.; Miller, H.C.; Wynne, C.D.L. How Dog Behavior Influences Pet Owner's Perceptions of Dog Preference for Dental Chews. *Animals* **2023**, *13*, 1964. https://doi.org/10.3390/ani13121964

Academic Editor: Betty McGuire

Received: 8 May 2023
Revised: 8 June 2023
Accepted: 10 June 2023
Published: 12 June 2023

1. Introduction

In 2022, around 69 million households in the United States owned a dog and spent 138.6 billion dollars in the pet industry [1]. Fifty-eight billion dollars were spent on food and treats [1], so understanding what a dog prefers to consume and how that might align with owner preference is economically important. The relationship between owner perceptions of dog preferences and dog behavior also raises basic questions of how well people understand their dogs' sources of pleasure and discomfort.

The National Research Council defined palatability for dog and cat food as the "physical and chemical properties of the diet which are associated with promoting or suppressing feeding behavior during the pre-absorptive or immediate post-absorptive period" [2] (p. 24). Yet, as argued by Aldrich and colleagues [3], palatability is much more than a substance's sensory properties and what makes it appealing to a dog, and researchers should seek a more holistic understanding that considers the animal, prior exposure to food, and human and environmental factors.

Within palatability, one can analyze a dog's mere acceptance of a food product or parse out preference by looking at a deliberate choice for one product over another [3].

Historically, acceptance of food has been analyzed through a single bowl test where the dog is presented with one bowl of food and the amount consumed is measured [3]. By switching between an old and a new food, one can compare acceptance of foods and how intake varies between the two [3]. This method is primarily useful in assessing true dislike of a product; to assess more graded choice, Aldrich and colleagues [3] recommended a two-choice design.

In a two-choice design, a dog is simultaneously given equal amounts of two food products in two different bowls [3] The dog's preference is assessed by observing the product the dog first approaches, first consumes, and, after a set time, comparing the amount of food remaining in the bowls [3]. This design has the drawback that pet dogs might not be able to discriminate between two items or may be indifferent to them [3]. Laboratory dogs, on the other hand, can be trained to be highly discriminatory, but their refined palates have been shown to be quite different than those of pet dogs [4].

To better understand motivation for a particular food, some modifications of the standard two-choice design have been attempted. Some researchers have added in operant tasks and compared food products using a dog's motivation to press a lever [5] or perform a certain trained behavior such as touching its nose to the experimenter's hand [6]. Working within the applied behavior analysis tradition, one study utilized a paired stimulus assessment and a subsequent reinforcer assessment to parse out dog preference [6]. Dogs were offered choices among six food items presented in pairs; the six were then ranked from least to most preferred. Based on those preferences, the least and most preferred food types were later presented pairwise on two different reinforcement schedules where the dogs had to press their noses to the experimenter's hand to assess preference. The study found that the most preferred food item in the paired stimulus assessment functioned as a reinforcer for both schedule conditions with dogs responding more to the most preferred food item than to the least preferred [6]. Additionally, in a survey, dogs' owners were usually accurate in predicting the dog's most preferred food item [6].

Another study merged a two-choice design with direct analysis of dog behavior. Dogs first sampled two food products (a meat treat compared to a bland kibble). Then, during testing, the dogs were prevented from accessing the food by placing it in locked containers. Instead, their behavior in attempting to gain access to the food—such as by pawing at the container or sniffing—was measured and compared [7]. These behaviors on the inaccessible test were then compared to how much of the same products was consumed when they were presented in food puzzles. The authors found that the meatier treat was sniffed and engaged with more often during the inaccessible test than the less-preferred kibble and this preference was maintained during the food puzzle assessment [7].

Beyond direct comparative presentation of food, dog preference is often assessed through owner reports. These studies often relate how the owner sees their dog holistically "enjoying" a food product to their preference for obtaining that product [8–10]. Several factors have been shown to influence owner motivation to feed a product. One study analyzed how a pet food's aroma, appearance, and color might relate to owner preference and owner-reported dog's preference; owner's overall liking for a product corresponded with their ratings of its physical qualities, which in turn correlated with owner-reported predicted dog liking [8].

Few studies have considered the specific dog behaviors that might relate to an owner's assessment of a dog's preference for one food over another. In a comparison of dogs who ate either a conventional food diet, a raw meat diet, or a vegan diet, owners were asked to report factors that contributed to why they fed the specific food to their dog and whether their dog presented any of the ten palatability behavioral indicators the researchers proposed [11]. These indicators were derived from owner reports from a prior study [12] and included behaviors such as speed of eating, jumping, vocalizing, sniffing the food, and tail wagging [11]. Knight and Satchell [11] determined that certain palatability indicators loaded onto what they identified as dog enthusiasm towards a food (associated with behaviors including tail wagging or eating food quickly). Alternatively, sniffing, as also reported by Di Donfrancesco and colleagues [12], was interpreted as indicating hesitancy in consuming the food [11].

Several studies have investigated dogs' preferences for different dog foods, but fewer have investigated dogs' preference and ingestion of items sold as dog treats. White et al. [13] reported that 70% of surveyed dog owners viewed treats as something additional given to their dog, not part of their standard, daily diet. Sixty-two percent of owners identified chews as a common, popular treat and many owners reported giving treats more than three times a week [13]. Owners felt that treats were important to make the dog happy but also noted that treats could be context-specific, such as rewards during training. Owners also indicated that different chews could be given for different reasons—such as dental chews for dental hygienic care [13].

Periodontal disease, which includes both gingivitis and periodontitis, can result from the buildup of plaque on dogs' teeth which may lead to bacteria that can affect the tissues of the mouth [14]. It has been estimated that up to 85% of adult dogs suffer from some form of periodontal disease and, as a result, many veterinarians recommend regular teeth brushing or veterinary professional teeth cleaning to remove harmful plaque. However, teeth brushing can be difficult for owners to maintain on a daily schedule and putting a dog under anesthetic to carry out teeth cleaning can be risky [14]. As such, giving a daily dental chew that might reduce plaque could be very helpful. In a comparison of three commercially available dental chews, all were shown to inhibit plaque and calculus growth on teeth and reduce halitosis [14].

Offering a dental chew to a dog to facilitate dental care can only be as effective as the dog's motivation to engage with the chew. Because a dog is unable to articulate its preference for a food, an objective method of assessing preference through behavior has great potential value. No prior study has analyzed how owners' preferences for different dog treats relate to their reports of their dogs' preferences and to objective assessment of dogs' behavior. The aim of this study was to analyze video recordings made by owners during in-home consumer testing and compare the outcome of a direct analysis of the duration and frequency of the dogs' behaviors to owners' reports of their own and their dogs' preferences.

2. Materials and Methods

2.1. Participants

A market research group with an internal panel of over a million potential participants was contracted to identify, recruit, and perform the consumer testing. Prospective participants were sent a screening questionnaire to determine eligibility based on several criteria relating to the household and owned dogs (Table 1). A target sample size of at least 60 participants was estimated to provide a minimum power of 0.80 to detect a medium effect size at an alpha level of 0.05.

Table 1. List of criteria for inclusion in the survey.

Name of Criterion	Criterion Eligibility
Age	25–54 years old
Household Composition	Exclude grown children living with parents
Household Income	Greater than $30,000
Level of Education	Minimum high school graduate
Employment	Full- or part-time employed, retired, homemaker
Occupation type	NOT employed in advertising, market research, sales promotion, media, veterinarian or animal hospital, pet store, manufacturer pet supplies, or new product development
Number of Dogs in home	One or two dogs
Relationship to Dog in home	Make all or most purchase decisions for dog, schedule, and attend more veterinarian visits

Table 1. *Cont.*

Name of Criterion	Criterion Eligibility
Age of dog	1–10 years old
Muzzle shape	Medium to very long snout, no brachycephalic dogs
Last visit to veterinarian	Within last 6–8 months
Treat purchase and offering to dog	Regularly purchase and offer dental chews daily to once a week, need to have purchased a dental chew within the last 4 months
Type of Dental Chew purchase	One of eight brands in study: Arknaturals, Blue Dental Bones, Purina Dentalife, Pedigree Dentastix, Greenies, Merrick Fresh Kisses, Whimzees, or Blue Wilderness Wild Bones
Medical history of dog	No history of digestive issues after giving treats or new food, no dental issues, no food allergies
Technology	iPhone or Android, download of survey application

Potential participants were asked to complete an implicit bias test through an online portal to confirm that they owned and utilized a smart phone or internet-connected tablet. Once identified, participants verified that they could commit to the study for a total of four hours over an estimated five weeks. Participants who completed all tasks for the study received $500.

2.2. Procedure

Participants were sent the commercially available product bags of eight target dental chews (Naturals Brushless Toothpaste, Arknaturals, Tampa, FL, USA; Blue Dental Bones, Blue Buffalo Company LTD, Joplin, MO, USA; Greenies, Mars Petcare US, Franklin, TN, USA; Merrick Fresh Kisses, Merrick Pet Care, Hereford, TX, USA; Pedigree Dentastix, Mars Petcare US, Franklin, TN, USA; Purina Dentalife, Neehah, WI, USA; Whimzees, Wellness Pet Company, Tewksbury, MA, USA; and Blue Wilderness Wild Bones, Blue Buffalo Company LTD, Joplin, MO, USA). Participants were asked to video film their dogs engaging with each dental chew. To ensure video quality, all participants were first provided written instructions and a sample video and asked to practice filming their dog. Instructions included the owner behaving neutrally towards the dog when the treat was presented and staying out of frame during the filming but present to observe the dog for safety. Ninety participants with the best practice recordings, 30 with small dogs (est. 8–25 lbs; 3.6–11.3 kg), 30 medium dogs (est. 25–45 lbs; 11.3–20.4 kg), and 30 large (45–70 lbs; 20.4–31.8 kg), were selected. All completed practice videos were reviewed by the first author.

Participants with two dogs only submitted responses for one dog and were instructed to keep the other dog separated while the focal dog engaged with the chew.

The study lasted from 9 April 2021 to 14 May 2021.

Each participant presented a single brand of dental chew to their dog for three consecutive days. This same procedure was completed for the remaining seven dental chews brands. The order of presentation of the different brands of dental chew was randomized and assigned across participants.

Participants were instructed to feed a dental chew between meals and to not implement any other dietary changes for their dog during the study duration. On Days 1 and 2, all participants offered the chew to the dog and then completed a series of survey questions relating to the dog's enjoyment and focus with the chew and its perceived efficacy and freshness. These questions were of interest to the broader scope of the overall project but were not of interest in this study and are not reported here. Participants completed survey questions on Days 1 and 2 and only video recorded their dog on the third day. On Day 3, they were instructed to film the dog from when the chew was offered until it was completely consumed or for at least two minutes. As in the practice filming, owners were asked to behave neutrally and to stay out of frame for the duration of the video. On

Day 4, no chew was offered, and participants completed a survey on their preference and perception of the dogs' preference for that chew. Five questions were selected for analysis; additional questions not relevant to this study are not reported here. All the response options were presented on a Likert scale (Table 2).

Table 2. List of selected five questions from Day 4 survey with scale labels where applicable.

Overall, how well do you like or dislike these dog dental chews?	1: Dislike Extremely 2: Dislike very much 3: Dislike moderately 4: Dislike slightly 5: Neither like nor dislike 6: Like slightly 7: Like moderately 8: Like very much 9: Like extremely
I felt happy giving these dental dog chews to my dog	1: Not at all happy 2: 3: 4: Neither happy or unhappy 5: 6: 7: Very happy
I felt indifferent when giving these dental dog chews to my dog	1: Very Indifferent 2: 3: 4: Neutral 5: 6: 7: Not at all indifferent
My dog felt happy when I offered him/her these dental dog chews	1: Dog felt not at all happy 2: 3: 4: Neutral 5: 6: 7: Dog felt very happy
My dog felt disappointed when I offered him/her these dog dental chews	1: Very disappointed 2: 3: 4: Neutral 5: 6: 7: Not at all disappointed

2.3. *Video Analysis*

A total of 704 videos, 8 per participant, were analyzed using the event-logging software, BORIS [15], operated by video coders trained to 80% or better accuracy. The video coders were blind to the aims of the study. The first author randomly selected and reviewed 20% of the videos to ensure consistency.

All behaviors were coded for their duration. (Table 3).

Total appetitive engagement was defined from the ethogram as the sum of the four appetitive behaviors (Carry, Chew, Crumbs, and Investigate) from the first 60 s of each video. We selected 60 s for analysis because of a need to standardize a time period during which most dogs were still interacting with the chew (video length varied from 36 to 897 s). Inspection of the video recordings showed that most active engagement with the chews was complete in one minute. Only 2.7% of videos (19 of 704) showed "carry" and 7.7% of videos (54 of 704) showed "investigate" beyond 60 s. There was extensive chewing and

predicted owner liking (b = 1.92, t (702) = 6.01, $p = 0.001$, and $R^2 = 0.05$), owner happiness in giving the chew (b = 1.59, t (702) = 3.70, $p = 0.001$, and $R^2 = 0.02$), and owner indifference in giving the chew (b = 0.95, t (702) = 2.42, $p = 0.016$, and $R^2 = 0.01$).

Table 5. Regression analysis for survey questions on dental chews with the predictor variable appetitive behavior.

	R	R Square	Adj.R Square	Unstandardized B	p
Overall liking	0.22	0.05	0.05	1.92	0.001
Human felt happy	0.14	0.02	0.02	1.59	0.001
Human felt indifferent	0.09	0.01	0.01	0.95	0.016
Dog felt happy	0.43	0.18	0.18	4.86	0.001
Dog felt disappointed	0.44	0.19	0.19	4.21	0.001

Appetitive behavior accounted for more variance in the two questions relating to owner-reported dog preference, qualifying as moderate effect sizes [16]. Dog behavior significantly predicted owner reporting that the dog liked the chew (b = 4.86, t (702) = 12.55, $p = 0.001$, and $R^2 = 0.18$) and owner reporting the dog being disappointed with the dental chew (b = 4.21, t (702) = 12.79, $p = 0.001$, and $R^2 = 0.19$).

4. Discussion

This study compared direct behavioral analysis of dogs' engagement with eight different dental chews to owner survey responses. Overall, there were differences in the amount of appetitive behavior in the first 60 s of a dog receiving the dental chews. Owner responses on survey questions relating to their preference and their reports of dog preference for a chew significantly correlated with each other. Coded appetitive behavior correlated significantly with owner preference and with owner-reported dog preference, but the correlations differed in magnitude: owner preference had only a small relationship with dog behavior, whereas the effect sizes of the relationship between owner-perceived dog preferences and dog behavior were medium [16].

Some palatability-related behaviors frequently reported in the literature were not displayed by the dogs in our sample. Jumping and vocalizing, reported as indicators of palatability and enthusiasm towards food by Knight and Satchell [11], occurred infrequently (jumping) or not at all (vocalizing) in our sample. Knight and Satchell's [11] study, however, relied solely on owner reports of behavior rather than objective coding. As the authors noted, behavior assessments by untrained owners are less reliable than analysis of video recordings by trained observers blind to the aims of the study. By comparing owner-reported dog preference to objective behavioral coding, participants in our sample did appear to appropriately use some of their dog's behavior to report their dog's preference.

Prato-Previde and colleagues [17] showed that owners can have a large impact on how dogs interact with food. When owners displayed interest in a bowl containing a few pieces of kibble compared to a bowl with more, dogs counterintuitively chose the bowl with less food. When there was no owner influence, dogs chose the bowl with more kibble [17]. We attempted to minimize owner influence by instructing owners not to interact with their dog while it was engaging with the chew during filming, but we were only able to view the interaction from one camera angle and did not have recording of the first two dental chew engagements. It is, thus, possible that the behavior we observed might have been influenced by the owner. For example, the instruction to remain neutral may have led to more subdued behavior in the dogs which could explain why we did not observe much jumping or vocalizing as reported by Knight and Satchell [11].

Total appetitive behavior coded from video recordings accounted for about 18% of the variance in owner responses about their dog's preferences for the dog chews. Similar trends of correlations of moderate magnitude are seen in studies that have compared parental reports to direct analysis of children's behavior [18]. Stifter and colleagues [18] noted that several studies found significant correlations between parent ratings and unbiased observations but that these correlations were often weak to moderate in magnitude ($r < 0.30$). A study by Root and Stifter [19] compared mothers' questionnaire responses of how they would react to their child's negative emotions to unbiased observations in laboratory and classroom settings. Mothers' reported behavioral responses of support correlated with the observed laboratory and classroom behavior, accounting for a moderate effect size, $R^2 = 0.14$.

It is interesting that, although dog behavior predicted a moderate amount of the variance in owners' reports of dogs' perception of dental chews, it predicted very little of owners' reports of their own preference for the chews. This could indicate that other dog behaviors not coded here, or unanalyzed aspects of the chew, such as its size, color, or branding, may influence owner responses. It is also possible that when asked about their own preference among chew brands, people focus solely on their own impressions of texture, odor, and possible flavor, independent of anything their dog is doing. In a comparison of parent and child food preferences, researchers found that while there was overall similarity in preference, parents would often focus their preferences on the healthiness of a food which was not a factor in the children's own preferences [20]. Similarly, when parents were asked to report their preference for children's books and then predict their child's preference, parents preferred books with cultural acclaim while factors such as whether their child's gender matched the book's protagonist or the number of words per page affected how they predicted their child's preference [21].

The evidence here that participants were able to competently record and upload video recordings of their dogs' behavior opens the door for objective behavioral observation of dogs in their home without the intervention of strangers and, thus, where they are most comfortable. With some simple instructions, participants in our study were able to record footage that was then analyzed by trained coders and compared to previous studies that had relied on owner reporting of their dog's behavior which may be subject to bias and misinterpretation [11,12]. By merging objective observation with survey data, we gained insight into what features owners might be using when making a purchasing decision. Future research should expand on the opportunity that in-home video recording provides to obtain an understanding of how dogs and owners interact with pet food products.

Limitations

One possible limitation of this study was the variety of durations with which the dogs interacted with the chews and, consequently, the varied video-recording durations provided by the owners. We attempted to control for this by only analyzing the first 60 s of each video and found that some appetitive behaviors such as "investigate" and "carry" occurred at relatively low frequencies after that time. Chewing persisted beyond 60 s, but most dogs did not chew for more than around 40 of the first 60 s. Interaction with crumbs also persisted beyond 60 s but crumb behavior could only occur if the chew was complete and soft enough to form crumbs. However, the nature of a dog's interaction with a dental chew is not necessarily just a function of the dog's enthusiasm for it, but also related to its size, hardness, and abrasiveness.

Additionally, while participants were instructed to feed their dog between meals so their dog would not be satiated, we could not know the feeding schedule for dogs and how that might affect chew interest.

It might have been beneficial to have owners record their dogs' behavior with the dental chew over all three days of chew exposure. This would have made it possible to observe possible changes in behavior as the dogs adjusted to the chews. Dogs tend to be more neophilic than other animals and prefer variety in their daily diet [22]. Callon et al. [23] and Vondran [24] noted that dogs showed enhanced interest in novel food products. However, dogs can also present neophobic behavior and, as Callon and colleagues [23] noted, dogs presented a novel food showed more hesitation and were slower to eat on the first day of food presentation than the last. Thus, it seems likely that behavioral interaction with the chews we presented would have changed over the course of three days, however, neophilic or neophobic behavior was not the focus of this study.

5. Conclusions

This study utilized owner-reported surveys and owner-recorded videos during an in-home test of eight commercially available dental chews to investigate how dogs' overt behavior corresponded to owner survey responses. Overall durations of appetitive engagement during the first 60 s across the different brands of dental chew were quite similar. We found the observed dogs' behavior only had a small impact on owner preference while the same behavior predicted owner-reported dog preference with a moderate effect size. Researchers should take advantage of in-home video recording to better understand how owners perceive their dog's liking of a product and how that ultimately may affect their preference and purchasing intent.

Author Contributions: Conceptualization, A.C.J., H.C.M. and C.D.L.W.; methodology, A.C.J., H.C.M. and C.D.L.W.; validation, A.C.J. and C.D.L.W.; formal analysis, A.C.J. and C.D.L.W.; investigation, A.C.J. and H.C.M.; resources, H.C.M.; data curation, H.C.M.; writing—original draft preparation, A.C.J.; writing—review and editing, A.C.J. and C.D.L.W.; supervision, C.D.L.W.; project administration, H.C.M.; funding acquisition, H.C.M. and C.D.L.W. All authors have read and agreed to the published version of the manuscript.

Funding: This study was funded by General Mills. Grant number: G06179.

Institutional Review Board Statement: This study was conducted through industry research and did not go through IACUC or IRB protocol. The study funder has an internal peer-review process and a committee that includes a veterinarian who ensures the safety of animal participants. This study also adhered to standards from ASTM international.

Informed Consent Statement: All participants consented to participation, could remove themselves or their dogs at any time, and received monetary compensation for their participation.

Data Availability Statement: The data presented in this study are available on request from the corresponding author. The data are not publicly available due to data pertaining to private industry.

Acknowledgments: We thank all the owners and their dogs that participated in the study and the staff of Sago International for their assistance in executing the study and provided the survey interface for data collection.

Conflicts of Interest: Funders had a role in data collection and study design. H.C.M declares that she is currently a full-time employee of General Mills. A.C.J. declares that she has been employed as independent contractor for General Mills.

References

1. American Pet Products Association (APPA). Pet Industry Market Size & Ownership Statistics. 2022. Available online: https://www.americanpetproducts.org/press_industrytrends.asp (accessed on 8 June 2023).
2. National Research Council. *Nutrient Requirements of Dogs and Cats*; The National Academies Press: Washington, DC, USA, 2006. [CrossRef]
3. Aldrich, G.C.; Koppel, K. Pet food palatability evaluation: A review of standard assay techniques and interpretation of results with a primary focus on limitations. *Animals* **2015**, *5*, 43–55. [CrossRef] [PubMed]
4. Griffin, R.W.; Scott, G.C.; Cante, C.J. Food preferences of dogs housed in testing-kennels and in consumers' homes: Some comparisons. *Neurosci. Biobehav. Rev.* **1984**, *8*, 253–259. [CrossRef] [PubMed]
5. Rashotte, M.E.; Foster, D.F.; Austin, T. Two-pan and operant lever-press tests of dogs' preference for various foods. *Neurosci. Biobehav. Rev.* **1984**, *8*, 231–237. [CrossRef] [PubMed]
6. Vicars, S.M.; Miguel, C.F.; Sobie, J.L. Assessing preference and reinforcer effectiveness in dogs. *Behav. Process.* **2014**, *103*, 75–83. [CrossRef] [PubMed]
7. Thompson, H.; Riemer, S.; Ellis, S.L.; Burman, O.H. Behaviour directed towards inaccessible food predicts consumption—A novel way of assessing food preference. *Appl. Anim. Behav. Sci.* **2016**, *178*, 111–117. [CrossRef]
8. Di Donfrancesco, B.; Koppel, K.; Swaney-Stueve, M.; Chambers, E. Consumer acceptance of dry dog food variations. *Animals* **2014**, *4*, 313–330. [CrossRef] [PubMed]

9. Kwak, M.K.; Cha, S.S. A study on the selection attributes affecting pet food purchase: After COVID-19 Pandemic. *Int. J. Food Prop.* **2021**, *24*, 291–303. [CrossRef]
10. Vinassa, M.; Vergnano, D.; Valle, E.; Giribaldi, M.; Nery, J.; Prola, L.; Bergero, D.; Schiavone, A. Profiling Italian cat and dog owners' perceptions of pet food quality traits. *BMC Vet. Res.* **2020**, *16*, 131. [CrossRef]
11. Knight, A.; Satchell, L. Vegan versus meat-based pet foods: Owner-reported palatability behaviours and implications for canine and feline welfare. *PLoS ONE* **2021**, *16*, e0260198. [CrossRef]
12. Di Donfrancesco, B.; Koppel, K.; Aldrich, C.G. Pet and owner acceptance of dry dog foods manufactured with sorghum and sorghum fractions. *J. Cereal Sci.* **2018**, *83*, 42–48. [CrossRef]
13. White, G.A.; Ward, L.; Pink, C.; Craigon, J.; Millar, K.M. "Who's been a good dog?"–Owner perceptions and motivations for treat giving. *Prev. Vet. Med.* **2016**, *132*, 14–19. [CrossRef] [PubMed]
14. Carroll, M.Q.; Oba, P.M.; Sieja, K.M.; Alexander, C.; Lye, L.; De Godoy, M.R.; He, F.; Smorak, A.J.; Keating, S.C.J.; Sage, A.M.; et al. Effects of novel dental chews on oral health outcomes and halitosis in adult dogs. *J. Anim. Sci.* **2020**, *98*, skaa274. [CrossRef] [PubMed]
15. Friard, O.; Gamba, M. BORIS: A free, versatile open-source event-logging software for video/audio coding and live observations. *Methods Ecol. Evol.* **2016**, *7*, 1325–1330. [CrossRef]
16. Cohen, J. *Statistical Power Analysis for the Behavioral Sciences*, 2nd ed.; Laurence Erlbaum Associates: Mahwah, NJ, USA, 1988. [CrossRef]
17. Prato-Previde, E.; Marshall-Pescini, S.; Valsecchi, P. Is your choice my choice? The owners' effect on pet dogs' (*Canis lupus familiaris*) performance in a food choice task. *Anim. Cogn.* **2008**, *11*, 167–174. [CrossRef] [PubMed]
18. Stifter, C.A.; Willoughby, M.T.; Towe-Goodman, N. Agree or agree to disagree? Assessing the convergence between parents and observers on infant temperament. *Infant Child Dev. Int. J. Res. Pract.* **2008**, *17*, 407–426. [CrossRef] [PubMed]
19. Root, A.K.; Stifter, C. Temperament and maternal emotion socialization beliefs as predictors of early childhood social behavior in the laboratory and classroom. *Parent. Sci. Pract.* **2010**, *10*, 241–257. [CrossRef] [PubMed]
20. Søndergaard, H.A.; Edelenbos, M. What parents prefer and children like–Investigating choice of vegetable-based food for children. *Food Qual. Prefer.* **2007**, *18*, 949–962. [CrossRef]
21. Wagner, L. Factors influencing parents' preferences and parents' perceptions of child preferences of picture books. *Front. Psychol.* **2017**, *8*, 1448. [CrossRef]
22. Thorne, C. Feeding behaviour of domestic dogs and the role of experience. In *The Domestic Dog: Its Evolution, Behaviour and Interactions with People*; Serpell, J., Ed.; Cambridge University Press: Cambridge, UK, 1995; pp. 103–113.
23. Callon, M.C.; Cargo-Froom, C.; DeVries, T.J.; Shoveller, A.K. Canine food preference assessment of animal and vegetable ingredient-based diets using single-pan tests and behavioral observation. *Front. Vet. Sci.* **2017**, *4*, 154. [CrossRef] [PubMed]
24. Vondran, J.C. A Two-Pan Feeding Trial with Companion Dogs: Considerations for Future Testing. Master's Thesis, Kansas State University, Manhattan, KS, USA, 2013. Available online: http://hdl.handle.net/2097/1589 (accessed on 21 September 2022).

animals

MDPI

Article

Are Hair Cortisol Levels of Humans, Cats, and Dogs from the Same Household Correlated?

Justyna Wojtaś [1], Aleksandra Garbiec [1,*], Mirosław Karpiński [1], Patrycja Skowronek [2] and Aneta Strachecka [2]

1 Department of Animal Ethology and Wildlife Management, University of Life Sciences in Lublin, Akademicka 13, 20-950 Lublin, Poland; justyna.wojtas@up.lublin.pl (J.W.); miroslaw.karpinski@up.lublin.pl (M.K.)

2 Department of Invertebrate Ecophysiology and Experimental Biology, University of Life Sciences in Lublin, Akademicka 13, 20-950 Lublin, Poland; patrycja.skowronek@up.lublin.pl (P.S.); aneta.strachecka@up.lublin.pl (A.S.)

* Correspondence: aleksandra.garbiec@up.lublin.pl; Tel.: +48-81-445-42-69

Simple Summary: Dogs and cats are animals that have been accompanying humans for many years. There is no doubt that they are emotionally connected with people, although each of them in their own way. The study attempts to assess the emotional relationship between humans, dogs, and cats living in one household on the basis of the correlations between the hair cortisol level. The study involved 25 women who had at least one dog and at least one cat at home. Based on the study conducted, no significant correlation was found between the level of cortisol in the hair of the owners and their pets. There were, however, some interesting differences depending on the degree of emotional connection and the frequency of interactions.

Abstract: Human–animal interactions and the emotional relationship of the owner with the pet are the subjects of many scientific studies and the constant interest of not only scientists but also pet owners. The aim of this study was to determine and compare the hair cortisol levels of dogs, cats, and their owners living in the same household. The owners were asked to complete a questionnaire concerning the frequency of their interactions with pets and emotional relationship with each of their cats and each of their dogs. The study involved 25 women who owned at least one dog and at least one cat. In total, 45 dogs and 55 cats from 25 households participated in the study. The average level of hair cortisol of the owners was 4.62 ng/mL, of the dogs 0.26 ng/mL, and in the hair of cats 0.45 ng/mL. There was no significant correlation between the hair cortisol level of the owner and dog or the owner and the cat and between dogs and cats living together. A significant positive correlation was observed between the hair cortisol level in the owner and the pet, for dogs in which the owner performs grooming treatments once a week and for cats which are never kissed. Although our study did not find many significant correlations, studies using other stress markers might have yielded different results.

Keywords: hair cortisol level; human-animal interaction; dog; cat

Citation: Wojtaś, J.; Garbiec, A.; Karpiński, M.; Skowronek, P.; Strachecka, A. Are Hair Cortisol Levels of Humans, Cats, and Dogs from the Same Household Correlated? *Animals* **2022**, *12*, 1472. https://doi.org/10.3390/ani12111472

Academic Editor: Betty McGuire

Received: 9 May 2022
Accepted: 4 June 2022
Published: 6 June 2022

1. Introduction

Human–animal interactions are one of the key issues of interest to psychologists, zoopsychologists, and scientists in the field of behavior and animal welfare, both with respect to companions [1–3] and farm animals [4]. The companion animals are now treated as family members. In every family, including interspecies ones, there are a number of relationships, interactions, and conflicts that affect the stress level, and thus the wellbeing of individual members of the interspecies "herd." The relationship between the dog's and its handler or owner's reaction to stress has been examined. Acute stress was assessed by Buttner et al. [5] and Wojtaś et al. [3]. Chronic stress was assessed by Sundman et al. [6].

The behavioral and physiological effects of dog–human interactions were described by Payne et al. [7], Petersson et al. [8], and Willen et al. [9]. The human–cat relationship has been extensively analyzed by Turner [10]. The effects of humans on cats based on oxytocin and cortisol levels in urine were analyzed by Nagasawa et al. [11]. The effect of cats on humans was investigated by Turner et al. [12].

There are at least a few reasons why hair is increasingly used and appreciated as biological material in research in many fields. It is a material of high durability and resistance to external factors. As hair does not appear suddenly but grows over weeks or months, analysis of its composition makes it possible to measure physiological changes over a given time scale. Research shows that the hair growth rate is about 1 cm/month [13–15], therefore it is assumed that the presence of a given substance in 1 cm of a hair length corresponds to a period of about a month's exposure to a given substance [16]. In this way, it is possible to evaluate the organism's exposure to toxins retrospectively or to detect hormonal changes related to disease or pregnancy [17].

Social interactions and group relationships are important modulators of the activity of the HPA (hypothalamic–pituitary–adrenal) axis [18]. The HPA axis is a major component of the physiological stress response in mammals. Its activity can be analyzed by assessing the level of cortisol. Cortisol, known as the stress hormone, has a wide range of effects on how the body works. It participates in the metabolism of glucose and lipids, and has immunosuppressive and anti-inflammatory effects [19]. It influences protein, calcium, and water–electrolyte metabolism as well as hematopoietic systems [20]. Chronic stress is thought to be accompanied by a hyperactive HPA axis [19]. Measuring hair cortisol is a potential method of measuring the long-term stress response and reflects the average level of circulating cortisol built into growing hair over time [21].

Since 2004, when the first report on the measurement of cortisol in hair appeared [22], the interest in its laboratory potential and possible applications in the assessment of chronic stress levels was gradually increasing [23]. Studies on the HCL of dogs were conducted, among others, by Accorsi et al. [24] and Direksin et al. [25]. The level of HCL of cats has been studied by Accorsi et al. [24] and Franchini et al. [26]. As for other animal species, cortisol has been studied in, among others, sheep wool [27], pig hair [28], dairy cows [29], horses [30], and captive wild animals [31,32].

The aim of this study was to determine the hair cortisol level (HCL) of dogs, cats, and their owners living in the same house. To our knowledge, hormonal relationships in dogs and cats, dogs and their owners, and cats and their owners living all together have not yet been evaluated. We assumed that there would be hormonal correlations between individuals living in one household. Correlations between the levels of stress hormones in humans and their pets have already been observed in our previous studies [3] on acute stress as well as in Sundman's studies [6] on chronic stress.

2. Materials and Methods

All methods used in the study were in line with the Act of 15 January 2015 on the protection of animals used for scientific or educational purposes (Journal of Laws of 2015, item 266) and the Directive of the European Parliament and of the Council of 22 September 2010 on the protection of animals used for scientific purposes (2010/63/EU). The tests performed were non-invasive, which means that, within the meaning of the Directive, they did not cause pain, suffering, distress, or permanent damage to an extent equal to, or more severe than, a needle-stick injury. All owners were over 18 years of age and gave their written consent to participate in the study voluntarily.

Twenty-five women who owned at least one dog and at least one cat were invited to take part in the study. The invitation to participate in the study was made available on the internet, on a social network, in groups associating animal lovers. The condition for participation in the study was to have at least one dog and at least one cat at home. In total, 45 dogs and 55 cats from 25 households participated in the study. The women belonged to two age groups: <25 (n = 10) and 25–50 (n = 15). More than half (n = 14) indicated a village as their place of residence, the others ($n = 11$) indicated the city. The study participants included 15 working women, 6 working students, and 4 non-working students. The group of dogs included 27 females and 18 males. Of all dogs, 31 were castrated. Three dogs were under 2 years of age, 22 dogs were aged 2–6 years, and 20 dogs were 7 years or older. Fifty-five cats took part in the study—31 females and 24 males. There were 50 cats after the castration procedure and only 5 animals were not castrated. The animals were divided into three age groups—less than 3 years old ($n = 14$), 3–10 years old ($n = 29$), and 11 and more years old ($n = 12$). All animals participating in the study, as well as their owners, did not undergo any surgery for at least 6 months preceding the study and did not receive treatment for chronic diseases.

The women were asked to complete a questionnaire that was a shortened modification of the MDORS scale—Monash Dog-Owner Relationship Scale [33] and CORS—Cat-Owner Relationship Scale [34]. The first part of the questions concerned basic information about the owner, cat, and dog, such as gender, age, and place of residence. The second part of the questionnaire concerned the frequency of individual owner interactions with each of her cats and each of her dogs. In the third part of the questionnaire, the owner rated her emotional relationship with each of her cats and each of her dogs. The second and third parts of the questionnaire are provided as Supplementary Materials.

The biological material in the study was hair. The hair was collected in a non-invasive way, by cutting right next to the skin. Hair was collected from the lumbosacral area from dogs and cats and from the occipital area of the owners. For the analysis, the centimeter closest to the scalp end was used. Taking a hair sample is a simple procedure that can be performed after a short briefing and does not require the presence of a professional. The cutting of the hair is minimally invasive and painless [35]. The hair samples were placed in foil bags and stored at room temperature until analysis [23].

The extraction methodology was modified from Koren et al. [36] and Accorsi et al. [24]. Hair was first minced into 1–2 mm length fragments and 20 mg of trimmed hair was put in a glass vial. Three-and-a-half methanol (Sigma-Aldrich, Poznań, Poland) was added, and vials were incubated at 50 °C with gentle shaking for 24 h. After incubation, the supernatant was filtrated to separate the liquid phase and put into disposable glass culture tubes. Following this, this supernatant was evaporated to dryness under an air-stream suction hood at 37 °C. Dry residue was then dissolved into 1 mL of phosphate-buffered saline (PBS) 0.05 M, pH 7.5. Samples were vortexed for one minute followed by another 30 s until they were well mixed. The cortisol levels in the samples were determined with the DRG Salivary Cortisol HS ELISA assay. The procedures followed the manufacturer's instructions. All samples were measured in triplets. Cortisol concentrations were expressed in ng/mL.

The statistical analysis was performed with the use of the Statistica 13.3 statistical package. The analysis of the correlation between the HCL in the pets and their owners was performed based on Spearman's rank correlation coefficient, due to the deviations of the cortisol level distribution from the normal distribution. The compliance of the distributions with the normal distribution was assessed with the Shapiro–Wilk test. The analysis of the significance of differences in HCL in the tested pets depending on the strength of the relationship with the owner was carried out using the Mann–Whitney U test when comparing two groups and the Kruskal–Wallis test when at least three groups were compared. The results were considered significant when $p \leq 0.05$.

Methodological Limitations

At the initial stage of planning the research, we assumed that women who had 1 dog and 1 cat at home would take part in the study. As it turned out, however, very often a woman who has both a dog and a cat at home constantly expands her "herd" with new individuals. For this reason, our research group consisted of a total of 25 owners, 55 cats, and 45 dogs. The unequal number of specimens in households made it slightly difficult for us to later perform statistical work on the results. Companion animals are usually treated by the owners as family members and as a separate individuals. The owner has a different kind of emotional relationship with each of the animals, as well as with the children. Therefore, we also treated these animals as separate entities, and did not want to compare the owner's cortisol level with the average cortisol level of the cats/dogs living on the farm. Therefore, we compared the owner's cortisol level with each dog separately and with each cat separately.

3. Results

Descriptive statistics for human and animal hair cortisol levels (HCL) are presented in Table 1. Spearman's rank correlation coefficients were determined to investigate the relationship between the caregiver's and the animal's cortisol levels. Based on the study conducted, no significant relationship can be found between the HCL of owners and pets (both dogs and cats), as well as between dogs and cats living together (Table 2).

Table 1. Descriptive statistics for human and animal hair cortisol levels (HCL).

Variable	Mean	Std. Dev.	Minimum	Maximum	Median	Lower Quartile	Upper Quartile
Human HCL (ng/mL)	4.62	1.87	1.88	7.89	4.31	2.99	6.09
Dog HCL (ng/mL)	0.26	0.12	0.12	0.58	0.23	0.19	0.33
Cat HCL (ng/mL)	0.45	0.36	0.18	2.69	0.33	0.26	0.54

Table 2. Spearman Rank Order Correlations for human and animal hair cortisol levels (HCL).

Pair of Variables	Spearman R	T (N-2)	*p*-Value
Human HCL (ng/mL) and animal HCL (ng/mL)	−0.033	−0.325	0.746
Human HCL (ng/mL) and dog HCL (ng/mL)	0.049	0.319	0.751
Human HCL (ng/mL) and cat HCL (ng/mL)	−0.102	−0.744	0.460
Cat HCL (ng/mL) and dog HCL (ng/mL)	0.115	1.299	0.196

Based on the study conducted, no significant correlation can be found between the HCL of the owner and the animal, broken down by species and sex, or between dogs and cats living together, broken down by gender. Only a significant negative correlation ($R = -0.461$, $p = 0.023$) was observed between the HCL in owners and male cats.

3.1. The Strength of the Correlation between the HCL in the Owner and the Animal Depends on the Frequency of Interactions

The study found no significant correlation between the owner and the dog's HCL in any of the groups depending on the frequency of kissing the dog. In the case of cats, a significant positive correlation ($R = 0.686$, $p = 0.0096$) was found between human and cat cortisol levels when cats are never kissed. For the remaining groups, depending on the frequency of kissing cats, no significant correlation could be found between the level of cortisol of the owners and cats.

There was no significant relationship between the HCL in the owner and the dog or cat in any of the groups depending on the frequency of playing with the animal. Likewise, for the frequency of giving your dog/cat treats, the frequency of hugging the animal, and the frequency of having the animal with you while you relax. The study did not show a significant relationship between the HCL in humans and cats, in any of the groups, depending on the frequency of grooming treatments. All analysis results that are not statistically significant can be found in Supplementary Materials. As for dogs, a significant positive correlation was observed between the HCL in the owner and the dog, for dogs in which the owner performs grooming treatments once a week ($R = 0.836$, $p = 0.005$).

3.2. The Strength of the Correlation between the HCL in the Owner and the Animal Depends on the Emotional Relationship

We can find a significant positive correlation between human and cat HCL in the group of people who do not have any special conviction that the cat will be with them even when others leave it (R = 0.576, p = 0.031). In the study, there was a tendency towards a negative correlation between human and cat cortisol levels, where the cat helps to survive difficult times (R = −0.335, p = 0.095). We observe a tendency for a negative correlation (R = −0.450, p = 0.092) in the case of people whom the cat definitely keeps company. There was also a tendency for a negative correlation between human and cat cortisol levels (R = −0.754, p = 0.084) in the group of people who want to have their cat constantly nearby.

There is a tendency for a positive correlation between the HCL in the owner and the dog when the dog is definitely next to the human when she needs comfort (R = 0.321, p = 0.090). We can find a significant positive correlation (R = 0.526, p = 0.036) between the HCL in the owner and the dog, in the case of people who definitely happen to tell their dog what they would not tell anyone else. We also observe a tendency for a positive correlation (R = 0.401, p = 0.099) in the case of people whom the dog definitely keeps company, and a tendency for the occurrence of a negative correlation (R = −0.772, p = 0.072) in the case of people who do not keep up a companionship with the dog. We can find a significant negative correlation (R = −0.741, p = 0.036) between human and dog HCL in the case of people who deny that the dog constantly observes them and focuses its attention on them.

The study showed a significant positive correlation (R = 0.583, p = 0.009) between the HCL in the owner and the dog, in the case of people who have a definite feeling that the dog gives them a reason to wake up each morning. However, quite the opposite is the case for cats. There was a significant negative correlation (R = −0.704, p = 0.007) between human and cat cortisol levels for people who feel that the cat gives them a reason to wake up each morning.

The study did not show a significant relationship between the HCL in dogs and their owners, depending on how traumatic the dog's death would be. However, in the case of cats, there is a tendency to observe a negative correlation between the level of cortisol of the cat and its owner when the owner firmly believes that the death of her cat will be a traumatic event for her (R = −0.303, p = 0.097).

4. Discussion

Cats and dogs are the most popular companion animals. They have fully adapted to the human social environment and are capable of establishing long-term social relationships with people [37]. The influence of the emotional connection on the hormonal interactions between humans and animals is the subject of much research. In our study, no significant relationship was found between the owner and pet's hair cortisol levels, both for dogs and cats. Similarly, in the studies by Höglin et al. [38] it has not been found that the HCL of the owner is mirrored by the level of this hormone in the dog. However, in previous studies by these authors, a significant correlation was observed between the levels of cortisol in the hair of dogs and owners. However, the dogs participating in that study [6] were shepherd dogs, and the observed interaction, as the authors themselves indicate, may have resulted from the selection of these dogs for cooperation with humans [38]. Mutual understanding is stronger in the human–animal relationship the more time they spend together performing the same tasks. Commitment to training and time spent training together are associated with experiencing a close relationship and this may cause a stronger hormonal dependence [3,39–41]. In our study, a significant positive correlation was observed between the HCL of the owner and the dog when dogs were groomed at least once a week, which is also related to spending time together.

In our study, we also did not observe a significant correlation between the levels of cortisol in dogs and cats living in the same house. In addition, we asked the owner if, according to them, the pets live in harmony and if there were any conflicts between the pets. We found no significant correlation depending on the answers to these questions.

The social skills of domestic cats in the context of human–animal interactions have not been studied as thoroughly as for dogs [37]. Perhaps it is related to their shorter period of domestication and living with humans, as well as their higher sense of independence [10]. Even if we consider that cats and dogs have different predispositions to interact with humans, both species are able to communicate effectively with humans in different situations, and perform it differently, because, among other things, humans have developed a completely different type of relationship with these pets [42].

The results of our research suggest that if a human is strongly emotionally connected with a dog, then we observe a different hormonal relationship between them than in the case of a human strongly emotionally connected with a cat. If the owner thinks her dog gives her a reason to wake up each morning, we see a significant positive correlation in their cortisol levels. The opposite is true for humans and cats. If the owner believes that the cat is important enough to her to give her a reason to wake up in the morning, this significant correlation is negative. The more the owner is emotionally connected with her cat (the cat definitely helps her to survive difficult times, wants to have the cat always nearby, and his death will be a highly traumatic event), the more frequent the tendency towards a negative correlation of cortisol levels. Is it possible that the more attention the owner pays to the cat, the more stressful it is for the animal? Might it be that the stronger the emotional relationship between a person and a cat, the more discomfort it causes for the cat?

When the dog is constantly accompanying its owner, there is a tendency for a positive correlation between cortisol levels. Again, the situation is quite the opposite for the cat. If in the opinion of the owner, the cat is still accompanying her, there is a tendency for a negative correlation. Perhaps it is worth considering whether, in this case, a cat following a person and watching him might not be a symptom of passive aggression, as in the case of a cat–cat interaction [43]?

In a study by González-Ramírez and Landero-Hernández [44] comparing the relationships between humans and dogs and humans and cats, the respondents indicated greater emotional closeness with their dogs than with cats (they noticed greater social support, companionship, and unconditional love in their dogs). The relationship of the owners with their cats was assessed as requiring less responsibility and associated with fewer restrictions in daily activities. In a study by Morris et al. [45], cat owners viewed their pets as less emotional and intellectual compared to dog owners. Could this be because humans just cannot read or misinterpret emotions in cats [46]? In our study, owners often felt the same way about a dog as they did about a cat, but it triggered a completely different type of correlation (positive and negative) in cortisol levels.

Most of the research on hormones and stress, both in nonhuman animals and humans, has focused on the sympathetic system and the HPA axis. Although the sympathetic-adrenal system and the HPA axis are widely regarded as the most important physiological systems activated during stress, there are many peptides involved in the stress response in addition to these so-called classical stress systems. These include, but are not limited to, corticotropic releasing hormone (CRH), vasopressin (AVP), adrenocorticotropic hormone, opioid peptides, oxytocin, and several appetite-regulating hormones such as orexin, neuropeptide Y, agouti-related peptide, leptin, and ghrelin [47]. Recently, the effects of the neuropeptides kisspeptin, dynorphin, and neurokinin B, known as KNDy peptides, have also been studied in stress assessment [Ralph et al. 2016]. To fully investigate potential mechanisms underlying the activity or control of the HPA axis, measurements of both HPA direct axis hormones and brain chemicals (e.g., serotonin) that contribute to HPA axis activity would be required [48]. Serotonin is thought to interact complexly with dopamine and cortisol, and in general, serotonin can be said to increase dopamine production and inhibit cortisol production [49]. The role of adrenal steroids involves many interactions with the neurochemical systems in the hippocampus, including GABA in addition to serotonin [50]. Stressful conditions lead to increased free radical production. The increase in free radicals strongly affects all body systems, therefore oxidants and antioxidants arouse wide interest in biological and medical research. Disorders in the pro-oxidative-antioxidant system have been defined as oxidative stress [51]. Although our study did not find many significant correlations between the hair cortisol levels in owners and their pets, studies using other stress markers might have yielded different results.

5. Conclusions

Today, there is no doubt that both dogs and cats can create social relations with their owners. They also create a specific emotional relationship with them. The existence of hormonal correlation between humans and pets under acute stress conditions has been repeatedly confirmed in other studies. In our research, no significant correlations were found in the hair cortisol level between humans and dogs, humans and cats, or between dogs and cats living in one household. However, hormonal dependencies regarding chronic stress markers still require in-depth analyses. Broadly understood, human–pet interactions will remain an area of interest for scientists and a wide field of research for a long time to come.

Supplementary Materials: The following supporting information can be downloaded at: https://www.mdpi.com/article/10.3390/ani12111472/s1, Questionnaire, Insignificant results.

Author Contributions: Conceptualization, J.W. and A.S.; methodology, J.W., P.S. and A.S.; software, M.K.; validation, P.S. and A.S.; formal analysis, J.W.; investigation, J.W. and A.G.; resources, J.W.; data curation, J.W. and A.G; writing—original draft preparation, J.W. and P.S.; writing—review and editing, J.W. and P.S.; visualization, A.G.; supervision, A.S. and M.K.; project administration, M.K.; funding acquisition, J.W. and M.K. All authors have read and agreed to the published version of the manuscript.

Funding: The research was part of a research project for young scientists (ZKE/MN-6/ZIR/20) University of Life Sciences in Lublin, Poland.

Institutional Review Board Statement: The study was conducted in accordance with the Act of 15 January 2015 on the protection of animals used for scientific or educational purposes (Journal of Laws of 2015, item 266) and the Directive of the European Parliament and of the Council of 22 September 2010 on the protection of animals used for scientific purposes (2010/63/EU).

Informed Consent Statement: All owners were over 18 years of age and gave their written consent to participate in the study voluntarily. Written informed consent for publication must be obtained from identifiable human participants. For studies involving client-owned animals, written informed consent must be obtained from the owner of the animals (or an authorized agent for the owner).

Data Availability Statement: The data presented in this study are available on request from the corresponding authors.

Acknowledgments: The authors extend a cordial thanks to all the dog and cat owners who participated in this study.

Conflicts of Interest: The authors declare no conflict of interest. The funders had no role in the design of the study; in the collection, analyses, or interpretation of data; in the writing of the manuscript, or in the decision to publish the results.

References

1. Bergamasco, L.; Osella, M.C.; Savarino, P.; Larosa, G.; Ozella, L.; Manassero, M.; Badino, P.; Odore, R.; Barbero, R.; Re, G. Heart Rate Variability and Saliva Cortisol Assessment in Shelter Dog: Human-Animal Interaction Effects. *Appl. Anim. Behav. Sci.* **2010**, *125*, 56–68. [CrossRef]
2. Herzog, H.A. Gender Differences in Human-Animal Interactions: A Review. *Anthrozoos* **2007**, *20*, 7–21. [CrossRef]
3. Wojtaś, J.; Karpiński, M.; Czyżowski, P. Salivary Cortisol Interactions in Search and Rescue Dogs and Their Handlers. *Animals* **2020**, *10*, 595. [CrossRef] [PubMed]
4. Hemsworth, P.H. Human-Animal Interactions in Livestock Production. *Appl. Anim. Behav. Sci.* **2003**, *81*, 185–198. [CrossRef]
5. Buttner, A.P.; Thompson, B.; Strasser, R.; Santo, J. Evidence for a Synchronization of Hormonal States between Humans and Dogs during Competition. *Physiol. Behav.* **2015**, *147*, 54–62. [CrossRef]

6. Sundman, A.S.; van Poucke, E.; Svensson Holm, A.C.; Faresjö, Å.; Theodorsson, E.; Jensen, P.; Roth, L.S.V. Long-Term Stress Levels Are Synchronized in Dogs and Their Owners. *Sci. Rep.* **2019**, *9*, 7391. [CrossRef]
7. Payne, E.; Bennett, P.C.; McGreevy, P.D. Current Perspectives on Attachment and Bonding in the Dog–Human Dyad. *Psychol. Res. Behav. Manag.* **2015**, *8*, 71–79. [CrossRef] [PubMed]
8. Petersson, M.; Uvnäs-Moberg, K.; Nilsson, A.; Gustafson, L.L.; Hydbring-Sandberg, E.; Handlin, L. Oxytocin and Cortisol Levels in Dog Owners and Their Dogs Are Associated with Behavioral Patterns: An Exploratory Study. *Front. Psychol.* **2017**, *8*, 1796. [CrossRef]
9. Willen, R.M.; Mutwill, A.; MacDonald, L.J.; Schiml, P.A.; Hennessy, M.B. Factors Determining the Effects of Human Interaction on the Cortisol Levels of Shelter Dogs. *Appl. Anim. Behav. Sci.* **2017**, *186*, 41–48. [CrossRef]
10. Turner, D.C. A Review of over Three Decades of Research on Cat-Human and Human-Cat Interactions and Relationships. *Behav. Processes* **2017**, *141*, 297–304. [CrossRef]
11. Nagasawa, T.; Ohta, M.; Uchiyama, H. The Urinary Hormonal State of Cats Associated With Social Interaction With Humans. *Front. Vet. Sci.* **2021**, *8*, 680843. [CrossRef]
12. Turner, D.C.; Rieger, G.; Gygax, L. Spouses and Cats and Their Effects on Human Mood. *Anthrozoos* **2003**, *16*, 213–228. [CrossRef]
13. Contreras, E.T.; Vanderstichel, R.; Hovenga, C.; Lappin, M.R. Evaluation of Hair and Nail Cortisol Concentrations and Associations with Behavioral, Physical, and Environmental Indicators of Chronic Stress in Cats. *J. Vet. Intern. Med.* **2021**, *35*, 2662–2672. [CrossRef]
14. Lee, D.Y.; Kim, E.; Choi, M.H. Technical and Clinical Aspects of Cortisol as a Biochemical Marker of Chronic Stress. *BMB Rep.* **2015**, *48*, 209–216. [CrossRef]
15. Russell, E.; Koren, G.; Rieder, M.; van Uum, S. Hair Cortisol as a Biological Marker of Chronic Stress: Current Status, Future Directions and Unanswered Questions. *Psychoneuroendocrinology* **2012**, *37*, 589–601. [CrossRef] [PubMed]
16. Cordero, R.; Lee, S.; Paterson, S. Distribution of Concentrations of Cocaine and ItsMetabolites in Hair Collected Postmortem from Caseswith Diverse Causes/Circumstances of Death. *J. Anal. Toxicol.* **2010**, *34*, 543–548. [CrossRef]
17. Acker, M.; Mastromonaco, G.; Schulte-Hostedde, A.I. The Effects of Body Region, Season and External Arsenic Application on Hair Cortisol Concentration. *Conserv. Physiol.* **2018**, *6*. [CrossRef]
18. Creel, S.; Dantzer, B.; Goymann, W.; Rubenstein, D.R. The Ecology of Stress: Effects of the Social Environment. *Funct. Ecol.* **2013**, *27*, 66–80. [CrossRef]
19. Manenschijn, L.; Koper, J.W.; Lamberts, S.W.J.; van Rossum, E.F.C. Evaluation of a Method to Measure Long Term Cortisol Levels. *Steroids* **2011**, *76*, 1032–1036. [CrossRef]
20. Musiała, N.; Hołyńska-iwan, I.; Olszewska-słonina, D. Kortyzol–Nadzór Nad Ustrojem w Fizjologii i Stresie Cortisol—Inspection in the Physiology and Stress. *Diagn. Lab.* **2018**, *54*, 29–36. [CrossRef]
21. Bryson, H.E.; Mensah, F.; Goldfeld, S.; Price, A.M.H.; Giallo, R. Hair Cortisol in Mother–Child Dyads: Examining the Roles of Maternal Parenting and Stress in the Context of Early Childhood Adversity. *Eur. Child Adolesc. Psychiatry* **2021**, *30*, 563–577. [CrossRef] [PubMed]
22. Raul, J.S.; Cirimele, V.; Ludes, B.; Kintz, P. Detection of Physiological Concentrations of Cortisol and Cortisone in Human Hair. *Clin. Biochem.* **2004**, *37*, 1105–1111. [CrossRef]
23. Greff, M.J.E.; Levine, J.M.; Abuzgaia, A.M.; Elzagallaai, A.A.; Rieder, M.J.; van Uum, S.H.M. Hair Cortisol Analysis: An Update on Methodological Considerations and Clinical Applications. *Clin. Biochem.* **2019**, *63*, 1–9. [CrossRef] [PubMed]
24. Accorsi, P.A.; Carloni, E.; Valsecchi, P.; Viggiani, R.; Gamberoni, M.; Tamanini, C.; Seren, E. Cortisol Determination in Hair and Faeces from Domestic Cats and Dogs. *Gen. Comp. Endocrinol.* **2008**, *155*, 398–402. [CrossRef]
25. Direksin, K.; Srirat, T.; Juiboot, S.; Khongthananet, S.; Chaiyasaen, N.; Nopwinyoowong, S. Preliminary Study of Hair Cortisol Concentration (HCC) of Dogs before and after Surgery. *J. Mahanakorn Vet. Med.* **2022**. Available online: https://li01.tci-thaijo.org/index.php/jmvm/article/view/249430 (accessed on 9 May 2022).
26. Franchini, M.; Prandi, A.; Filacorda, S.; Pezzin, E.N.; Fanin, Y.; Comin, A. Cortisol in Hair: A Comparison between Wild and Feral Cats in the North-Eastern Alps. *Eur. J. Wildl. Res.* **2019**, *65*. [CrossRef]
27. Fürtbauer, I.; Solman, C.; Fry, A. Sheep Wool Cortisol as a Retrospective Measure of Long-Term HPA Axis Activity and Its Links to Body Mass. *Domest. Anim. Endocrinol.* **2019**, *68*, 39–46. [CrossRef]
28. Heimbürge, S.; Kanitz, E.; Tuchscherer, A.; Otten, W. Within a Hair's Breadth—Factors Influencing Hair Cortisol Levels in Pigs and Cattle. *Gen. Comp. Endocrinol.* **2020**, *288*, 113359. [CrossRef] [PubMed]
29. Vesel, U.; Pavič, T.; Ježek, J.; Snoj, T.; Starič, J. Welfare Assessment in Dairy Cows Using Hair Cortisol as a Part of Monitoring Protocols. *J. Dairy Res.* **2020**, *87*, 72–78. [CrossRef] [PubMed]
30. Sauveroche, M.; Henriksson, J.; Theodorsson, E.; Svensson Holm, A.C.; Roth, L.S.V. Hair Cortisol in Horses (*Equus caballus*) in Relation to Management Regimes, Personality, and Breed. *J Vet. Behav.* **2020**, *37*, 1–7. [CrossRef]
31. Salas, M.; Temple, D.; Abáigar, T.; Cuadrado, M.; Delclaux, M.; Enseñat, C.; Almagro, V.; Martínez-Nevado, E.; Quevedo, M.Á.; Carbajal, A.; et al. Aggressive Behavior and Hair Cortisol Levels in Captive Dorcas Gazelles (*Gazella dorcas*) as Animal-Based Welfare Indicators. *Zoo Biol.* **2016**, *35*, 467–473. [CrossRef]
32. Yamanashi, Y.; Teramoto, M.; Morimura, N.; Hirata, S.; Suzuki, J.; Hayashi, M.; Kinoshita, K.; Murayama, M.; Idani, G. Analysis of Hair Cortisol Levels in Captive Chimpanzees: Effect of Various Methods on Cortisol Stability and Variability. *MethodsX* **2016**, *3*, 110–117. [CrossRef]

33. Dwyer, F.; Bennett, P.C.; Coleman, G.J. Development of the Monash Dog Owner Relationship Scale (MDORS). *Anthrozoos* **2006**, *19*, 243–256. [CrossRef]
34. Howell, T.J.; Bowen, J.; Fatjó, J.; Calvo, P.; Holloway, A.; Bennett, P.C. Development of the Cat-Owner Relationship Scale (CORS). *Behav. Processes* **2017**, *141*, 305–315. [CrossRef]
35. Heimbürge, S.; Kanitz, E.; Otten, W. The Use of Hair Cortisol for the Assessment of Stress in AnimalsGen. *Comp. Endocrinol.* **2019**, *270*, 10–17. [CrossRef]
36. Koren, L.; Mokady, O.; Karaskov, T.; Klein, J.; Koren, G.; Geffen, E. A Novel Method Using Hair for Determining Hormonal Levels in Wildlife. *Anim. Behav.* **2002**, *63*, 403–406. [CrossRef]
37. Merola, I.; Lazzaroni, M.; Marshall-Pescini, S.; Prato-Previde, E. Social Referencing and Cat–Human Communication. *Anim. Cogn.* **2015**, *18*, 639–648. [CrossRef]
38. Höglin, A.; van Poucke, E.; Katajamaa, R.; Jensen, P.; Theodorsson, E.; Roth, L.S.V. Long-Term Stress in Dogs Is Related to the Human–Dog Relationship and Personality Traits. *Sci. Rep.* **2021**, *11*, 8612. [CrossRef]
39. Bennett, P.C.; Rohlf, V.I. Owner-Companion Dog Interactions: Relationships between Demographic Variables, Potentially Problematic Behaviours, Training Engagement and Shared Activities. *Appl. Anim. Behav. Sci.* **2007**, *102*, 65–84. [CrossRef]
40. Dotson, M.J.; Hyatt, E.M. Understanding Dog-Human Companionship. *J. Bus. Res.* **2008**, *61*, 457–466. [CrossRef]
41. Meyer, I.; Forkman, B. Dog and Owner Characteristics Affecting the Dog-Owner Relationship. *J. Vet. Behav.* **2014**, *9*, 143–150. [CrossRef]
42. Miklósi, Á.; Pongrácz, P.; Lakatos, G.; Topál, J.; Csányi, V. A Comparative Study of the Use of Visual Communicative Signals in Interactions between Dogs (*Canis familiaris*) and Humans and Cats (*Felis catus*) and Humans. *J. Comp. Psychol.* **2005**, *119*, 179–186. [CrossRef]
43. Hargrave, C. When Does Inter-Cat Communication Become Inter-Cat Aggression? *Vet. Nurse* **2016**, *7*, 336–343. [CrossRef]
44. González-Ramírez, M.T.; Landero-Hernández, R. Pet–Human Relationships: Dogs versus Cats. *Animals* **2021**, *11*, 2745. [CrossRef] [PubMed]
45. Morris, P.; Doe, C.; Godsell, E. Secondary Emotions in Non-Primate Species? Behavioural Reports and Subjective Claims by Animal Owners. *Cogn. Emot.* **2008**, *22*, 3–20. [CrossRef]
46. Arahori, M.; Kuroshima, H.; Hori, Y.; Takagi, S.; Chijiiwa, H.; Fujita, K. Owners' View of Their Pets' Emotions, Intellect, and Mutual Relationship: Cats and Dogs Compared. *Behav. Processes* **2017**, *141*, 316–321. [CrossRef]
47. Tilbrook, A.J.; Ralph, C.R. Hormones, Stress and the Welfare of Animals. *Anim. Prod. Sci.* **2018**, *58*, 408–415. [CrossRef]
48. King, S.L.; Hegadoren, K.M. Stress Hormones: How Do They Measure Up? *Biol. Res. Nurs.* **2002**, *4*, 92–103. [CrossRef]
49. Field, T.; Hernandez-Reif, M.; Diego, M.; Schanberg, S.; Kuhn, C. Cortisol Decreases and Serotonin and Dopamine Increase Following Massage Therapy. *Int. J. Neurosci.* **2005**, *115*, 1397–1413. [CrossRef]
50. McEwen, B.S. Central Effects of Stress Hormones in Health and Disease: Understanding the Protective and Damaging Effects of Stress and Stress Mediators. *Eur. J. Pharmacol.* **2008**, *583*, 174–185. [CrossRef]
51. Megahed, G.A.; Anwar, M.M.; Wasfy, S.I.; Hammadeh, M.E. Influence of Heat Stress on the Cortisol and Oxidant-Antioxidants Balance during Oestrous Phase in Buffalo-Cows (*Bubalus bubalis*): Thermo-Protective Role of Antioxidant Treatment. *Reprod. Domest. Anim.* **2008**, *43*, 672–677. [CrossRef] [PubMed]

animals

MDPI

Article

Impact of the Dog–Human Bond on Canine Social Evaluation: Attachment Predicts Preference toward Prosocial Actors

Emily M. Richards *, Zachary A. Silver and Laurie R. Santos

Department of Psychology, Yale University, New Haven, CT 06520, USA; zsilver@oxy.edu (Z.A.S.)
* Correspondence: e.richards@yale.edu

Simple Summary: The human species naturally judges whether other agents are nice or mean from a young age. Recent research has suggested that such social judgments are influenced by the way humans form attachment bonds with others. Given dogs' rich evolutionary history alongside humans, researchers have become interested in whether dogs make similar evaluations of human social interactions, for instance, by distinguishing between someone who is helpful or unhelpful. However, this concept, to date, has shown mixed results. In the present study, we explore whether dogs' attachment bonds impact their ability to form these judgments. Specifically, the present study sought to investigate whether dogs' attachment bonds to their owners could predict the extent to which they successfully evaluated unfamiliar humans who interacted with their owners. We found that dogs with stronger attachment bonds to their owners were more likely to prefer people who helped their owners but were no more likely to avoid people who refused to help their owners. These results suggest that, as in humans, a dog's attachment may impact the way that they evaluate potential social partners.

Abstract: Scholars have argued that social evaluation, the capacity to evaluate different potential social partners, is an important capacity not just for humans but for all cooperative species. Recent work has explored whether domesticated dogs share a human-like ability to evaluate others based on prosocial and antisocial actions toward third parties. To date, this work has shown mixed results, suggesting that individual differences may play a role in dogs' capacity to evaluate others. In the present study, we test whether attachment—an individual difference that affects human social evaluation performance—can explain the mixed pattern of social evaluation results observed in dogs. We first tested dogs on a social evaluation task in which an experimenter either helped or refused to help the dog's owner open a container. We then assessed dogs' attachment strength using a subset of the C-BARQ. We found that attachment was a statistically significant predictor of dogs' preference toward the prosocial actor but was not a predictor in antisocial or control conditions. This finding provides early evidence that attachment may drive positivity biases in dogs and that attachment might explain mixed results within canine social evaluation literature.

Keywords: social evaluation; attachment; domestication; *Canis familiaris*

Citation: Richards, E.M.; Silver, Z.A.; Santos, L.R. Impact of the Dog–Human Bond on Canine Social Evaluation: Attachment Predicts Preference toward Prosocial Actors. *Animals* **2023**, *13*, 2480. https://doi.org/10.3390/ani13152480

Academic Editor: Betty McGuire

Received: 29 June 2023
Revised: 26 July 2023
Accepted: 28 July 2023
Published: 1 August 2023

1. Introduction

The ability to recognize and evaluate the actions of others is an important skill for any cooperative species. Many scholars have argued that the capacity to distinguish prosocial others—individuals who are likely to be helpful or cooperative in the future—from antisocial others—individuals who may be selfish and uncooperative—could be beneficial for the survival of social animals, e.g., [1,2]. Much research has shown that social evaluation is a critical skill in the human species and is one that develops very early in life. Human infants as young as 3 months in age prefer novel agents that behave prosocially (e.g., helping another agent to achieve this goal) relative to agents that behave neutrally and choose to avoid novel agents that behave antisocially (e.g., preventing an agent from achieving his goal) relative to neutral agents, see [3–7]. In one classic study [3], infants watched as a puppet tried and failed to make it up a steep hill. Infants were then introduced to two new puppets: one who acted prosocially, helping the first character up the hill, and a second who acted antisocially, hindering the first character by pushing him down the hill. When given a choice between the two puppets, infants reliably preferred to interact with the prosocial over the antisocial puppet. Results like these suggest that some capacity to evaluate the actions of agents is present within the first few months of human life.

The early emergence of social evaluation in the human species has prompted comparative researchers to explore whether similar capacities exist in other non-human species or whether such abilities are instead unique to humans. To test this question, researchers first explored whether non-human primates possessed the ability to socially evaluate novel agents. Krupenye and Hare [8] presented bonobos with a task similar to the ones used to test human infants and found that, in contrast to the performance of human infants, bonobos preferred antisocial humans. However, not all primates appear to show this antisocial preference. For example, Kawai et al. [9] found that marmoset monkeys avoided third parties who did not reciprocate during a social exchange. Similarly, Anderson and colleagues found that tufted capuchin monkeys tended to avoid antisocial humans who explicitly refused to help a third party [10] or failed to reciprocate goods with another actor [11]. Other studies have observed that non-human primates appear to make some social judgments when socially eavesdropping on prosocial or antisocial actors, but only sometimes and often with varying effects [12]. For example, Herrmann et al. [13] directly compared the social evaluative capacities of human children and non-human great apes (chimpanzees, bonobos, and orangutans) and found that human children and orangutans preferred a prosocial human actor when they themselves were the direct recipient of the actor's actions, while chimpanzees and bonobos did not exhibit any preference. Interestingly, Russell et al. [14] found nearly the opposite pattern of results; they observed that chimpanzees tended to prefer a prosocial actor—who gave food to a begging experimenter—compared to an antisocial actor but found that orangutans, gorillas, and bonobos exhibited no preference [14]. Taken together, studies on primate social evaluation to date show a rather mixed pattern of results, suggesting that non-human primates might possess a more limited capacity for social evaluation than developing human infants and children.

Other researchers have begun to explore whether non-human animals share human-like social evaluation capacities by focusing on a different group of non-human subjects, ones that have more experience interacting with human agents: domesticated pet dogs (*Canis familiaris*). Many researchers have argued that dogs might be an especially good species to test for social evaluation, given the close domestication history that dogs have shared with humans [1]. However, such canine social evaluation studies have also yielded mixed results to date: see the review in Silver et al. [15]. Some studies have found that dogs prefer prosocial over antisocial humans [15–18], whereas many other studies have found that dogs show no significant preferences when choosing between prosocial and antisocial individuals [19–24].

Researchers have now begun to investigate why dogs show such mixed performance on social evaluation tasks. Some researchers have begun testing whether specific methodological factors can explain the extent to which dogs are able to socially evaluate agents. Freidin et al. [22], for example, found that dogs are more likely to distinguish between prosocial and antisocial agents when they are given more explicit body language and verbal reactions to help them distinguish between the behavior of different agents. Similarly, Carballo et al. [17] found that dogs were successfully able to distinguish prosocial and antisocial human actors when those actors were of different genders but not when the two actors were of the same gender. In another example, Chijiiwa and colleagues [25] investigated whether dogs might show a more human-like pattern of social evaluation when they have a close relationship with the third-party individual who is being helped or hurt. To test this question, Chijiiwa and colleagues presented dogs with people who directed prosocial and antisocial actions not toward strangers, as in most studies, but instead toward the dogs' owners. Interestingly, Chijiiwa and colleagues found that dogs did not prefer individuals who helped their owners over neutral individuals who did not interact with their owner and instead showed a bias against antisocial individuals that no previous studies had found [25]. Other studies have taken a different approach to understanding dogs' mixed performance in social evaluation tasks, examining whether individual differences in dogs' backgrounds or training could explain the pattern of effects observed in social evaluation studies. Silver et al. [15], for example, found that trained agility dogs showed a human-like pattern of preferring prosocial to antisocial experimenters, whereas untrained pet dogs showed no preference.

The present study aims to explore whether another stable individual difference can explain the mixed pattern of results observed in canine social evaluation studies. Specifically, this study explores whether the way that dogs form relationships with humans can serve as a factor in determining whether dogs show preferences for prosocial or antisocial actors. Researchers in human psychology have found that attachment bonds—the emotional bond from one individual to another—emerge early in human development and form during early interactions between infants and their primary caregivers [26]. The nature of a young child's attachment to their primary caregiver has wide-reaching impacts on their development. For instance, the nature of this bond has a strong impact on the child's feelings of safety and security in the presence and absence of their primary caregiver as well as on their willingness to engage with novel stimuli and social partners [26].

Furthermore, these early attachment experiences form a relatively stable foundation for how we approach, develop, and maintain close relationships even through adulthood [27,28]. Research in adult humans has shown that a person's attachment style—one of several predictable patterns of attachment—can impact their preference for cognitive closure and how likely they are to use new information during social evaluations [29]. Similarly, research has shown that attachment style can predict a person's level of social curiosity, e.g., [29,30], and their sensitivity to social expressions [31]. Adults with different attachment styles also show differences in neural activation during social appraisals [31] and in their level of attentional control during non-social tasks [32]. Importantly for the purposes of the present experiment, new work has also shown that a person's early attachment affects their responses to prosocial versus antisocial behaviors during social evaluation paradigms, e.g., [33–35]. These findings suggest that there may be an important connection between attachment and social evaluation in humans.

Given that attachment appears to be a meaningful individual difference in human social preferences, this study aimed to explore whether similar individual differences in dog attachment could explain the mixed performance that dogs exhibit in standard social evaluation studies. While much work has examined the nature of human attachment (including that of human-to-dog attachment, see [36,37]), less work has tested the nature of dog-to-human attachment and non-human–animal-to-human attachment more broadly. However, a growing body of work has hinted that dogs may exhibit stable individual differences in the attachments they set up with others, e.g., [38–49], with the characteristics of dogs' patterns of attachment to their owners closely resembling those of human infants to their primary caregivers [45,46]. Additionally, emerging evidence has suggested that dogs' relationships with their owners seem to impact their behavior in cognitive tests. For example, the nature of this relationship appears to impact dogs' performances during problem-solving tasks [48,50] and their heart rate responses during threatening situations [51]. Taken together, these results provide evidence that, like humans, dogs' attachment bond to their owners appears to be a stable individual difference impacting their performance in cognitive tasks [48] and their willingness to engage with novel social partners [52].

The goal of the present study was to test whether a dog's attachment also affects the dog's success in social evaluations. To test this question, dogs were presented with the social evaluation task used by Chijiiwa and colleagues [25]. In this task, dogs first watched as a novel experimenter acted either prosocially, antisocially, or neutrally (i.e., did not interact) toward their owner. Then, dogs were released and could choose to take a high-value food reward from that actor or from a second neutral experimenter. Based on Chijiiwa and colleagues' findings [25], we hypothesized that dogs would exhibit a negativity bias after witnessing an antisocial interaction with their owner. In contrast with Chijiiwa et al. [25], however, we also hypothesized that dogs could exhibit a positivity bias after witnessing a prosocial interaction with their owner. We made this prediction because, although the literature on social evaluation in dogs provides mixed results, at least some studies, e.g., [16,17,19,20,22], have provided evidence of a positivity bias in dogs in certain contexts and under certain conditions. Note that we specifically chose to use a social evaluation method that involved dogs' owners since we hypothesized that attachment would be most likely to affect the dog's performance when the recipient of the observed interaction was the individual most connected to the dogs.

After assessing the dogs' performance on this task, we then assessed each dog's attachment bonds to its owner and tested whether this predicted the dogs' behavioral performance. Previous research has typically used one of two different methods to assess attachment relationships between dogs and their owners: behavioral assessments and owner-survey methods. Behavioral tests, such as the classic Strange Situation Test originally developed for research in human infants [53], typically use observational data to classify dogs' individual attachment styles based on the changes in dogs' behavior when their owner is present versus when their owner is absent. These behavioral tests, however, have several limitations. First, these tests often place dogs in intentionally stressful situations (e.g., by separating dogs from their owners when in an unfamiliar location, as in the Strange Situation Test or the Secure-Base Test, e.g., [45,47], or perhaps by placing dogs in the presence of a threatening individual, as in the Threatening Stranger Procedure [45,51]) which sometimes causes ethical concerns with dog owners. Secondly, these behavioral tests are often relatively long in duration, with the Strange Situation Test, for example, taking over twenty minutes to complete. Given the stress-inducing nature of many behavioral tests, in addition to the fact that the tests at our center are often shorter in duration, we worried that a long behavioral test like the Strange Situation Test could increase dogs' frustration and anxiety during their visits. As a result, we instead opted to assess dogs' attachment using an owner-survey method.

While various scales have been developed to assess owners' attachment to their dogs—such as the Dog Attachment Questionnaire [54], the Lexington Attachment to Pets Scale [55], or the Monash Dog Owner Relationship Scale [56]—fewer owner-survey methods have been developed to assess dogs' attachment to their owners. One questionnaire-based behavioral evaluation tool known as the Canine Behavioral Assessment & Research Questionnaire (C-BARQ), however, is widely used to explore a variety of canine personality traits, including attachment [57]. This standardized, 100-item validated assessment includes six internally consistent ($\alpha = 0.74$) questions that explore the degree to which a dog displays attachment and attention-seeking behaviors toward their human owner (see Table 1) [57]. Notably, the attachment and attention-seeking behaviors assessed in the C-BARQ closely resemble the types of behavior examined during standard behavioral canine attachment assessments. For instance, both the C-BARQ and behavioral tests such as the Strange Situation Test or Secure-Base test examine the degree and invasiveness of proximity-seeking behavior that a dog demonstrates toward their owner as well as whether dogs display a preference for their attachment figure [46,47,57]. Additionally, the C-BARQ examines dogs' reactions to interactions between the dog's owner and an unfamiliar person, as in the Strange Situation Test [45,46] or the Threatening Stranger Procedure [45,51]. Thus, the attachment and attention-seeking subset of the C-BARQ appears to be a promising owner-based survey method for investigating the nature of dogs' relationships with their owners that could be an optimal alternative when behavioral assessments are not feasible. As a result, we opted to use the attachment and attention-seeking subset of the C-BARQ to investigate whether attachment might impact dogs' performance on an owner-based social evaluation task.

Table 1. C-BARQ attachment and attention-seeking questions, see [57].

Question Number	Thinking Back over the Recent Past, How Often Has Your Dog Shown the Following Signs of Attachment or Attention-Seeking on a Scale from 0 (Never) to 4 (Always):
68	Displays a strong attachment for one particular member of the household.
69	Tends to follow you (or other members of the household) about the house, from room to room.
70	Tends to sit close to, or in contact with, you (or others) when you are sitting down.
71	Tends to nudge, nuzzle, or paw you (or others) for attention when you are sitting down.
72	Becomes agitated (whines, jumps up, tries to intervene) when you (or others) show affection for another person.
73	Becomes agitated (whines, jumps up, tries to intervene) when you (or others) show affection for another dog or animal.

2. Part A: Do Dogs Socially Evaluate Individuals Who Interact with Their Owners

2.1. Methods

2.1.1. Participants

Thirty-seven domesticated pet dogs (17 female, M_{age} = 5.68 years, SD_{age} = 2.82, $range_{age}$ = 1–13) were tested alongside their owners (36 unique owners, 1 owner tested two pet dogs of the same household) at the Canine Cognition Center at Yale University (see Table S1 for additional demographic information). An additional 4 dogs (2 female) were excluded due to either owner (N = 2) or experimenter (N = 2) errors during the demonstration. Note that we had initially planned to test a full sample of 60 dogs, but due to a 19-month lapse in testing during the COVID-19 pandemic, we could test only 12–13 dogs in each of the conditions rather than the 20 that we had originally pre-registered. To underscore the reliability of our existing data, however, it was important to note that our final sample size was comparable to the original sample size in Chijiiwa et al. [25].

2.1.2. Procedure

We began by testing dogs in a social evaluation study closely modeled after the one used by Chijiiwa and colleagues [25] (see Figure 1 for the set-up). Owners sat in the middle of a testing room with two female experimenters (hereafter the *actors*), one on either side. One of the actors (hereafter the *target actor*, see three conditions below) interacted with the owner during the presentation, whereas the other remained neutral. The owner and the actors faced the dog subject, who was located in the corner of the testing room and handled by a third experimenter. Owners held a closed transparent Tupperware container with a red block (the target object) inside and were cued by the third experimenter to begin the demonstration. The demonstration consisted of the owner unsuccessfully attempting to open the lid of the container for approximately 10 s while the actors focused their gaze on the ground in front of them *(initial attempt period)*. After the initial attempt period, the dogs then saw one of three different test conditions. In the *prosocial condition* (N = 13), the owner turned to present the container to one of the actors (the prosocial target actor) while the other actor remained neutral. The prosocial actor would then hold the base of the container so that the owner could successfully open the container. The owner then turned back to the dog, removed and held out the red block, then secured it back in the container. For the *antisocial condition* (N = 12), after the initial attempt period, the owner similarly turned to one of the actors (the antisocial target actor) while the other actor remained neutral. Rather than helping the owner, the antisocial actor looked at and made eye contact with the owner, then behaved uncooperatively by turning their entire body away from the owner for approximately 2 s. After this, the owner made 3 more seconds of failed attempts to open the container. In the *control condition* ($N = 12$), the owner did not turn to either actor to recruit help but instead paused for 2 s, looking straight at the container held out in front of them, while one actor (the control target actor) turned away. The owner then made a failed attempt for 3 more seconds to open the container. This allowed us to control whether the movement of the target actor alone affected the dogs' choices. In all three conditions, the owner placed the container on the ground at the end of the demonstration.

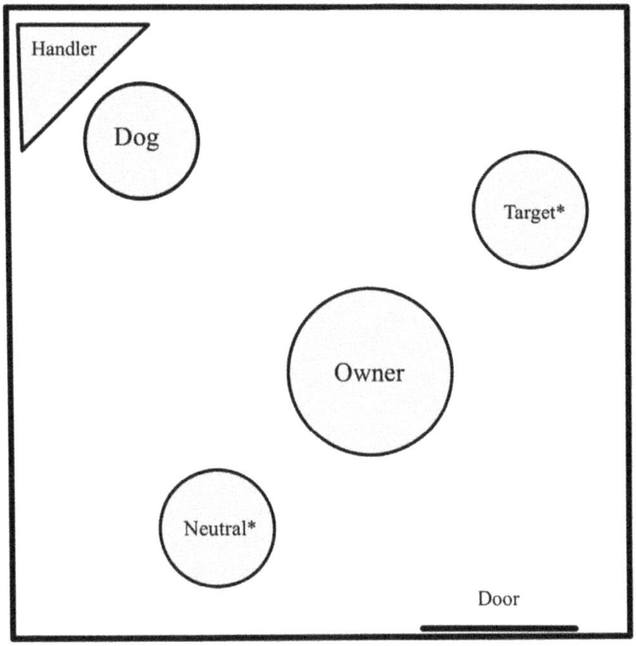

Figure 1. The test set up for our social evaluation study. The identities of the two actors (denoted by *) were counterbalanced between dogs and their positions were counterbalanced across trials.

The dogs were then presented with a choice between the two actors. During this *choice phase*, the target and neutral actors, still looking down, each extended their hands with a piece of high-value food (i.e., freeze-dried beef liver or, for dogs with food sensitivities, a comparable owner-provided treat), and the dog was released. The dog was allowed to receive only 1 reward, and our primary dependent measure was which actor the dogs approached first to receive their reward.

All dogs completed four trials of the same condition. Of the full sample, four trials were excluded due to either an owner error ($N = 3$) or the failure of the dog to approach an experimenter within 30 s of the trial's onset ($N = 1$). The actors' roles were counterbalanced between the dogs, and their positions were counterbalanced across the trials. Our secondary dependent measures were the total duration that the dogs spent looking at each of the actors during all four of the demonstrations as well as their average latency to choose either the target or neutral actor.

2.2. Results

A Kruskal–Wallis test revealed the strong effect ($\varepsilon^2 = 0.41$) of the condition on dogs' target actor choices ($H(2) = 14.58$, $p < 0.001$) (see Figure 2). Pairwise Mann–Whitney U tests between the conditions (with Bonferroni adjusted p-value of $0.05/3 = 0.017$) revealed that there were statistically significant differences between the control and prosocial conditions ($W = 28$, $p = 0.004$), as well as between the antisocial and prosocial conditions ($W = 18$, $p < 0.001$). There was no significant difference between the antisocial and control conditions

($W = 84$, $p = 0.428$). Additional pairwise Wilcoxon Signed-Rank tests were performed to determine if dogs preferred the target actor, as opposed to the neutral actor, in each condition (with Bonferroni adjusted $p = 0.017$). We found that dogs only had a statistically significant preference for the target actor in the prosocial condition ($V = 66$, $p = 0.003$). There were no statistically significant preferences toward the actor in the antisocial ($V = 3.5$, $p = 0.713$) nor the control conditions ($V = 8$, $p = 0.357$).

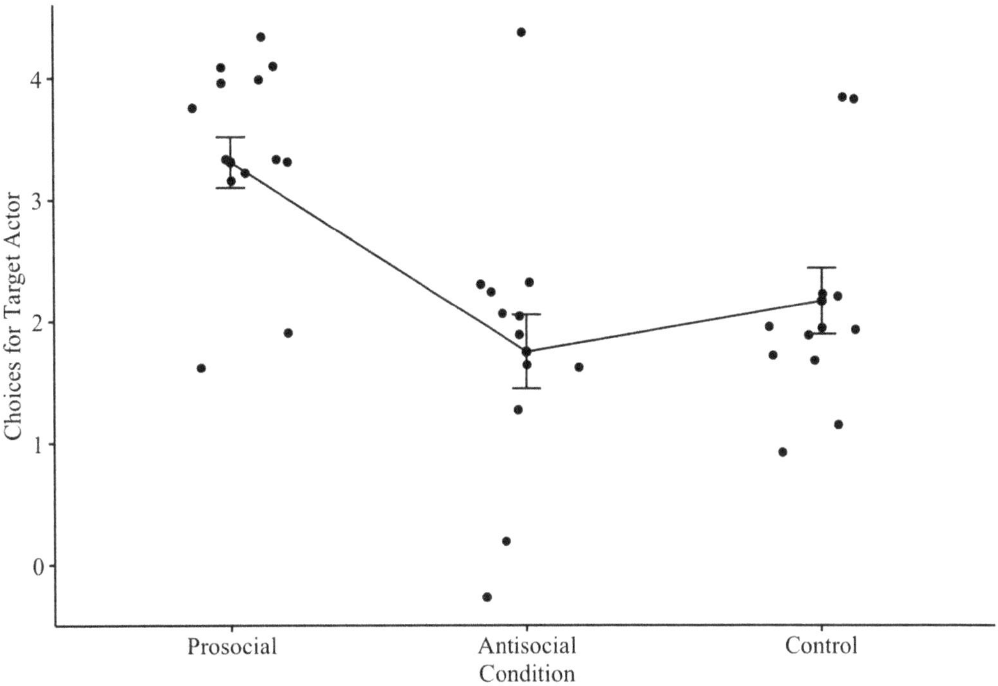

Figure 2. Jitter plot of dogs' mean choices for target actor by condition. Notably, dogs chose the target actor at above chance levels in the prosocial condition but not in the antisocial and control conditions.

Additional pairwise Wilcoxon Signed-Rank tests were performed to determine if the dogs looked longer at the target actor than the neutral actor during the demonstrations (Bonferroni adjusted $p = 0.017$) (see Table S2 for descriptive statistics). We found in all three conditions that the dogs looked significantly longer at the target actor than they did at the neutral actor (prosocial: $V = 88$, $p = 0.001$; antisocial: $V = 76$, $p = 0.001$; control: $V = 74$, $p = 0.003$), suggesting that dogs successfully visually distinguished between the two female actors in each condition (see Figure 3).

Lastly, we examined whether condition impacted dogs' average latency per actor choice. The results of pairwise Wilcoxon Signed-Rank tests (with Bonferroni adjusted $p = 0.017$) revealed that dogs in the prosocial condition were significantly faster to choose to accept a food reward from the neutral actor than the prosocial actor ($V = 74$, $p = 0.007$), but no significant effect was found within antisocial ($V = 29$, $p = 0.470$) nor control ($V = 47$, $p = 0.569$) conditions (see Table S3 for descriptive statistics, see also Figure S1). Given that dogs significantly preferred the target actor in the prosocial condition, this finding that dogs were faster to make a choice for the neutral actor suggests that the dogs might have been thinking more critically about their choice when choosing the prosocial actor, therefore taking longer, and perhaps choosing quickly on impulse when choosing the neutral actor.

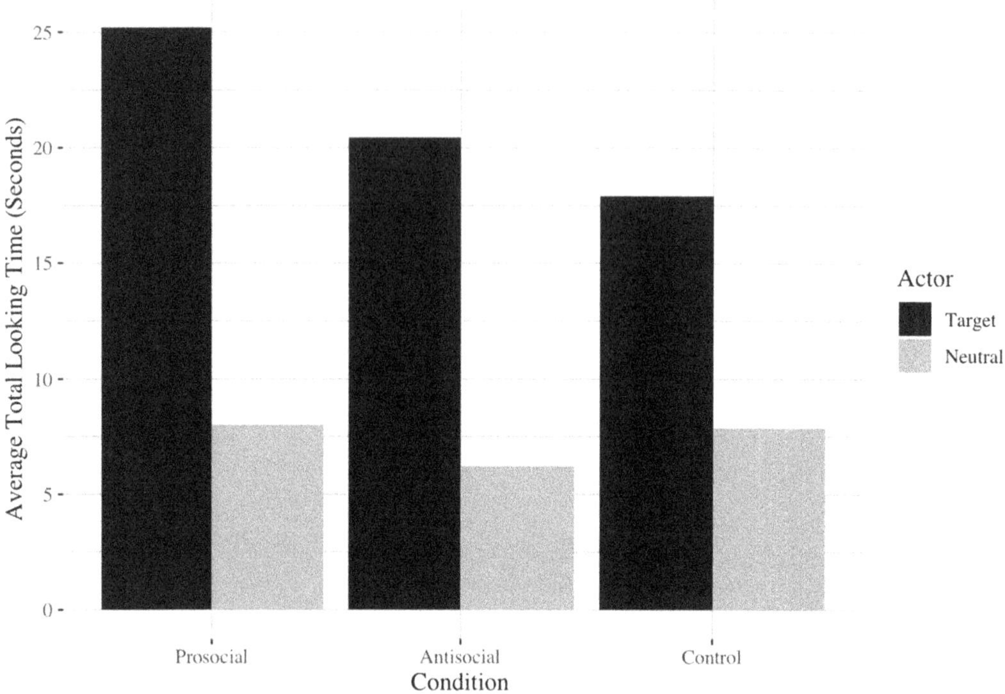

Figure 3. Dogs' average looking time in seconds by actor and condition. Notably, dogs looked significantly longer at the target actor than the neutral actor in all three conditions.

2.3. Discussion

Dogs successfully distinguished between the target and neutral actors across all three conditions, and dogs in the prosocial condition exhibited a significant preference for the prosocial actor compared to the neutral actor, while dogs in the antisocial and control conditions exhibited no preference. These results suggest that dogs prefer to interact with individuals who help their owners but show no clear evaluation of individuals who fail to help their owners. Note that our results differ from the previous findings observed by Chijiiwa et al. [25] in two ways. First, our results revealed a positivity bias that was not found in Chijiiwa et al. [25]—dogs in our study showed a significant bias toward the prosocial actor, whereas no such preference was observed by Chijiiwa and colleagues [25]. Second, we failed to observe the negativity bias pattern that Chijiiwa and colleagues [25] observed; in contrast to their results, our dog participants showed no avoidance of the antisocial actor relative to the neutral actor. The lack of a negativity bias found in our study is somewhat surprising, particularly given that research has shown that a negativity bias is privileged in human infant development, with infants as young as 3 months old avoiding antisocial experimenters, e.g., [4]. However, our results are consistent with other studies that show dogs as sometimes indicating a positivity bias for helpful experimenters, e.g., [15].

Given the variance in dogs' evaluation performance in our study, the second part of our study then went on to test whether dogs' attachment to their owners could explain the variation observed in their social evaluation performance. Historically, researchers have used two different methods to assess dogs' attachment to their owners: measuring their behavioral performance, e.g., the classic Strange Situation Test [45], and owner survey methods, e.g., the C-BARQ [57]. We chose to use the C-BARQ survey method rather than a Strange Situation Test for a few reasons. First, the Strange Situation Test sometimes causes concerns with dog owners because it does cause some stress during the period when dogs are isolated. Second, the Strange Situation Test takes over twenty minutes to complete; since many of the tests we run at our center are shorter in duration, we worried that a long test like the Strange Situation Test would increase dogs' frustration and anxiety during their visits. For these reasons, we chose to assess dogs' attachment using the attachment questions developed in the C-BARQ instead.

We hypothesized that dogs who were shown by the C-BARQ to display more attachment and attention-seeking behaviors toward their owners (i.e., with stronger attachment bonds) would be more likely to prefer an actor who helped their owner than a neutral experimenter compared to dogs who displayed fewer attachment and attention-seeking behaviors (i.e., had weaker attachment bonds) to their owner. Additionally, we predicted that dogs with stronger attachment bonds to their owner would be more likely to prefer a neutral experimenter than an actor who did not help their owners compared to dogs who had weaker attachment bonds to their owners. Finally, we predicted that attachment would not affect dogs' choices in the control condition. Taken together, we hypothesized that dogs with stronger attachment bonds to their owner would exhibit both stronger positivity and negativity biases compared to dogs with weaker attachment bonds to their owners.

3. Part B: Attachment as a Predictor of Actor Choices during Social Evaluation

3.1. Methods

3.1.1. Participants

Twenty-six of the domesticated pet dogs from Part A (11 female, M_{age} = 5.87 years, SD_{age} = 2.74, $\text{range}_{\text{age}}$ 1–13) were assessed by their owners for analysis in Part B after previously participating in Part A (see Table S1 for additional demographic information). Eleven dogs were excluded from the original thirty-seven due to the owner's failure to complete the survey.

3.1.2. Procedure

All owners who had participated in Part A were later administered a questionnaire-based assessment for research on domestic dogs called the C-BARQ [57]. The C-BARQ contains a variety of questions on dog behavior and temperament, including a section on attachment and attention-seeking behaviors specifically. We chose to use an average score on a scale from 0 (never) to 4 (always) from Questions 68–73 (see Table 1) of the C-BARQ [57] to measure the strength of dogs' attachment bond to their owner. Thus, each dog received a numeric score from 0 to 4, detailing the strength of their attachment bond to their owner. Dogs who displayed more attachment and attention-seeking behaviors (i.e., displayed evidence of stronger attachment bonds) received higher scores.

3.2. Results

Simple linear regression was used to evaluate whether average attachment scores predicted the number of actor choices in each condition of Part A (using Bonferroni adjusted $p = 0.017$). In the prosocial model ($N = 11$), the results of the linear regression indicated that average attachment was a significant predictor of actor choices ($F(1, 9) = 11.86$, $p = 0.007$, $R^2_{adjusted} = 0.52$). The results of a Pearson correlation verified this result, as there was a strong significant positive association between the average attachment and choices for the target actor in the prosocial condition ($r(9) = 0.75$, $p = 0.007$) (see Figure 4A).

The results of the antisocial model ($N = 7$), however, revealed no significant prediction of attachment on actor choices ($F(1,5) = 0.20$, $p = 0.672$, $R^2_{adjusted} = -0.15$) (see Figure 4B). Additionally, a Pearson correlation test verified that there was no association between average attachment and actor choices in the antisocial condition ($r(5) = 0.20$, $p = 0.672$). Similarly, there was no significant prediction of attachment on actor choices in the control condition ($N = 8$, $F(1,6) = 0.33$, $p = 0.589$, $R^2_{adjusted} = -0.11$), and this result was confirmed by a Pearson correlation test ($r(6) = -0.23$, $p = 0.589$) (see Figure 4C). In other words, as the strength of attachment increased, dogs became more likely to prefer a prosocial actor compared to a neutral actor. On the other hand, the strength of attachment did not appear to have any relationship with the dogs' preferences toward an antisocial actor or a control actor.

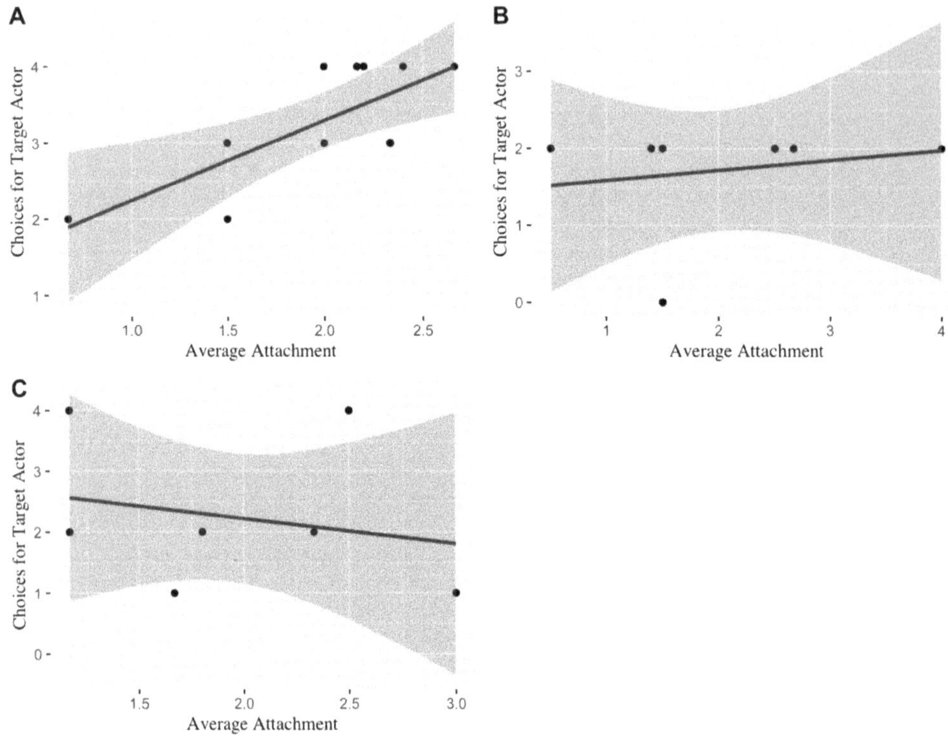

Figure 4. The relationship between the strength of dogs' attachment bonds to their owners and their choices for the target actor in the prosocial (**A**), antisocial (**B**), and control conditions (**C**).

Additionally, simple linear regression was used to determine whether average attachment scores predicted how long dogs spent looking at the target actor within each condition (using Bonferroni adjusted $p = 0.017$). The results of the prosocial, antisocial, and control models revealed that attachment scores did not significantly predict how long dogs spent looking at the target actor (prosocial: $F(1,9) = 1.18$, $p = 0.305$, $R^2_{adjusted} = 0.02$; antisocial: $F(1,5) = 0.10$, $p = 0.768$, $R^2_{adjusted} = -0.18$; control: $F(1,6) = 0.06$, $p = 0.817$, $R^2_{adjusted} = -0.16$). These results were also confirmed using Pearson correlation tests (prosocial: $r(9) = 0.34$, $p = 0.305$; antisocial: $r(5) = 0.14$, $p = 0.768$; control: $r(6) = -0.10$, $p = 0.817$).

Lastly, simple linear regression was used to determine whether average attachment scores predicted how quickly the dogs chose to accept a reward from the target actor within each condition (using Bonferroni adjusted $p = 0.017$). The average attachment was not a significant predictor of dogs' average choice latency for the target actor in any of the three conditions (prosocial: $F(1,9) = 0.004$, $p = 0.951$, $R^2_{adjusted} = -0.11$; antisocial: $F(1,5) = 0.20$, $p = 0.671$, $R^2_{adjusted} = -0.15$; control: $F(1,6) = 3.53$, $p = 0.109$, $R^2_{adjusted} = 0.27$). These results were confirmed using Pearson correlation tests (prosocial: $r(9) = -0.02$, $p = 0.951$; antisocial: $r(5) = 0.20$, $p = 0.671$; control: $r(6) = -0.61$, $p = 0.109$).

3.3. Discussion

Dogs' average attachment was a significant predictor of preference toward prosocial actors in an owner-centered social evaluation paradigm. Specifically, dogs with stronger attachment bonds were significantly more likely to prefer an actor who helped their owner compared to dogs with weaker attachment bonds. Interestingly, dogs with stronger attachment bonds did not have a stronger aversion to the antisocial actor. Indeed, attachment did not seem to have any effect on whether dogs chose to approach the antisocial or neutral actor in the antisocial condition in contrast with our predictions. Our findings suggest that dogs who have stronger attachment bonds to their owners are more likely to positively socially evaluate people who help their owners but do not seem to evaluate and avoid people who refuse to help their owners.

4. General Discussion

The goal of the present study was to determine whether we could explain dogs' mixed performance in social evaluation studies with a factor known to affect how individuals relate to one another: attachment bonds. To examine this question, we first replicated a previous test to determine dogs' social evaluation [25] and then tested whether the dogs' performance was mediated by the strength of their attachment bonds to their owner as measured by the attachment and attention-seeking subset of an owner-administered C-BARQ survey. In Part A of our study, we found that dogs exhibited a positivity bias but not a negativity bias in their social evaluation. Dogs significantly preferred to interact with people who helped their owners, but showed no avoidance of individuals who actively refused to help their owners. In Part B, we found evidence that dogs' attachment bonds to their owner significantly predicted their performance in the prosocial condition but did not predict their choices in the antisocial and control conditions. Taken together, our findings provide early evidence that attachment may predict meaningful individual differences across dogs during social evaluation tasks.

Our canine findings align nicely with previous results in humans, which have demonstrated that attachment impacts how people behave in [29–32] and physiologically respond to [31] a variety of social situations. Given the evidence that a similar attachment system may be present in dogs, e.g., [38–49,51,52], it makes sense that we observed attachment playing a comparable role in dogs' evaluations of actors who help their owners.

Our findings not only provide support that attachment affects dogs' performance in social evaluation studies but also may shed some light on the mixed results that have been observed to date within existing social evaluation research. Researchers have long observed inconsistent patterns in dogs' performance on canine social evaluation tasks, with some studies finding that dogs successfully evaluated prosocial and antisocial actors [15–18,25] and others finding no evidence for successful evaluation [19–24]. Our results provide a hint on why researchers may have observed such varying patterns of performance. Specifically, since the nature of a dog's relationship to their owner can vary substantially, not just between individual dogs but also between populations of dogs, it is possible that studies showing stronger social evaluation effects may have happened to test dogs who have stronger attachment bonds to their owners.

Indeed, there are hints in the existing literature that attachment may play more of a role in dogs' mixed performance than previous works have recognized. Consider, for example, the mixed results of Silver et al. [15], which found that trained agility dogs exhibited a strong positivity bias during social evaluation, but untrained pet dogs did not. Silver et al. [15] initially argued that training might impact dogs' capacity for social evaluation. Our results provide a new take on this interpretation—suggesting that highly trained dogs may develop stronger attachment bonds compared to untrained dogs. This hypothesis is supported by the research of Fallani and colleagues [38], showing that highly trained guide dogs tend to have stronger attachment bonds to their owner than untrained pet dogs. In this way, future research should consider investigating the relationship between attachment and training in dogs and the role that both of these factors play in dogs' social evaluation performance.

Interestingly, our findings in Part A stand in contrast to the results of Chijiiwa and their colleagues' study [25], which used the same social evaluation paradigm that we used in our study. Chijiiwa and colleagues [25] observed that dogs showed a negativity bias in their task but no positivity bias; however, we found an opposite pattern of performance. Although we hypothesized in Part B that dogs who displayed more attachment and attention-seeking behaviors (i.e., had stronger attachment bonds) would result in a stronger aversion to antisocial behavior toward their owner, it is possible that we did not observe an effect of attachment in this condition since we did not see any behavioral evidence of a negativity bias in Part A. Future work should attempt to alter the antisocial condition to make the act of avoidance more salient. For example, work in human infants has shown that participants often show a stronger avoidance of antisocial actors in scenarios in which an antisocial actor takes an object or actively hinders another individual [58]—it is possible that we would observe more robust social evaluations in the antisocial condition if we used a more salient antisocial action.

One limitation of our design concerns our use of the C-BARQ to operationalize attachment. Given that we found a significant effect of attachment, it can be reasonably assumed that our method had sound construct validity. It is important to note, however, that the previous literature on attachment in dogs has primarily measured this construct using behavioral assessments such as the Strange Situation Test. As a result, future research should aim to employ this test. A second limitation of the current design is that we may not have seen any effect of attachment on an antisocial scenario because we did not isolate specific attachment styles. The C-BARQ allowed us to assess dogs who displayed fewer attachment and attention-seeking behaviors and who we interpreted to have weaker attachment bonds, which appeared behaviorally similar to dogs with avoidant attachment styles. However, more attachment and attention-seeking behaviors could possibly be an indicator of either secure or anxious attachment; however, these attachment styles typically result in very different behavior. For instance, work in humans has shown that anxious children are far less likely to explore a novel environment than secure children [53]. Future work could thus profit from using more fine-grained measures of dog attachment to test the role that this individual difference plays in social evaluation.

5. Conclusions

The present results suggest that, just as in humans, a dog's pattern of attachment may be an important individual difference that affects their social evaluation skills and behavior. We found that dogs not only tended to prefer actors that helped their owner over neutral actors but also that dogs who displayed more attachment and attention-seeking behaviors (i.e., appear to have stronger attachment bonds) toward their owners were significantly more likely to exhibit this positivity bias. These results provide promising evidence that attachment may be a meaningful variable to analyze in future canine social evaluation research and open avenues for new work on the relationship between attachment and social cognition in non-human species more broadly.

Supplementary Materials: The following supporting information can be downloaded at: https://www.mdpi.com/article/10.3390/ani13152480/s1, Table S1: Demographic information for Part A (participants 1–37) and Part B (participants 1–26), Table S2: Descriptive statistics for average total looking time (seconds) per actor per condition, Table S3: Descriptive statistics for average response latency (seconds) per actor per condition, Figure S1: Dogs' average response latency in seconds by actor and condition.

Author Contributions: E.M.R. served as lead for conceptualization, methodology, formal analysis, investigation, writing—original draft preparation, and writing—review and editing. Z.A.S. served as lead for supervision and served in a supporting role for conceptualization, methodology, formal analysis, and writing—review and editing. L.R.S. served in a supporting role for methodology, writing—review and editing, and supervision. All authors have read and agreed to the published version of the manuscript.

Funding: This research received no external funding.

Institutional Review Board Statement: The animal study protocol was approved by the Institutional Animal Care and Use Committee of Yale University (#2017-11448).

Informed Consent Statement: Informed consent was obtained from all owners involved in the study.

Data Availability Statement: The data presented in this study are available upon reasonable request from the corresponding author. The data are not publicly available due to privacy and ethical reasons.

Acknowledgments: We would like to thank the members of the Canine Cognition Center at Yale for their help in data collection.

Conflicts of Interest: The authors declare no conflict of interest.

References

1. Abdai, J.; Miklósi, Á. The Origin of Social Evaluation, Social Eavesdropping, Reputation Formation, Image Scoring or What You Will. *Front. Psychol.* **2016**, *7*, 1772. [CrossRef] [PubMed]
2. Melis, A.P.; Hare, B.; Tomasello, M. Chimpanzees Recruit the Best Collaborators. *Science* **2006**, *311*, 1297–1300. [CrossRef]
3. Hamlin, J.K.; Wynn, K.; Bloom, P. Social Evaluation by Preverbal Infants. *Nature* **2007**, *450*, 557–559. [CrossRef] [PubMed]
4. Hamlin, J.K.; Wynn, K.; Bloom, P. Three-Month-Olds Show a Negativity Bias in Their Social Evaluations. *Dev. Sci.* **2010**, *13*, 923–929. [CrossRef] [PubMed]
5. Hamlin, J.K.; Wynn, K. Young Infants Prefer Prosocial to Antisocial Others. *Cogn. Dev.* **2011**, *26*, 30–39. [CrossRef]
6. Hamlin, J.K. The Case for Social Evaluation in Preverbal Infants: Gazing toward One's Goal Drives Infants' Preferences for Helpers over Hinderers in the Hill Paradigm. *Front. Psychol.* **2015**, *5*, 1563. [CrossRef]
7. Hamlin, J.K. Context-Dependent Social Evaluation in 4.5-Month-Old Human Infants: The Role of Domain-General versus Domain-Specific Processes in the Development of Social Evaluation. *Front. Psychol.* **2014**, *5*, 614. [CrossRef]
8. Krupenye, C.; Hare, B. Bonobos Prefer Individuals That Hinder Others over Those That Help. *Curr. Biol.* **2018**, *28*, 280–286.e5. [CrossRef]
9. Kawai, N.; Yasue, M.; Banno, T.; Ichinohe, N. Marmoset Monkeys Evaluate Third-Party Reciprocity. *Biol. Lett.* **2014**, *10*, 20140058. [CrossRef]
10. Anderson, J.R.; Kuroshima, H.; Takimoto, A.; Fujita, K. Third-Party Social Evaluation of Humans by Monkeys. *Nat. Commun.* **2013**, *4*, 1561. [CrossRef]
11. Anderson, J.R.; Takimoto, A.; Kuroshima, H.; Fujita, K. Capuchin Monkeys Judge Third-Party Reciprocity. *Cognition* **2013**, *127*, 140–146. [CrossRef]
12. Subiaul, F.; Vonk, J.; Okamoto-Barth, S.; Barth, J. Do Chimpanzees Learn Reputation by Observation? Evidence from Direct and Indirect Experience with Generous and Selfish Strangers. *Anim. Cogn.* **2008**, *11*, 611–623. [CrossRef] [PubMed]
13. Herrmann, E.; Keupp, S.; Hare, B.; Vaish, A.; Tomasello, M. Direct and Indirect Reputation Formation in Nonhuman Great Apes (Pan Paniscus, Pan Troglodytes, Gorilla Gorilla, Pongo Pygmaeus) and Human Children (Homo Sapiens). *J. Comp. Psychol.* **2013**, *127*, 63–75. [CrossRef]
14. Russell, Y.I.; Call, J.; Dunbar, R.I.M. Image Scoring in Great Apes. *Behav. Process.* **2008**, *78*, 108–111. [CrossRef] [PubMed]
15. Silver, Z.A.; Furlong, E.E.; Johnston, A.M.; Santos, L.R. Training Differences Predict Dogs' (Canis Lupus Familiaris) Preferences for Prosocial Others. *Anim. Cogn.* **2021**, *24*, 75–83. [CrossRef] [PubMed]
16. Kundey, S.M.A.; De Los Reyes, A.; Royer, E.; Molina, S.; Monnier, B.; German, R.; Coshun, A. Reputation-like Inference in Domestic Dogs (Canis Familiaris). *Anim. Cogn.* **2011**, *14*, 291–302. [CrossRef]
17. Carballo, F.; Freidin, E.; Putrino, N.; Shimabukuro, C.; Casanave, E.; Bentosela, M. Dog's Discrimination of Human Selfish and Generous Attitudes: The Role of Individual Recognition, Experience, and Experimenters' Gender. *PLoS ONE* **2015**, *10*, e0116314. [CrossRef]
18. Anderson, J.R.; Bucher, B.; Chijiiwa, H.; Kuroshima, H.; Takimoto, A.; Fujita, K. Third-Party Social Evaluations of Humans by Monkeys and Dogs. *Neurosci. Biobehav. Rev.* **2017**, *82*, 95–109. [CrossRef]
19. Marshall-Pescini, S.; Passalacqua, C.; Ferrario, A.; Valsecchi, P.; Prato-Previde, E. Social Eavesdropping in the Domestic Dog. *Anim. Behav.* **2011**, *81*, 1177–1183. [CrossRef]
20. Nitzschner, M.; Kaminski, J.; Melis, A.; Tomasello, M. Side Matters: Potential Mechanisms Underlying Dogs' Performance in a Social Eavesdropping Paradigm. *Anim. Behav.* **2014**, *90*, 263–271. [CrossRef]
21. Nitzschner, M.; Melis, A.P.; Kaminski, J.; Tomasello, M. Dogs (Canis Familiaris) Evaluate Humans on the Basis of Direct Experiences Only. *PLoS ONE* **2012**, *7*, e46880. [CrossRef] [PubMed]
22. Freidin, E.; Putrino, N.; D'Orazio, M.; Bentosela, M. Dogs' Eavesdropping from People's Reactions in Third Party Interactions. *PLoS ONE* **2013**, *8*, e79198. [CrossRef]
23. McAuliffe, K.; Bogese, M.; Chang, L.W.; Andrews, C.E.; Mayer, T.; Faranda, A.; Hamlin, J.K.; Santos, L.R. Do Dogs Prefer Helpers in an Infant-Based Social Evaluation Task? *Front. Psychol.* **2019**, *10*, 591. [CrossRef]
24. Jim, H.-L.; Plohovich, M.; Marshall-Pescini, S.; Range, F. Wolves and Dogs Fail to Form Reputations of Humans after Indirect and Direct Experience in a Food-Giving Situation. *PLoS ONE* **2022**, *17*, e0271590. [CrossRef] [PubMed]
25. Chijiiwa, H.; Kuroshima, H.; Hori, Y.; Anderson, J.R.; Fujita, K. Dogs Avoid People Who Behave Negatively to Their Owner: Third-Party Affective Evaluation. *Anim. Behav.* **2015**, *106*, 123–127. [CrossRef]
26. Bowlby, J. Attachment and Loss: Retrospect and Prospect. *Am. J. Orthopsychiatry* **1982**, *52*, 664–678. [CrossRef]

27. Benoit, D.; Parker, K.C.H. Stability and Transmission of Attachment across Three Generations. *Child Dev.* **1994**, *65*, 1444–1456. [CrossRef] [PubMed]
28. Waters, E.; Hamilton, C.E.; Weinfield, N.S. The Stability of Attachment Security from Infancy to Adolescence and Early Adulthood: General Introduction. *Child Dev.* **2000**, *71*, 678–683. [CrossRef]
29. Mikulincer, M. Adult Attachment Style and Information Processing: Individual Differences in Curiosity and Cognitive Closure. *J. Pers. Soc. Psychol.* **1997**, *72*, 1217–1230. [CrossRef]
30. Renner, B. Curiosity About People: The Development of a Social Curiosity Measure in Adults. *J. Pers. Assess.* **2006**, *87*, 305–316. [CrossRef]
31. Vrtička, P.; Andersson, F.; Grandjean, D.; Sander, D.; Vuilleumier, P. Individual Attachment Style Modulates Human Amygdala and Striatum Activation during Social Appraisal. *PLoS ONE* **2008**, *3*, e2868. [CrossRef]
32. Gillath, O.; Giesbrecht, B.; Shaver, P.R. Attachment, Attention, and Cognitive Control: Attachment Style and Performance on General Attention Tasks. *J. Exp. Soc. Psychol.* **2009**, *45*, 647–654. [CrossRef]
33. Holvoet, C.; Scola, C.; Arciszewski, T.; Picard, D. Infants' Preference for Prosocial Behaviors: A Literature Review. *Infant Behav. Dev.* **2016**, *45*, 125–139. [CrossRef]
34. Johnson, S.C.; Dweck, C.S.; Chen, F.S. Evidence for Infants' Internal Working Models of Attachment. *Psychol. Sci.* **2007**, *18*, 501–502. [CrossRef] [PubMed]
35. Johnson, S.C.; Dweck, C.S.; Chen, F.S.; Stern, H.L.; Ok, S.-J.; Barth, M. At the Intersection of Social and Cognitive Development: Internal Working Models of Attachment in Infancy. *Cogn. Sci.* **2010**, *34*, 807–825. [CrossRef]
36. Beck, L.; Madresh, E.A. Romantic Partners and Four-Legged Friends: An Extension of Attachment Theory to Relationships with Pets. *Anthrozoös* **2008**, *21*, 43–56. [CrossRef]
37. Kurdek, L.A. Pet Dogs as Attachment Figures for Adult Owners. *J. Fam. Psychol.* **2009**, *23*, 439–446. [CrossRef]
38. Fallani, G.; Previde, E.P.; Valsecchi, P. Do Disrupted Early Attachments Affect the Relationship between Guide Dogs and Blind Owners? *Appl. Anim. Behav. Sci.* **2006**, *100*, 241–257. [CrossRef]
39. Fallani, G.; Prato Previde, E.; Valsecchi, P. Behavioral and Physiological Responses of Guide Dogs to a Situation of Emotional Distress. *Physiol. Behav.* **2007**, *90*, 648–655. [CrossRef]
40. Gácsi, M.; Topál, J.; Miklósi, Á.; Dóka, A.; Csányi, V. Attachment Behavior of Adult Dogs (Canis Familiaris) Living at Rescue Centers: Forming New Bonds. *J. Comp. Psychol.* **2001**, *115*, 423–431. [CrossRef]
41. Palestrini, C.; Previde, E.P.; Spiezio, C.; Verga, M. Heart Rate and Behavioural Responses of Dogs in the Ainsworth's Strange Situation: A Pilot Study. *Appl. Anim. Behav. Sci.* **2005**, *94*, 75–88. [CrossRef]
42. Parthasarathy, V.; Crowell-Davis, S.L. Relationship between Attachment to Owners and Separation Anxiety in Pet Dogs (Canis Lupus Familiaris). *J. Vet. Behav.* **2006**, *1*, 109–120. [CrossRef]
43. Prato-Previde, E.; Custance, D.M.; Spiezio, C.; Sabatini, F. Is the Dog-Human Relationship an Attachment Bond? An Observational Study Using Ainsworth's Strange Situation. *Behaviour* **2003**, *140*, 225–254. [CrossRef]
44. Rehn, T.; McGowan, R.T.S.; Keeling, L.J. Evaluating the Strange Situation Procedure (SSP) to Assess the Bond between Dogs and Humans. *PLoS ONE* **2013**, *8*, e56938. [CrossRef] [PubMed]
45. Solomon, J.; Beetz, A.; Schöberl, I.; Gee, N.; Kotrschal, K. Attachment Security in Companion Dogs: Adaptation of Ainsworth's Strange Situation and Classification Procedures to Dogs and Their Human Caregivers. *Attach. Hum. Dev.* **2019**, *21*, 389–417. [CrossRef]
46. Topál, J.; Miklósi, Á.; Csányi, V.; Dóka, A. Attachment Behavior in Dogs (Canis Familiaris): A New Application of Ainsworth's (1969) Strange Situation Test. *J. Comp. Psychol.* **1998**, *112*, 219–229. [CrossRef]
47. Udell, M.A.R.; Brubaker, L.; Thielke, L.E.; Wanser, S.S.H.; Rosenlicht, G.; Vitale, K.R. Dog–Human Attachment as an Aspect of Social Cognition: Evaluating the Secure Base Test. In *Comparative Cognition: Commonalities and Diversity*; Anderson, J.R., Kuroshima, H., Eds.; Springer: Singapore, 2021; pp. 305–320, ISBN 9789811620287.
48. Brubaker, L.; Udell, M.A.R. Does Pet Parenting Style Predict the Social and Problem-Solving Behavior of Pet Dogs (Canis Lupus Familiaris)? *Anim. Cogn.* **2023**, *26*, 345–356. [CrossRef]
49. Valsecchi, P.; Previde, E.P.; Accorsi, P.A.; Fallani, G. Development of the Attachment Bond in Guide Dogs. *Appl. Anim. Behav. Sci.* **2010**, *123*, 43–50. [CrossRef]
50. Horn, L.; Huber, L.; Range, F. The Importance of the Secure Base Effect for Domestic Dogs—Evidence from a Manipulative Problem-Solving Task. *PLoS ONE* **2013**, *8*, e65296. [CrossRef]
51. Gácsi, M.; Maros, K.; Sernkvist, S.; Faragó, T.; Miklósi, Á. Human Analogue Safe Haven Effect of the Owner: Behavioural and Heart Rate Response to Stressful Social Stimuli in Dogs. *PLoS ONE* **2013**, *8*, e58475. [CrossRef]
52. Topál, J.; Gácsi, M.; Miklósi, Á.; Virányi, Z.; Kubinyi, E.; Csányi, V. Attachment to Humans: A Comparative Study on Hand-Reared Wolves and Differently Socialized Dog Puppies. *Anim. Behav.* **2005**, *70*, 1367–1375. [CrossRef]
53. Ainsworth, M.D.S.; Blehar, M.C.; Waters, E.; Wall, S.N. *Patterns of Attachment: A Psychological Study of the Strange Situation*; Psychology Press: New York, NY, USA, 2015; ISBN 978-1-135-01618-0.
54. Archer, J.; Ireland, J.L. The Development and Factor Structure of a Questionnaire Measure of the Strength of Attachment to Pet Dogs. *Anthrozoös* **2011**, *24*, 249–261. [CrossRef]
55. Johnson, T.P.; Garrity, T.F.; Stallones, L. Psychometric Evaluation of the Lexington Attachment to Pets Scale (Laps). *Anthrozoös* **1992**, *5*, 160–175. [CrossRef]

56. Dwyer, F.; Bennett, P.C.; Coleman, G.J. Development of the Monash Dog Owner Relationship Scale (MDORS). *Anthrozoös* **2006**, *19*, 243–256. [CrossRef]
57. Hsu, Y.; Serpell, J. Development and Validation of a Questionnaire for Measuring Behavior and Temperament Traits in Pet Dogs. *J. Am. Vet. Med. Assoc.* **2003**, *223*, 1293–1300. [CrossRef] [PubMed]
58. Margoni, F.; Surian, L. Infants' Evaluation of Prosocial and Antisocial Agents: A Meta-Analysis. *Dev. Psychol.* **2018**, *54*, 1445–1455. [CrossRef]

animals

MDPI

Article

What Is Written on a Dog's Face? Evaluating the Impact of Facial Phenotypes on Communication between Humans and Canines

Courtney L. Sexton [1,2,*], Colleen Buckley [2], Jake Lieberfarb [3], Francys Subiaul [2,4], Erin E. Hecht [5] and Brenda J. Bradley [2]

1 Department of Population Health Sciences, Virginia-Maryland College of Veterinary Medicine, Virginia Polytechnic Institute and State University, Blacksburg, VA 24061, USA
2 Center for the Advanced Study of Human Paleobiology, Department of Anthropology, The George Washington University, Washington, DC 20052, USA
3 Independent Data Analyst, Washington, DC 20052, USA
4 Department of Speech, Language and Hearing Sciences, The George Washington University, Washington, DC 20052, USA
5 Department of Human Evolutionary Biology, Harvard University, Cambridge, MA 02138, USA
* Correspondence: csexton@gwu.edu; Tel.: +1-609-709-3717

Simple Summary: As dogs evolve to fill a new and increased number of roles in human societies, it is critical that we understand how they communicate with people. Here, we investigate whether markings on dogs' faces influence how expressive they are perceived to be by humans. Using standardized systems to analyze dogs' facial complexity and behaviors, we find that dogs with plainer faces (fewer markings) objectively score as more behaviorally expressive. Age and skill or training level also impact expressivity, with adult dogs being more expressive than senior dogs and dogs that are highly skilled being more expressive than those who have had no training or working experience. Interestingly, dogs tend to use their face more "wholistically" during highly social interactions with owners than when presented with ambiguous cues, and owners of adult dogs with plainer faces tend to be more accurate at judging their dog's expressivity. These data are important to consider as the human–dog relationship continues to develop, both from an evolutionary perspective and especially in the context of canine training and welfare.

Abstract: Facial phenotypes are significant in communication with conspecifics among social primates. Less is understood about the impact of such markers in heterospecific encounters. Through behavioral and physical phenotype analyses of domesticated dogs living in human households, this study aims to evaluate the potential impact of superficial facial markings on dogs' production of human-directed facial expressions. That is, this study explores how facial markings, such as eyebrows, patches, and widow's peaks, are related to expressivity toward humans. We used the Dog Facial Action Coding System (DogFACS) as an objective measure of expressivity, and we developed an original schematic for a standardized coding of facial patterns and coloration on a sample of more than 100 male and female dogs ($N = 103$), aged from 6 months to 12 years, representing eight breed groups. The present study found a statistically significant, though weak, correlation between expression rate and facial complexity, with dogs with plainer faces tending to be more expressive ($r = -0.326$, $p \leq 0.001$). Interestingly, for adult dogs, human companions characterized dogs' rates of facial expressivity with more accuracy for dogs with plainer faces. Especially relevant to interspecies communication and cooperation, within-subject analyses revealed that dogs' muscle movements were distributed more evenly across their facial regions in a highly social test condition compared to conditions in which they received ambiguous cues from their owners. On the whole, this study provides an original evaluation of how facial features may impact communication in human–dog interactions.

Keywords: human–canine interaction; canine communication; dogs; domestication; canine behavior; facial communication; facial markings; human–animal interaction

Citation: Sexton, C.L.; Buckley, C.; Lieberfarb, J.; Subiaul, F.; Hecht, E.E.; Bradley, B.J. What Is Written on a Dog's Face? Evaluating the Impact of Facial Phenotypes on Communication between Humans and Canines. *Animals* **2023**, *13*, 2385. https://doi.org/10.3390/ani13142385

Academic Editor: Betty McGuire

Received: 1 July 2023
Revised: 15 July 2023
Accepted: 20 July 2023
Published: 22 July 2023

1. Introduction

Dogs have acquired behavioral and anatomical traits that engender successful social interaction with humans. The unique interspecies relationship between humans and dogs seems an evolutionary anomaly, differing in biologically and cognitively significant ways from other instances of heterospecific mutualism and commensalism [1–3].

This unique relationship has influenced the bounty of the current research aiming to uncover and define the spectrum of cognitive abilities of dogs living in human societies [4], as well as how these abilities compare to those of dogs' close taxonomic relatives (namely, wolves and other wild canids). Dogs are remarkably socially attuned to humans, attend to direct human signals, speech, and ostensive cues, and are highly trainable for a variety of tasks (e.g., herding, scent detection, medical detection, search and rescue, etc.). Investigations of the genetic bases for such abilities indicate that these traits are highly heritable between breeds and breed groups [5,6]. From a neuroanatomical perspective, historical selection by humans for working skills influences the brain structure in individual breeds [7]. Still, significant gaps in knowledge of how and why dogs arrived at their current state remain, which places us at a disadvantage in understanding the role of a critical character in our own evolutionary tale, and yet, some answers may be staring us in the face.

For highly social humans, faces are useful for identifying individuals but are also the key to understanding, analyzing, and modifying behavior in response to the perceived thoughts, intentions, and feelings of others [8,9]. When looking at another human, our faces exhibit minute shifts in position that unconsciously mimic the feeling meant to be depicted in the other's expression [10,11]. This automatic mimicry allows our brains to process the emotion in addition to the visual signal presented [12,13]. Additionally, according to Wood and colleagues [14], it allows us to better empathize with the emotional state of the "other".

Indeed, across human cultures and populations, a mechanism for establishing the connections necessary for social learning and, arguably, survival comes in the form of a "universal" language grounded in visual cues that are especially reliant on the face [15–17], but also see Jack et al. [18].

Living closely with humans, dogs have not only evolved the ability to distinguish familiar human faces and process human facial cues, but they have also developed a propensity for responding in kind [19–21]. In particular, dogs make and maintain eye contact and use a variety of facial gestures to effectively communicate with human companions [21–23] and may even have developed facial expressions in response to non-human stimuli, such as pain [24]. They likewise understand the emotional valence of human faces [25,26]. Nagasawa and colleagues [27] show that "human-like modes of communication, including mutual gaze, in dogs may have been acquired during domestication with humans".

While dogs may be unique in their readiness to make and keep eye contact with humans, gaze behavior is not insignificant for other canids, especially dogs' wolf relatives. Gaze among conspecifics is typically regarded as an agonistic signal, though other facial expressions are also generally relevant for canids in some similar ways as for primates, especially those facial signals related to play [28,29]. Unlike other canids, however, wolves have facial color patterns in which gaze direction can be easily identified, which Ueda et al. [30] suggest is related to obligate group living and cooperative hunting—not unlike the adaptation of white sclerae in humans [31].

Receptive features of communication (e.g., markings and coloration patterns) on the faces of dogs' close relatives (wolves) and human close relatives (other primates) may be shaped by similar forces, namely the effect they have on the intended signal receiver. Santana et al. [32] and Santana et al. [33] find that while more social primate species tend to have more facial coloration, additional research finds that among those species, individuals with plainer faces display a broader repertoire of facial expressions ("plain face phenomenon") [34]. This pattern among fixed traits likely evolved to aid the conspecific comprehension of information contained in those more flexible, productive features of facial communication, gestures, and expressions [35]. Dogs, however, have developed highly expressive facial behaviors, including paedomorphic expressions that potentially increase the likelihood of receiving human care [36] due to social interactions with humans [21,37,38] more so than with other dogs.

Therefore, in this study, we ask: are dogs with more complex facial features more or less behaviorally expressive toward humans than those with plainer faces? Given the heritability of temperament (including communication-related traits) across breeds [5,39], we also explored whether age, breed group, or work status impacts the dogs' objective scores of facial expressions. Finally, we ask how accurately humans subjectively perceive canine facial expressivity compared to the objective scores. The overall aim of this study is to investigate the variations in facial markings, expressivity, and behaviors of domesticated dogs integrated into human homes and communities in order to determine whether there is a connection between superficial facial phenotypes and behavioral adaptations for communication with human social partners.

We hypothesize that the selective pressures on the physical facial features observed in primates and wolves will be disrupted in dogs, given the history of domestication and intentional breeding in dogs. That is, due to multiple breeds with unique physical phenotypes maintaining similarly close social relationships with humans, the "plain face phenomenon" [34] observed in primates should not apply to dogs.

As an alternative hypothesis, superficial markings could serve to enhance performance and, thus, desirability for specific breed-to-task orientation. For example, on the one hand, markings on the faces of herding dogs may help to mask the visibility of facial expressions that might otherwise give away behavioral intention [40,41], which would be suitable for working sheep or cattle. On the other hand, a plain face would be more advantageous to a retriever primarily directing signals toward human hunting partners if expressions are, in fact, clearer on a plain-faced dog. However, because of the physical variation within breeds and breed groups, this alternative hypothesis would predict that the differences in behavioral rates of expression will only be observed *between* breed groups. Of course, there is also the possibility that facial markings could be unrelated to actual or perceived expressivity.

2. Materials and Methods

2.1. Ethical Considerations

The Harvard University-Area Committee approved the experiment on the Use of Human Subjects under the protocol title: Cognition, motivation, and emotion in domestic dog breeds; Harvard Principal Investigator: Erin Hecht; Protocol #: IRB19-0476/SITE20-0061. The above-named committee approved the George Washington University as a relying institution; George Washington Principal Investigator Courtney Sexton; Federal-wide Assurance: FWA00005945.

2.2. Subjects

Recruitment. Volunteers and their dogs were recruited for this project (titled "What is Written on a Dog's Face?") personally and via a robust outreach plan, including online and social media platforms and fliers advertising the custom-made project website. Participation from human companions entailed recording and submitting a series of videos of dogs in the home (see below). Human companions gave their written informed consent prior to voluntary participation in the study and were provided updates and opportunities to further engage with the project over the course of the data collection and analyses.

Participants. One hundred and eight (108) pet dogs living in households with human companions in North America and Europe were tested. A total of 5 of the 108 dogs submitted were excluded from analyses due to one or more uncodable videos, giving a final total N of 103 (Table 1). The canine subjects included various "purebred" and mix-breed dogs, as reported by their human companions. Per the American Kennel Club (AKC) breed group designation [42], there were 11 Working, 7 Toy, 7 Terrier, 24 Sporting, 9 Non-sporting, 16 Herding, 6 Hound, and 23 Mutt/Mixed-breed dogs.

Table 1. Summary of dogs included in study sample.

	Classification	N Dogs in Sample
AKC Breed Group	Working	11
	Toy	7
	Terrier	7
	Sporting	24
	Non-sporting	9
	Herding	16
	Hound	6
	Mixed-breed	23
Age Bin	Young (6 months–2 years)	20
	Adult (2.1–6.9 years)	49
	Senior (7+ years)	34
Sex	Male	50
	Female	53
Reproductive Status	De-sexed	88
	Intact	15
Training/Skill Level	Unskilled/No training	40
	Basic obedience	31
	Skilled	32

The dogs had to be at least six months of age at the testing time for inclusion in an effort to reduce potential confounds related to the dogs' early social developmental window (3–4 months of age). The chronological ages of each participant were collected and binned into developmental/cognitive age grades based on the six-category system proposed by Harvey [43]. There were 20 dogs classified as "young" (6 months–2 years), 49 dogs classified as "adult" (2.1–6.9 years), and 34 dogs classified as "senior" (7 years and above). The mean age was 5.2 years (SD = 3.22). The dogs included both males (50) and females (53) and those who were both de-sexed (88) and reproductively intact (15). Forty (40) dogs had no formal training/work status; 31 dogs had basic obedience-level skills; and 32 dogs were highly skilled and/or working dogs, as reported by their human companions.

2.3. Experimental Procedure

The data collection period for this study occurred during the COVID-19 pandemic, and thus in-person experimentation was generally not feasible as per the university's public health protocols. Data collection for a small number (N = 20) of subjects was conducted in person, outdoors, though not all of these individuals were included in the final sample (see above). To maximize the total number of study participants and engage the public in community science efforts, the majority of human companions, recruited through interpersonal networking and social media, were given the opportunity to participate remotely via video upload of the dog(s) living in their homes. At-home participants were provided with a study protocol and instructions for uploading their images and videos to secure remote (Dropbox) storage. A potential benefit of in-home data collection is that the test may have more ecological validity by virtue of taking place in the dogs' natural, day-to-day social and physical environments.

After filling out a brief demographics survey and behavioral assessment for each canine subject, the dogs' human companions were instructed to take a photo of their dog(s)' face(s) and to record four 30-second-long videos of the dog(s) in the following conditions, in the specified order:

Condition 1: Asocial/Dog at rest—Dog at rest without eye contact from human.

Condition 2: Eye contact only—Human making eye contact with the dog without speaking, gesturing, or otherwise encouraging a social response.

Condition 3: Eye contact + Unfamiliar words—Human looking at the dog and speaking in a neutral tone, repeating an unfamiliar phrase twice, slowly.

Condition 4: Eye contact + Familiar words—Human looking at the dog and speaking in a normal to slightly excited tone, using words and/or phrases familiar to the dog, attempting to encourage a social response.

The conditions were designed to elicit the maximum Objective Behavioral Sum (OBS) score (see below) from each individual whose responses to human communication would likely depend on previous experience. The participants recorded their dog(s) in each condition only once unless the dog's face moved completely out of the frame for more than a third of the video.

For conditions 3–4, humans spoke to the dogs in the language the dog was most used to hearing. The unfamiliar phrase used by all participants was, "*Ancient Egyptians built enormous pyramids to honor the pharaohs. Ruins from many of these sites have been excavated over the years, unearthing mummies, art and relics*".

The participants were instructed to keep the front of the dog's face clearly visible for the duration of each session and to complete filming of all four conditions within 72 h, when possible, allowing at least 30 min between recording different conditions. The participants were asked to locate a quiet, well-lit area of the home to conduct the recording sessions and, where possible, to avoid distractions, such as other humans, dogs, animals, etc. All videos included in the final analysis observed these general instructions. Unfortunately, because some participants uploaded their videos in bulk (i.e., after completing all four conditions), we cannot verify (e.g., using time stamps) the length of time between the condition recordings.

No experimental training phases were required for this study.

2.4. Analyses

Statistical analyses were performed using JASP [Version 0.17.1] and Jupyter Notebook for Python [3.7.15].

The study used a mixed within-between design consisting of 4 conditions (1–4) repeated within-subjects and breed group, age, and sex as the between-subjects variables.

2.5. Dependent Measures

Physical Score (PS): The objective measure of physical markings. To assess the complexity of facial physical phenotype and assign a corresponding score, each dog's face was evaluated using an original matrix that accounted for both pigmentation and perceptible marks/patterning. Perceptible facial marks/patterns included but were not limited to patches (eye or otherwise), "eyebrows", masks, spots, ticking, "widow's peak", and chin strips (Figure 1). The dogs' facial phenotypes were scored by humans unfamiliar with the individual dogs (i.e., not the human participants).

Per the complexity matrix, a minimum physical score of one (1) would indicate a solid-coated or hyper- "plain-faced" dog; a maximum physical score of nine (9) would indicate a dog with more than two coat colors visible on the face, and at least two markings in each of three facial regions: head/ears, eye area, and mid-lower face.

Objective Behavior (OB): Objective behavioral measures of facial movements (expressivity). The dogs were assessed by independent coders using the Dog Facial Action Coding System DogFACS [44] (see below) in each condition. This measure ranged from 4 (the lowest in any condition across the sample) to 71 (the highest in any condition across the sample).

Objective Behavioral Sum (OBS): The score for each dog was calculated as the sum of the behavioral expressivity scores (OB), as coded according to DogFACS (see description below) across all four conditions (1–4), including the movements for all facial regions indicated (see below). This measure ranged from 41 (the lowest across the sample) to 258 (the highest across the sample). Because the physical score (PS) did not change across conditions, this collapsed measure was used in comparison to the PS.

Behavioral Bin (Bin): The percentile rank of the OBS score for each dog (1–10). This was used in order to compare the OBS to the owner's subjective expressivity score (see below).

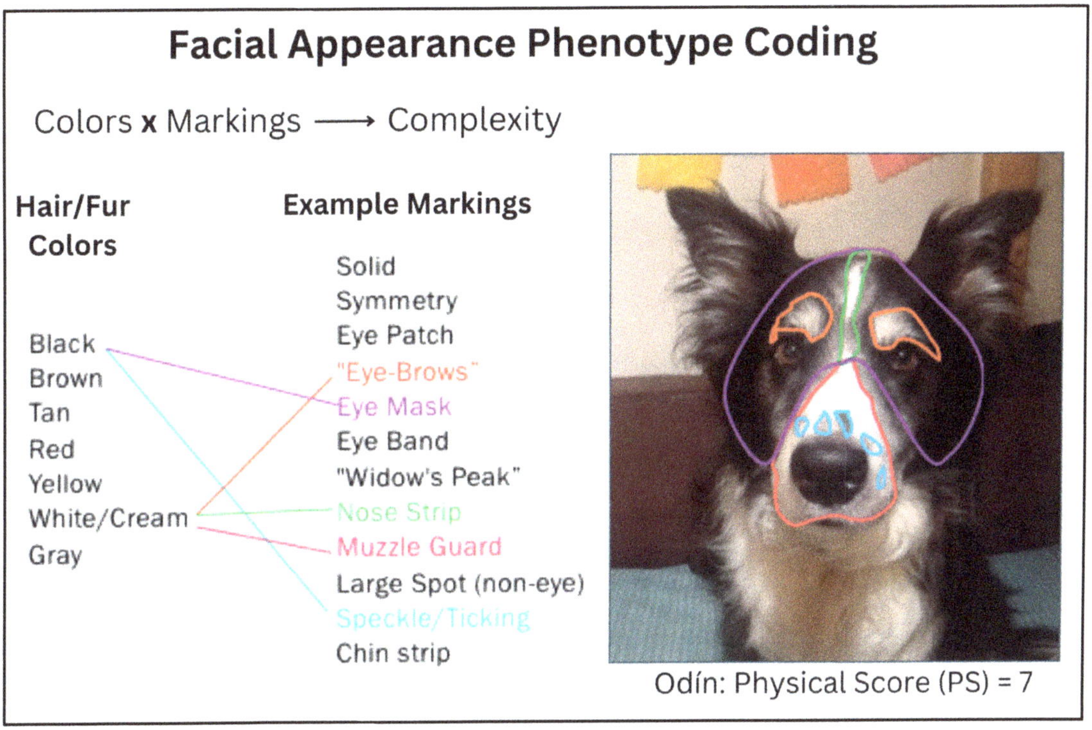

Figure 1. Facial features that contributed to a dog's overall complexity score (PS) included both color and markings, such as those highlighted here. Odín, a border collie who participated in the study, provides an example of a dog with a physical score (PS) 7.

Expressivity: The owner's subjective evaluation of their own dog's expressiveness. The owners were asked to rate on a scale of 1–10 their dog(s)' level of non-vocalizing expression, with 1 = does not seem expressive at all and 10 = very expressive.

Agreement: Composite measures consisting of the difference between the owner's subjective rank of their dog's expressivity (Expressivity) and the binned objective OBS measure of expressivity, where a score of "0" means the objective and subjective measures are in complete agreement, a negative score indicates the owner ranked expressivity lower than the objective binned OBS, and a positive score indicates the owner ranked expressivity higher than the objective binned OBS.

2.6. Independent Variables

In addition to the dog's breed group, sex, and age, we performed exploratory analyses on how the PS, OBS, and expressivity measures varied by the following:

Eyebrows: The presence or absence of a physically colored "eyebrow" marking on the face.

Time-in-Home: The duration of time in years the dog had lived in the home with the owner.

Origin: The last known place of origin of the dog, as reported by the owner (shelter, rescue, breeder, re-homed, or self-bred).

Work Status: Owners reported the level of training/work experience their dog had achieved at test time. Those who had never taken a formal training class were categorized as "unskilled"; those who had completed at least a basic obedience class were categorized as "obedience" level; and those who had one or more training certificates/titles/working dog statuses were considered "skilled". This included skills in the following areas: agility, rally, conformation, scent work/detection, herding, fieldwork, search and rescue, and service.

2.7. DogFACS

The Facial Action Coding System (FACS) [17] is the most widely used and well-regarded tool for measuring facial expressions in human research. FACS is an anatomically based system that describes observable movements of the face in the context of the underlying muscles responsible for the movements. Numbered action units, or Aus, correspond with the visible movements. The system is an effective approach for minimizing experimenter biases related to human emotions and expressions. FACS has been adapted for several animal species (www.animalfacs.com, accessed on 1 November 2020), including dogs. The Dog Facial Action Coding System, or DogFACS [44], is similarly reliable and useful in reducing biases, especially those potentially introduced through anthropomorphizing. Using DogFACS in research requires practice, testing, and certification. Two certified DogFACS coders manually coded the video samples independently, according to the DogFACS manual [44].

The facial areas coded included:

Upper Face action units (Inner Brow Raiser (AU10), Eye Closure (AU143), Blink (AU145));

Mid and Lower Face action units (Nose Wrinkler (AU109), Upper Lip Raiser (AU110), Lip Corner Puller (AU12), Lower Lip Depressor (AU116), Lip Pucker (AU118), Lips Part (AU25), Jaw Drop (AU26), Mouth Stretch (AU27));

Mouth action descriptors (Tongue Show (AD19), Blow (AD33), Suck (AD35), Lip Wipe (AD37), Nose Lick (AD137));

Ear action descriptors (Ears Forward (AD101), Ear Adductor (AD102), Ear Flattener (AD103), Ear Rotator (AD104), Ears Downward (AD105));

Head/Eye action descriptors (Head Turn L/R (AD51/52), Head Up/Down (AD53/54), Head Tilt L/R (AD55/56), Eyes Turn L/R (AD61/62), Eyes Up/Down (AD63/64)). Lip Wipes (AD37) and Nose Licks (AD137) were coded in addition to Tongue Show (AD19), not in place of, where applicable.

All the above action units and descriptors were included in the calculation of the objective behavioral sum (OBS) score, individual condition scores (OB), and facial region subscores. The miscellaneous behaviors, including sniffing, vocalizing, panting, chewing, licking, and body shakes, were noted but not included in the analyses. An OBS of zero would indicate there were no discernable facial movements with the corresponding DogFACS units in any of the four conditions. The highest behavioral score for anyone canine participant recorded was 258; the lowest OBS for any single canine participant was 41.

2.8. Score Validation

All images and video recordings from each canine subject were coded independently by two different DogFACS-certified coders for reliability. There was, generally, concordance between the two scorers; however, if there was an intercoder difference of greater than 5 points (behavioral) or 2 points (physical), those videos/images were rescored. No subjects had to be thrown out due to scorer discordance.

SUMMARY OF SIGNIFICANT FINDINGS

QUESTION	VARIABLES	RESULTS	INTERPRETATIONS
Does dog facial appearance impact their expressivity?	Objective Behavioral Sum (OBS) *Total DogFACS-coded movements* **X** Physical Score (PS) *Complexity of Facial Markings*	Pearson's r (r = -0.326, p = < 0.001)	The higher the PS score, the lower the OBS score. • Dogs with plainer faces were more behaviorally expressive. • Though statistically significant, facial markings and coloration probably do not have a real biological effect on dogs' capacity for facial movements.
Does age influence how expressive dogs' faces are?	Objective Behavioral Sum (OBS) **X** Age Bin *Young, Adult, Senior*	ANOVA [F(2,94) = 5.5, p = 0.005, η2= 0.10; pbonf = 0.004] *Adult vs. Senior*	Adult dogs (ages 2.1-6.9 yrs) were more behaviorally expressive than senior dogs (7 years and older). • Senior dogs may be less physically mobile/physically uncomfortable. • Diminished cognition in older dogs slows response • Older dogs may feel they don't need to "try as hard" with familiar human partners.
Does training/skill influence how expressive dogs' faces are?	Objective Behavioral Sum (OBS) **X** Training/Skill Level *Unskilled, Obedience, Skilled*	ANOVA [F(1,66) = 4.02, p = 0.049, η2= 0.050; pbonf = 0.008] *Unskilled vs. Skilled*	Dogs reported as skilled (having one or more advanced training certificate) are more behaviorally expressive than those with no skill or only basic obedience. • Attention and response to humans demanded of working dogs would suppose increased use of facial gestures.
How do expressions compare across facial regions between conditions?	Objective Behavior (OB) / Condition **X** Objective Behavior / Face Region *Ears, Mid-face, Upper-face*	2-factor Rep. Meas. ANOVA [F(2,204) = 84.99, p = < 0.001, η2 = 0.231] *Main Effect - Face Region* [F(3,306) = 8.65, p = < 0.001, η2= 0.013] *Main Effect - Condition* [F(6,612) = 11.24, p = < 0.001, η2 = 0.032] *Face region X condition interaction*	Movements of the Upper Face were higher in less directly social conditions (2 and 3). • Dogs may have been confused or awaiting further instruction during Conditions 2/3, and thus reduced movements of other parts of the face. • Condition 4 may have provoked a more equal distribution of movements across facial regions because it was unambiguously highly social.
How do owners perceive their own dog's expressivity?	Physical Score (PS) **X** Expressivity *Owner's subjective perception*	Pearson's r (r = -0.167, p = 0.092)	No significant correlation between dogs' facial complexity and owners' subjective expressivity ranking, but appearance did impact how *well* owners ranked expressivity. Owners of adult dogs were more accurate compared to objective scores (OBS) if the dog had a plain face than if the dog's face was more complex. • Plain faces may reduce "noise" introduced by multiple markings and pigments.

3. Results

The preliminary analyses evaluating how sex, the presence of eyebrows, and origin might have impacted the OBS, PS, and expressivity scores found no significant effect. Consequently, these variables were excluded from additional analyses.

3.1. Behavior and Physical Score

To evaluate the effects of various demographic variables on the PS and OBS, a Pearson's r correlation was used. It included the objective behavioral sum (OBS) score, physical score (PS), age (un-binned), expressivity, time-in-home, agreement, and work status. Several relationships were significant, including OBS and PS (r = −0.326, $p \leq 0.001$); OBS and age (r = −0.283, $p = 0.004$); OBS and work status (r = 0.289, $p = 0.003$); OBS and time-in-home (r = −0.313, $p = 0.001$); OBS and agreement (r = −0.726, $p \leq 0.001$); age and work status (r = −0.268, $p = 0.006$); age and agreement (r = 0.268, $p = 0.006$); age and time-in-home (r = 0.876, $p \leq 0.001$); time-in-home and agreement (r = 0.276, $p = 0.005$). There was a negative trend toward significance between expressivity and PS (r = −0.167, $p = 0.092$).

The higher the PS score, the lower the OBS score (r = −0.326, $p \leq 0.001$) (Figure 2); the higher the OBS score, the lower the age (r = −0.283, $p = 0.004$) and the shorter time the dog had lived in their home (r = −0.313, $p = 0.001$). The lower the age, the shorter amount of time in the home (r = 0.876, $p \leq 0.001$); and the higher the OBS, the less disagreement between expressivity and the OBS (r = −0.726, $p \leq 0.001$). The more skilled/more training dogs had, the younger they tended to be (r = −0.268, $p = 0.006$) and the higher their OBS was (r = 0.289, $p = 0.003$).

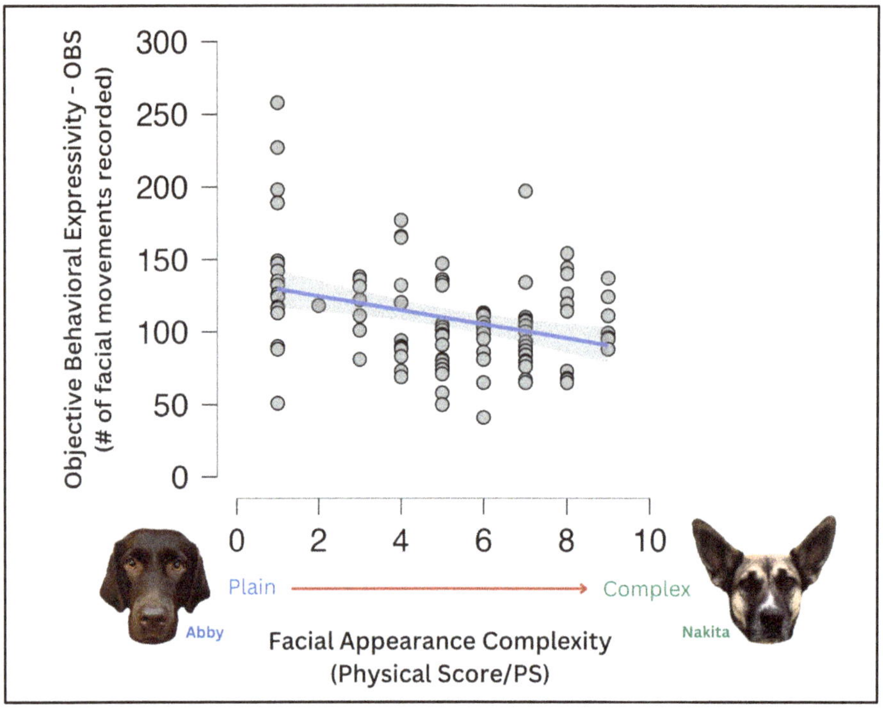

Figure 2. Relationship between OBS and PS. There was a slight though statistically significant correlation between the objective behavioral sum (OBS) score and appearance or physical score (PS). Plainer-faced dogs had slightly higher OBS scores than those with more complex faces.

3.2. Differences in Behavior across Age and Training/Skill Level Groups

An ANOVA further evaluating the differences in the OBS and age groups and the OBS and work status levels was significant for the age bin [F(2, 94) = 5.5, $p = 0.005$, $\eta^2 = 0.10$], and the post hoc comparisons revealed significant differences between adult and senior dogs, ($p_{bonf} = 0.004$). Despite the significant correlation noted above, the main effect for work status at three levels (no skill, obedience, skill) was not significant. To increase statistical power, we ran another ANOVA with a two-level version of this factor (no skill–skill). There was a marginally significant main effect with a medium effect size [F(1, 66) = 4.02, $p = 0.049$, $\eta^2 = 0.050$; $p_{bonf} = 0.008$] (Figure 3). The age-by-skill interaction was not significant [F(2, 66) = 0.227, $p = 0.797$, $\eta^2 = 0.006$].

3.3. Behavior between Conditions

To evaluate the changes in the DogFACS (OB) scores across conditions (1–4), a repeated-measures ANOVA that included a condition (four levels: 1–4) as a repeated measure and breed group as a between-subjects factor produced a main effect for the condition [F(2.8, 285) = 9.06, $p \leq 0.001$, $\eta^2 = 0.082$], and a between-subjects effect for breed group [F (7, 95) = 2.28, $p = 0.034$, $\eta^2 = 0.144$]. Post hoc pairwise comparisons using the Bonferroni correction indicated that the OB for Condition 1 was significantly lower than for Condition 3 ($p_{bonf} = 0.008$) and Condition 4 ($p_{bonf} \leq 0.001$). The OB score for Condition 2 was significantly lower than Condition 4 ($p_{bonf} = 0.001$). No other contrast was statistically significant after correction.

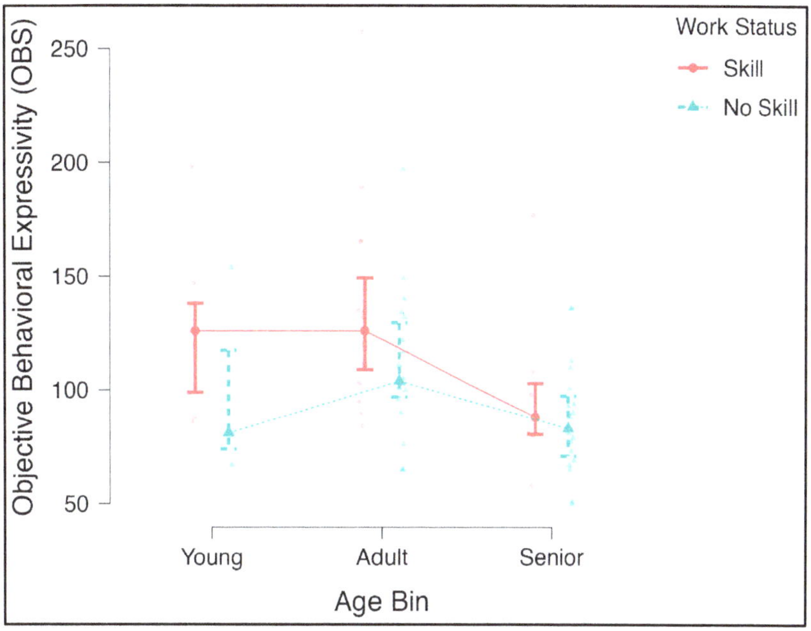

Figure 3. OBS and Age Groups. On average, adult dogs had a significantly higher objective behavioral sum (OBS) score than senior dogs. Dogs who were reported by owners to be "skilled", that is, had one or more training classifications or advanced work status above basic obedience, received a higher OBS than those dogs who had no training/status.

Post hoc comparisons indicated that the breed group population numbers in the sample were insufficient to determine a significant relationship between breed groups.

3.4. Movements across Facial Regions between Conditions

To evaluate the differences in movements within multiple regions of the face across conditions (1–4), a two-factor repeated-measures ANOVA that included face parts (three levels: Ears, Upper Face, Mid Face) and condition (four levels: 1–4) produced the main effects for the face parts [$F(2, 204) = 84.99$, $p \leq 0.001$, $\eta^2 = 0.231$], condition [$F(3, 306) = 8.65$, $p \leq 0.001$, $\eta^2 = 0.013$], and the face parts X condition interaction [$F(6, 612) = 11.24$, $p \leq 0.001$, $\eta^2 = 0.032$]. The number of movements in the Upper Face decreased from Conditions 2–3 to Condition 4, whereas the number of movements in the Ears and Mid Face increased from Conditions 2–3 to Condition 4 (Figure 4).

Conditions 3 and 4 (eye contact with humans speaking unfamiliar words; eye contact with humans speaking familiar words) accounted for a higher percentage of the OBS score than Conditions 1 or 2 (baseline; eye contact/no words).

Interestingly, the majority of action units and action descriptors coded in the Head/Eye and Upper Face regions combined decreased in the percentage of movements per region compared to the OB from Condition 2 to Condition 4 (Condition 2 = 81%; Condition 3 = 76%; Condition 4 = 64%). That is, as the percentage of movements increased across the conditions, so too did the spread of movements across the facial regions in which they were being made (Figure 5).

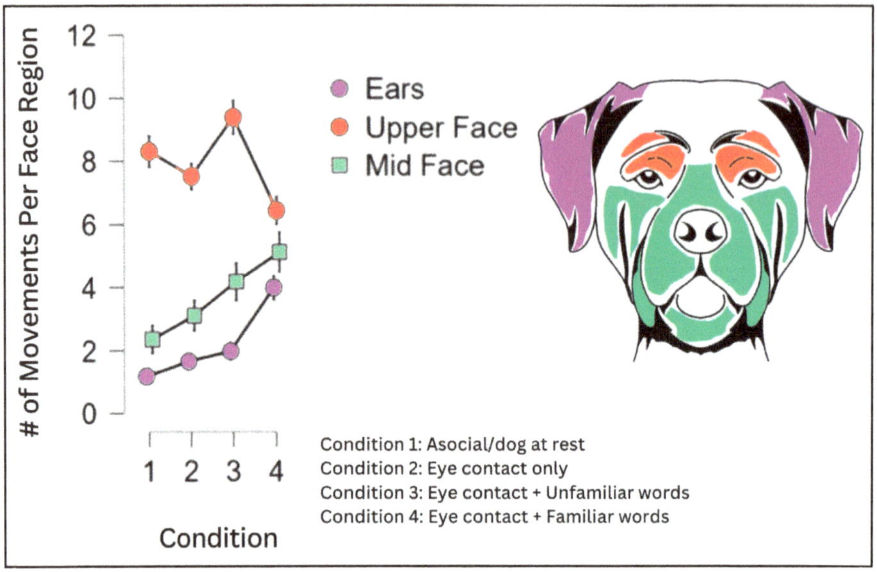

Figure 4. Face Movements Across Conditions. Movements of the Upper Face were higher in Conditions 2 and 3, wherein humans stared at dogs and said nothing (Condition 2), or else used unfamiliar words (Condition 3), than in Condition 4, where humans made eye contact and used familiar words. Condition 4 provoked a more equally distributed spread of movements across facial regions.

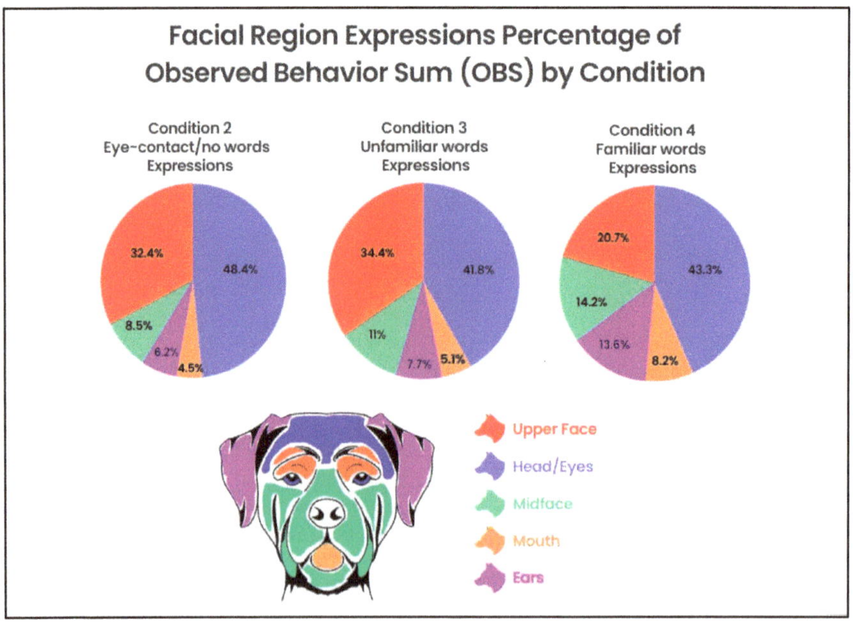

Figure 5. In Conditions 2 and 3, humans stared at dogs and said nothing (Condition 2) or else used unfamiliar words (Condition 3), provoking movements in the head/eye region, which may be related to anticipatory gaze holding. Dog facial movements are more broadly distributed across facial regions when responding to humans who are speaking in familiar words and tones (Condition 4).

3.5. Objective vs. Subjective Measures of Expressivity

Human companions were asked to rank their dog(s)' level of non-vocal expressivity on a scale of 1–10 (1 = "does not seem expressive at all"; 10 = "very expressive"). Nearly half of the people were within a 2-point score deviation (48.5%), and more than two-thirds were within a 4-point score deviation (68.9%) from the dogs' objective behavior scores. Those who were less accurate (31%) deviated at 5 points difference or more. Notably, the presence/absence of "eyebrow"-like physical markings did not influence human ranking.

Among the most accurate owners (the upper third who were in complete objective–subjective score agreement or with 1 point deviating), eight dogs, or 30% of the dogs, were plain-faced (a PS score of 1), whereas, among the least accurate owners (the lower third, who were 5 points or more deviating), only one dog, or 3% of the dogs had a solid face (Figure 6).

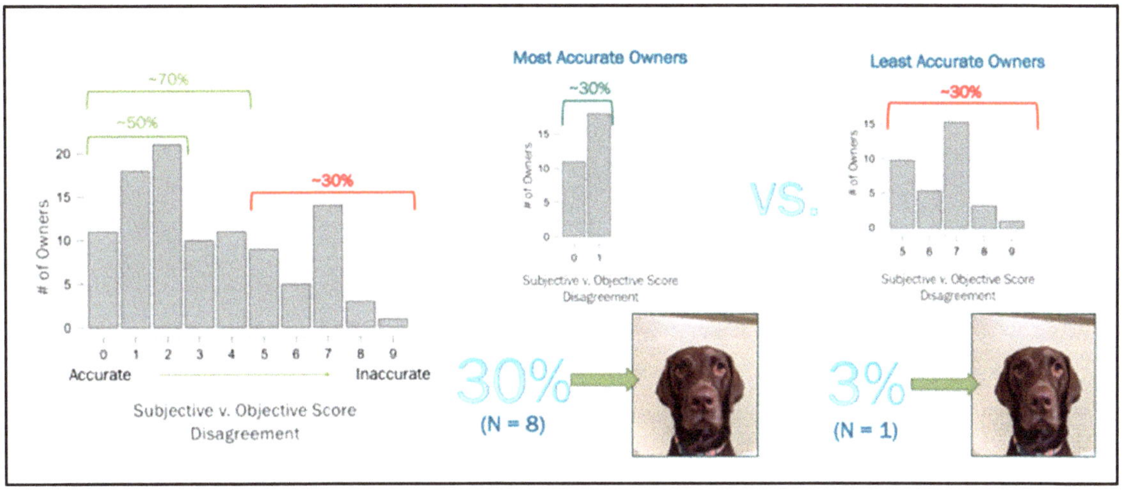

Figure 6. There was a higher percentage of human companions who had a plain-faced dog and were accurate when subjectively judging their dog's expressivity than those who had a plain-faced dog and were inaccurate.

However, as noted above, there was no significant correlation between the human owners' subjective measures of their dogs' expressivity (i.e., expressivity score) and PS and OBS scores. But, there was a trend toward significance between expressivity and the PS, with the higher the subjective expressivity score, the lower the PS ($r = -0.167$, $p = 0.092$), which we hypothesized may be indicative of an indirect effect of the PS on agreement (the relationship between the objective OBS and subjective expressivity)—the OBS was higher for dogs with lower PSs, while objective and subjective scores of dogs with higher OBSs were more closely aligned. For adult dogs only, owners' subjective assessments of their dogs' expressivity were more aligned with their dogs' objective behavioral expressivity scores when the dogs had a lower PS or fewer physical markings on the face. That is, owners of adult dogs gauged their dogs' expressivity more accurately if the dog had a plain face than if the dog's face was more complex (Figure 7).

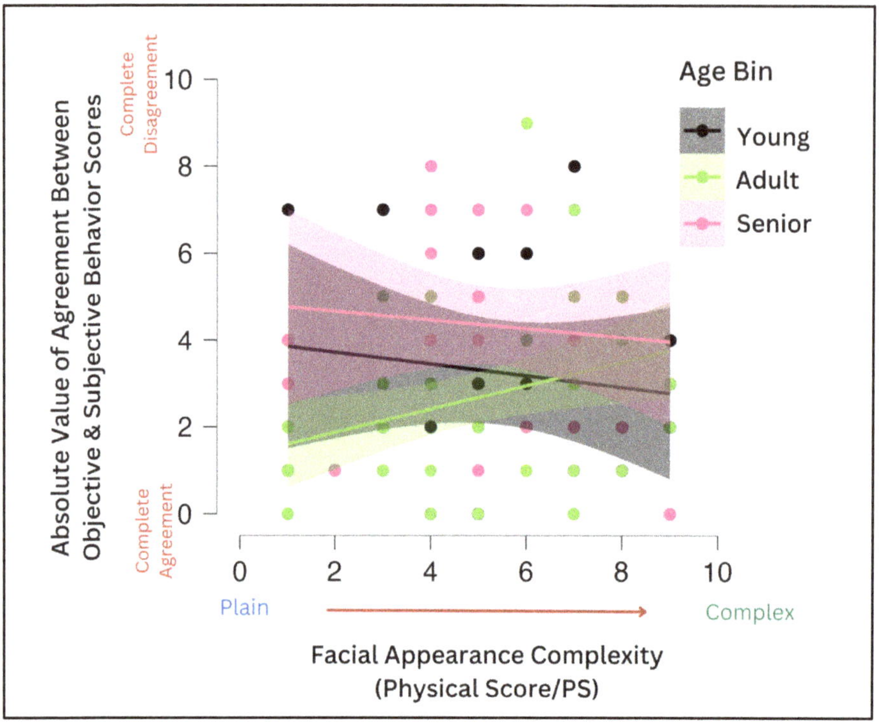

Figure 7. Age as a Determinant of Agreement. While there was no correlation between the physical score and subjective expressivity ranking, owners of adult dogs gauged their dogs' expressivity more accurately compared to objective scores (OBS) if the dog had a plain face than if the dog's face was more complex.

4. Discussion

It is somewhat surprising that this study finds that the markings and coloration on dogs' faces have a similar effect on the perception of their facial expressions as do the markings on the faces of social primates (that is, plainer faces are seen as more expressive). Dogs display a striking convergent evolution with non-human primates in regard to the diversity of facial hair patterns and ornamentation, including and especially such markings as "eyebrows" and "widow's peaks" [45], color(s), and furnishings. However, while these phenotypes are naturally occurring in non-human primates, they are artificially selected in dogs. The history and nature of intentional breeding for dogs who are adept at performing distinct tasks within human society would, by default, also necessitate that dogs of all different physical phenotype variations have the ability to communicate well with humans (including and especially attending to human faces).

Therefore, while statistically significantly correlated in this study, facial markings and coloration probably do not have a real biological effect on dogs' capacities for facial movements; rather, the significance may be the result of movement being the salient signal to human observers. Indeed, dogs seem to have adapted their behavioral features of the face significantly to communication with humans, regardless of the influence of physical features, and have also developed early emerging social skills to prepare and allow for cooperative communication with humans [46–49].

Because the physical features of faces may not be as important to conspecific communication in domestic dogs as they are for group-living/hunting wolves and non-human primates [30,34]—see Gergely et al. [50] and Mongillo et al. [51]—it is reasonable to assume that dogs of any breed/mix would behaviorally overcome any natural and/or artificial selection by humans for specific physical traits. All breeds possessing the same potential primacy of productive rather than passive (e.g., markings) communication toward humans can be considered a novel adaptation unique to the demands of being dependent in an interspecies relationship. Our findings that dogs who were skilled and certified working dogs were more behaviorally expressive in their facial movements than those with no skill/training (Figure 3) support this hypothesis—attention and response to humans during episodes of social learning, cooperation, human–dog coordinated action, and other such communication-heavy activities that working dogs engage in would suppose an increased use of facial gestures.

Likewise, differences in the objective expression scores (OBS) between adult and senior dogs (with senior dogs making significantly fewer expressions) could be the result of senior dogs being generally less physically mobile/physically uncomfortable or may suggest that (a) diminished cognition slows response [52,53], or (b) older dogs have learned that a higher rate of gestural expression may not be necessary to convey their intended signals to a familiar human partner (they do not need to "try" as hard).

Although we lacked the power to determine what the significant breed group differences were, it is unsurprising that they should exist. Given the dramatically different working and companionship roles of the dogs included in the groups, it is reasonable that their facial communication strategies would differ [54–57]. For example, while sporting breeds have historically been bred to work alongside hunters in the field, pointing and retrieving (mostly without auditory signaling so as not to alert their prey), non-sporting dogs breed more diverse social/working backgrounds, and toy breeds are those that have been selected strictly as companion animals.

Regarding the within-subject variation, we found that as the information from the human companion changed, the response from the dogs changed—different parts of the face seemed to serve different functions in different conditions of human attention (Figure 5).

As Conditions 3 and 4 were conditions under which human companions were speaking to the dogs, it is reasonable to assume that the increase in expressive behaviors in these conditions was related to the increased attention and responsiveness from the dogs toward the human companions in a more social context.

Discounting the resting state condition (Condition 1), the majority of movements compared to the OB in Conditions 2 and 3 occur in the Head/Eye + Upper Face (including Brow Raiser, Blink, Eye Closure, Eye Movements Up, Down and Left and Right, and Head Movements). Recall that Condition 2 involved a human companion making eye contact but not speaking, and Condition 3 involved a human speaking unfamiliar words. These were conditions under which the canine subjects may have been confused and/or awaiting further instruction/clarification. Infant developmental literature suggests that a similar phenomenon occurs with human infants when faced with similarly ambiguous cues or an attentive but still face from an adult caregiver, wherein the infants typically decrease expressive behavior and even gaze and often become stressed when presented with a still face [58,59]. In a recent pilot study examining the still-face paradigm in dogs, Barrera et al. [60] reported a decrease in affiliative behaviors in dogs toward humans during the still-face phase.

Condition 4, on the other hand, has a much greater distribution of movements across facial regions. In this condition, in addition to paying attention to the humans, dogs may have been provoked by familiar and exciting words and phrases to respond using a broader gestural repertoire (Figures 4 and 5).

Finally, humans characterized their own dog(s)' level of expression with moderate accuracy, though, of particular interest to those hoping to enhance the human–dog bond, human companions in this study tended to overestimate their dogs' expressivity—only 22 respondents (~20%) scored their dogs as less expressive than the OBS score indicated—indicating perhaps a confound between dogs' responses during experimental conditions and their "everyday" behavior, or else over-eager interpretation. However, according to Sullivan et al. [61], humans are better at categorizing canine facial displays of emotion than they are at categorizing those of chimpanzees or bonobos. Those authors attribute this skill discrepancy to the fact that, although *Pan* looks more human-like, we are more *socially familiar* with dogs.

In reviewing the data here, human companions may characterize dogs' rates of facial expressivity with more accuracy if the dog has a plainer face independent of emotional valence (we found this was the case for owners of adult dogs, see Figure 7). Conversely, Bloom et al. [62] suggest that people are able to identify emotions across all breeds at a rate higher than expected by chance, except for Dobermans (a plain-faced breed), which they hypothesize is due to the breed's dark facial color, which may obscure expressions.

Among additional future studies, our results encourage further investigation of human perception of dogs' facial expressivity as humans may be transferring an entrained preference for reading the relatively plain faces of human conspecifics to our interactions with canines. Although human facial morphology and relative muscular innervation support some of the most complex facial expressions [63,64], we are relatively plain-faced compared to many social, group-living primates (e.g., guenons, callitrichids) [34], and individual differences in superficial facial features (though not movements) may be used more for identification than communication [65,66]. Markings and patterning potentially obscure the behavioral features on dogs' faces for the humans looking at them, and thus a solid-coated face would seem to be more expressive simply because there is less visual "noise".

Of course, it would also be worth repeating this study without some of the limitations imposed by pandemic-era data collection. Primarily, a larger sample size, including an equal number of participants from each breed and age group (especially given the behavioral differences observed among these groups), would be of value. A larger sample size would also aid in controlling for commonly observed canine facial features, such as ear position, brachycephaly, wrinkling (e.g., one participant was a Shar Pei), long/shaggy hair around the eyes, and variations in muscular robusticity, which may have contributed to skew. Comparing the results to those using data collected in a controlled laboratory setting where processes could be standardized would also be of interest, as technical challenges arising from at-home recordings could skew the visibility of facial expressions. This would be especially pertinent as community science solicitations become more widely used for data collection.

5. Conclusions

In our study of analyzing the facial expressivity and physical characteristics of more than 100 companion dogs ($N = 103$), we found that dogs with plainer faces (fewer markings and/or colors) appear to be more behaviorally expressive in objective measures. Among the age groups, adult dogs are more expressive than senior dogs, and dogs that are highly skilled are more expressive than those who have had no training or working experience. Especially relevant to interspecies communication and cooperation, dogs respond with movements more evenly distributed across multiple facial regions when responding to familiar words and tones from humans than from ambiguous or asocial cues; humans tend to be more accurate at judging the expressivity of dogs with plainer faces.

The domestication of dogs and their coexistence with humans has influenced the biological and social development of both species. While the suite of physical changes that now separates dogs from extant wolves has largely been selected for by humans, studies like this one suggest that some changes may not have been as deliberately cultivated as others. The results from this study suggest that there may even be underlying, conserved preferences for certain facial features that humans have unwittingly selected for similar reasons that we may find one human more or less attractive, trustworthy, "easy to read", or any number of other traits.

Understanding how and to what degree biases such as these and other interactions with humans (including the potential projection of human biases onto dogs) impacts the development of novel modes of communication in dogs could provide valuable insight into what shaped early human culture. Likewise, and perhaps more importantly, by gaining a fuller view of how dogs communicate with humans and how we receive and perceive their efforts, we can be better equipped to support them in the critical roles they fill within our society.

Indeed, as the field of canine science expands, findings from studies such as this offer new insight into understanding and navigating the continuously evolving relationship between humans and dogs and will hopefully also prove useful in exploring new avenues of research among a myriad of other taxa and social systems.

Author Contributions: C.L.S.: Conceptualization, Methodology, Formal Analysis, Investigation, Data Curation, Writing—Original Draft, Writing—Review and Editing, Visualization, Funding Acquisition, and Project Administration; C.B.: Methodology, Validation, Data Curation, and Writing—Review and Editing; J.L.: Software and Formal Analysis; F.S.: Formal Analysis, Resources, and Writing—Review and Editing; E.E.H.: Resources, Supervision, and Writing—Review and Editing; B.J.B.: Resources, Supervision, and Writing—Review and Editing. All authors have read and agreed to the published version of the manuscript.

Funding: This research received no external funding.

Institutional Review Board Statement: The experiment was approved by the Harvard University-Area Committee on the Use of Human Subjects under the protocol title: Cognition, Motivation, and Emotion in Domestic Dog Breeds; Harvard Principal Investigator: Erin Hecht; Protocol #: IRB19-0476/SITE20-0061. The George Washington University was approved by the above-named committee as a relying institution; George Washington Principal Investigator: Courtney Sexton; Federal-wide Assurance: FWA00005945.

Informed Consent Statement: Informed consent was obtained from all owners of the canine participants.

Data Availability Statement: Video data may be available upon request in accordance with privacy considerations. Subject coding data are available publicly via Mendeley Data: Sexton, Courtney (2023), "Canine Facial Phenotypes and Expressive Behaviors", Mendeley Data, V1, doi:10.17632/br92x9768y.1.

Acknowledgments: We would like to thank Carson Murray, Emily Bray, and the members of the Canine Brains Project at Harvard University for their invaluable input and support throughout the research process; Katie Auerswald and Dasha Ordynat for their recruitment support; and Haley McKey for designing the logo for our outreach materials. We would also like to thank the Awesome Foundation, DC Chapter; the Lewis N. Cotlow Research Fund; The Columbian Women of the George Washington University; the Shenkman Fund, and the Ann Gordon Webster Endowment/Shirley H. & Robert L. Richards Fund at the George Washington University for project funding. Finally, we would like to extend our sincere thanks to all the participants in our study (human and canine) who generously shared their time with us. Their willingness to engage with our research was essential to the success of this project, and we are deeply grateful for their participation.

Conflicts of Interest: The authors declare no conflict of interest.

References

1. Horowitz, A.; Hecht, J. Examining dog-human play: The characteristics, affect, and vocalizations of a unique interspecific interaction. *Anim. Cogn.* **2016**, *19*, 779–788. [CrossRef]
2. Sipple, N.; Thielke, L.; Smith, A.; Vitale, K.R.; Udell, M.A.R. Intraspecific and Interspecific Attachment between Cohabitant Dogs and Human Caregivers. *Integr. Comp. Biol.* **2021**, *61*, 132–139. [CrossRef]
3. Smith, T.D.; Van Valkenburgh, B. The dog-human connection. *Anat. Rec.* **2021**, *304*, 10–18. [CrossRef]
4. Aria, M.; Alterisio, A.; Scandurra, A.; Pinelli, C.; D'Aniello, B. The scholar's best friend: Research trends in dog cognitive and behavioral studies. *Anim. Cogn.* **2021**, *24*, 541–553. [PubMed]
5. MacLean, E.L.; Snyder-Mackler, N.; vonHoldt, B.M.; Serpell, J.A. Highly heritable and functionally relevant breed differences in dog behaviour. *Proc. R. Soc. B.* **2019**, *286*, 20190716. [CrossRef] [PubMed]
6. Shan, S.; Xu, F.; Brenig, B. Genome-Wide Association Studies Reveal Neurological Genes for Dog Herding, Predation, Temperament, and Trainability Traits. *Front. Vet. Sci.* **2021**, *8*, 693290. [CrossRef]
7. Hecht, E.E.; Smaers, J.B.; Dunn, W.D.; Kent, M.; Preuss, T.M.; Gutman, D.A. Significant Neuroanatomical Variation Among Domestic Dog Breeds. *J. Neurosci.* **2019**, *39*, 7748–7758. [CrossRef]
8. Andrew, R.J. Evolution of Facial Expression. *Science* **1963**, *142*, 1034–1041. [CrossRef]

9. Oberman, L.M.; Winkielman, P.; Ramachandran, V.S. Face to face: Blocking facial mimicry can selectively impair recognition of emotional expressions. *Soc. Neurosci.* **2007**, *2*, 167–178. [CrossRef]
10. Dimberg, U.; Thunberg, M.; Elmehed, K. Unconscious facial reactions to emotional facial expressions. *Psychol. Sci.* **2000**, *11*, 86–89. [CrossRef]
11. Lundqvist, L.; Dimberg, U. Facial Expresssions are Contagious. *J. Psychophysiol.* **1995**, *9*, 203–211.
12. Atkinson, A.P.; Adolphs, R. Visual Emotion Perception: Mechanisms and Processes. In *Emotion and Consciousness*; Barrett, L.F., Ed.; The Guilford Press: New York, NY, USA, 2005; Volume 420, pp. 150–182.
13. Seibt, B.; Mühlberger, A.; Likowski, K.U.; Weyers, P. Facial mimicry in its social setting. *Front. Psychol.* **2015**, *6*, 1122. [CrossRef] [PubMed]
14. Wood, A.; Rychlowska, M.; Korb, S.; Niedenthal, P. Fashioning the Face: Sensorimotor Simulation Contributes to Facial Expression Recognition. *Trends Cogn. Sci.* **2016**, *20*, 227–240. [CrossRef] [PubMed]
15. Curio, C.; Bülthoff, H.H.; Giese, M.A. (Eds.) *Dynamic Faces: Insights from Experiments and Computation*; Boston Review: Boston, MA, USA, 2011.
16. Darwin, C.R. *The Expression of the Emotions in Man and Animals*, 1st ed.; John Murray: London, UK, 1872.
17. Ekman, P.; Friesen, W. *Facial Action Coding System: Manual*; Consulting Psychologists Press: Palo Alto, CA, USA, 1978.
18. Jack, R.E.; Garrod, O.G.B.; Yu, H.; Caldara, R.; Schyns, P.G. Facial expressions of emotion are not culturally universal. *PNAS* **2012**, *109*, 7241–7244. [CrossRef] [PubMed]
19. Caeiro, C.; Guo, K.; Mills, D. Dogs and humans respond to emotionally competent stimuli by producing different facial actions. *Sci. Rep.* **2017**, *7*, 15525. [CrossRef] [PubMed]
20. Huber, L.; Racca, A.; Scaf, B.; Virányi, Z.; Range, F. Discrimination of familiar human faces in dogs (*Canis familiaris*). *Learn Motiv.* **2013**, *44*, 258–269. [CrossRef]
21. Kaminski, J.; Hynds, J.; Morris, P.; Waller, B.M. Human attention affects facial expressions in domestic dogs. *Sci. Rep.* **2017**, *7*, 12914. [CrossRef]
22. Johnston, A.M.; Turrin, C.; Watson, L.; Arre, A.M.; Santos, L.R. Uncovering the origins of dog–human eye contact: Dingoes establish eye contact more than wolves, but less than dogs. *Anim. Behav.* **2017**, *133*, 123–129. [CrossRef]
23. Savalli, C.; Resende, B.; Gaunet, F. Eye Contact Is Crucial for Referential Communication in Pet Dogs. *PLoS ONE* **2016**, *11*, e0162161. [CrossRef]
24. Mota-Rojas, D.; Marcet-Rius, M.; Ogi, A.; Hernández-Ávalos, I.; Mariti, C.; Martínez-Burnes, J.; Mora-Medina, P.; Casas, A.; Domínguez, A.; Reyes, B.; et al. Current Advances in Assessment of Dog's Emotions, Facial Expressions, and Their Use for Clinical Recognition of Pain. *Animals* **2021**, *11*, 3334. [CrossRef]
25. Albuquerque, N.; Guo, K.; Wilkinson, A.; Savalli, C.; Otta, E.; Mills, D. Dogs recognize dog and human emotions. *Biol. Lett.* **2016**, *12*, 20150883. [CrossRef] [PubMed]
26. Buttelmann, D.; Tomasello, M. Can domestic dogs (*Canis familiaris*) use referential emotional expressions to locate hidden food? *Anim. Cogn.* **2013**, *16*, 137–145. [CrossRef]
27. Nagasawa, M.; Mitsui, S.; En, S.; Ohtani, N.; Ohta, M.; Sakuma, Y.; Onaka, T.; Mogi, K.; Kikusui, T. Social evolution. Oxytocin-gaze positive loop and the coevolution of human-dog bonds. *Science* **2015**, *348*, 333–336. [CrossRef] [PubMed]
28. Fox, M.W. A comparative study of the development of facial expressions in canids; Wolf, coyote and foxes. *Behaviour* **1970**, *36*, 49–73. [CrossRef]
29. Hobkirk, E.R. Through the Eyes of a Wolf: Using Non-Invasive Methods to Quantify and Classify the Facial Signalling of Wolves (*Canis lupus*) and Domestic Dogs (*Canis lupus* familiaris). Master's Thesis, Durham University, Durham, UK, 2019.
30. Ueda, S.; Kumagai, G.; Otaki, Y.; Yamaguchi, S.; Kohshima, S. A comparison of facial color pattern and gazing behavior in canid species suggests gaze communication in gray wolves (*Canis lupus*). *PLoS ONE* **2014**, *9*, e98217. [CrossRef]
31. Kobayashi, H.; Kohshima, S. Unique morphology of the human eye. *Nature* **1997**, *387*, 767–768. [CrossRef]
32. Santana, S.E.; Lynch Alfaro, J.; Alfaro, M.E. Adaptive evolution of facial colour patterns in Neotropical primates. *Proc. R. Soc. B Biol. Sci.* **2012**, *279*, 2204–2211. [CrossRef] [PubMed]
33. Santana, S.E.; Alfaro, J.L.; Noonan, A.; Alfaro, M.E. Adaptive response to sociality and ecology drives the diversification of facial colour patterns in catarrhines. *Nat. Commun.* **2013**, *4*, 2765. [CrossRef]
34. Santana, S.E.; Dobson, S.D.; Diogo, R. Plain faces are more expressive: Comparative study of facial colour, mobility and musculature in primates. *Biol. Lett.* **2014**, *10*, 20140275. [CrossRef]
35. Rincon, A.V.; Waller, B.M.; Duboscq, J.; Mielke, A.; Pérez, C.; Clark, P.R.; Micheletta, J. Socially tolerant macaques use more complex facial behavior than intolerant macaques. *bioRxiv* **2022**. [CrossRef]
36. Waller, B.M.; Peirce, K.; Caeiro, C.C.; Scheider, L.; Burrows, A.M.; McCune, S.; Kaminski, J. Paedomorphic Facial Expressions Give Dogs a Selective Advantage. *PLoS ONE* **2013**, *8*, e82686. [CrossRef]
37. Benz-Schwarzburg, J.; Monsó, S.; Huber, L. How Dogs Perceive Humans and How Humans Should Treat Their Pet Dogs: Linking Cognition With Ethics. *Front. Psychol.* **2020**, *11*, 584037. [CrossRef]
38. Bhattacharjee, D.; Sau, S.; Das, J.; Bhadra, A. Free-ranging dogs prefer petting over food in repeated interactions with unfamiliar humans. *J. Exp. Biol.* **2017**, *220 Pt 24*, 4654–4660. [CrossRef]
39. Gnanadesikan, G.E.; Hare, B.; Snyder-Mackler, N.; MacLean, E.L. Estimating the heritability of cognitive traits across dog breeds reveals highly heritable inhibitory control and communication factors. *Anim. Cogn.* **2020**, *23*, 953–964. [CrossRef] [PubMed]

40. Ancillotto, L.; Mori, E. Adaptive significance of coat colouration and patterns of Sciuromorpha (Rodentia). *Ethol. Ecol. Evol.* **2017**, *29*, 241–254. [CrossRef]
41. Dutrow, E.V.; Serpell, J.A.; Ostrander, E.A. Domestic dog lineages reveal genetic drivers of behavioral diversification. *Cell* **2022**, *185*, 4737–4755.e18. [CrossRef]
42. American Kennel Club. *The Complete Dog Book: The Photograph, History, and Official Standard of Every Breed Admitted to AKC Registration, and the Selection, Training, Breeding, Care, and Feeding of Pure-Bred Dogs*; Howell Book House: New York, NY, USA, 1979.
43. Harvey, N.D. How Old Is My Dog? Identification of Rational Age Groups in Pet Dogs Based Upon Normative Age-Linked Processes. *Front. Vet. Sci.* **2021**, *8*, 643085. [CrossRef]
44. Waller, B.; Caeiro, C.C.; Peirce, K.; Burrows, A.; Kaminski, J. DogFACS: The Dog Facial Action Coding System. 2013. Available online: https://www.semanticscholar.org/paper/DogFACS%3A-the-dog-facial-action-coding-system-Waller-Caeiro/8ba6d22adfd3efb1fe3b6a1d65095d3863e704fc (accessed on 1 November 2020).
45. Bradley, B.J.; Mundy, N.I. The primate palette: The evolution of primate coloration. *Evol. Anthropol.* **2008**, *17*, 97–111.
46. Bray, E.E.; Gnanadesikan, G.E.; Horschler, D.J.; Levy, K.M.; Kennedy, B.S.; Famula, T.R.; MacLean, E.L. Early-emerging and highly heritable sensitivity to human communication in dogs. *Curr. Biol.* **2021**, *31*, 3132–3136.e5. [CrossRef] [PubMed]
47. Byosiere, S.-E.; Mundry, R.; Range, F.; Virányi, Z. Selective responding to human ostensive communication is an early developing capacity of domestic dogs. *Dev. Sci.* **2022**, *26*, e13361. [CrossRef]
48. Kujala, M.V.; Kauppi, J.P.; Törnqvist, H.; Helle, L.; Vainio, O.; Kujala, J.; Parkkonen, L. Time-resolved classification of dog brain signals reveals early processing of faces, species and emotion. *Sci. Rep.* **2020**, *10*, 19846. [CrossRef] [PubMed]
49. Salomons, H.; Smith, K.C.M.; Callahan-Beckel, M.; Callahan, M.; Levy, K.; Kennedy, B.S.; Bray, E.E.; Gnanadesikan, G.E.; Horschler, D.J.; Gruen, M.; et al. Cooperative Communication with Humans Evolved to Emerge Early in Domestic Dogs. *Curr. Biol.* **2021**, *31*, 3137–3144.e11. [CrossRef] [PubMed]
50. Gergely, A.; Petró, E.; Oláh, K.; Topál, J. Auditory–Visual Matching of Conspecifics and Non-Conspecifics by Dogs and Human Infants. *Animals* **2019**, *9*, 17. [CrossRef]
51. Mongillo, P.; Eatherington, C.; Lõoke, M.; Marinelli, L. I know a dog when I see one: Dogs (*Canis familiaris*) recognize dogs from videos. *Anim. Cogn.* **2021**, *24*, 969–979. [CrossRef]
52. Bray, E.E.; Raichlen, D.A.; Forsyth, K.K.; Promislow, D.E.L.; Alexander, G.E.; MacLean, E.L.; Dog Aging Project Consortium. Associations between physical activity and cognitive dysfunction in older companion dogs: Results from the Dog Aging Project. *GeroScience* **2023**, *45*, 645–661. [CrossRef]
53. Watowich, M.M.; MacLean, E.L.; Hare, B.; Call, J.; Kaminski, J.; Miklósi, Á.; Snyder-Mackler, N. Age influences domestic dog cognitive performance independent of average breed lifespan. *Anim Cogn.* **2020**, *23*, 795–805. [CrossRef]
54. Mehrkam, L.R.; Wynne, C.D.L. Behavioral differences among breeds of domestic dogs (*Canis lupus* familiaris): Current status of the science. *Appl. Anim. Behav. Sci.* **2014**, *155*, 12–27.
55. Serpell, J.A.; Duffy, D.L. Dog Breeds and Their Behavior. In *Domestic Dog Cognition and Behavior: The Scientific Study of Canis familiaris*; Horowitz, A., Ed.; Springer: Berlin/Heidelberg, Germany, 2014; pp. 31–57.
56. Turcsán, B.; Kubinyi, E.; Miklósi, Á. Trainability and boldness traits differ between dog breed clusters based on conventional breed categories and genetic relatedness. *Appl. Anim. Behav. Sci.* **2011**, *132*, 61–70. [CrossRef]
57. Van Poucke, E.; Höglin, A.; Jensen, P.; Roth, L.S.V. Breed group differences in the unsolvable problem task: Herding dogs prefer their owner, while solitary hunting dogs seek stranger proximity. *Anim. Cogn.* **2022**, *25*, 597–603. [CrossRef]
58. D'Entremont, B.; Muir, D.W. Five-months-olds' attention and affective responses to still-faced emotional expressions. *Infant. Behav. Dev.* **1997**, *20*, 563–568. [CrossRef]
59. Gunning, M.; Halligan, S.L.; Murray, L. Contributions of maternal and infant factors to infant responding to the still face paradigm: A longitudinal study. *Infant. Behav. Dev.* **2013**, *36*, 319–328. [CrossRef] [PubMed]
60. Barrera, G.; Guillén-Salazar, F.; Bentosela, M. Still-face Effect in Dogs (*Canis familiaris*): A Pilot Study. *J. Appl. Anim. Welf. Sci.* **2021**, *26*, 271–284. [CrossRef] [PubMed]
61. Sullivan, S.K.; Kim, A.; Vinicius Castilho, L.; Harris, L.T. Comparing emotion inferences from dogs (*Canis familiaris*), panins (Pan troglodytes/Pan paniscus), and humans (Homo sapiens) facial displays. *Sci. Rep.* **2022**, *12*, 13171. [CrossRef]
62. Bloom, T.; Trevathan-Minnis, M.; Atlas, N.; MacDonald, D.A.; Friedman, H.L. Identifying facial expressions in dogs: A replication and extension study. *Behav. Process* **2021**, *186*, 104371. [CrossRef] [PubMed]
63. Burrows, A.M.; Waller, B.M.; Parr, L.A.; Bonar, C.J. Muscles of facial expression in the chimpanzee (Pan troglodytes): Descriptive, comparative and phylogenetic contexts. *J. Anat.* **2006**, *208*, 153–167. [CrossRef] [PubMed]
64. Powell, V.; Esteve-Altava, B.; Molnar, J.; Villmoare, B.; Pettit, A.; Diogo, R. Primate modularity and evolution: First anatomical network analysis of primate head and neck musculoskeletal system. *Sci. Rep.* **2018**, *8*, 2341. [CrossRef]

65. Lander, K.; Butcher, N. Independence of face identity and expression processing: Exploring the role of motion. *Front. Psychol.* **2015**, *6*, 255. [CrossRef]
66. Sheehan, M.J.; Nachman, M.W. Morphological and population genomic evidence that human faces have evolved to signal individual identity. *Nat. Commun.* **2014**, *5*, 4800. [CrossRef]

animals

MDPI

Article

Reactivation of a Hospital-Based Therapy Dog Visitation Program during the COVID-19 Pandemic

Lisa Townsend [1,2,3,4,*], Jennifer K. Heatwole [4] and Nancy R. Gee [2,4]

1 Department of Pediatrics, Division of Adolescent Medicine, School of Medicine, Virginia Commonwealth University, Richmond, VA 23298, USA
2 Department of Psychiatry, School of Medicine, Virginia Commonwealth University, Richmond, VA 23298, USA; nancy.gee@vcuhealth.org
3 Children's Hospital of Richmond (CHoR), Richmond, VA 23298, USA
4 Center for Human-Animal Interaction, Virginia Commonwealth University, Richmond, VA 23298, USA; heatwolejk@vcu.edu
* Correspondence: lisa.townsend@vcuhealth.org

Simple Summary: Negative mental health outcomes have affected healthcare workers, patients, and community members following pandemics: most recently, the SARS-CoV-2 (COVID-19) outbreak. Although therapy dog visitation programs are associated with reduced stress, most hospital-based programs were placed on hiatus during the COVID-19 pandemic. This study examined human–animal interactions during the reactivation of a hospital-based therapy dog program during the pandemic. Characteristics of the interactions and the participants involved were recorded and analyzed. Findings indicated that most visit recipients were healthcare workers, while the longest interaction times occurred with adult and pediatric patients. High levels of adherence to human and animal safety protocols indicate that human–dog therapy teams can safely return to hospital visitation work.

Abstract: This study examined human–animal interactions during the reactivation of a hospital-based therapy dog program during the COVID-19 pandemic. Data were collected from human–dog interactions at an academic medical center in Virginia. Interaction length, participant role, age group (pediatric or adult), and observed gender were recorded. Handler adherence to human and animal safety protocols (donning personal protective equipment (PPE), using hand sanitizer, and limiting visit length) was measured. Observations from 1016 interactions were collected. *t*-tests and analysis of variance were conducted. Most visit recipients were healthcare workers (71.69%). Patients received longer visits than other participants ($F(4880) = 72.90$, $p = <0.001$); post hoc Bonferroni analyses ($p = 0.05/4$) showed that patients, both adult ($M = 2.58$ min, $SD = 2.24$) (95% C.I = 0.35–1.68) and pediatric ($M = 5.81$, $SD = 4.38$) (95% C.I. 3.56–4.97), had longer interaction times than healthcare workers ($M = 1.56$, $SD = 1.92$) but not visitors ($p = 1.00$). Gender differences were not statistically significant ($t(552) = -0.736$), $p = 0.462$). Hand sanitizer protocols were followed for 80% of interactions. PPE guidelines were followed for 100% of visits. Most interactions occurred with healthcare workers, suggesting that therapy dog visits are needed for this population. High adherence to COVID-19 safety protocols supports the decision to reactivate therapy animal visitation programs in hospitals. Challenges to safety protocol adherence included ultra-brief interactions and crowds of people surrounding the dog/handler teams. Program staff developed a "buddy system" mitigation strategy to minimize departures from safety protocols and reduce canine stress.

Keywords: therapy dog visitation; hospital; COVID-19; healthcare workers; safety protocols; animal welfare

Citation: Townsend, L.; Heatwole, J.K.; Gee, N.R. Reactivation of a Hospital-Based Therapy Dog Visitation Program during the COVID-19 Pandemic. *Animals* **2022**, *12*, 1842. https://doi.org/10.3390/ani12141842

Academic Editor: Betty McGuire

Received: 21 June 2022
Accepted: 19 July 2022
Published: 20 July 2022

1. Introduction

Hospital-based therapy dog programs provide important relief from physical and emotional discomfort for many types of patients. Hospitalization is often associated with anxiety, discomfort, loneliness [1–3], and unpleasant or distressing sensory experiences [4]. These experiences may increase vulnerability to anxiety, depression, and reductions in well-being [5]. Therapy dog visits have been associated with improvements in ratings of stress, anxiety, fear [5], pain, depression, well-being [6], loneliness, and boredom [4,5]. Interacting with a dog has also been associated with improvements in certain physiological parameters, such as blood pressure, heart rate, cardiovascular reactivity, exercise, and motor functioning [4,5,7,8]. Benefits have been demonstrated among adult [9] and child inpatients [10].

The hospital environment may also place strain on healthcare workers, who are vulnerable to poor mental health outcomes [11] due to job-specific stressors such as long shifts with heightened psychological demands. Compassion fatigue and burnout are common sequelae of healthcare work [12]. Physician burnout has been reported in 55–70% of emergency healthcare workers and 45–50% of non-emergency workers [13–15]. This is not only concerning for the well-being of healthcare workers themselves, but also has implications for patient care. Decreases in empathy and compassion for patients have been associated with healthcare worker stress; furthermore, chronically stressed nurses are more likely to make medical errors [16].

The recent COVID-19 pandemic has heightened existing psychological demands on healthcare workers. During the pandemic, healthcare workers have experienced sleep disturbances and insomnia [17–21], lack of personal protective equipment (PPE) [17,19,22], burnout, and mental exhaustion [17,23]. Healthcare workers may experience trauma (or vicarious traumatization) by watching patients suffer or pass away on a frequent basis [17]. The combination of these major and frequent stressors has increased the prevalence of PTSD [24] and suicidal thoughts and behaviors among healthcare workers [17,20,22].

Hospital-based therapy dog programs have been shown to alleviate stress among healthcare workers. Barker [25] found that 5 min with a therapy dog produces the same amount of stress reduction as 20 min of quiet rest. A study by Kline [26] revealed that healthcare workers rate their stress levels lower after spending 5 min with a therapy dog than following 5 min of coloring. Jensen [11] showed that healthcare workers reported less work-related burnout, less job-related depression, and less intention to leave one's job after interacting with a therapy dog. A recent systematic review including 12 studies suggested that it is feasible to implement such programs in healthcare settings and that they may be associated with reductions in healthcare worker stress [27]. Due to therapy dogs' unique ability to provide significant stress relief within relatively short periods of time, hospital-based therapy dog programs could provide significant amelioration of healthcare worker stress during and following the pandemic.

1.1. Barriers to Program Reactivation

Concerns regarding COVID-19 transmission have augmented existing concerns about infection prevention for therapy animal visits in healthcare settings. Consequently, most hospital-based therapy animal visitation programs were suspended during the height of the pandemic, a time when the benefits of such visits may have been sorely needed. Some programs, such as the PAWS Your Stress Therapy Dog Program of the University of Saskatchewan and St. John Ambulance, transitioned to an online format, where therapy dog visits were conducted virtually [28]. Although early findings indicate that such online programs are well-received and important sources of social connection [28], no data exist as yet to support the comparative effectiveness of virtual vs. in-person visits in their effects on patients and healthcare workers.

As knowledge accumulated regarding the SARS-CoV-2 virus, the Centers for Disease Control and Prevention (CDC) issued a statement that animals do not significantly contribute to the spread of COVID-19 [29], especially if standard and pandemic-specific infection prevention protocols are followed. For example, face masks and eye protection can significantly reduce the transmission of airborne infections. This finding altered the risk landscape for hospital and program administrators, paving the way for hospital-based therapy animal programs to consider reactivation. Programs have successfully managed risk for transmission of other common infections, such as methicillin-resistant staphylococcus (MRSA), by implementing appropriate infection control protocols (such as the use of hand sanitizer before and after touching a dog) [30]. However, there are no data on infection control protocol adherence among human–animal therapy dog couplets in hospitals during the COVID-19 pandemic.

1.2. Animal Welfare during Reactivation

Reactivation during the pandemic offered a unique opportunity to examine the impact of program reactivation on aspects of canine welfare. The program's therapy dogs had been on hiatus from hospital visitation for a year, and significant changes were made to how humans navigated the hospital environment. For example, only four people were allowed in an elevator car at once, which caused bottlenecks and longer wait times in the main hallways. Patients, visitors, and staff wore masks and face shields or goggles that altered their appearance and non-verbal social cues. One way of reducing canine stress during reactivation was to adhere strictly to visit time limits [4,10], which can vary greatly depending upon program and setting [31]. Furthermore, individual interactions that take place during visits can vary greatly in time, activity (such as petting vs. talking to a dog), and location within the hospital [30,31]. COVID-19 and associated risk reduction strategies may change the way people interact with dogs—for example, more people interacting at one time or how they appear to the dog. There are no data that characterize the behavior of therapy dogs, handlers, and visit recipients in a hospital setting during the pandemic. Detailed exploration of therapy dog program delivery in a hospital during a pandemic may help similar institutions to make well-informed decisions about implementation that promote human and animal welfare.

1.3. Purpose of Current Study

The COVID-19 pandemic offered a unique opportunity to examine characteristics of human–animal interactions in a hospital setting as teams were reactivated and implementing new safety protocols. Detailed examination of human–animal interaction characteristics and the implementation of infection prevention protocols in a hospital-based therapy animal program can inform program development and improvement efforts for all hospital-based AAI programs. Furthermore, the variability in the execution of human–animal interaction programs in hospitals makes it difficult to effectively draw consistent/generalizable conclusions about canine-assisted interactions (CAIs) in a hospital setting [30]. Interactions vary in time, location, and frequency, and visit recipients differ along a wide range of characteristics. In addition, there is significant variation in infection prevention protocols and adherence to them. Although the literature contains examples of model programs and protocols [4], there are few finely grained descriptions of human–animal interactions inside a hospital, particularly during the global COVID-19 pandemic. Information regarding hospital-based therapy dog interactions and the implementation of infection prevention protocols can provide essential information regarding meeting service needs while maintaining appropriate safety precautions during a global pandemic.

This study aimed to provide a behaviorally based description of canine-assisted human–animal interactions among a large sample of visit recipients inside a hospital during a global pandemic. By doing so, the authors hope to provide data that inform decision-making about the deployment of therapy dog services, challenges to safety protocol implementation, and strategies for program improvement. This study examined characteristics of human–animal interactions and adherence to human and animal safety and welfare protocols during the reactivation of a hospital-based therapy dog visitation program. This information may also be helpful in future situations that may require hospital administrators to decide whether to put a program on hiatus or keep it active. The protocol was deemed exempt from review by the university Institutional Review Board and the Institutional Animal Care and Use Committee. Exemption criteria were met given the quality assurance purpose of the study and that no data were collected directly from visit recipients or animals.

2. Materials and Methods

2.1. *Therapy Dog/Handler Teams*

Dogs on Call is a therapy dog visitation program established in the Center for Human–Animal Interaction at Virginia Commonwealth University (VCU) School of Medicine in 2001. Each Dogs on Call team consists of one dog and one handler. All handlers must provide documentation of external therapy dog registration (Pet Partners or Alliance of Therapy Dogs). Handlers must also complete VCU Medical Center volunteer services training (such as a background check and HIPAA education), Dogs on Call training, and adhere to the center's policies and procedures, including human health screenings and vaccinations. In total, 20 handlers and 20 dogs were observed during the execution of this study. Table 1 details the standard (pre-COVID-19) and enhanced (during COVID-19) health requirements and safety protocols for handlers and their dogs.

Table 1. Handler and dog health and safety requirements.

	Pre-COVID-19	**Additions during COVID-19**
Handler	Varicella (vaccine or titer) MMR (measles, mumps, rubella) (vaccine or titer) Annual flu vaccine Tuberculosis screening	COVID-19 vaccine COVID-19 booster Level 3 face mask Face shield or goggles Temperature measurement and respiratory symptom checklist upon hospital entry
Dog	Registration with Pet Partners or Alliance of Therapy Dogs w/Canine Good Citizen Test Annual veterinary exam Vaccine or titer for: rabies, distemper, and parvovirus Negative annual fecal exam Two-hour visit limit	Reactivation shadowing Canine stress evaluation by program staff Three one-hour reactivation visits for reacclimation
Visit Protocol	Hand sanitizer before/after touching dog Contact tracing	No entry into COVID+ ("Hot") zones Remain at home if exposed to COVID-19 virus or experiencing respiratory symptoms

Handler and Therapy Dog Characteristics

Handlers are routinely asked to provide demographic information about themselves and their dog(s) for administrative purposes to the Center for Human–Animal Interaction. These data were accessed for the dog/handler teams that participated in this quality assurance investigation and are presented in the results below.

2.2. *Measures*

2.2.1. Participant Role

Participants were individuals in the hospital who interacted with the dog/handler teams. Participant roles were classified as adult patient (a hospitalized person visibly over the age of 18), pediatric patient (a hospitalized person visibly under the age of 18), public adult (any person not employed by the hospital or receiving treatment who was visibly over the age of 18), public child (any person not employed by the hospital or receiving treatment who was visibly under the age of 18), or healthcare worker (HCW). Determining whether a patient was an adult versus a child was facilitated by the location of the patient in the hospital, because children are typically treated in pediatric units. Public adult or public child status was determined by the absence of an employee badge and/or uniform or the presence of either a visitor wristband or a visitor name tag. HCWs were defined as any person employed by VCU Health as indicated by a badge depicting the VCU Health logo, staff member name, and department. HCWs included nurses, doctors, social workers, administrative faculty, maintenance workers, and medical students as well as volunteer services (VS) staff.

2.2.2. Observed Gender

Observed gender data collection began on the 24th visit. The term "observed gender" is used because there was no way for the researchers to confirm individuals' gender identity without asking them directly. Participants were defined as "male" if they displayed a traditionally masculine appearance and "female" if they presented a traditionally feminine appearance.

2.2.3. Total Visit Time

The observer started a timer at the beginning of the visit (when the Dogs on Call team opened the door to walk into the VS office). The timer continued to run as teams interacted with people in the hospital and was stopped when teams left VS (the door closed) after checking out. Total visit time was recorded for each visit. The timer was located at the top of the researcher's clipboard so that times could be noted at a glance.

2.2.4. Time Spent in Volunteer Services (VS)

The time spent during volunteer check-in and check-out was recorded; check-in start time began when the door to VS opened and ended when the door closed and hospital visitation began. The same recording strategy was used during check-out. Total check in and check out times were added together to determine the total time spent in the VS office.

2.2.5. Interaction Characteristics

Interaction characteristics were recorded on a pre-defined checklist on which the researcher made tick marks (see Figure A1). This checklist was developed by the authors based on the human–animal interaction expertise of the third author (N.R.G.) and feedback from handlers regarding their experiences in the hospital. Behavior was classified as an interaction if a person engaged with a therapy dog for four seconds or more and paused to visit the handler or dog. Four seconds was used as the cut off based on preliminary casual observations of a subset of interactions that our team labeled "drive-by" interactions in which an individual would walk by and run their hand along the dog's body as the dog passed by but did not stop and spend time engaging in an interaction with the dog/handler. Interactions were classified in one of three categories: (1) Talk; a person talking to a dog, (2) Pet; petting/touching a dog, or (3) Talk and Pet; talking while touching and petting the dog. Totals of Talk, Pet, or Talk and Pet interactions were recorded. The observer recorded how long each individual interaction took by looking at the running timer attached to a clipboard. When the first interaction began, the observer recorded the start time. End time was recorded when the subject was no longer talking to or petting the dog. *Total interaction time* was later calculated in seconds.

It was possible for multiple people to talk and/or pet the dog at the same time, meaning that multiple people could participate in one interaction. The *total number of people* involved in each interaction was recorded. *Participant role* was recorded for each person who interacted with the dog–handler team. As described above, roles were defined as adult patient, pediatric patient, public adult, public child, HCW, or a member of the volunteer services staff. Groupings of people from multiple roles were defined as a "mixed" population. Each interaction was coded as taking place with either male (only males participated in the interaction), female (only females participated in the interaction), or mixed group (both males and females participated in the interaction). *Observed Gender Total* was used to obtain a running total of males and females who participated in the interactions. Observed gender was used as a categorical variable to examine gender differences in interaction characteristics.

2.2.6. "Love Bombing"

The term "love bombing" was developed by the authors following initial feedback from handlers as they returned to hospital visitation. It was defined as an interaction consisting of three or more people that created crowding. Crowding was coded positively when a team's ability to move throughout the hospital was impeded by the number of people present during an interaction. Each interaction was coded as "yes" if the interaction met qualifications of a love bomb or "no" if the interaction did not meet those criteria. The number of people who participated in a love bomb was also recorded.

2.2.7. Floor

Visits took place on various floors of the VCU Medical Center and Children's Hospital of Richmond with the following exceptions: teams did not visit areas that required handlers to don extra personal protective equipment, rooms where patients tested positive for SARS-CoV-2 (COVID-19), areas where food is served, and active labor and delivery rooms. Table 2 provides a description of each floor and the services provided in those locations.

Table 2. Description of hospital floors and services provided.

Common Areas	Areas which all persons in the hospital (staff, visitors, volunteers, etc.) are free to use (with the exception of food service areas where teams do not visit)
Inpatient/Inpatient Support	Floors that provide general medical care and an array of services such as respiratory therapy, trauma treatment, cardiac care, orthopedics, intensive care, etc.
Pediatric	Floors that specialize in the treatment of pediatric patients including the Children's Hospital of Richmond
ICU	Floors that specialize in the treatment of patients with critical illness or injury
Volunteer Services	Volunteer service office where Dogs on Call teams sign in and out before and after hospital visits
Gateway	The Gateway Building serves as VCU Medical Center's "front door" and houses some of its outpatient services. Check-in and waiting areas for surgical services are located on the 5th floor of this building
Emergency Department	Department that provides immediate treatment for life threatening or time-sensitive health concerns
Nelson Clinic	Various outpatient services such as OB/GYN & Women's Health, Outpatient Eye Clinic, and dental care are housed here

Table 2. *Cont.*

West Hospital	West Hospital houses clinical, administrative, and support services for VCU Medical Center, as well as academic and administrative offices of VCU's School of Medicine and College of Health Professions
Psychiatric, Palliative Care	These departments share the same floor. Psychiatry treats those suffering from mental illness. Palliative care refers to end-of-life treatment

2.2.8. Location

Interaction location was also recorded. An interaction could take place in a hall (an area not bound by four walls and/or a door including common areas such as lobbies and elevator waiting areas), a patient room (a room designated for patient treatment only), or an office (a room with four walls with a door that designates space for employee functions, elevator, or the volunteer services office.

2.2.9. Hand Sanitizer Use

The Dogs on Call program adheres to infection prevention guidelines recommended by the American Veterinary Medical Association [32] and the Society for Healthcare Epidemiology America [33]. These guidelines indicate that all people who touch a therapy dog should use hand sanitizer before and after each interaction. Handlers are responsible for ensuring that these hand sanitizer guidelines are followed by providing the sanitizer to individuals who wish to interact with the dogs from small bottles they carry with them. Hand sanitizer behavior was coded as before-only (hand sanitizer was used before the interaction), after-only (hand sanitizer was used after the interaction), before and after (hand sanitizer was used before and after the interaction), not applicable (when the participant only talked to a dog or when a patient was unable to pet a dog due to immobility or contact restrictions), or none (when no hand sanitizer was used and physical contact between a human and dog occurred).

2.2.10. "Drive-Bys"

"Drive-by" interaction definitions were developed for this project by the study team. A drive-by interaction was coded if the interaction lasted 3 s or less and the individual did not pause near the handler or dog. Drive-bys were defined as verbal (talking only), physical (petting only), or both (Talk and Pet). The total number of drive-bys, as well as totals of each type of drive-by (Talk, Pet, Talk and Pet) were recorded for each visit. This definition was developed to distinguish ultra-brief, spontaneous interactions that usually occurred in hallways and other public spaces and were conducted in passing.

2.3. Procedure

2.3.1. Data Collection

All visit recipients were either employed by, receiving treatment from, or visiting/accompanying someone at VCU Medical Center. Program protocol requires that visit recipients provide assent before being approached by a therapy dog team. All recipients are free to decline or postpone a visit. As part of the reactivation, the Dogs on Call program implemented extra infection prevention precautions to reduce the transmission of COVID-19 in addition to the program's standard use of hand sanitizer before and after touching the dog. These precautions included the use of Level 3 face masks and face shields by human handlers, mandatory temperature and respiratory symptom screenings upon entering the hospital and refraining from visiting areas of the hospital that would require donning additional personal protective equipment, such as rooms housing COVID-positive patients or the burn unit. All Dogs on Call handlers are also required to be fully vaccinated against COVID-19 in order to participate in hospital visits. Handlers who did not wish to receive the vaccine were offered the opportunity to participate with their dogs in a virtual visitation program that was not a part of this study. Handler–dog teams are given a maximum time limit of two hours in the hospital to minimize canine stress and fatigue.

The observer accompanied dog/handler teams on visits throughout the hospital during a three-and-a-half-month period between June and September 2021. She walked next to the handler and remained near the handler/dog team throughout their hospital visit. Teams were eligible to visit all inpatient units except those requiring the use of additional personal protective equipment, such as the burn unit, rooms with patients who tested positive for SARS-CoV-2 (COVID-19), and food service areas. Table 2 provides a list of hospital units visited during data collection. The observer met teams in the VS office at the beginning of the visit and followed the team throughout the duration of the hospital visit, ending data collection when the door to volunteer services closed at check-out. The observer documented the details of every human–animal interaction that took place from the time the team clocked in until the time they clocked out. This allowed the observer to observe and record details that are typically only observed by handlers during their visitation time. Data were recorded on an observation sheet in real time, as each interaction took place. Tick marks were used to record visit characteristics on a standardized checklist containing the variables describe above. A stopwatch was attached to the top of the observer's clipboard for ease of time notation. Given that the data were collected for program quality assurance purposes, researchers did not obtain consent from participants and no personal information was collected from them.

2.3.2. Study Design

The study design was observational and descriptive. The observer made no attempt to engage in teams' interactions with participants. Handlers were aware of the observer's role and that interaction characteristics were being recorded for quality assurance purposes.

2.3.3. Data Analysis

All analyses were conducted using SPSS (version 26, IBM, Armonk, NY, USA) and Stata (version 15, StataCorp, College Station, TX, USA). Univariate statistics were used to examine the frequencies and distributions of categorical and continuous variables and to ensure that distributional characteristics of continuous variables were appropriate for planned analyses.

Role x Time Analysis. The participant role variable was used to obtain a running total of how many members of each group participated in an interaction. Role data were used to create a categorical variable to examine group differences in interaction characteristics. For example, a common interaction consisted of a pediatric patient and their parent/guardian (a public adult). This interaction would fall into the "mixed" category for interactions but would be counted as one pediatric patient and one public adult. A one-way analysis of variance (ANOVA) was performed to examine differences in length of interaction time between individuals of different roles (e.g., pediatric vs. adult patients). Bonferroni's correction was used to account for multiple comparisons. Post hoc Tukey HSD comparisons were conducted to examine specific group differences in interaction time.

Gender Differences in Interaction Time. A one-sample, independent groups t-test was conducted to explore gender differences in interaction time.

Differences in Number of Interactions per Floor. Floor was used as a categorical variable in analyses of interaction characteristics. A chi-squared test using the equiprobability model was used to evaluate differences in the number of visits received by the various floors. Pearson's standardized residuals were used to determine whether differences in number of visits between floors were statistically significant using a cut-off value of $+/-2.0$ [34].

3. Results

The observer collected data from 57 visits starting on 2 June 2021 and ending on 15 September 2021. Data were gathered from 69.25 h of observation. There were

1016 interactions recorded, involving 1783 participants. Observed gender information was collected for 1182 participants.

3.1. Handler and Dog Characteristics

The majority of handlers (15/20) were female, and all identified as White. Their mean age was 65 years, while the mean age of the dogs was 8 years. On average, dogs were approximately 58.56 cm tall (at the shoulders) and weighed 20.85 kg. Table 3 lists the breeds represented on the teams observed for this study.

Table 3. Characteristics of Dogs on Call therapy dogs observed during the quality assurance study *.

Dog	Age (Years)	Sex	Breed	Height (cm)	Weight (kg)
1	-	Female	Labradoodle	-	-
2	4	Female	Golden Retriever	71.12	27.22
3	10	Male	Mixed Breed (Large Terrier/Wolfhound)	78.74	27.22
4	11	Male	English Cream Golden Retriever	71.12	26.76
5	3	Male	English Cream Golden Retriever	71.12	29.94
6	9	Female	Leonberger	88.90	41.73
7	4	Male	Mixed Breed (Terrier x)	30.48	7.26
8	13	Male	Irish Setter	66.04	29.48
9	9	Female	Irish Setter	68.58	29.48
10	10	Female	Pembroke Welsh Corgi	38.10	11.34
11	12	Male	Mixed Breed (Lab/Pug/Boxer)	60.96	21.77
12	7	Female	Golden Doodle	76.20	27.22
13	8	Male	Shih Tzu	38.10	8.16
14	7	Male	Miniature Schnauzer	35.56	3.40
15	13	Female	Jack Russell Terrier	30.48	7.26
16	5	Female	English Cream Golden Retriever	71.12	29.48
17	-	Female	Chocolate Labrador Retriever	-	-
18	8	Male	Standard Wire Hair Dachshund	40.64	12.70
19	10	Female	Maltipoo	25.40	2.27
20	2	Male	English Cream Golden Retriever	91.44	32.66

* Some therapy dog information is missing because although handlers are asked to provide this information, they are not required to do so.

3.2. Participant Role and Observed Gender

When using participant role as the unit of measure, most visit recipients were healthcare workers (71.69%), and the remainder consisted of 9.30% adult patients, 9.08% pediatric patients, and 9.87% public adults (see Figure 1). When examining interactions as the unit of measure (see Figure 2), 57.2% (581) of interactions occurred with healthcare workers, 12.9% (131) with a mixed population (involving participants from multiple roles), 11.6% (118) with adult patients, 9.6% (98) with pediatric patients, 5.0% (51) with public adults (visitors), and 3.6% (37) with volunteer services staff. No public children were observed. Analyses revealed significant differences in interaction length by participant role ($F(4880) = 72.90$, $p = <0.001$); post hoc Bonferroni analyses using a p value of 0.05/4 showed that patients, both adult (M = 2.58 min, SD = 2.24) (95% C.I = 0.35–1.68) and pediatric (M = 5.81, SD = 4.38) (95% C.I. 3.56–4.97), had longer interaction times than healthcare workers (M = 1.56, SD = 1.92) but not visitors ($p = 1.00$) (see Figure 3). Pediatric patients had significantly longer interaction times than any other group ($p = 0.001$ for all comparisons).

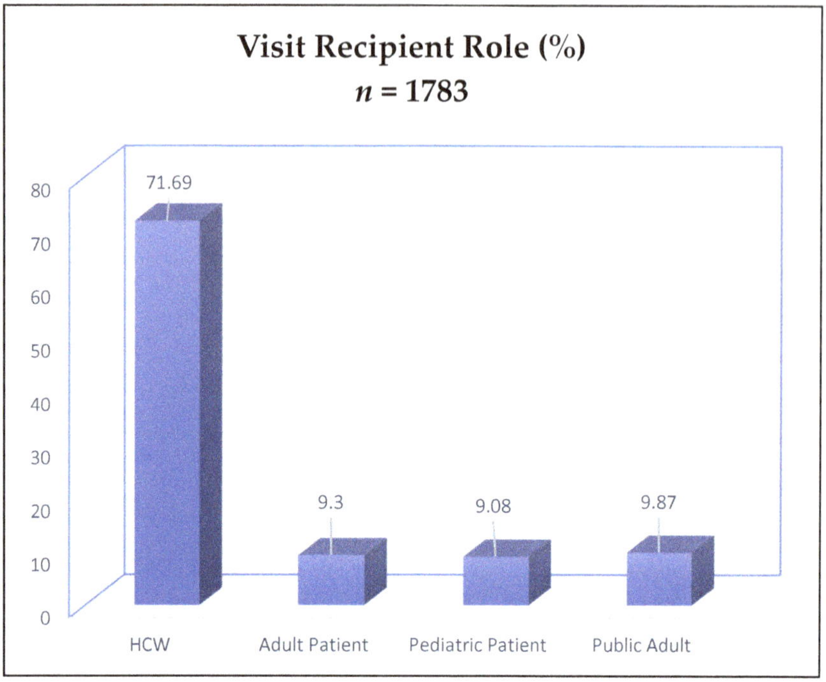

Figure 1. Visit recipient roles as percentage of visit recipients.

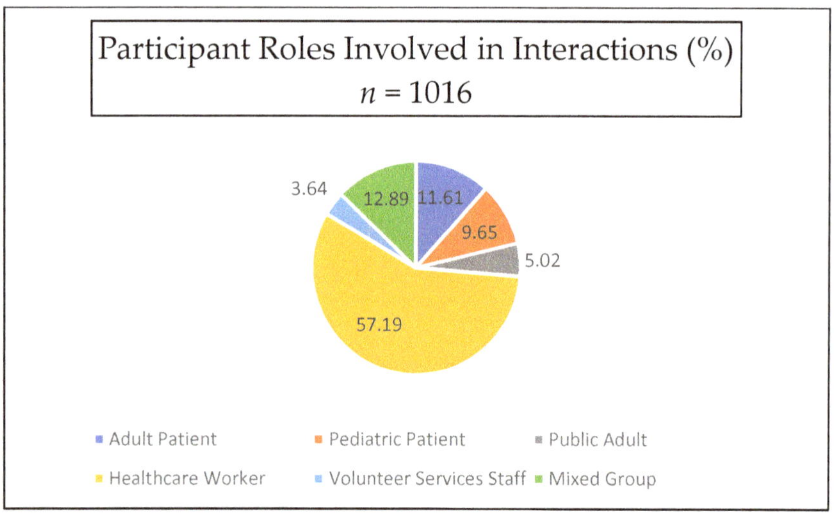

Figure 2. Visit recipient roles as percentages of interactions.

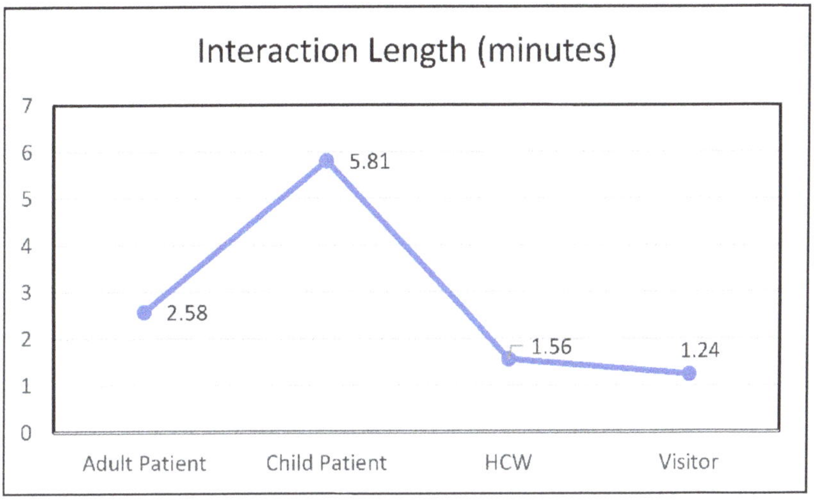

Figure 3. Interaction length by participant role.

Observed gender data collection started on 18 July 2021, which was the 24th out of 57 visits. In total, 661 interactions contained information on gender. A total of 928 females (78.40%) participated in canine-assisted interventions. Interactions with females accounted for 64.8% (428) of interactions. Males accounted for 19.1% (126) of interactions, and the remaining interactions involved both genders, accounting for 16.2% (107) of all interactions. Interaction time did not differ significantly by gender ($t(552) = -0.736$, $p = 0.462$).

3.3. Total Visit Time

The average visit time in the hospital for each team was 87 min (SD = 24.69). Total visit times were divided into groups based on 30 min intervals. As illustrated in Figure 4, 16.67% of visits (8) were less than 60 min, 31.25% (15) were 60–90 min, 45.83% (22) were 90–120 min, and 6.25% of visits (3) were more than 120 min. Visit length was missing for 9 visits due to teams arriving earlier than expected or visiting in a pediatric location that does not have a volunteer services office for check-in. Overall, 93.8% of total visit times were less than 120 min (45).

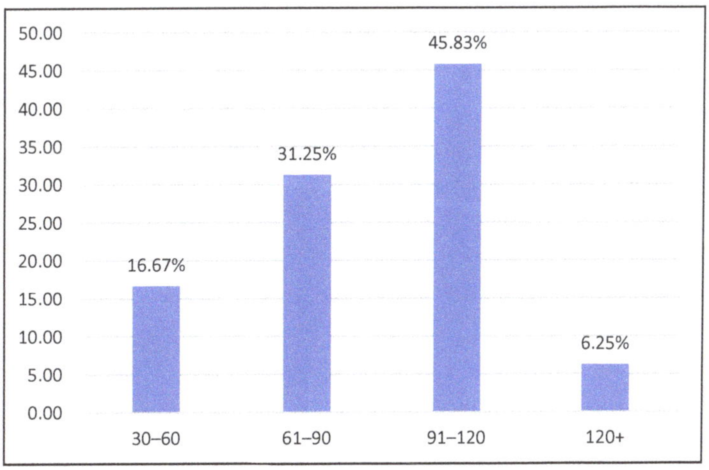

Figure 4. Distribution of total visit times across 30 min intervals.

3.4. Interaction Characteristics

3.4.1. Interaction Type

The majority of interactions (95.9%) involved a participant both talking to and petting a dog, 2.7% (27) involved a participant only talking to a dog, and 1.5% (15) involved a participant only petting a dog.

3.4.2. Interaction Time

Hospital interaction data were available for all 57 visits. During the visits, a total of 58.97% (40.77 h) of visit time was spent interacting with people. On average, each interaction lasted 2.408 min (SD = 2.96). To better examine the distribution of interaction times, interaction times were divided into one-minute interval groups. As illustrated in Figure 5, 40.26% of interactions (409) were one minute or less, 63.88% (649) were two minutes or less, and 75.49% (767) were three minutes or less. The remaining 24.51% (249) of interactions lasted longer than three minutes. Approximately 2.46% of interactions (25) lasted longer than ten minutes.

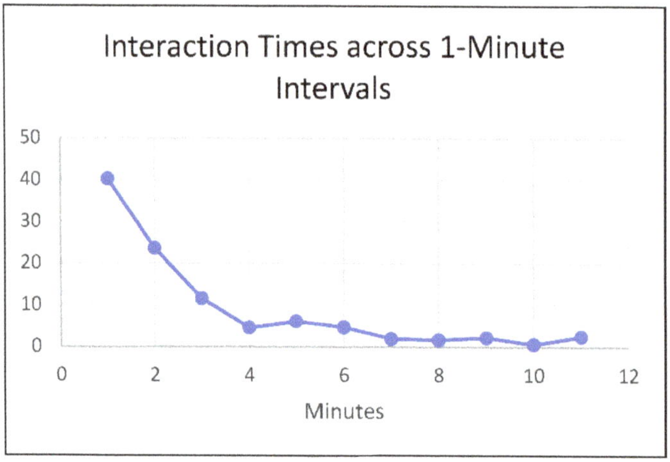

Figure 5. Distribution of interaction times across 1 min intervals.

3.4.3. Number of People per Interaction

On average, 1.76 (SD = 1.266) people were involved in an interaction (range = 1–14). Over half of interactions (59%) involved one person, 23.7% involved two people, and 17.3% (175) involved three or more people.

3.5. Love Bombing

Love bombs made up 16.33% (166) of all interactions. On average, 4.09 people (SD = 1.498) were involved in a love bomb (range = 3–14).

3.6. Floors

As shown in Table 4, most interactions took place on Inpatient and Inpatient Support (n = 286, 29.39%) as well as Critical Care Inpatient floors (n = 254, 26.10%). Pediatric Inpatients represented the floor with the third highest number of interactions (n = 235, 24.15%).

Table 4. Frequency of interactions by hospital location/floor.

Floor	Interaction Frequency (*n*, %)	Pearson's Residuals
Common Areas	123, 12.64	−3.08 *
Inpatient/Inpatient Support	286, 29.39	9.72 *
Pediatric Inpatient	235, 24.15	5.72 *
Critical Care	254, 26.10	7.21 *
Non-emergency Outpatient	27, 2.77	−10.61 *
Emergency Department	48, 4.93	−8.97 *
Total	973, 100.0	

* Forty-three interactions were excluded because they occurred in volunteer services or administrative support areas, which do not serve patients.

Chi-squared tests based on the equiprobability model (equal distribution of interactions across hospital floors) were conducted, and Pearson's residuals were used to examine whether certain floors received significantly greater or fewer interactions than expected. The expected frequency of interactions under conditions of equiprobability in this case was 162 per floor. Results indicated that the observed distribution of interactions differed significantly from that expected under the equiprobability assumption [χ^2 (7) = 381.77, p = 0.000], using a significance value of +/−2.00 [33]. As shown in Table 4, results indicated that Inpatient/Inpatient Support, Pediatrics, and Critical Care units received significantly more interactions than other units. In contrast, common areas, non-emergency outpatient centers, and the emergency department received significantly fewer interactions than expected.

3.7. Location

Just over half (57%) of interactions (579) occurred in a hallway, 32% of interactions (325) occurred in a patient room, 5.7% of interactions (58) occurred in an office, 1.3% of interactions (13) occurred in an elevator, and 3.9% of interactions (40) occurred in the volunteer services office.

3.8. Hand Sanitizer Use

In total, 79.9% of interactions (812) were carried out with proper hand-sanitizing behavior. For 2.7% of interactions (27), hand sanitizer was not required because no physical contact was made with a dog. Hand sanitizer was applied according to protocol (before and after touching a dog) in 77.3% of interactions (785). Participants applied hand sanitizer before touching a dog but not afterwards in 5.3% of interactions (54). In 5.4% of interactions (55), participants applied hand sanitizer after touching a dog, but not before. In 9.4% of interactions (95), no hand sanitizer was applied.

3.9. Drive-Bys

A total of 65.8% (669) interactions were considered drive-bys. Approximately 75% (503) of drive-bys were Talk only, 16.74% (112) were Pet only, and 8.07% (54) were Talk and Pet drive-bys. On average, 11.74 (SD = 5.453) drive-bys occurred per visit.

4. Discussion

Data from this study provide valuable insights regarding the impact of a hospital-based therapy dog program on patients and healthcare workers, the challenges of implementing therapy animal visitation during the COVID-19 pandemic, and the effectiveness of strategies to maintain animal welfare during a period of intense stress and high demand for therapy animal interventions.

4.1. Program Impact on Visit Recipients

Findings suggest that a hospital-based therapy dog program is a highly efficient way to reach large numbers of patients and hospital staff. This may have been particularly important given the potential for increased social isolation faced by patients as a result of restrictions on the number of hospital visitors due to COVID-19. Our data collection period spanned three and a half months (2 June–15 September 2021) and represents observations from 57 therapy dog visits from twenty different therapy dog teams. During these visits, Dogs on Call teams reached a total of 1783 people, including patients, visitors, healthcare workers, and other staff members. Over 70% of visit recipients were healthcare workers; it is important to note that our program reactivated amidst ongoing waves of COVID-19 surges, a period of intense stress for medical providers. A variety of interventions have been mobilized to support those who provide medical care; these include crisis intervention hotlines [17], digital support groups [35], wellness programs [36], emotion regulation training [37], and "nature-inspired recharge rooms" [38]. A therapy dog visitation program may offer unique benefits for busy providers, such as the flexibility of program delivery and stress reduction benefits from ultra-brief interactions [25]. Visits from Dogs on Call can be requested by staff at any time, scheduled for specific employee wellness and stress reduction events, or can occur spontaneously with handler–dog teams deployed throughout the hospital. In addition, our results show that most interactions lasted two minutes or less. This counters the concern that the presence of therapy dogs takes excessive amounts of time and cannot be carried out without significantly impeding clinical care. Further research is needed to evaluate ways in which hospital-based therapy dog programs can be expanded or adapted to address unmet needs among healthcare workers.

Patients and their families may also receive significant benefits from interacting with therapy dogs. Various studies have shown that hospital-based therapy animal programs are associated with a range of physical and mental health benefits for patients, including distraction from pain [39], decreased psychological distress [40], and higher activity levels [41]. Our study highlights that inpatient/inpatient support, critical care, and pediatric units receive significantly more therapy dog visits than other hospital departments. These units care for patients with potentially life-threatening illnesses and injuries. Findings could suggest that healthcare workers and/or volunteer handlers recognize a greater level of need among those patients; further study is needed to determine whether healthcare workers request more therapy dog visits for those patients or whether handlers simply tend to prefer visiting those units.

4.2. Strategies for Maintaining Human and Animal Welfare

Findings indicate a high level of adherence to human safety and animal welfare standards. The average age of our human volunteers was 65 years; this age group may be particularly vulnerable to negative sequelae associated with COVID-19 infection, with death rates increasing exponentially with age during the initial outbreak in 2020 and decreasing significantly among this population following introduction of the vaccine [42]. All hospital volunteers were required to receive full doses of the FDA-approved COVID-19 vaccine and provide documentation of vaccination prior to returning to active volunteer status in the hospital, including Dogs on Call handlers. Those handlers who did not wish to receive the vaccine or were unable to receive it due to religious beliefs or pre-existing health conditions were offered the opportunity to participate in a virtual visitation program. Handlers were also required to don Level 3 masks and a face shield or goggles upon entering the hospital to further minimize COVID-19 infection. Hand sanitizer was applied according to protocol in 80% of interactions that involved touching a dog. Additionally, our program asks handlers to systematically track all requested and spontaneous visits by location. Forms designed for this purpose are available when handlers check in for their hospital visits and are entered into a database that can be queried in the event that contact tracing is needed. These strategies were highly effective and no COVID-19 infections were reported among handlers, despite them being present in the hospital during several

COVID-19 surges. These infection prevention findings are important given that 96% of interactions in this study involved individuals talking to and petting the therapy dogs in close proximity to handlers.

Our study showed consistent fidelity to animal welfare safety guidelines. The majority (93.8%) of visits were two hours or less, indicating a high level of adherence to animal welfare visit length recommendations. Average visit length for reactivating teams fell under the two-hour time limit required by Pet Partners [43] and the VCU Center for Human–Animal Interaction [4] at 87 min. This visit length was likely shorter than visit lengths under non-pandemic conditions given that teams were instructed to limit their initial three reactivation visits to approximately one hour as their dogs reacclimated to the hospital following their year-long hiatus. Nevertheless, it is encouraging to see that handlers adhered to this recommendation despite the high levels of patient, staff, and healthcare worker need for relief from pandemic-related stress. Anecdotal observations of reactivation visits suggest that people wanted to spend longer amounts of time with the teams, which could have created social pressure for them to remain past the recommended one-hour time limit during reactivation.

4.3. Challenges during Program Reactivation

A number of challenges to maintaining high human safety and animal welfare standards presented themselves during our reactivation study. These included "love bombs", in which groups of people crowded handlers and their dogs, as well as "drive-bys", interactions in which people would touch the dogs and leave too quickly for handlers to offer hand sanitizer. Although the average number of people per interaction was approximately two people, love bombs could involve as many as fourteen people at once. We developed a "buddy system" to mitigate these departures from protocol; program staff or additional volunteer handlers accompanied teams to assist with crowd control, animal monitoring, and hand sanitizer application. We recommend this system for all unusual situations, such as the activation of a new visitation program, reactivation of an existing program that has been on hiatus, or special events at which crowds of people may congregate.

4.4. Visit Recipient Characteristics

Over 75% of visit recipients were female and over 60% of interactions took place among females. An explanation for this difference may lie in the large percentage of women present in healthcare professions such as nursing [44]. There were no significant differences between male and female interaction times, suggesting that men and women are equally willing to engage in human–animal interaction in a hospital. There is a lack of data within other HAI studies that examine spontaneous interactions with therapy dogs across genders. Unlike previous studies, this study was solely observational, where participation in human–animal interaction occurred spontaneously as the dogs and their handlers made themselves available for interactions throughout the hospital. This allowed analysis of what types of participants actively seek therapy dog interactions in a hospital setting. Future research should investigate gender-specific factors that influence the length of engagement with a therapy dog (ex. gender of dog, gender of handler). We also suggest that future researchers examine other issues of diversity, equity, and inclusion to better understand populations served and under-served by hospital-based therapy animal visitation programs.

A major finding of this study was the large percentage of healthcare workers who participated in canine-assisted interaction. Healthcare workers comprised 71.69% of visit recipients and 57.2% of interactions occurred solely between healthcare workers. This may indicate a strong need for stress relief among medical staff due to effects of the COVID-19 pandemic, given that participants voluntarily sought interactions. Recent studies indicate that hospital staff may experience elevated rates of PTSD, suicidal ideation, extreme burnout, and fear during and after a pandemic [20,22,24,45]. The literature shows that interacting with a therapy dog significantly reduces the stress and burnout of healthcare workers [11,25,26]. In-hospital therapy dog programs have potential to significantly combat the negative psychological consequences of COVID-19 among hospital personnel. However, many of these programs were shut down during the pandemic, making utilization of these resources scarce. Our study suggests that hospital-based therapy dog programs may meet stress-reduction needs among hospital staff safely during pandemic conditions. Barker's (2005) randomized cross-over trial suggests that brief (5 min) interactions with a therapy dog confers similar benefits to longer (20 min) interactions [25]. Additional research is needed to evaluate the effects of ultra-brief (<5 min) therapy dog interactions on stress responses in healthcare workers and to explore the range of these effects more fully.

This study also investigated differences in interaction times by visit recipient roles. Two main findings emerged from this analysis. First, both adult patients and pediatric patients had significantly longer interaction times than other groups. Patients may spend more time with a therapy dog because patients are typically free of time constraints or other work-related obligations. The second finding was that pediatric patients had significantly longer interactions than any other group, including adult patients. Children may spend more time with a dog due to higher levels of stress, greater need for stimulation, or greater levels of excitement when interacting with a dog. Future research should investigate the relationships between age (child vs. adult), hospital role (e.g., healthcare worker, patient, visitor), and stress-related outcomes to maximize the targeted deployment of animal-assisted hospital visitation programs.

4.5. Limitations

Several specific limitations warrant consideration when interpreting our findings. First, gender data were not collected until the second month of the study, and given the quality assurance nature of the study, could not be collected by participant interview. Characterizing participants by observation rather than by self-report introduces potential error in gender findings and certainly excludes individuals whose gender identity may not match their physical appearance. Second, the total visit length was unavailable for nine visits; this occurred when the researcher was unable to meet teams directly upon check-in to the hospital. There is no reason to suspect that these nine visits differed in significant ways from visits in which the researcher accompanied the teams from start to finish. Third, recorded interaction times were close approximations rather than precise values and sometimes depended upon the reaction time of the observer during ultra-brief interactions. Relatedly, only one observer counted the number of people involved in interactions, which could have become challenging when crowds formed during love bombs. With that said, error was minimized by having a one-page checklist, recording interaction length using a running timer attached to a clipboard, and using tick marks for counting; these techniques allowed the researcher to rapidly note times and record data with rapid strokes of a pen. In addition, all data were collected by the same person using the same approach; thus, it is reasonable to assume that any measurement error was consistent across interactions and is unlikely to represent a systematic bias in the recording of timings. It is important to point out that video-taping interactions and having multiple raters score each video recording is not possible in this setting due to patient privacy laws and COVID-19-related restrictions on social distancing. Lastly, without testing every handler and visit recipient, definitive conclusions regarding the transmission of COVID-19 (or the lack thereof) cannot be drawn. When these data were collected, rapid testing for SARS-CoV-2 was not widely available and tests were reserved for patients with respiratory symptoms and healthcare workers, rendering daily testing for handlers infeasible. However, temperature checks and respiratory symptom screenings were conducted for all handlers entering the hospital; furthermore, the program's contact tracing system was consistent with that used by the health system to track potential exposure to the virus.

5. Conclusions

The findings indicate that the Dogs on Call hospital-based therapy dog program reached large numbers of patients, staff, and healthcare workers efficiently and safely during successive waves of the COVID-19 pandemic. Strict adherence to human and animal welfare standards allowed the program to serve others with no reported cases of COVID-19 transmission associated with human–animal or human–volunteer contact, despite the highly contagious nature of the virus. To the best of our knowledge, this is the only study of its kind to evaluate such outcomes during the pandemic; the findings can inform policies and procedures for the development and reactivation of other human–animal visitation programs in similar contexts.

Author Contributions: N.R.G. conceptualized the study and its design. L.T. supervised data analysis and interpretation. J.K.H. assisted with data collection and analysis. All authors contributed to the content of the manuscript. All authors have read and agreed to the published version of the manuscript.

Funding: This research received no external funding.

Institutional Review Board Statement: Ethical review and approval were not applicable for this study because no data were collected directly from humans or animals. The Institutional Review Board at Virginia Commonwealth University made this determination.

Informed Consent Statement: Not applicable.

Data Availability Statement: Data sharing is not applicable to this article.

Conflicts of Interest: Lisa Townsend's spouse receives or has received research support, acted as a consultant and/or has received honoraria from Acadia, Adamas, Aevi, Afecta, Akili, Alkermes; Allergan, American Academy of Child & Adolescent Psychiatry, American Psychiatric; Press, Arbor, Axsome, Daiichi-Sankyo, Emelex, Gedeon Richter, Genentech, Idorsia; Intra-Cellular Therapies, Kempharm, Luminopia, Lundbeck, MedAvante-ProPhase, Merck; MJH Life Sciences, NIH, Neurim, Otsuka, PaxMedica, PCORI, Pfizer, Physicians; Postgraduate Press, Q BioMed, Receptor Life Sciences, Roche, Sage, Signant Health; Sunovion, Supernus Pharmaceuticals, Syneos, Syneurx, Takeda, Teva, Tris, and Validus. Nancy R. Gee receives or has received research support, or acts/acted as a consultant for, and/or received honoraria from Animal-Assisted Interventions Singapore, the Annenberg Foundation, the British Society of Animal Studies, Human–Animal Bond Research Institute, Mars Petcare, Nestle Purina. Jennifer K. Heatwole has no conflicts of interest to disclose.

Appendix A

CHAI Dogs on Call Research Observation Form

Team (Handler/Dog):				Date:	
VS Check-In		VS Check Out		Total VS Time	

Interaction #	Population (see legend)	# of people	Type of Interaction (see legend)	Floor	Locale	Int Start Time	Int End Time	Total Time	Hand Sanitizer	Love Bomb (√)	Additional Comments	Drive-By Tally
1									B / A			
2									B / A			
3									B / A			
4									B / A			
5									B / A			
6									B / A			
7									B / A			
8									B / A			
9									B / A			
10									B / A			
11									B / A			
12									B / A			
13									B / A			
14									B / A			
15									B / A			

Population	Type of Interaction
PtA – Patient Adult; PubA – Public Adult; HC – Staff Adult	T – Talking; P – Pet; T/P – Talking and Pet
PtC – Patient Child; PubC – Public Child	DB – Drive-by (verbal hello or pet of the dog without initiating team stop = < 1 sec.)
	Love Bomb – 3 or more individuals who arrive at the same time for an unspecified period of time.

Figure A1. Dogs on Call Visit Observation Checklist.

References

1. Chahraoui, K.; Laurent, A.; Bioy, A.; Quenot, J.P. Psychological experience of patients 3 months after a stay in the intensive care unit: A descriptive and qualitative study. *J. Crit. Care* **2015**, *30*, 599–605. [CrossRef] [PubMed]
2. Eakin, M.N.; Patel, Y.; Mendez-Tellez, P.; Dinglas, V.D.; Needham, D.M.; Turnbull, A.E. Patients' outcomes after acute respiratory failure: A qualitative study with the PROMIS framework. *Am. J. Crit. Care* **2017**, *26*, 456–465. [CrossRef]
3. Hosey, M.M.; Jaskulski, J.; Wegener, S.T.; Chlan, L.L.; Needham, D.M. Animal-assisted intervention in the ICU: A tool for humanization. *Crit. Care* **2018**, *22*, 1–4. [CrossRef] [PubMed]
4. Barker, S.B.; Gee, N.R. Canine-Assisted Interventions in Hospitals: Best Practices for Maximizing Human and Canine Safety. *Front. Vet. Sci.* **2021**, *8*, 615–730. [CrossRef]
5. Perkins, A. The benefits of pet therapy. *Nurs. Made Incred. Easy* **2020**, *18*, 5–8. [CrossRef]
6. Carey, B.; Dell, C.A.; Stempien, J.; Tupper, S.; Rohr, B.; Carr, E.; Cruz, M.; Acoose, S.; Butt, P.; Broberg, L.; et al. Outcomes of a controlled trial with visiting therapy dog teams on pain in adults in an emergency department. *PLoS ONE* **2022**, *17*, e0262599. [CrossRef] [PubMed]
7. Allen, K.; Shykoff, B.E.; Izzo, J.L. Pet ownership, but not ACE inhibitor therapy, blunts home blood pressure responses to mental stress. *Hypertension* **2001**, *38*, 815–820. [CrossRef] [PubMed]
8. Shiloh, S.; Sorek, G.; Terkel, J. Reduction of state-anxiety by petting animals in a controlled laboratory experiment. *Anxiety Stress Coping* **2003**, *16*, 387–395. [CrossRef]
9. Barker, S.B.; Dawson, K.S. The effects of animal-assisted therapy on anxiety ratings of hospitalized psychiatric patients. *Psychiatr Serv.* **1998**, *49*, 797–801. [CrossRef]
10. McCullough, A.; Ruehrdanz, A.; Jenkins, M.A. HABRI Central Briefs: The Use of Dogs in Hospital Settings. Available online: https://habricentral.org/resources/54871/download/hc_brief_dogsinhospitals20160115Access.pdf. (accessed on 4 June 2022).
11. Jensen, C.L.; Bibbo, J.; Rodriguez, K.E.; O'Haire, M.E. The effects of facility dogs on burnout, job-related well-being, and mental health in paediatric hospital professionals. *J. Clin. Nurs.* **2021**, *30*, 1429–1441. [CrossRef]
12. Graystone, R. Prevent compassion fatigue and burnout with a magnet culture. *J. Nurs. Adm.* **2019**, *49*, 231–233. [CrossRef] [PubMed]
13. Bragard, I.; Dupuis, G.; Fleet, R. Quality of work life, burnout, and stress in emergency department physicians: A qualitative review. *Eur. J. Emerg. Med.* **2015**, *22*, 227–234. [CrossRef] [PubMed]
14. Takayesu, J.K.; Ramoska, E.A.; Clark, T.R.; Hansoti, B.; Dougherty, J.; Freeman, W.; Weaver, K.R.; Chang, Y.; Gross, E. Factors associated with burnout during emergency medicine residency. *Acad. Emerg. Med.* **2014**, *21*, 1031–1035. [CrossRef]
15. Shanafelt, T.D.; Boone, S.; Tan, L.; Dyrbye, L.N.; Sotile, W.; Satele, D.; West, C.P.; Sloan, J.; Oreskovich, M.R. Burnout and satisfaction with work-life balance among US physicians relative to the general US population. *Arch. Intern. Med.* **2012**, *172*, 1377–1385. [CrossRef] [PubMed]
16. Cross, L.A. Compassion fatigue in palliative care nursing: A concept analysis. *J. Hosp. Palliat. Nurs.* **2019**, *21*, 21–28. [CrossRef]
17. Feinstein, R.E.; Kotara, S.; Jones, B.; Shanor, D.; Nemeroff, C.B. A health care workers mental health crisis line in the age of COVID-19. *Depress. Anxiety* **2020**, *37*, 822–826. [CrossRef] [PubMed]
18. Xiao, H.; Zhang, Y.; Kong, D.; Li, S.; Yang, N. The effects of social support on sleep quality of medical staff treating patients with Coronavirus Disease 2019 (COVID-19) in January and February 2020 in China. *Med. Sci. Monit.* **2020**, *26*, e923549-1–e923549-8. [CrossRef]
19. Muller, A.E.; Hafstad, E.V.; Himmels, J.; Smedslund, G.; Flottorp, S.; Stensland, S.Ø.; Stroobants, S.; Van de Velde, S.; Vist, G.E. The mental health impact of the COVID-19 pandemic on healthcare workers, and interventions to help them: A rapid systematic review. *Psychiatry Res.* **2020**, *293*, 113441. [CrossRef]
20. Mortier, P.; Vilagut, G.; Ferrer, M.; Serra, C.; Molina, J.D.; Lopez-Fresnena, N.; Puig, T.; Pelayo-Teran, J.M.; Pijoan, J.I.; Emparanza, J.I.; et al. Thirty-day suicidal thoughts and behaviors among hospital workers during the first wave of the spain COVID-19 outbreak. *Depress. Anxiety* **2021**, *38*, 528–544. [CrossRef]
21. Pappa, S.; Ntella, V.; Giannakas, T.; Giannakoulis, V.G.; Papoutsi, E.; Katsaounou, P. Prevalence of depression, anxiety, and insomnia among healthcare workers during the COVID-19 pandemic: A systematic review and meta-analysis. *Brain Behav. Immun.* **2020**, *88*, 901–907. [CrossRef]
22. Jahan, I.; Ullah, I.; Griffiths, M.D.; Mamun, M.A. Covid-19 suicide and its causative factors among the healthcare professionals: Case study evidence from press reports. *Perspect. Psychiatr. Care* **2021**, *57*, 1707–1711. [CrossRef] [PubMed]
23. Cheng, P.; Xiz, G.; Pang, P.; Wu, B.; Jiang, W.; Li, Y.T.; Wang, M.; Ling, Q.; Chang, X.; Wang, J.; et al. COVID-19 epidemic peer support and crisis intervention via social media. *Community Ment. Health J.* **2020**, *56*, 786–792. [CrossRef]
24. Yuan, K.; Gong, Y.M.; Liu, L.; Sun, Y.K.; Tian, S.S.; Wang, Y.J.; Zhong, Y.; Zhang, A.Y.; Su, S.Z.; Liu, X.X.; et al. Prevalence of posttraumatic stress disorder after infectious disease pandemics in the twenty-first century, including COVID-19: A meta-analysis and systematic review. *Mol. Psychiatry* **2021**, *26*, 4982–4998. [CrossRef] [PubMed]
25. Barker, S.B.; Knisely, J.S.; McCain, N.L.; Best, A.M. Measuring stress and immune response in healthcare professionals following interaction with a therapy dog: Pilot study. *Psychol. Rep.* **2005**, *96*, 713–729. [CrossRef] [PubMed]
26. Kline, J.A.; Van Ryzin, K.; Davis, J.C.; Parra, J.A.; Todd, M.L.; Shaw, L.L.; Haggard, B.R.; Fisher, M.A.; Petit, K.L.; Beck, A.M. Randomized trial of therapy Dogs Versus deliberative Coloring (Art therapy) to reduce stress in emergency Medicine Providers. *Acad. Emerg. Med.* **2020**, *27*, 266–275. [CrossRef]

27. Acquadro Maran, D.; Capitanelli, I.; Cortese, C.G.; Ilesanmi, O.S.; Gianino, M.M.; Chirico, F. Animal-Assisted Intervention and Health Care Workers' Psychological Health: A Systematic Review of the Literature. *Animals* **2022**, *12*, 383. [CrossRef]
28. Dell, C.; Williamson, L.; McKenzie, H.; Carey, B.; Cruz, M.; Gibson, M.; Pavelich, A. A commentary about lessons learned: Transitioning a therapy dog program online during the COVID-19 pandemic. *Animals* **2021**, *11*, 914. [CrossRef]
29. Centers for Disease Control and Prevention. Animals and COVID-19. 5 October 2021. Available online: https://www.cdc.gov/coronavirus/2019-ncov/daily-life-coping/animals.html (accessed on 5 June 2022).
30. Bert, F.; Gualano, M.R.; Camussi, E.; Pieve, G.; Voglino, G.; Siliquini, R. Animal Assisted Intervention: A systematic review of benefits and risks. *Eur. J. Integr. Med.* **2016**, *8*, 695–706. [CrossRef]
31. Chur-Hansen, A.; McArthur, M.; Winefield, H.; Hanieh, E.; Hazel, S. Animal-assisted interventions in children's hospitals: A critical review of the literature. *Anthrozoös* **2014**, *27*, 5–18. [CrossRef]
32. American Veterinary Medical Association. Animal Assisted Interventions: Guidelines. Available online: https://www.avma.org/resources-tools/avma-policies/animal-assisted-interventions-guidelines (accessed on 5 June 2022).
33. Murthy, R.; Bearman, G.; Brown, S.; Bryant, K.; Chinn, R.; Hewlett, A.; Glenn George, B.; Goldstein, E.J.C.; Holzmann-Pazgal, G.; Rupp, M.E.; et al. Animals in healthcare facilities: Recommendations to minimize potential risks. *Infect. Control. Hosp. Epidemiol.* **2015**, *36*, 495–516. [CrossRef]
34. Agresti, A. *Categorical Data Analysis*, 2nd ed.; Wiley: New York, NY, USA, 2002; p. 81.
35. Viswanathan, R.; Myers, M.F.; Fanous, A.H. Support groups and individual mental health care via video conferencing for frontline clinicians during the COVID-19 pandemic. *Psychosom. J. Consult. Liaison Psychiatry* **2020**, *61*, 538–543. [CrossRef] [PubMed]
36. Gonzalez, A.; Cervoni, C.; Lochner, M.; Marangio, J.; Stanley, C.; Marriott, S. Supporting health care workers during the COVID-19 pandemic: Mental health support initiatives and lessons learned from an academic medical center. *Psychol. Trauma Theory Res. Pract. Policy* **2020**, *12*, S168–S170. [CrossRef] [PubMed]
37. Landa-Ramírez, E.; Domínguez-Vieyra, N.A.; Hernández-Nuñez, M.E.; Díaz-Vásquez, L.P.; Santana-García, I.A. Psychological support during COVID-19 death notifications: Clinical experiences from a hospital in Mexico. *Psychol. Trauma Theory Res. Pract. Policy* **2020**, *12*, 518–520. [CrossRef] [PubMed]
38. Putrino, D.; Ripp, J.; Herrera, J.E.; Cortes, M.; Kellner, C.; Rizk, D.; Dams-O'Connor, K. Multisensory, nature-inspired recharge rooms yield short-term reductions in perceived stress among frontline healthcare workers. *Front. Psychol* **2020**, *11*, 560833. [CrossRef] [PubMed]
39. Nepps, P.; Stewart, C.N.; Bruckno, S.R. Animal-Assisted Activity: Effects of a Complementary Intervention Program on Psychological and Physiological Variables. *J. Evid. Based Complementary Altern. Med.* **2014**, *19*, 211–215. [CrossRef]
40. Gagnon, J.; Bouchard, F.; Landry, M.; Belles-Isles, M.; Fortier, M.; Fillion, L. Implementing a hospital-based animal therapy program for children with cancer: A descriptive study. *Can. Oncol. Nurs. J.* **2004**, *14*, 217–222. [CrossRef]
41. Abate, S.V.; Zucconi, M.; Boxer, B.A. Impact of canine-assisted ambulation on hospitalized chronic heart failure patients' ambulation outcomes and satisfaction: A pilot study. *J. Cardiovasc. Nurs.* **2011**, *26*, 224–230. [CrossRef]
42. Centers for Disease Control. Weekly Counts of Deaths by Age Group. Available online: https://www.cdc.gov/nchs/nvss/vsrr/covid19/excess_deaths.htm (accessed on 5 May 2022).
43. Pet Partners. Pet Partners Brief Information. Available online: https://petpartners.org/wp-content/uploads/2019/12/PP_Info-General-Information-w-terms.pdf (accessed on 7 May 2022).
44. Smiley, R.A.; Ruttinger, C.; Oliveira, C.M.; Hudson, L.R.; Allgeyer, R.; Reneau, K.A.; Silvestre, J.H.; Alexander, M. The 2020 National Nursing Workforce Survey. *J. Nurs. Regul.* **2021**, *12*, S1–S96. [CrossRef]
45. Ruiz, M.A.; Gibson, C.A.M. Emotional impact of the COVID-19 pandemic on U.S. health care workers: A gathering storm. *Psychol. Trauma Theory Res. Pract. Policy* **2020**, *12*, S153–S155. [CrossRef]

animals

MDPI

Article

Patient and Caregiver Perceptions of Animal Assisted Activity in Orthodontics

Katelyn Cass [1,2], Clare Bocklage [1,3], Taylor Sulkowski [1], Christina Graves [3], Nare Ghaltakhchyan [1,3], Allen Rapolla [1], Tate Jackson [1], Kimon Divaris [4], Chris Wiesen [5], Timothy Strauman [6] and Laura Jacox [1,3,*]

[1] Orthodontics Group, Division of Craniofacial and Surgical Care, Adams School of Dentistry, University of North Carolina, 270 Brauer Hall, CB #270, Chapel Hill, NC 27599, USA; katie13cass@gmail.com (K.C.); eclare@live.unc.edu (C.B.); taylor.sulkowski@unc.edu (T.S.); nare@live.unc.edu (N.G.); arapolla@live.unc.edu (A.R.); tatejackson@unc.edu (T.J.)
[2] Summers Orthodontics, 4207 E North St, Greenville, SC 29615, USA
[3] Division of Oral and Craniofacial Health Sciences, Adams School of Dentistry, University of North Carolina, 385 S Columbia St, CB #7455, Chapel Hill, NC 27599, USA; cxg@email.unc.edu
[4] Division of Pediatric and Public Health, Adams School of Dentistry, University of North Carolina, 385 S Columbia St, CB #7455, Chapel Hill, NC 27599, USA; kimon_divaris@unc.edu
[5] Howard W. Odum Institute for Research in Social Science, Davis Library, University of North Carolina, 208 Raleigh St, CB #3355, Chapel Hill, NC 27514, USA; chris_wiesen@unc.edu
[6] Psychology and Neuroscience Department, Trinity College of Arts and Sciences, Duke University, 417 Chapel Dr, CB #90086, Durham, NC 27514, USA; tjstraum@duke.edu
* Correspondence: ljacox@live.unc.edu; Tel.: +1-(919)-537-3768

Simple Summary: Dental anxiety impacts a significant fraction of children and adults, leading to lifelong avoidance of the dentist and increased emergency dental care. Animal-assisted activity (AAA) is widely used in medicine to reduce anxiety and pain, with promise in dentistry. However, dentistry has been slow to adopt AAA, with a state dental board banning therapy animals in dental clinics due to patient concerns over dog safety, allergies, and cleanliness. Our goal was to determine how orthodontic patients and their caregivers viewed canine therapists in dental clinics to see whether AAA would be welcomed by most families. (No dog therapy occurred as part of this study, so the efficacy of AAA for dental anxiety management was not evaluated). Orthodontic patients and parents/caregivers were asked to fill out a survey about their dental anxiety and their desire for and concerns regarding therapy animals in dental clinics. More than a third of patients had moderate or greater anxiety related to dental care. A vast majority of participants believed that therapy dogs would make dental experiences more enjoyable and reduce fear, with a small minority raising concerns about cleanliness, allergies, and safety. Among patients and caregivers, there is broad acceptance and desire for AAA in dental and orthodontic settings. Future research should be aimed at determining how AAA can improve the experiences of dental patients.

Abstract: Dental anxiety affects up to 21% of children and 80% of adults and is associated with lifelong dental avoidance. Animal assisted activity (AAA) is widely used to reduce anxiety and pain in medical settings and has promise in dentistry. The primary objective of this study was to evaluate caregiver and patient perceptions of canine AAA in orthodontics. A cross-sectional survey consisting of pre-tested and validated questions was conducted (n = 800) including orthodontic patients (n = 352 minors, n = 204 adults) and parents/caregivers (n = 244) attending university orthodontic clinics. In this study, AAA and dog therapy were not used or tested for dental anxiety management. More than a third of orthodontic patients (37%) had moderate or greater anxiety related to care. Participants believed that therapy animals would make dental experiences more enjoyable (75%) and reduce anxiety (82%). There was little to no concern expressed regarding cleanliness (83%), allergies (81%), and safety (89%) with a therapy animal in dental settings. Almost half of the participants would preferentially select an orthodontic office offering AAA. In light of the COVID-19 pandemic, we assessed whether perceptions of AAA changed before and after the shutdown of dental offices, with no significant differences. Across patients and caregivers, the responses support the use of AAA in orthodontic settings with minimal concerns.

Citation: Cass, K.; Bocklage, C.; Sulkowski, T.; Graves, C.; Ghaltakhchyan, N.; Rapolla, A.; Jackson, T.; Divaris, K.; Wiesen, C.; Strauman, T.; et al. Patient and Caregiver Perceptions of Animal Assisted Activity in Orthodontics. *Animals* **2022**, *13*, 1862. https://doi.org/10.3390/ani12141862

Academic Editors: Betty McGuire and Mónica Teresa González-Ramírez

Received: 10 June 2022
Accepted: 20 July 2022
Published: 21 July 2022

Keywords: dogs; dental anxiety; dental fear; anxiety; orthodontists; orthodontics; dentistry; animal therapy; animal assisted therapy; COVID-19

1. Introduction

Dental anxiety (DA) affects 50–80% of adults and 6–21% of children [1,2]. DA commonly emerges during childhood due to traumatic experiences and often results in lifelong distress and care avoidance [3,4]. DA presents a major challenge to optimizing oral health outcomes and is associated with increased incidence of caries, infection, and urgent care [4,5]. To care for anxious children, the American Academy of Pediatric Dentistry advocates the use of pharmacological and non-pharmacological behavior guidance techniques [6]. Pharmacological sedation, which is required for highly anxious patients, carries a low risk of respiratory depression, neurological injury, and death [7]. Due to these risks, parents may elect against the use of sedation and medications. This is especially true for orthodontics, which is elective and often delayed until patients can comply; however, this delay can cause patients to miss optimal treatment timing [8]. As a result, non-pharmacological approaches are needed for managing anxious patients in orthodontic settings.

Animal assisted activity (AAA) is a promising intervention in which a certified, trained animal is introduced by a trained professional to interact with an individual to enhance their quality of life [9]. AAA is utilized to reduce anxiety, stress, and the perception of pain; it usually involves dogs that are trained to be obedient, calm, and comforting, and is an option for behavior management in dentistry (Figure 1) [10,11]. AAA distracts patients and is effective at reducing stress hormones, increasing endorphins, and activating mirror neurons [10–12]. AAA has been deployed successfully in inpatient and outpatient medical settings [12–14]. Data regarding AAA's positive effects are abundant in medicine, however, the use of AAA in dentistry is in its nascent stages, with promising early findings [13,15–19]. Dental patients with a therapy animal exhibited decreased discomfort, lower blood pressure, and improvement in experience and compliance [16,17]. Among children verbalizing distress, AAA decreased their physiological arousal [17].

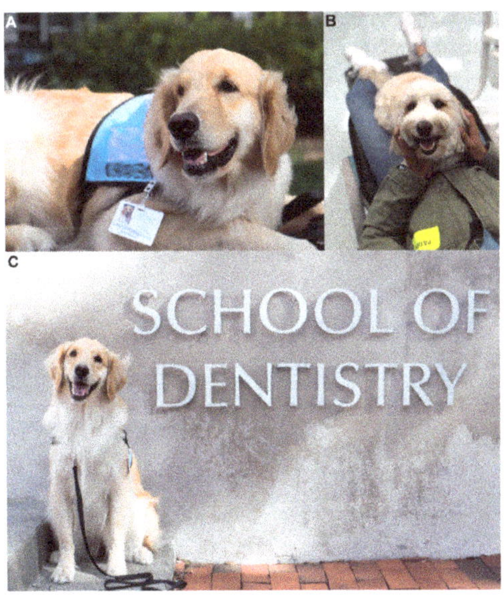

Figure 1. Therapy animals in dental clinics. (**A**) Certified canine therapist, Grayson Siggi. (**B**) Farley Cass comforting orthodontic patients. (**C**) Grayson welcoming visitors, as one of the first dental facility therapy dogs.

Though research indicates AAA's promise for DA management, clinical adoption by practitioners depends on patient interest and acceptance [15–18]. This study aims to determine the patient perceptions of AAA in dentistry and orthodontics to inform its adoption. A review proposed that hazards of therapy animals in dental offices included safety risks, cleanliness, and allergens [20]. Safety concerns included the risk of dog bites, disease transmission (zoonosis), dog entanglement in instruments, and accidental tripping over the dog, with the potential for fall injuries [20]. Cleanliness concerns relate to waste removal (dog urine and feces) and dirt dispersion from the paws. Allergy concerns refer to airborne dander, hair shedding, and facial licking [20]. Health care protocols are used to mitigate these risks in medical and dental settings, as detailed in the guidelines for animal assisted interventions [21]. However, it is unknown whether the concerns for these hazards are held by the patients and parents/caregivers, and whether families will accept therapy animals in orthodontic clinics. To address this, we evaluated the perceptions and concerns of AAA in an orthodontic setting using pre-tested and validated survey scales. We hypothesized that AAA is acceptable to orthodontic patients and caregivers, with a majority (>70%) believing that therapy animals would make dental experiences more enjoyable, with infrequent concerns for allergies, cleanliness, and safety (<30% with medium to large concerns). Dog therapy was not performed, nor did we evaluate the efficacy of AAA for anxiety management in this study. The survey results can inform practitioners regarding AAA implementation in orthodontic contexts.

2. Materials and Methods

We conducted a cross-sectional survey of orthodontic patients and their parents/caregivers to determine perceptions of AAA at the University of North Carolina (UNC) Orthodontics Graduate and Faculty Clinics. Our sample (n = 800) included consecutively enrolled minor patients (under 18 [12–17 years old]), adult patients (>18 years old), and adult caregivers of minor patients (Table 1 and Table S1). Survey questions (File S1) were developed using pre-tested or validated questions on anxiety and AAA (data in Tables 2–4, Supplementary Tables S2–S7). The validated Corah Dental Anxiety Scale (DAS) was included as a widely used, reliable measure of anxiety; answers are scored and summed to determine anxiety level (<8 limited, 9–12 moderate, 13–14 high, 15–20 severe) (Table 4 and Table S3) [16,22,23]. Under the guidance of a survey expert, we adapted the Corah DAS to suit the orthodontic setting with minor changes, and then pre-tested and revised the questions. For topics with no published tools, the team developed, pre-tested, and revised the questions. Prior surveys and position pieces on AAA were referenced for theme inclusion [15,17–19]. Pre-testing was performed with iterative revisions until a final draft was approved by the investigators. Pre-testers included seven laypeople (four adults; three minors), seven residents, two private orthodontists, and two faculty. Of the pre-testers, eight owned dogs and 13 had orthodontic treatment. Among the pre-testers, individuals who did and did not own dogs were included to represent these perspectives, as dog owners were compared to non-dog owners in the survey results.

Table 1. Descriptive information of study participants.

Category	Group	Frequency (%) and Number (n) per Group *
Participant groups	Patients who are minors under 18	44.0% (n = 352)
	Adult patients over 18	25.5% (n = 204)
	Caregivers	30.5% (n = 244)

Table 1. *Cont.*

Category	Group	Frequency (%) and Number (n) per Group *
Race	Caucasian	68.8% (n = 391)
	Black	21.7% (n = 123)
	Asian	7.6% (n = 43)
	Other	1.9% (n = 11)
	Prefer to not answer	n = 232
Ethnicity	Hispanic	18.8% (n = 150)
	Not Hispanic	81.3% (n = 650)
Sex	Female	65.7% (n = 460)
	Male	34.3% (n = 240)
	Prefer to not answer	n = 100
Dog Allergy	Diagnosed allergy to dogs	4.1% (n = 30)
	No allergy	95.9% (n = 706)
Fear of dogs	Not at all afraid of dogs	77.1% (n = 566)
	Only a little afraid of dogs	17.0% (n = 125)
	Somewhat afraid of dogs	4.1% (n = 30)
	Very afraid	1.8% (n = 13)
Dog Presence **	Dog present **	41.0% (n = 328)
	No dog present **	59.0% (n = 472)
COVID ^	Pre-shutdown ^	16.1% (n = 129)
	Post-shutdown ^	83.9% (n = 671)
Pet at home ^^	Pet(s) at home of any species	71.0% (n = 512)
	No pet at home	29.0% (n = 288)
	Pet dog(s) ^^	60.1% (n = 434)
	No pet dog ^^	39.9% (n = 288)
		Total n = 800

* Participants who did not answer the questions on the demographic variables (e.g., race, gender) were not considered in the frequency calculations. ** Participants who completed the survey in the presence of a dog in the waiting room (Dog). Participants who responded to the survey without a dog in the waiting room (No dog). ^ Responses collected before the pandemic shutdown (pre-shutdown) or after the shutdown (post-shutdown). ^^ Participants with a pet dog at home (pet dog) or without a pet dog at home (no pet dog).

Table 2. The concerns related to animal assisted activity (AAA).

When Thinking about a Therapy Dog in a Dental Setting, How Much Concern Would You Have for Each of the Following? (Q10)						
			Little to No Concern	Medium to Large Concern	n ^	p-Value *
Age Groups	Cleanliness	Overall	83% 599 ^	17% 122 ^	721 ^	0.5106
		Under 18 **	83.8% 244 ^	16.2% 47 ^		
		Over 18 **	80.9% 157	19.1% 37		
		Caregivers **	83.9% 198	16.1% 38		

Table 2. *Cont.*

When Thinking about a Therapy Dog in a Dental Setting, How Much Concern Would You Have for Each of the Following? (Q10)						
			Little to No Concern	**Medium to Large Concern**	***n*** ^	***p*-Value** *
	Allergies	Overall	81% 583	19% 135	718	0.3436
		Under 18	79.3% 230	20.7% 60		
		Over 18	84.4% 162	15.6% 30		
		Caregivers	80.9% 191	19.1% 45		
	Safety	Overall	89% 642	11% 76	718	0.9829
		Under 18	89.7% 260	10.3% 30		
		Over 18	89.1% 171	10.9% 21		
		Caregivers	89.4% 211	10.6% 25		
Dog vs. No Dog in Clinic **	Cleanliness	Dog ***	87% 256	13% 39	721	0.0213 *
		No Dog ***	81% 343	19% 83		
	Allergies	Dog	83% 246	17% 49	718	0.2877
		No Dog	80% 337	20% 86		
	Safety	Dog	93% 274	7% 21	718	0.0185 *
		No Dog	87% 368	13% 55		
Pet dog vs. No pet dog ^^	Cleanliness	Pet ^^	86.8% 382	13.2% 58	721	0.00005 *
		No Pet Dog ^^	77.2% 217	22.8% 64		
	Allergies	Pet	84.2% 369	15.8% 69	718	0.0040 *
		No Pet Dog	76.4% 214	23.6% 66		
	Safety	Pet	92.4% 404	7.6% 33	718	0.00009 *
		No Pet Dog	84.7% 238	15.3% 43		

* $p < 0.05$ statistical significance criterion. ** Orthodontic patients who were minors under 18 (Under 18). Adult orthodontic patients over 18 (Over 18). Caregivers included parents and legal guardians of orthodontic patients who were minors (Caregivers). *** Respondents who filled out the survey while a dog was in the clinic waiting room (Dog). Respondents who filled out the survey without a dog in the clinic waiting room (No Dog). ^ The number of respondents for each subgroup. ^^ Respondents with a pet dog at home (Pet dog) or without a pet dog at home (No pet dog).

Table 3. The AAA's impact on patient and caregiver's orthodontic office selection.

If You Were Making a Choice Between Two Similar Orthodontic Practices, Would the Presence of a Dog Matter to You? (Q19)				
	Yes	**No**	n ^	p-**Value**
Overall	48.1% 334 ^	51.9% 361 ^	695 ^	
Under 18 **	54.5% 152	45.5% 127	279	0.0210 *
Over 18 **	43.2% 80	56.8% 105	185	
Caregivers **	44.2% 102	55.8% 129	231	
Pet Dog ^^	53.8% 227	46.2% 195	422	0.0017 *
No Pet Dog ^^	39.2% 107	60.8% 166	273	
If Answered "Yes" Above, Which Practice Would You Pick? (Q20)				
	With Dog	**Without Dog**	n ^	p-**Value**
Overall	92.2% 308^	7.8% 26^	334 ^	
Under 18	96.1% 146	3.9% 6	152	0.0523 *
Over 18	90.0% 72	10.0% 8	80	
Caregivers	88.2% 90	11.8% 12	102	
Pet Dog	95.6% 217	4.4% 10	227	0.0080 *
No Pet Dog	85.0% 91	15.0% 16	107	

* $p < 0.05$ statistical significance criterion. ** Orthodontic patients who were minors under 18 (Under 18). Adult orthodontic patients over 18 (Over 18). Caregivers included the parents and legal guardians of orthodontic patients who were minors (Caregivers). ^ Number of respondents for each subgroup. ^^ Respondents with a pet dog at home (pet dog) or without a pet dog at home (no pet dog).

Table 4. Corah Dental and Orthodontic Modified Dental Anxiety Scales by group (Q21–28).

		Corah Dental Anxiety Category ^^					**Orthodontic Anxiety Category ^^**					
Group	**Total** n **^**	**Limited**	**Moderate**	**High**	**Severe**	**Group** p-**Value**	**Limited**	**Moderate**	**High**	**Severe**	**Group** p-**Value**	p-**Value Dent v. Ortho ^^**
All	664 ^	364 ^	221 ^	39 ^	40 ^		416 ^	191 ^	30 ^	27 ^		0.0030 *
		54.8%	33.4%	5.9%	5.9%		62.65%	28.76%	4.52%	4.07%		
Gender						0.0001 *					0.0066 *	
Males	224	146	70	6	8		150	63	9	2		0.2360
Females	432	63.5% 220	30.4% 155	2.6% 34	3.5% 32		63.5% 261	30.4% 125	2.6% 21	3.5% 25		0.0001 *
Age group		49.9%	35.2%	7.7%	7.3%	0.0079 *	60.4%	28.9%	4.9%	5.8%	0.0007 *	
Patients Under 18	266	153	94	12	13		172	74	13	7		0.0059 *
		56.3%	34.6%	4.4%	4.8%		64.7%	27.8%	4.9%	2.6%		

Table 4. *Cont.*

Group	Total *n* ^	Corah Dental Anxiety Category ^^ Limited	Moderate	High	Severe	Group *p*-Value	Orthodontic Anxiety Category ^^ Limited	Moderate	High	Severe	Group *p*-Value	*p*-Value Dent v. Ortho ^^
Patients Over 18	174	109	53	9	8		127	35	6	6		0.0341 *
		60.9%	29.6%	5.0%	4.5%		73.0%	20.1%	3.5%	3.5%		
Caregiver	224	110	80	19	19		117	82	11	14		0.0907
		48.3%	35.1%	8.3%	8.3%		52.2%	36.6%	4.9%	6.2%		
Dog/No Dog						0.1586					0.0389 *	
Dog present	278	114	93	26	15		155	88	14	13		0.1664
		51.8%	33.5%	9.4%	5.4%		57.4%	32.6%	5.2%	4.8%		
No dog present	401	228	134	14	2		261	104	16	14		0.0007 *
		56.9%	33.4%	3.5%	6.2%		66.2%	26.1%	4.1%	3.6%		

* $p < 0.05$ statistical significance criterion. ^ Number of participants for each subgroup. ^^ Corah Dental Anxiety DAS score (dental). Orthodontic modified anxiety DAS score (ortho). Dental Anxiety Scales (DAS) are summed (four questions, 1–5 points each) to determine the anxiety level (<8 limited, 9–12 moderate, 13–14 high, and 15–20 severe).

Surveys were administered using Qualtrics (Qualtrics XM, Inc., Provo, UT, USA). Potential participants were discretely approached in the clinic's reception by study staff who screened and enrolled subjects, at 2 h intervals to ensure turnover (Table S1). Patient participants were orthodontic patients, 12–65 years old, and treated at the UNC Graduate or Faculty Orthodontics Practices (Table 1 and Table S1). Parent participants were caregivers of a minor orthodontic patient treated at the UNC Graduate or Faculty Orthodontics Practices. Study coordinators did not enroll parent–child pairs to limit any effect of their relationship on the responses. To enroll children, we would meet with the parent and child to gain parental consent and minor participant assent on IRB-approved digital forms; for these minors, we did not enroll their caregiver. To enroll caregivers, we approached them on their own, once their child was taken into the clinic for their appointment. Subjects who met the enrollment criteria (Table S1) and verbally agreed to participate were given consent forms, and then the survey to complete while they were in the orthodontics department. A screening question excluded repeat responses.

Our dental clinics host therapy dogs during clinic sessions, as part of their normal operations. A therapy dog is present during routine care in our pediatric and orthodontic clinics during most workdays; in orthodontics, the dog is seated in the reception area with her handler, or walked in the clinic by her handler to greet patients before or after their dental visit. Routine care in orthodontics includes bonding braces, removing braces, or replacing wires. Seeing a dog has the potential to influence survey participants. To control for this potential confounder and evaluate whether seeing a therapy dog influenced the perceptions of AAA, data were collected from participants with and without a therapy dog in the reception area; the dog was present on alternating weeks. The therapy dog is a 3-year-old, female, medium-sized goldendoodle who underwent therapy dog training and certification with her handler, as specified by the university with the observation of animal welfare policies (Dog: Farley Cass; Handler: Dr. Katelyn Cass, DDS, MS). During the study, the dog was seated with her handler in the reception area in an open air pen (size: 7 ft by 7 ft) positioned next to the front desk. Participants saw the dog when checking in and waiting for their appointment and could pet the dog if desired. No dog therapy occurred; we did not evaluate the efficacy of AAA for anxiety management in this study.

The survey distribution began before the COVID-19 pandemic and continued after reopening (2/2020–10/2020). Participants who took the survey prior to the shutdown

(n = 105 minor patients, no dog) did not have questions about COVID-19. After re-opening, pandemic-related questions were added (File S1, Q29–35).

Statistics: Descriptive statistics are reported in the tables of the response frequencies. A row mean difference test was used to determine the differences between groups for ordered categorical variables. This test of equality was also used to evaluate the differences in mean scores of the outcome variable (e.g., anxiety scale value) among the grouping variable (e.g., gender, patient group). Multiple comparison tests (MCT) were conducted to evaluate the differences among groups containing three categories (e.g., under 18, over 18, and caregivers). Additionally, the Holm method was used to adjust the p-values of multiple comparisons to reduce the chance of type I error. McNemar's test was used to compare dental versus orthodontic anxiety. Subjects responded to each item and the null hypothesis was that the marginal distribution of item responses was the same for both items. Matching was within subjects. Significance was defined as $p < 0.05$. Statistical analyses were conducted using SAS 9 software (SAS Institute, Cary, NC, USA). Graphs were made using Prism 9 Software (GraphPad Software Inc., La Jolla, CA, USA) and figures were created using Adobe Suite (Adobe Inc., San Jose, CA, USA).

This research was approved by the UNC Institutional Review Board (IRB #19-1908) with the protection of human subjects and their rights.

3. Results

3.1. *Aaa Concerns and Benefits*

To determine the perceptions of AAA, questions probed the concerns, desired experiences, and anxiety. The response rate was 84.9% with 800 participants including 204 adult patients, 352 minor patients, and 244 caregivers (Table 1). Data showed that a large majority of patients and caregivers reported "little" or "no concern" regarding cleanliness (83%), exposure to allergens (81%), or safety (89%) with a therapy animal in a dental setting (Figure 2A,B, Table 2). Participants with pet dogs and those who filled out the survey with a dog present more frequently reported "little" or "no concern" than participants without pet dogs and without a dog present, which is consistent with our hypotheses (Table 2).

Three quarters (75%) of participants and 85% of minors ($p = 0.00003$) indicated that having a therapy dog in a dental office would create a more enjoyable patient experience and 82% selected that the therapy animal would reduce dental anxiety (Figure 2E, Tables S2 and S8). Roughly half of the caregivers (44%) and patients (55%) under 18 indicated that the presence of a therapy animal in an orthodontic office would be important to decide which office they selected for care (Figure 2C, Table 3). Of these participants, 88% of caregivers and 96% of patients under 18 would preferentially select an office offering AAA (Figure 2D, Table 3). Dog owners were even more likely to choose an office with AAA than non-dog owners, however, 85% of those without a pet dog would still choose a practice offering AAA (Table 3). The overwhelming majority of patients and caregivers indicated that therapy animals would reduce anxiety and increase enjoyment in orthodontic settings with minimal concerns.

3.2. *Dental Anxiety and Aaa*

Almost half of the participants (45%) suffered from some level of anticipatory dental anxiety and 37% had orthodontic-related anxiety at a level of moderate or greater (Figure 3A, Tables 4 and S3). Adult and minor patients reported a higher severity of dental anxiety than orthodontic anxiety (Figure 3B,C, Tables 4 and S3). Moreover, the caregivers indicated more dental and orthodontic anxiety than adults and minor patients despite not having appointments (Figure 3B,C, Tables 4, S3 and S8). Females reported higher dental than orthodontic anxiety, along with more dental and orthodontic anxiety than males (Figure 3D, Tables 4 and S3). Patients and caregivers who took the survey with a therapy dog had minimal change in the anticipated dental or orthodontic anxiety when compared to participants who did not see the dog (Tables 4 and S3).

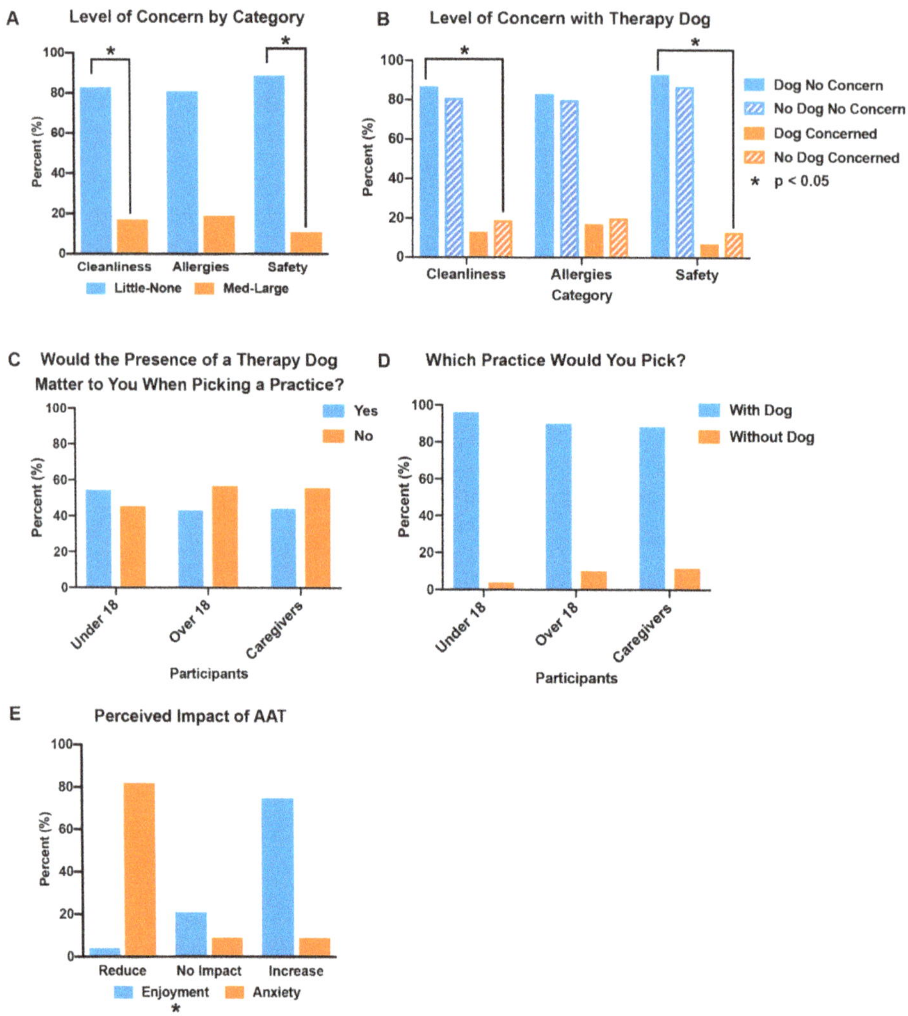

Figure 2. The patient and caregiver perceptions of AAA. (**A**) Frequency of participants responding "little concern" or "no concern" (pooled data, blue) versus "medium concern" and "large concern" (orange) with regard to cleanliness, allergies, and safety when having a therapy animal in a clinical dental setting. (Table 2); (**B**) Level of concern (no concern—blue; concerned—orange) about having a therapy dog present with and without a dog in the waiting area (dog present—solid; no dog present—hatched) in regard to cleanliness, allergies, and safety; (**C**) Participants responding whether the presence of a therapy dog matters (yes—blue; no—orange) to patients under 18, patients over 18, and caregivers when selecting between two similar orthodontic practices. (Table 3); (**D**) Participants (under 18 patients, over 18 patients, caregivers) responding to which practice they would pick (with a dog—blue; without a dog—orange); (**E**) Perceived impact (reduce, no impact, increase) of AAA on enjoyment (blue) and anxiety (orange) (Table S2). Statistically significant at the $p < 0.05$ level. Survey questions are in File S1.

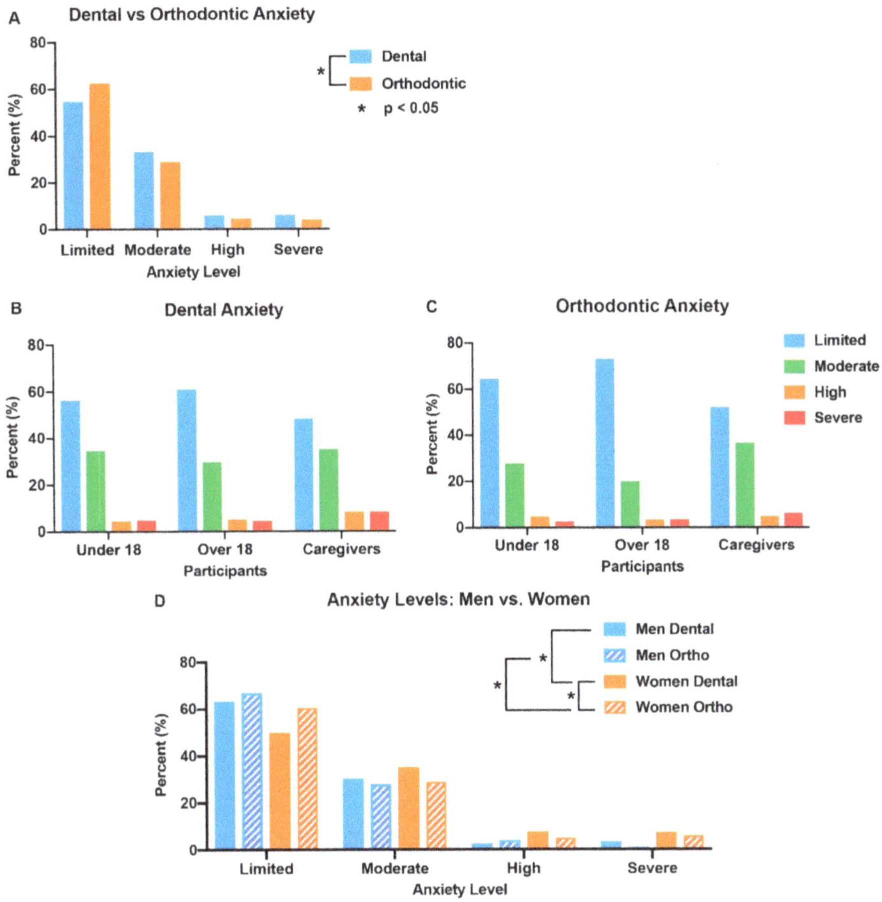

Figure 3. Dental and orthodontic anxiety. (**A**) Percentage of participants with limited, moderate, high, or severe anxiety (dental anxiety—blue; orthodontic anxiety—orange); (**B**) Participants (patients under 18, patients over 18, caregivers) with dental anxiety (limited—blue; moderate—green; high—orange; severe—red); (**C**) Participants (patients under 18, patients over 18, caregivers) with orthodontic anxiety (limited—blue; moderate—green; high—orange; severe—red); (**D**) Levels of dental (solid) and orthodontic (hatched) anxiety in males (blue) and females (orange). Dental and orthodontic anxiety determined by the Corah Dental Anxiety Scale (DAS) and modified orthodontic DAS, respectively (Tables 4 and S3).

3.3. Impact of the Covid-19 Pandemic

After the COVID-19 pandemic, anxiety levels increased due to pervasive uncertainty and fear of disease [24,25]. Despite reports that SARS-CoV-2 can be spread by dogs, concerns of contracting COVID-19 from a dog was low, with 88% of participants reporting "little" to "no concern" (Figure 4A, Table S4) [26–36]. Though most patients had little to no concern of contracting COVID-19 from dogs and the risk of zoonotic transmission is considered low, it is important for handlers and facilities with therapy animals to adhere to proper health protocols and the use of personal protective equipment, to minimize the risk of spread and ensure safety for both the canine and patients [27–36].

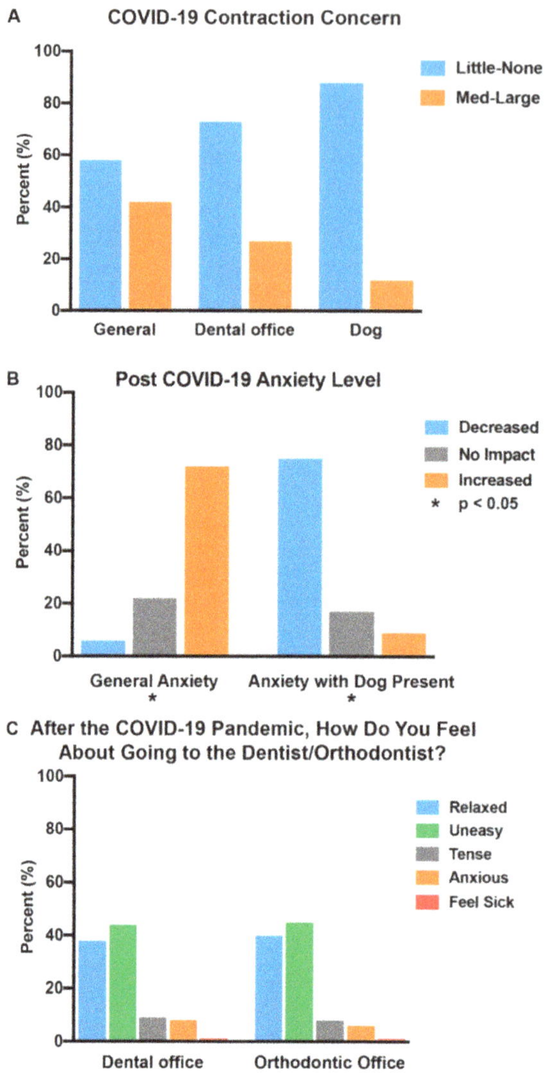

Figure 4. Concerns regarding COVID-19. (**A**) Percentage of participants responding "little concern" or "no concern" (pooled data, blue) versus "medium concern" and "large concern" (orange) with regard to contracting COVID-19 in general, at the dentist's office, or from a dog (Table S4); (**B**) Perceived impact of COVID-19 on general anxiety and anxiety with a dog present (decreased- blue; no impact—grey; increase—orange) (Table S5); (**C**) Concern regarding dental professionals after COVID-19 in a dental office or orthodontic office (relaxed—blue; uneasy—green; tense—grey; anxious—orange; feel sick—red) (Table S6).

While 72% of participants reported an increase in general anxiety during the pandemic, with adults and caregivers having greater reported anxiety than minors, 75% of respondents indicated that their anxiety would be reduced with a therapy dog in day-to-day life (Figure 4B, Tables S5 and S8). A sizable minority (42%) reported a "moderate" to "large" concern of contracting the virus in day-to-day life while only a quarter (27%) were concerned about contracting SARS-CoV-2 at a dental or orthodontic office (Figure 4A, Table S4). Meanwhile 40% of subjects felt "relaxed" about going to an orthodontist, while a plurality of participants reported feeling "uneasy" (44%) or "tense," "anxious" or "sick" (16%) going to a dentist or orthodontist after the outbreak (Figure 4C, Table S6). There were no differences in concern for the safety, allergies, and cleanliness of AAA, before and during the pandemic (Table S7).

4. Discussion

We found that most patients and caregivers would welcome AAA in orthodontics, consistent with AAA's widespread adoption in medicine [13]. The majority of participants indicated that AAA would alleviate anxiety and offer an enjoyable experience. Furthermore, the vast majority of caregivers and patients had "little" or "no concern" regarding cleanliness, allergens, or the safety of therapy animals in dental clinics, even during the pandemic. These results are consistent with our hypothesis and the medical literature, pointing to widespread acceptance of therapy animals in diverse health care settings [19,37]. For nearly half of the participants, an office with a therapy dog would influence their choice of provider, with the vast majority (92%) choosing the practice with a dog. Taken together, our findings suggest that incorporating therapy dogs in orthodontic practice could improve the patient experience, reduce barriers to care, and provide practice growth potential.

Dog owners and participants that took the survey with a dog present were more likely to select an office offering AAA and less likely to indicate concerns about therapy dogs, suggesting dog interactions allay fears of AAA. However, the dog's presence in the reception was unrelated to the anticipated anxiety scales for dentistry and orthodontics. This is possibly because participants had no structured interaction with the dog and did not undergo animal therapy; participants could choose to ignore the dog, and some were seated farther from the dog than others.

Across the patient groups, dental anxiety was more severe than orthodontic-related anxiety, consistent with the literature [38]. This may be due to the greater use of injections, hand pieces, and involved procedures in general dentistry [38]. Of those participants reporting orthodontic anxiety, 29% had moderate anxiety, and may benefit from anxiety-reducing interventions such as AAA. Moderate anxiety patients have stressors that can be managed in the dental clinic, while high anxiety patients require significant intervention such as anesthesia or medications [6]. Although adult patients reported lower orthodontic and dental-related anxiety than minor patients, one in five adults and one in three minors reported orthodontic-related anxiety, which is a significant fraction of patients.

When comparing across groups, minors indicated significantly higher anticipated levels of enjoyment in the presence of a therapy dog than the adults or caregivers, suggesting that the use of AAA in dental settings could be especially beneficial for pediatric patients.

Because the pandemic caused a marked increase in anxiety and depression, stress, mood disorders, and suicidal ideation, questions were added to assess the pandemic's effects [25,39]. Most participants (72%) reported an increase in general anxiety and thought a therapy animal would reduce the day-to-day stress and dental-related anxiety, suggesting that the expanded presence of therapy animals could be beneficial in difficult times, particularly for adult patients and caregivers who reported anxiety more often compared to minors, since the pandemic [25]. Similarly, pet owners reported increases in animal engagement for emotional support during the pandemic [40]. However, the pandemic interfered with AAA delivery, with marked reductions in therapy animal visits and team availability due to paused health care programming and fewer volunteers and dogs [40]. Health care providers need to enact protocols for re-launching AAA services post-pandemic, with careful attention paid to minimize the risks posed by dogs and SARS-CoV-2 [40,41]. Though the risk of zoonotic transmission involving dogs is low, it is important for therapy dog handlers and health care facilities to enact protocols to minimize the risk of SARS-CoV-2 spread; the Centers for Disease Control (CDC) and the American Veterinary Medical Association (AVMA) have issued useful guidelines for safe practices, cleaning, and personal protective equipment usage for canine and patient safety (27–36). Specific attention needs to also be directed at designing AAA protocols for dental clinics including accommodations for children with severe dog allergies or cynophobia.

We found no differences regarding the AAA concerns of cleanliness, allergies, or safety pre- and post-pandemic, with a small proportion (12%) of participants reporting concerns of contracting SARS-CoV-2 from dogs. This finding may be related to articles regarding transmission through pets, despite the CDC stating that the risk of human–canine transmission is low [26,36,42]. Furthermore, there is no evidence that the virus can spread from the skin, fur, or hair of pets [36].

Our response rate was 84.9%, in line with similar surveys (71%–84%), suggesting that there was no undue respondent burden [23,37]. Consistent with the general population, our sample reported a 4.1% prevalence of canine allergies and 22.9% prevalence of cynophobia including 5.9% that were "somewhat" or "very" afraid of dogs (Table 1) [43].

Limitations: Participants were consecutively enrolled during a fixed time window, with no power calculation guiding sample size. However, the sample size (n = 800) was within a range judged as "very good" (>500) and "above the acceptable range" (300–450) for surveys [44,45]. Data on dental anxiety may have been subject to recall bias, as patients were visiting an orthodontist, and not a general dentist. There was potential for selection bias, namely volunteer bias. Data were collected at one university with its therapy dog; this sampling bias may influence generalizability to other regions and private practices. Private practices were not included due to the pandemic's state-wide ban on non-facility dogs. Response bias may have occurred, specifically social desirability bias, whereby participants reported the desirable outcome of positive feelings toward dogs. It was infeasible for our team to re-validate the orthodontic DAS, but modifications to the DAS were pre-tested, revised, and guided by a survey expert.

Future directions include evaluating provider perspectives and enrolling patients from other regions, specialties, and private practice. Studies investigating the effects of AAA in dentistry are needed to guide protocol development and implementation. AAA has been widely adopted in medicine; dentistry and orthodontics are the next frontier due to the high prevalence of dental anxiety and AAA's potential to mitigate stress with few perceived risks and broad patient acceptance [2–5,38].

5. Conclusions

- Over a third of patients under 18 have a level of orthodontic anxiety that could benefit from interventions such as AAA.
- The majority of patients and caregivers believe dental AAA will reduce anxiety and boost enjoyment.
- For nearly half of the participants, an office with a therapy dog would influence their choice of provider, with most (92%) choosing the dog.
- The majority of participants were unconcerned with the potential allergies (81%), safety risks (89%), and cleanliness (83%) of the therapy dogs.
- AAA could be a valuable practice builder and promising anxiety-management tool welcomed by most families.

Supplementary Materials: The following supporting information can be downloaded at: https://www.mdpi.com/article/10.3390/ani12141862/s1, Table S1: Inclusion and exclusion criteria; Table S2: AAA's impact on anxiety and enjoyment; Table S3: Corah Dental and Orthodontic Modified Anxiety Scales Numerical; Table S4: COVID-19 concerns regarding the contraction of the virus; Table S5: The impact of COVID-19 on stress and AAA; Table S6: Feelings toward going to a dental office after the COVID-19 pandemic; Table S7: Concerns related to animal assisted activity (AAA) pre- and post-pandemic shutdown; Table S8: Multiple comparison test (MCT) results by groups; File S1: Qualtrics survey questions and text for patients and caregivers; File S2: Dental anxiety indices.

Author Contributions: Project administration, K.C.; Investigation, K.C., C.B., and L.J.; Funding acquisition, K.C., C.G., A.R., and L.J.; Conceptualization, K.C., T.J., T.S. (Timothy Strauman), and L.J.; Formal analysis, K.C., C.W., and L.J.; Visualization, K.C., C.B., T.S. (Taylor Sulkowski), and C.G.;

Writing—original draft, K.C., C.B., and L.J.; Writing—review and editing, K.C., C.B., T.S. (Taylor Sulkowski), C.G., N.G., A.R., T.J., K.D., C.W., T.S. (Timothy Strauman), and L.J.; Supervision, L.J. All authors have read and agreed to the published version of the manuscript.

Funding: This research was funded by the Southern Association of Orthodontics Research Award [to K.C. and A.R.], the NIH NIDCR K08 grant to L.J. (K08DE030235), the NIH NCATS grant to L.J. (UL1TR002489), and the American Association of Orthodontics Foundation (AAOF) Junior Faculty Fellowship [to L.J.] and the Resident Research Award [to K.C. and A.R.].

Institutional Review Board Statement: The study was conducted in accordance with the Declaration of Helsinki, and approved by the Institutional Review Board of the University of North Carolina (IRB #19-1908, 10/29/2019).

Informed Consent Statement: Informed consent was obtained from all subjects involved in the study.

Data Availability Statement: The data presented in this study are available within this article and the Supplementary Materials.

Acknowledgments: We thank Teresa Edwards at the UNC Odum Institute for Research in Social Science for her valuable guidance in the survey development. We appreciate the advice of Eric Hodges and Siggi Saemundsson, DDS. We thank Samantha Jhingree for her support and engagement in the project. We acknowledge our furry friends who have kept us sane and loved throughout this pandemic.

Conflicts of Interest: The authors declare no conflict of interest. The funders had no role in the design of the study; in the collection, analyses, or interpretation of data; in the writing of the manuscript, or in the decision to publish the results.

References

1. White, A.M.; Giblin, L.; Boyd, L.D. The Prevalence of Dental Anxiety in Dental Practice Settings. *J. Dent. Hyg.* **2017**, *91*, 30–34. [PubMed]
2. Grisolia, B.M.; dos Santos, A.P.P.; Dhyppolito, I.M.; Buchanan, H.; Hill, K.; Oliveira, B.H. Prevalence of Dental Anxiety in Children and Adolescents Globally: A Systematic Review with Meta-Analyses. *Int. J. Paediatr. Dent.* **2021**, *31*, 168–183. [CrossRef] [PubMed]
3. Locker, D.; Liddell, A.; Dempster, L.; Shapiro, D. Age of Onset of Dental Anxiety. *J. Dent. Res.* **1999**, *78*, 790–796. [CrossRef] [PubMed]
4. Klingberg, G.; Broberg, A.G. Dental Fear/Anxiety and Dental Behaviour Management Problems in Children and Adolescents: A Review of Prevalence and Concomitant Psychological Factors. *Int. J. Paediatr. Dent.* **2007**, *17*, 391–406. [CrossRef] [PubMed]
5. McGrath, C.; Bedi, R. The Association between Dental Anxiety and Oral Health-Related Quality of Life in Britain. *Community Dent. Oral. Epidemiol.* **2004**, *32*, 67–72. [CrossRef]
6. American Academy of Pediatric Dentistry Behavior Guidance for the Pediatric Dental Patient. *Pediatr. Dent.* **2018**, *40*, 254–267.
7. Coté, C.J.; Wilson, S.; American Academy of Pediatrics; American Academy of Pediatric Dentistry. Guidelines for Monitoring and Management of Pediatric Patients Before, During, and After Sedation for Diagnostic and Therapeutic Procedures. *Pediatrics* **2019**, *143*, e20191000. [CrossRef]
8. Fleming, P.S. Timing Orthodontic Treatment: Early or Late? *Aust. Dent. J.* **2017**, *62* (Suppl. 1), 11–19. [CrossRef]
9. American Veterinary Medical Association. Animal-Assisted Interventions: Definitions. Available online: https://www.avma.org/resources-tools/avma-policies/animal-assisted-interventions-definitions (accessed on 7 July 2022).
10. Marcus, D.A. The Science behind Animal-Assisted Therapy. *Curr. Pain Headache Rep.* **2013**, *17*, 322. [CrossRef]
11. Braun, C.; Stangler, T.; Narveson, J.; Pettingell, S. Animal-Assisted Therapy as a Pain Relief Intervention for Children. *Complement. Ther. Clin. Pract.* **2009**, *15*, 105–109. [CrossRef]
12. Marcus, D.A.; Bernstein, C.D.; Constantin, J.M.; Kunkel, F.A.; Breuer, P.; Hanlon, R.B. Animal-Assisted Therapy at an Outpatient Pain Management Clinic. *Pain Med.* **2012**, *13*, 45–57. [CrossRef] [PubMed]
13. Kamioka, H.; Okada, S.; Tsutani, K.; Park, H.; Okuizumi, H.; Handa, S.; Oshio, T.; Park, S.-J.; Kitayuguchi, J.; Abe, T.; et al. Effectiveness of Animal-Assisted Therapy: A Systematic Review of Randomized Controlled Trials. *Complement. Ther. Med.* **2014**, *22*, 371–390. [CrossRef]
14. Filan, S.L.; Llewellyn-Jones, R.H. Animal-Assisted Therapy for Dementia: A Review of the Literature. *Int. Psychogeriatr.* **2006**, *18*, 597–611. [CrossRef] [PubMed]
15. Nammalwar, R.B.; Rangeeth, P. A Bite out of Anxiety: Evaluation of Animal-Assisted Activity on Anxiety in Children Attending a Pediatric Dental Outpatient Unit. *J. Indian Soc. Pedod. Prev. Dent.* **2018**, *36*, 181–184. [CrossRef] [PubMed]
16. Cruz-Fierro, N.; Vanegas-Farfano, M.; González-Ramírez, M.T. Dog-Assisted Therapy and Dental Anxiety: A Pilot Study. *Animals* **2019**, *9*, 512. [CrossRef] [PubMed]

17. Havener, L.; Gentes, L.; Thaler, B.; Megel, M.E.; Baun, M.M.; Driscoll, F.A.; Beiraghi, S.; Agrawal, S. The Effects of a Companion Animal on Distress in Children Undergoing Dental Procedures. *Issues Compr. Pediatr. Nurs.* **2001**, *24*, 137–152. [CrossRef]
18. Charowski, M.; Wells, M.H.; Dormois, L.; Fernandez, J.A.; Scarbecz, M.; Maclin, M. A Randomized Controlled Pilot Study Examining Effects of Animal Assisted Therapy in Children Undergoing Sealant Placement. *Pediatr. Dent.* **2021**, *43*, 10–16.
19. Gupta, N.; Yadav, T. Parents' Acceptance and Their Children's Choice of Pet for Animal-Assisted Therapy (A.A.T.) in 3- to 12-Year-Old Children in the Dental Operatory—A Questionnaire-Based Pilot Study. *Int. J. Paediatr. Dent.* **2018**, *28*, 373–379. [CrossRef]
20. Gussgard, A.M.; Weese, J.S.; Hensten, A.; Jokstad, A. Dog-Assisted Therapy in the Dental Clinic: Part A-Hazards and Assessment of Potential Risks to the Health and Safety of Humans. *Clin. Exp. Dent. Res.* **2019**, *5*, 692–700. [CrossRef]
21. Lefebvre, S.L.; Golab, G.C.; Christensen, E.; Castrodale, L.; Aureden, K.; Bialachowski, A.; Gumley, N.; Robinson, J.; Peregrine, A.; Benoit, M.; et al. Guidelines for Animal-Assisted Interventions in Health Care Facilities. *Am. J. Infect. Control* **2008**, *36*, 78–85. [CrossRef]
22. Corah, N.L.; Gale, E.N.; Illig, S.J. Assessment of a Dental Anxiety Scale. *J. Am. Dent. Assoc.* **1978**, *97*, 816–819. [CrossRef] [PubMed]
23. Humphris, G.; Crawford, J.R.; Hill, K.; Gilbert, A.; Freeman, R. UK Population Norms for the Modified Dental Anxiety Scale with Percentile Calculator: Adult Dental Health Survey 2009 Results. *BMC Oral Health* **2013**, *13*, 29. [CrossRef] [PubMed]
24. Coelho, C.M.; Suttiwan, P.; Arato, N.; Zsido, A.N. On the Nature of Fear and Anxiety Triggered by COVID-19. *Front. Psychol.* **2020**, *11*, 581314. [CrossRef]
25. Xiong, X.; Wu, Y.; Fang, X.; Sun, W.; Ding, Q.; Yi, Y.; Huang, Y.; Gong, J.; Liu, J.; Wang, J. Mental Distress in Orthodontic Patients during the Coronavirus Disease 2019 Pandemic. *Am. J. Orthod. Dentofac. Orthop.* **2020**, *158*, 824–833. [CrossRef]
26. Jason, D. Can You Get Covid-19 From Your Dog? Here's What Disease Experts Say. Available online: https://www.wsj.com/articles/can-you-get-covid-19-from-your-dog-heres-what-disease-experts-say-11599494410 (accessed on 5 April 2022).
27. World Organization for Animal Health. Infection with SARS-CoV-2 in Animals. Available online: https://www.woah.org/app/uploads/2021/11/en-factsheet-sars-cov-2-20211025.pdf (accessed on 11 July 2022).
28. Sila, T.; Sunghan, J.; Laochareonsuk, W.; Surasombatpattana, S.; Kongkamol, C.; Ingviya, T.; Siripaitoon, P.; Kositpantawong, N.; Kanchanasuwan, S.; Hortiwakul, T.; et al. Suspected Cat-to-Human Transmission of SARS-CoV-2, Thailand, July–September 2021. *Emerg. Infect. Dis.* **2022**, *28*, 1485–1488. [CrossRef] [PubMed]
29. Centers for Disease Control and Prevention. Animals and COVID-19. Available online: https://www.cdc.gov/coronavirus/2019-ncov/daily-life-coping/animals.html (accessed on 11 July 2022).
30. American Veterinary Medical Association. SARS-CoV-2 in Animals. Available online: https://www.avma.org/resources-tools/one-health/covid-19/sars-cov-2-animals-including-pets (accessed on 11 July 2022).
31. Sit, T.H.C.; Brackman, C.J.; Ip, S.M.; Tam, K.W.S.; Law, P.Y.T.; To, E.M.W.; Yu, V.Y.T.; Sims, L.D.; Tsang, D.N.C.; Chu, D.K.W.; et al. Infection of Dogs with SARS-CoV-2. *Nature* **2020**, *586*, 776–778. [CrossRef]
32. Bosco-Lauth, A.M.; Hartwig, A.E.; Porter, S.M.; Gordy, P.W.; Nehring, M.; Byas, A.D.; VandeWoude, S.; Ragan, I.K.; Maison, R.M.; Bowen, R.A. Experimental Infection of Domestic Dogs and Cats with SARS-CoV-2: Pathogenesis, Transmission, and Response to Reexposure in Cats. *Proc. Natl. Acad. Sci. USA* **2020**, *117*, 26382–26388. [CrossRef]
33. Patterson, E.I.; Elia, G.; Grassi, A.; Giordano, A.; Desario, C.; Medardo, M.; Smith, S.L.; Anderson, E.R.; Prince, T.; Patterson, G.T.; et al. Evidence of Exposure to SARS-CoV-2 in Cats and Dogs from Households in Italy. *Nat. Commun.* **2020**, *11*, 6231. [CrossRef]
34. Calvet, G.A.; Pereira, S.A.; Ogrzewalska, M.; Pauvolid-Corrêa, A.; Resende, P.C.; de Tassinari, W.S.; de Costa, A.P.; Keidel, L.O.; da Rocha, A.S.B.; da Silva, M.F.B.; et al. In-vestigation of SARS-CoV-2 Infection in Dogs and Cats of Humans Diagnosed with COVID-19 in Rio de Janeiro, Brazil. *PLoS ONE* **2021**, *16*, e0250853. [CrossRef]
35. Centers for Disease Control and Prevention. Guidance for Handlers of Service and Therapy Animals. Available online: https://www.cdc.gov/healthypets/covid-19/service-therapy-animals.html (accessed on 11 July 2022).
36. Centers for Disease Control and Prevention. What You Should Know about COVID-19 and Pets. Available online: https://www.cdc.gov/healthypets/covid-19/pets.html#:~{}:text=Pets%20worldwide%2C%20including%20cats%20and,19%20to%20people%20is%20low (accessed on 11 July 2022).
37. Vincent, A.; Easton, S.; Sterman, J.; Farkas, K.; Heima, M. Acceptability and Demand of Therapy Dog Support Among Oral Health Care Providers and Caregivers of Pediatric Patients. *Pediatr. Dent.* **2020**, *42*, 16–21.
38. Roy, J.; Dempster, L.J. Dental Anxiety Associated with Orthodontic Care: Prevalence and Contributing Factors. *Semin. Orthod.* **2018**, *24*, 233–241. [CrossRef]
39. Murata, S.; Rezeppa, T.; Thoma, B.; Marengo, L.; Krancevich, K.; Chiyka, E.; Hayes, B.; Goodfriend, E.; Deal, M.; Zhong, Y.; et al. The Psychiatric Sequelae of the COVID-19 Pandemic in Adolescents, Adults, and Health Care Workers. *Depress. Anxiety* **2021**, *38*, 233–246. [CrossRef] [PubMed]
40. Ng, Z.; Griffin, T.C.; Braun, L. The New Status Quo: Enhancing Access to Human–Animal Interactions to Alleviate Social Isolation & Loneliness in the Time of COVID-19. *Animals* **2021**, *11*, 2769. [CrossRef]
41. Meers, L.L.; Contalbrigo, L.; Stevens, V.A.; Ulitina, O.M.; Laufer, S.J.; Samuels, W.E. The State of Animal-Assisted Interventions: COVID-19 Safety Protocols and Ethical Considerations. *J. Appl. Anim. Ethics Res.* **2021**, *3*, 103–125. [CrossRef]

42. Kiros, M.; Andualem, H.; Kiros, T.; Hailemichael, W.; Getu, S.; Geteneh, A.; Alemu, D.; Abegaz, W.E. COVID-19 Pandemic: Current Knowledge about the Role of Pets and Other Animals in Disease Transmission. *Virol. J.* **2020**, *17*, 143. [CrossRef] [PubMed]
43. Brunekreef, B.; Groot, B.; Hoek, G. Pets, Allergy and Respiratory Symptoms in Children. *Int. J. Epidemiol.* **1992**, *21*, 338–342. [CrossRef] [PubMed]
44. Comrey, A.L.; Lee, H.B. *A First Course in Factor Analysis*, 2nd ed.; Lawrence Erlbaum: Hillsdale, NJ, USA, 1992; Volume XII, ISBN 9781315827506.
45. Guadagnoli, E.; Velicer, W.F. Relation of Sample Size to the Stability of Component Patterns. *Psychol. Bull.* **1988**, *103*, 265–275. [CrossRef]

animals

MDPI

Article

"An Aid with Soul"—Understanding the Determinants of Guide Dog-Owner Compatibility from Qualitative Interviews

Yana Bender [1,2,*], Tim Matschkowski [1], Stefan R. Schweinberger [2] and Juliane Bräuer [1,2]

1 Max Planck Institute of Geoanthropology, DogStudies, Kahlaische Strasse 10, 07745 Jena, Germany; matschkowski@shh.mpg.de (T.M.); juliane.braeuer@uni-jena.de (J.B.)

2 Department of General Psychology and Cognitive Neuroscience, Friedrich Schiller University, Leutragraben 1, 07743 Jena, Germany; stefan.schweinberger@uni-jena.de

* Correspondence: bender@shh.mpg.de

Simple Summary: Guide dogs can help visually impaired persons to feel more confident and independent. Twenty-one guide dog owners reported the following factors to be important for a good match between a dog and an owner: sharing hobbies, similar activity levels or higher activeness in dogs, similar expressions of calmness; happiness; greediness; and friendliness. Owners also felt like a good match with their dog when they were both open or their dogs were more open than themselves and when they were dominant personalities and their dogs were more submissive. Moreover, the relationship to a former guide dog can have a big impact on the next relationship. Owners who felt similar in their personality to their dogs, as well as owners who felt like a good match with their dogs, reported positive aspects such as a strong bond and less influence from previous relationships. However, a strong bond might sometimes also have negative effects. The findings can help to understand what makes a dog and an owner a good match and improve the matching processes of guide dogs and handlers.

Abstract: Guide dogs hold the potential to increase confidence and independence in visually impaired individuals. However, the success of the partnership between a guide dog and its handler depends on various factors, including the compatibility between the dog and the handler. Here, we conducted interviews with 21 guide dog owners to explore determinants of compatibility between the dog and the owner. Experienced compatibility between the dog and the owner was associated with positive relationship aspects such as feeling secure with the dog. Certain characteristics emerged as subjective determinants of compatibility, including shared hobbies, high levels of openness in both or only the dog, similar activity levels and higher activeness in dogs, similar expressions of calmness; happiness; greediness; friendliness; and a complementary dominance–submissiveness relationship. Owners who perceived themselves to be similar in their personality to their dogs often reported to have a strong bond, to feel secure with their dog and to be less influenced by previous relationships. However, our results suggest that a strong bond between the dog and the owner does not exclusively yield positive effects. Moreover, prior dog ownership seems to have a potentially strong impact on the subsequent relationship. Our results contribute to the understanding of dog–owner compatibility and may improve the matching process of guide dogs and their prospective handlers.

Keywords: human–animal bond; guide dogs; dog–owner compatibility; personality

Citation: Bender, Y.; Matschkowski, T.; Schweinberger, S.R.; Bräuer, J. "An Aid with Soul"—Understanding the Determinants of Guide Dog-Owner Compatibility from Qualitative Interviews. *Animals* **2023**, *13*, 2751. https://doi.org/10.3390/ani13172751

Academic Editor: Lynette A. Hart

Received: 25 July 2023
Revised: 18 August 2023
Accepted: 24 August 2023
Published: 29 August 2023

1. Introduction

The bond between humans and dogs is widely recognized as a unique and significant relationship [1]. Over the past decade, research has shed light on the remarkable cooperative tendencies displayed by dogs. Dogs show several prosocial behaviors, such as sharing food and informing, when they receive cues signaling the need for help [2]. This motivation for cooperation, paired with reward-based training, has fostered dogs helping humans in different areas of life, such as smelling scents of missing persons, assisting disabled people with opening doors and picking up objects or guiding blind people. However, those dog–human dyads are not always successful in their work, and neither are owners always satisfied with the relationship they share with their (working) dogs. This is reflected in the current high number of relinquishment of pet dogs in Western countries [3]. This is not only the case for family dogs but also for working dogs where the relationship between the dog and the handler may be unsuccessful. Lloyd and colleagues [4], for example, found that 36% of all dogs were returned to the guide dog training establishment before reaching retirement age (in a sample of N = 118 teams). This high return rate is associated with both economic and personal costs. The intensive training that the dogs have undergone is very expensive; assistance dogs can cost up to USD 50,000 (according to the National Service Animal Registry, 2019 [5]). In these dyads, where humans rely on and trust their dogs' abilities to solve particular tasks (such as in assistance dogs but also police or rescue dogs), one can imagine that failure also may have tremendous social and personal consequences, including life-threatening ones in extreme cases. In addition to a sound education of the dogs and constant training, we here consider the matching of a working dog to its prospective handler as a factor for the development of functionality and satisfaction in a dyad [6].

1.1. *Determinants of Success in Dog–Owner Dyads*

The general literature background on the determinants of success in dog–owner dyads is still considerably small. Existing papers tend to stress uncontrollable aspects like the age and profession of the owner, the area of residence or the number of household members [7–9], which generally impact ownership in either a negative or a positive way. Aspects that can be used to actively form functioning dog–human dyads especially include personality aspects [10,11]. There is a substantial body of evidence suggesting that dog personality can be assessed along similar dimensions or traits that have been established to assess human personality traits [12,13]. In addition, we already know that the majority of dog–owner pairs resemble each other in their personality traits [14,15]. However, not much is known about the consequences this might have on functionality and satisfaction in the relationship. While it seems plausible that pairings of those similar individuals are especially functional, an alternative possibility could be that complementary traits favor the development of a functional team. Moreover, the particular attachment style of the dog and the owner, as well as the general quality of the bond, are factors that interact with these personality aspects and influence the performance of a dyad [10,16,17]. To date, no study has investigated these five potential aspects of a successful match (performance, dog and owner personality traits, quality of the bond and attachment aspects) and their interaction.

1.2. *Functionality in Guide Dog–Owner Pairs*

Besides this knowledge on general dog–human relationships, additional aspects might play a role when analyzing the success of working dogs and their handlers. The functionality of cooperation to solve different tasks determines their success on many levels, potentially also their bond and satisfaction within the general relationship. Thus, the importance of assigning the "right" dog to the "right" owner might even play a more significant role than in family dogs and their owners. Guide dogs are probably the type of working dogs that people trust their lives to the most, as reflected in their training, which lasts several months and, unlike for many other working dogs, usually is not performed by the owners [18]. While it is clear that the degree of compatibility between the human and the (guide) dog impacts the therapeutic value of the partnership [19], the few existing studies on the compatibility of guide dog–owner teams lack a viewpoint that includes detailed consideration of the personality traits of both counterparts [19–21]. Lloyd and colleagues [22] recently showed that within a sample of 50 New Zealand guide dog owners, the average subjective compatibility was very high. However, the determinants of this compatibility remain largely unexplored. Problems in research on guide dogs resemble those on family dog–owner matching (see above), including a small volume of research.

Existing studies emphasize contextual factors and mobility factors [6,23]. Mobility is certainly important and is already considered by guide dog trainers; even evaluated processes to match guide dogs to their handlers, like the *Orientation and Mobility Outcomes (OMO)* tool, exist. [20]. Concerning contextual factors, a recent study has identified four essential areas: societal, social support, environmental and personal factors [23]. Many of these factors were outside the handler and the guide dog organization's control, emphasizing the importance of research focusing on more modifiable issues of the dog–human relationship.

The widespread belief that unsuccessful relationships are caused by aspects of the dog (e.g., inadequate training or behavior in general) can be disproved by the observation that mismatched dogs can often be successfully rematched [4]. This suggests that a successful relationship depends on the interaction between the handler and the dog, instead of solely being one party's fault. We therefore focus the interviews in the current study on the subjective fit that handlers feel between themselves and their dogs, thus exploring the potential power of matching processes. Our focus furthermore lies on aspects other than immutable contextual or already extensively studied mobility factors, such as personality traits, which represent a promising approach based on family dog research. Another aspect that will be considered is that former relationships have been shown to impact the subsequent relationships with the next guide dogs [24]. As an additional point, we included the potential influence of an existing or non-existing dog affinity (meaning that persons have general positive feelings toward and a preference for dogs) when acquiring the guide dog, which has not yet been investigated.

1.3. Research Gap and Study Rationale

A functional relationship between the owner and the guide dog increases subjective confidence and independence in blind persons and leads to better social relationships [25]. Consequently, there is a high relevance of research on what favors this functionality, especially in processes that are influenceable, such as the matching process. Compatibility between dogs and humans in these dyads is not yet well researched, as discussed above. There is a specific lack of studies that focus on the personality of both partners. Our aim is therefore to find determinants of compatibility between the dog and the owner. In order to do so, we conducted semi-structured interviews with guide dog owners. Guide dog owners often have many years of experience with different dogs and can be expected to have good experiential knowledge regarding the determinants that have affected the quality of different relationships. The methodological approach of conducting semi-structured interviews furthermore allows for freely exploring other psychological determinants of compatibility.

2. Materials and Methods

2.1. Participant Recruitment and Demographics

We conducted expert interviews with 21 guide dog owners on what makes guide dogs for the blind and their owners a compatible, successful team. The interviews were realized in the setting of an extensive quantitative study whose subjects also include guide dog owners. We recruited participants through expert contacts (assistance and guide dog trainers, club chairmen) of the DogStudies Lab at the MPI GEA in Jena, who spread word of the study via newsletters, WhatsApp and Facebook groups, as well as calls in the local press. Participants had to be visually impaired or blind guide dog owners over 18 years old. After interested parties had contacted us by mail, a telephone appointment was arranged. The interviewer (Y.B.) explained the procedure and the possibility of participating in an interview. Of the 24 participants in the quantitative project, 21 owners agreed to take part and were interviewed in German between July 2022 and January 2023. Men constituted 19% (N = 4) of the sample, and ages ranged from 28 to 69 years (M = 54.1 years; for demographical details, see Table 1).

Table 1. Participants demographics.

Number	Gender	Age (Years)	Impairment	Interview Location	Dog's Age (Years)	Dog's Sex	Time Together	Guide Dog History
1	F	57	Fully blind	Telephone	9	F	7 years	3rd
2	F	67	Fully blind	Telephone	8	F	7.5 years	2nd
3	F	47	Fully blind	In person	3	M	1.5 years	2nd
4	F	45	Partially blind	Telephone	3	M	1.5 years	3rd
5	F	69	Partiallly blind	Telephone	2	M	8 months	4th
6	F	36	Fully blind	In person	6.5	F	4.5 years	2nd
7	F	28	Fully blind	In person	4	F	2 years	1st
8	F	61	Partiallly blind	In person	7	M	5 years	1st
9	F	67	Fully blind	Telephone	10	M	8 years	3rd
10	F	51	Partially blind	In person	3	M	1.5 years	3rd
11	M	55	Partiallly blind	Telephone	4	F	2 years	1st
12	F	64	Partiallly blind	In person	5	F	3 years	1st
13	M	46	Partiallly blind	In person	8	F	6 years	2nd
14	F	36	Fully blind	In person	6.5	F	3.5 years	2nd
15	M	65	Partiallly blind	Telephone	5	F	2.5 years	1st
16	F	59	Partiallly blind	Telephone	2.5	F	11 months	3rd
17	F	51	Partiallly blind	In person	7.5	F	3.5 years	4th
18	M	59	Partially blind	In person	11	M	9 years	1st
19	F	56	Fully blind	In person	3	M	5 months	1st
20	F	65	Partially blind	Telephone	6	M	4 years	1st
21	F	52	Partially blind	In person	7	F	2 years	1st

2.2. Ethics

The study received approval by the Max Planck Ethics Council on 27 June 2022 (Application No: 2022_12).

2.3. Realization of the Interviews

All interviews started with an introduction by the interviewer. She explained that the topic will be the relationship between dog and human and the subjectively experienced match between the two. Furthermore, she stated that the interviewee could give their information in as much detail or as concisely as they felt comfortable. A privacy statement was signed at the beginning of the quantitative behavioral study. We recorded the interviews via the Apple app "Voice Memos". Twelve of the interviews could be realized in face-to-face live setting, while nine had to be conducted via telephone due to logistic and time reasons. The interview length ranged from 12.20 min to 45.15 min, with an average time of 25.19 min.

2.4. Questionnaire Design

Based on the state of research and our background knowledge, we focused on four aspects:

- The subjectively experienced match with their dog (based on personality traits);
- Relationship parameters such as satisfaction, initial expectations and problems;
- The comparison to and influence of other (former) dog–human relationships;
- The influence of dog affinity.

According to Kuckartz [26], we tried to balance the semi-structured questionnaire between questions based on the state of research and an open question format, to profit from the participants' expertise and not influence them in their answers. This approach resulted in a minimum of nine and a maximum of twelve questions, depending on whether participants had owned another guide dog before and experienced problems within the relationship (see Table 2). The initial questions were more general and subsequently got more detailed, but note that the order could also be varied according to the interview flow, due to the semi-structured format. At the end of the interview, the participants also had the opportunity to add further aspects relating to the topic they felt were important. The interviewer sometimes asked more detailed questions about the differences between off- and at-work relationship, as major differences in these distinctive settings exist [27].

Table 2. Questions of the semi-structured interview.

Question 1	Please describe your relationship with your current guide dog and how satisfied you are with it.
Question 2	Do you think you and your guide dog are a good match?
Question 3	Which of your personality traits do you think are a particularly good match? Example: openness to new experiences, agreeableness, and extraversion (with short explanations).
Question 4	Which of your personality traits do you think are less compatible?
Question 5	Are there any characteristics you would change in your dog if you could?
Question 6	What problems do you experience in your daily life with your dog that have to do with your dog's behavior or the relationship between you?
Question 7	Do you have any idea where these problems might be coming from?
Question 8	Have the expectations you had for your guide dog been met?
Question 9	Did you have another guide dog before your current guide dog, and if so, please tell me about your relationship with him.
Question 10	Do you feel you matched better or worse with your previous guide dog?—What could have been the reason for this?
Question 11	Do you feel that the previous relationship affected your current relationship?
Question 12	Before living with your guide dog, would you have described yourself as a dog person, or did you like dogs?

2.5. Data Processing

All interviews were transcribed via the software f4transcript [28]. Afterward, transcripts were manually checked and adapted according to Kuckartz [26]. For all analyses, the original German language was kept. The data were evaluated using qualitative content analysis according to Kuckartz [26], which ties in with Mayring's content analysis [29]. The individual phases and the illustration in the form of a cycle enable a comprehensible and clear data evaluation (see Figure 1). Qualitative content analysis is characterized by the formation of categories that are worked out both inductively and deductively. Throughout the entire process as shown in Figure 1, an iterative and cyclical approach was employed to ensure a comprehensive and high-quality analysis of the interview texts.

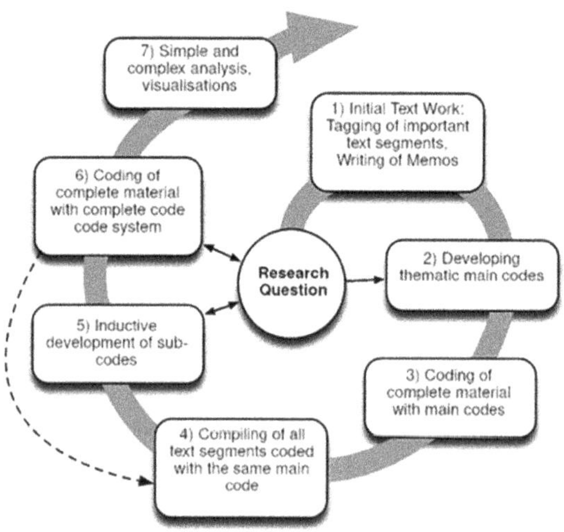

Figure 1. Process of a content structuring content analysis as presented by Kuckartz [30].

T.M. coded all interviews two times, firstly based on the a priori evaluated codes based on the questionnaires. T.M. and Y.B. then discussed the sub-codes based on the participants' answers and defined the coding criteria. T.M. coded the data a second time, and afterward, Y.B. went through all of them again. Differing cases (about 20) were discussed within the research team. Table 3 shows the eight main and 13 sub-codes that resulted from this approach. All text passages relevant to the research question were assigned to one of these codes.

Table 3. Main- and sub-codes used for coding the interviews.

1 Compatibility	**4 Compatibility with previous guide dogs**
1.1 Specific personality traits	4.1 Specific personality traits
1.2 General temperament	4.2 General temperament
1.3 General activity/energy level	4.3 General activity/energy level
1.4 General similarity	4.4 General similarity in personality
1.5 General difference	4.5 General difference in personality
2 Bond and relationship aspects	**5 Influence of previous guide dog relationship**
2.1 Positive attributes of the dog	**6 Dog-affinity**
2.2 Shared experiences with the dog	**7 Effects on social life**
2.3 Overall importance of the dog	**8 Problems in everyday life**
3 Expectations and fulfillments	

The analyses were again performed against the background of the methodology by Kuckartz [26] and with the analysis software f4analyse [31]. For this, the following analyses were carried out and written up in a logical order: category-based evaluation of main categories, correlations of subcategories within a main category and correlations between main categories. Besides analyses between the categories, other demographic aspects and shared features were considered in the next analysis step as suggested by Kuckartz [26].

3. Results and Discussion

3.1. Compatibility—Most Owners Feel like a Good Match for Various Reasons

Table 4 presents an overview of the main results of compatibility. Nineteen of twenty-one owners felt they were a good match with their dog, whereas two owners felt themselves and their dog were rather not a good match. Out of the 19 owners with a well-matched dog, most felt they and their dog were similar in their personality (14/19), and about one-quarter of them felt they were different (5/19). Of all participants, six owners (6/21) reported their dogs would adapt to them in some aspects. Four owners (4/21) reported their relationship with their dog was "love at first sight", all of them continued to believe they were a good match (4/4), and most of them felt more similar in their personality (3/4). These four participants also reported that their expectations for their guide dog were met or exceeded.

Table 4. Overview of compatibility.

Good match: $N = 19$	Different: 5/19
	Similar: 14/19
Mismatch: $N = 2$	Different: $\frac{1}{2}$ Similar: $\frac{1}{2}$
Love at first sight: $N = 4$	Good match: 4/4 Similar: $\frac{3}{4}$ Expectations met or exceeded: 4/4

Three owners (3/21) did not name any specific personality traits that they identified as matching or not matching with their dog. Research classifies these difficulties as a common phenomenon and that assessing one's own personality is biased by egoistic and moralistic biases and is not always easy, or accurate [32–34]. Most of those owners who could not name precise personality traits (2/3) did not feel like a good match with their dog.

Many owners reported that they share certain characteristics that make them and their dog a good match. These read as follows: both are friendly (mentioned by three), greedy (mentioned by two), happy (mentioned by two), reserved, interested, cuddly, humorous, physical, loyal, scattered or both show no pain. For example, participant 6 said "My dog is friendly, greedy, and likes to swim. All characteristics that I have as well." Comparable to the result regarding friendliness and happiness, Bauer and Woodward [16] found an expression of the trait "warmth" to predict owners' attachment to the pet and satisfaction with the human–animal bond. Moreover, a study by Curb et al. [14] supports these results, in which four out of eight similarity characteristics between the owner and the dog were associated with owner satisfaction. Similarly, strong openness resp. extraversion in both was important for six participants. Four of them said they and their dog were open and approached other people together, while two described that they themselves approached other people openly, and their dogs did the same with other dogs. Participant 18 described it like this: "When someone comes up to us in town, he first approaches everyone in a friendly manner. He also offers his friendship to every dog. And that's actually how I am, too." Indeed, research supports the importance of openness in owners—it was found to be generally associated with greater attachment between the dog and the owner [35]. Two owners described that their dogs' openness was a door opener to social contact with others for them, as they tended not to be open themselves, and another said it was a good fit that her dog was more open and curious than she was. This positive impact of high openness in dogs on relationship satisfaction has been observed by Cavanaugh and colleagues [36] before. Also, an explanatory approach can be found within the social support hypothesis [37], which proposes that companion animals act as facilitators of social interactions between other human beings and provide social support themselves [38]. If the dog expresses an open attitude toward other humans, this effect could be reinforced.

In terms of complementary personality traits, the following results emerged. Two participants said they had the dominance their dog needed. This also is supported by previous research as Bauer and Woodward [16] reported the combination of submissiveness and dominance to be linked to higher attachment of the dog and the owner. Another participant reported that his dog and he complemented each other well because his dog grasped situations more quickly than he did. One owner described that she herself was often nervous and that her dog's calm nature helped her to calm down more quickly. Similarly, participant 21 reported that her dog's calm manner helped her overcome her anxiety: "Well, I am naturally a very fearful person (...) but (name of the dog) was supposed to show me that it can be done differently. And we achieved that."

As a non-matching trait, one participant mentioned that her dog was too meticulous, while another reported that she was very fond of music and that her dog disliked music; both participants still felt like an overall good match with their dog. One owner who felt she was not a good match with her dog (1/2) indicated that her dog was too temperamental for her, while the other owner who felt she was not a good match could not provide specifics.

For an overview of temperament and activity level, see Table 5. Five participants felt they were a good match with their dog because they were similarly active and three because they were similarly calm. Three of all participants stated it was a good fit that their dog was more active than they were and thus carried them along (participant 10: "On days when I'm in a worse mood and maybe want to go for smaller walks only, he goes: 'No, but we still have to do more. That's good for you', (. . .) Yes, he then brings you back on the right track."). This has also been the case in a study by Chopik and Weaver [39], in which owners reported higher relationship quality when dogs were more active than themselves. These three participants also feel that themselves and their dog are generally different, with one of them not feeling like an overall good match with her dog.

Participants reported the importance of shared hobbies within their relationship. Two owners said they and their dogs equally loved water. Another participant loved ball sports as much as her dog. Three owners felt the shared hobby of walking was a great fit, and two participants felt it matched their active lifestyles that their dogs could go everywhere with them and were always relaxed. One participant reported the matching common weather preference for winter. This is supported by former studies that identified shared hobbies to increase owner satisfaction [14,40] and decrease risk factors of problematic dog behavior [39].

Table 5. Overview of temperament and activity.

Good match because both active: $N = 5$	
Good match because both happy: $N = 2$	
Good match because both calm: $N = 2$	
Good match that dog more active: $N = 3$	Different: 3/3 Overall Good Match: 2/3
Enjoying shared hobbies: $N = 6$	

3.2. Relationship Parameters—Can Dog and Owner Have a Too Close Bond?

Table 6 shows an overview of the results regarding relationship aspects. No owner reported being generally unsatisfied. Fourteen of all participants reported to have a subjectively strong bond with their dog (this was not asked explicitly but considered if owners stated how intimate/intense/close the relationship was or that the dog was their best friend/partner). From this group, most owners felt similar to their dog (12/14), and all felt they were a good match with their dog (14/14). Five participants felt very secure with their dog and stated that they could rely on them. All of these (5/5) had a strong bond and felt similar to their dog. Out of six teams in which the dog and the owner shared a high expression of the openness trait, almost all (5/6) had a strong bond.

Table 6. Overview of quality of the bond.

Strong bond: $N = 14$	Similar personality: 12/14 Good match: 14/14
Feeling secure: $N = 5$	Strong bond: 5/5 Similar personality: 5/5

A variety of statements were made about the value of the dog, such as feeling dependent on it or seeing it as a family member. Participant 3 described the importance of her dog as follows: "(...) aid with soul. So, of course, a family member, friend, but just also an aid." Three participants said their dogs brought a lot of relief and freedom in everyday life; two said they were partners and best friends. Two owners described that even in times when things were not going well with the dog, they still did not want to give it up. Two participants described a very close proximity to their dogs and that they had nothing else and found their lives no longer worth living without them (participant 20: "If he's no longer there, then... then you can put me in the coffin right away. There is nothing more then.").

Previous studies have shown that benefits of pet ownership show when owners are moderately attached to their dogs [41,42]. Very high or extreme expressions of attachment are associated with the development of mental health issues (at least in elderly women, which are also strongly represented in our study [41]). Three of the strongly attached owners also felt negative social impacts on their life, and two owners found their lives no longer worth living without their dogs. These findings and the fact that we did not investigate the general satisfaction with life or mental status of our participants lead to the conclusion that negative psychological impacts on strongly attached owners cannot be excluded and need to be further explored.

3.3. Strong Bonds Can Develop Even if Expectations Are Not Met

Nine participants (9/21) indicated that the expectations they had for their future guide dog were met (for an overview, see Table 7). Three participants (3/21) said they had no expectations. Five (5/21) stated their expectations were even exceeded, and they all now share a strong bond with their dog. Two participants stated that their expectations had not been met. While existing research underlines the negative consequences of high expectations, such as increased returning rates [43], these two participants in our sample still have a strong bond with their dogs. One participant was ambivalent about whether her expectations were met. In addition, participants expressed expectations on the levels of mobility and, in private, fears about not being a good match and expectations of the general abilities of a guide dog in first-time owners. Participant 11 said, "As someone who has never had a guide dog (...) you don't have expectations, but you have dreams about what the dog could do. But it was really hard work (...) where I thought to myself in between: "Yes, why do I need the guide dog, if I have to do everything myself anyway?".

Table 7. Overview of expectations.

Expectations met: $N = 9$	
No expectations: $N = 3$	
Expectations exceeded: $N = 5$	Strong bond: 5/5
Expectations not met: $N = 2$	Strong bond: 2/2

3.4. Former Guide Dog Relationships Can Have Strong Influence on the Subsequent Ones

Of 12 participants who had one or more guide dogs before their current one, three stated they were just as well matched with the previous one as they were with their current one (for an overview of compatibility in previous relationships, see Table 8). None of these (0/3) were disappointed in their expectations of a guide dog, and two (2/3) had a strong bond with their dog.

Table 8. Overview of compatibility in previous relationships.

Good match before and now: $N = 3$	Expectations disappointed: 0/3
	Strong bond: 2/3
Better match before: $N = 2$	Good match: 0/2 Strong bond: 0/2 Feeling secure: 0/2
Better match now: $N = 5$	

One participant reported that her previous dog was a better match for her because he was more generally sensitive and temperamental when playing, and they were generally more similar. Participant 4 reported that she had a closer bond with her previous dog because "I sometimes felt myself that he notices and thinks what I feel and when I wasn't feeling well or I was upset or something, he felt and sensed that exactly." These two participants who experienced their previous dog as a better fit are the ones who experienced their current dog as a mismatch, did not have a strong bond or felt very secure with their dog. Participant 2 described it as follows: "So you always have a sweetheart dog, unfortunately, and I'm sorry about that, that was just her predecessor. That's maybe like first love too." This might also explain why they could not name precise personality traits as determinants of compatibility: The individuals may have had certain expectations about the compatibility based on their previous relationship. However, when faced with the actual interactions and behaviors of their dogs, they may have found discrepancies in comparison to their former dog. This incongruence could lead to difficulties in identifying and reporting specific traits as determinants of compatibility. It is possible that the high expectations due to former relationships themselves influenced the feeling of incompatibility. A study by Powell and colleagues [44] furthermore showed that previous dog owners had reduced odds of expecting challenges than non-owners. This indicates bias through selective recall of positive experiences from previous ownership. The same could be the case for the two interviewees in our study. What should be mentioned when looking at this possible explanation, though, is that none of the two mismatching owners reported to be disappointed in their expectations (one was ambivalent though, and both described their answer to the question more on a mobility level).

Two participants (2/12) reported their previous dog was very active, which suited them well at the time; they are now somewhat calmer, and the current, less temperamental dog is a better match. Five owners (5/12) reported their current dog was a better match than previous ones, one because his calmness suited her, one because the chemistry was just better, one said the previous dog was too reserved, and another said he had too much anxiety. One participant felt her previous dog was unsuitable for her but believed this was also due to his training with punishment, which she herself did not use. Participant 13 reported that his previous dog was good for beginners and his recent one a little too temperamental: "So I no longer need to be completely carefree. But I don't necessarily need quite as much stress as with this one right away. Something in between would be quite good".

Four prior guide dog owners (4/12) felt their current relationship was influenced by their previous one (for an overview, see Table 9). Two of them (2/4) felt the match with their current dog was better, and two (2/4) felt it was worse than with their previous one. The fact that none of those owners thought the dogs equally matched might hint at a lack of neutral evaluation due to an extremely positive or negative prior relationship. This underlines the power of the influence of prior relationships as already found by Lloyd and colleagues [24].

Table 9. Overview of influence of previous relationship.

Influenced by previous relationship: $N = 4$	Match with current dog better: 2/4 Match with previous dog better: 2/4
Not influenced by previous relationship: $N = 5$	Influence on social life 1/5 Expectations disappointed 0/5

Five owners (5/12) said they were not affected by the relationship with their previous guide dog, of those only one (1/5) felt any impact on their social life, and none (0/5) were disappointed in their expectations of a guide dog. Two (2/12) participants did not give specific statements about being influenced, and one person was ambivalent. There was no apparent connection between being influenced and second-, third- or fourth-time ownership. Two individuals said their expectations had been very high due to good previous relationships; these two were also the two participants who experienced themselves as not matching with their dog. For these two participants, also a higher attachment to their prior dog could be the case that led to more grief and sorrow, in turn negatively influencing the subsequent relationship [45]. One participant said she had known exactly what she did not want because of a previous mismatch. Four participants (4/12) said it was important to them not to compare the dogs to each other. Two owners (2/12) said the transition was difficult for them because the dogs had very different personalities. One first-time owner commented that she would have liked to have had some dog experience and believed this would have positively influenced the current relationship.

3.5. Dog Affinity Is Not Crucial for Compatibility

To our knowledge, no study has yet analyzed the impact of dog affinity on success in (guide) dog–owner relationships. It has been shown that perceived cuteness (which might be higher in dog-affine persons) predicts the relationship quality though [46]. In our sample, sixteen participants (16/21) said they had an affinity for dogs before getting their guide dog (for an overview, see Table 10). Five owners (5/21) said they had no previous connection to dogs, were afraid of dogs or would not have acquired a dog but for their disability. Eleven owners (11/21) had a family dog before their first guide dog. One participant grew up with dogs but would not have acquired a dog without her visual impairment; this was counted in the second group of non-dog-affine participants. Two owners expressed the opinion that one should only get a guide dog if one would otherwise acquire a dog. Of the five owners who said they had not been dog-affinitive before, all felt like a good match with their current dog, and three of them (3/5) had a strong bond with their dog. Being non-dog-affine therefore did not seem to negatively impact the relationship. Moreover, even a potentially positive aspect emerged: even though four owners (4/5) had another guide dog before, none of them (0/4) felt influenced by their previous relationship.

Table 10. Overview of dog affinity.

Dog affinity: N = 16	
No dog affinity: N = 5	Good match: 5/5 Strong bond: 3/5 Influenced by previous relationship: 0/4
Owned/lived with family dog before: N = 11	

3.6. Owners Experience Positive and Negative Effects on Their Social Life

Twelve participants (12/21) mentioned the effects of living with a guide dog on their social life. Of these, two owners (2/12) reported exclusively negative experiences, and one (1/12) reported both positive and negative experiences, such as the dog getting too much attention from others and rejection of the dog (participant 7: "And there is also sometimes a lot of rejection in my private environment, where I (...) thought, was it right (to get the dog)?"). The most frequently mentioned positive effect was increased social contacts. For example, participant 16 reported, "When you have a dog and you're a little more open-minded, you have a lot of friends who have dogs." This is supported by literature that suggests that dog ownership can increase opportunities for social contact and even new friendships [47], as well as specifically increase and change social interaction for guide dog owners [25]. Of the twelve owners who experienced impacts, nine had a strong bond with their dogs, including the three persons who experienced negative impacts of the social environment. This reflects that negative consequences do not necessarily impair the relationship. The negative social consequences might even be a result of the potentially "too close" bond itself, as discussed above.

3.7. Problems in Everyday Life Are Diverse

Nineteen of twenty-one participants reported problems in everyday life with their dog. Most of these problems were undesirable characteristics of the dog, such as greediness (named by ten participants) and being too open to other people (named by four participants). The other problems mentioned were diverse, including allergies of the dogs, yapping and over-excitement. The two participants who did not name any problems felt that they and their dog were a good match, and one shared a strong bond with her dog.

3.8. No Meaningful Differences in Off-/At-Work Relationship

Although inquired at some points by the interviewer, the owners reported only little difference between the at-work and off-work relationship throughout their interviews. Based on theory, it was expected to find major differences between those two modes and that the working relationship has a bigger impact on whether or not handlers consider the match to be a successful one [22]. We even found contrary results: both owners who felt like a mismatch with their dogs reported that the guide work was well functioning.

4. What Is Better—Similar or Different Teams?

Within the comparison of different and similar teams, the prominent differences in the following areas evolved: sharing a strong bond, feeling secure with the dog, being influenced by the previous relationship and feeling the current dog matches better than the previous one. As shown in Table 11, the difference in compatibility between different and similar teams was not striking. One participant in each group felt like a mismatch with their dog.

Table 11. Prominent differences in the comparison of different and similar teams.

Aspect	Percentage in Different Teams (N = 6)	Percentage in Similar Teams (N = 15)
Good match	5/6 (83.3%)	14/15 (93.3%)
Strong bond	2/6 (33.3%)	13/15 (86.7%)
Feeling secure	0/6 (0%)	5/15 (33.3%)
Influenced by previous relationship [1]	3/5 (60%)	1/7 (14.3%)
Match with current dog better [1]	1/5 (20%)	4/7 (57.1%)

[1] in participants that owned guide dogs before (N = 5 in different teams, N = 7 in similar teams).

Our results show that similarity in ownership is not clearly indicative of higher compatibility but is associated with positive relationship characteristics (the similar teams more often shared a strong bond, felt secure with their dog, felt their dog matches better than the one before and were less often influenced by their previous relationship). This is also supported by the state of research. Studies showed several advantages of similar expressions of traits, such as more positive attitudes toward their dog when they were similar on the dimension of warmth [48] and owner satisfaction when they were similar to their dogs in sharing possessions or the enjoyment of running outside [14]. But there are also studies that emphasize the positive impact of different combinations of traits, such as higher relationship satisfaction when dogs are more open, agreeable and neurotic than their owners [36]. The positive impact of being similar in the dimension of openness, as well as high openness in the dog, has also been described by participants in our study, as discussed above. Interestingly, a too high openness of the dog was one of the most often named problems of the participants, which probably plays a more dominant role in guide dogs than in family dogs, where they should not be distracted from external stimuli when guiding their owners. This discrepancy in our results, as well as in general research, underlines the need for extensive quantitative studies on the dog–owner compatibility based on their personality traits.

5. Limitations

Choosing a qualitative approach allowed us to freely explore possible determinants of guide-dog–owner compatibility and include owners' subjective perspective. It needs to be mentioned, though, that this type of research always has interpretative parts [49] and contains the possibility of social desirability bias (a tendency to present reality as what is perceived to be socially acceptable [50,51]). Additionally, nearly half of the interviews were conducted via telephone. We did not see any systematic differences between the groups of persons interviewed via telephone vs. those interviewed in person in our study. Still, telephone interviews are discussed to have advantages such as cost and time efficacy but also disadvantages such as potential effects on the content of the responses due to anonymity created by spatial separation [52,53].

Another bias that probably occurred in our study is the volunteer bias [54]. Accordingly, unsatisfied owners who experience major problems in their guide dog relationship will be less likely to expose those in an extensive study. This could contribute to the fact that our sample only included a very small number of incompatible teams. Another characteristic of the sample that limits generalizability is the fact that mainly older and female owners, as typical for animal research [55], participated in the study.

6. Conclusions and Future Research

Owners who perceived themselves and their dogs as a good match were more likely to identify personality traits as determinants of compatibility, experience a strong bond and feel secure in their relationship with their dog. They also expressed satisfaction in their expectations and were rarely influenced by previous relationships, underlining the importance and positive potential outcomes of a good match.

Certain traits emerged as subjective determinants of compatibility, including shared hobbies, shared high openness and high openness in the dog, similar activity levels and higher activity in dogs and the combination of dominant owners with submissive dogs. Additionally, similarities in the expressions of calmness, happiness, greediness and friendliness were deemed important. While similar teams tended to have a stronger bond, feel more secure and be less influenced by previous relationships, the differences in compatibility between different and similar teams were not particularly salient. Our study therefore points toward a positive influence of similarity on the relationship, but future research is necessary to confirm this assumption.

Dog affinity, expectations not being met and differences between the on- and off-work relationship did not influence compatibility or other relationship parameters in our sample. Prior dog ownership, however, seems to have a potentially strong impact, as participants who reported to be influenced due to a former very positive guide dog relationship did not feel compatible with their current dog. The majority of owners reported positive social consequences associated with guide dog ownership, with only a few reporting negative effects. Nevertheless, all of them maintained a strong bond with their dog, indicating that these consequences did not impair the overall relationship. It is important to consider, however, that our results, in line with the existing literature, suggest that a strong bond between the dog and the owner does not exclusively yield positive effects.

Having taken advantage of a qualitative approach to freely explore possible aspects of compatibility, it is necessary to further study the parameters found. What is needed, then, is a comprehensive study that captures personality traits of participants and their dogs using validated questionnaires, thus circumventing the problem that many participants had difficulty assessing them freely. In addition, it would be important to further explore the differentiation between the off- and at-work relationship, as well as to include a measurable outcome in the area of mobility of guide dogs, such as an obstacle course. With respect to our results, it could also be revealing to further investigate the impact of high attachment between the dog and the owner on the owners' mental status or consequences in their social life.

However, the above-named conclusions contribute to the understanding of dog–owner compatibility and can already be used to improve the matching process of guide dogs and their prospective owners. This furthermore can potentially increase the success rate of compatible matches and thus the animal welfare, as well as the mobility of visually impaired persons.

Author Contributions: Conceptualization, J.B., S.R.S. and Y.B.; methodology, Y.B. and T.M.; software, Y.B. and T.M.; validation, Y.B. and T.M.; formal analysis, Y.B. and T.M.; data curation, Y.B. and T.M.; writing—original draft preparation, Y.B.; writing—review and editing, J.B., S.R.S. and T.M.; supervision, J.B. and S.R.S.; project administration, Y.B. All authors have read and agreed to the published version of the manuscript.

Funding: Yana Bender was supported by a scholarship of the "Stiftung der Deutschen Wirtschaft". Juliane Bräuer was supported by the DFG grant BR 3601/7–1, and Stefan R. Schweinberger was supported by DFG grant SCHW 511/24–1.

Institutional Review Board Statement: This study was conducted according to the guidelines of the Declaration of Helsinki and approved by the Max Planck Ethics Council (protocol code 2022_12, approved on 27 June 2022).

Informed Consent Statement: Informed consent was obtained from all subjects involved in the study.

Data Availability Statement: The data presented in this study are not publicly or otherwise available due to the need to protect the anonymity of the participants.

Acknowledgments: The authors wish to warmly thank the participants of this study. We also want to thank the Alliance for Assistance Dogs—Pfotenpiloten e.V. for their collaboration, support and mediation of contacts. Further thanks go to the Max Planck Institute of Plant Breeding Research, the Association for the Blind and Visually Impaired District of Soest, TU Dortmund University, Department of Rehabilitation Sciences, the General Association for the Blind and Visually Impaired Berlin and the Association for the Blind and Visually Impaired Lower Saxony e. V. for their interest in our project and the provision of premises.

Conflicts of Interest: The authors declare no conflict of interest. The funders had no role in the design of the study; in the collection, analyses or interpretation of data; in the writing of the manuscript or in the decision to publish the results.

References

1. Kaminski, J.; Marshall-Pescini, S. *The Social Dog: Behavior and Cognition*, 1st ed.; Elsevier: Amsterdam, The Netherlands, 2014.
2. Bräuer, J. I do not understand but I care: The prosocial dog. *Interact Stud.* **2015**, *16*, 341–360. [CrossRef]
3. Jensen, J.B.; Sandøe, P.; Nielsen, S. Owner-Related Reasons Matter more than Behavioural Problems—A Study of Why Owners Relinquished Dogs and Cats to a Danish Animal Shelter from 1996 to 2017. *Animals* **2020**, *10*, 1064. [CrossRef]
4. Lloyd, J.; Budge, C.; La Grow, S.; Stafford, K. An Investigation of the Complexities of Successful and Unsuccessful Guide Dog Matching and Partnerships. *Front. Vet. Sci.* **2016**, *3*, 114. [CrossRef] [PubMed]
5. National Service Animal Registry. How Much Does a Service Dog Cost: A Buyer's Guide for Your Service Dog. 2019. Available online: https://www.nsarco.com/blog/service-dog-buyers-guide.html (accessed on 18 August 2023).
6. Lloyd, J.; Stafford, K.; La Grow, S.; Budge, R.C. Matching guide dogs to people: Assessing the relationship between 50 people and their first guide dogs. In Proceedings of the First Conference of the Australian Working Dog Alliance, Sydney, NSW, Australia, 4–5 November 2013; pp. 26–27.
7. Defelipe, R.; Savalli, C.; Otta, E. Demographics and self-reported well-being of Brazilian adults as a function of pet ownership: A pilot study. *Heliyon* **2020**, *6*, e04069. [CrossRef]
8. Lord, M.S.; Casey, R.A.; Kinsman, R.H.; Tasker, S.; Knowles, T.G.; Da Costa, R.E.; Woodward, J.L.; Murray, J.K. Owner perception of problem behaviours in dogs aged 6 and 9-months. *Appl. Anim. Behav. Sci.* **2020**, *232*, 105147. [CrossRef]
9. Pitteri, E.; Mongillo, P.; Adamelli, S.; Bonichini, S.; Marinelli, L. The quality of life of pet dogs owned by elderly people depends on the living context, not on the owner's age. *J. Vet. Behav.* **2014**, *9*, 72–77. [CrossRef]
10. Bender, Y.; Bräuer, J.; Schweinberger, S.R. What makes a good dog-owner team?—A systematic review about compatibility in personality and attachment. *Appl. Anim. Behav. Sci.* **2023**, *260*, 105857. [CrossRef]
11. Canejo-Teixeira, R.; Armelim Almiro, P.; Baptista, L.V.; Niza, M.M.R.E. Predicting Dysfunctional Human–Dog Dyads. *Anthrozoös* **2020**, *33*, 743–758. [CrossRef]
12. Gosling, S.D.; Kwan, V.S.; John, O.P. A dog's got personality: A cross-species comparative approach to personality judgements in dogs and humans. *J. Pers. Soc. Psychol.* **2003**, *85*, 1161–1169. [CrossRef]
13. Hsu, Y.; Serpell, J.A. Development and validation of a questionnaire for measuring behaviour and temperament traits in pet dogs. *J. Am. Vet. Med. Assoc.* **2003**, *223*, 1293–1300. [CrossRef]
14. Curb, L.A.; Abramson, C.I.; Grice, J.W.; Kennison, S.M. The Relationship between Personality Match and Pet Satisfaction among Dog Owners. *Anthrozoös* **2013**, *26*, 395–404. [CrossRef]
15. Turcsan, B.; Range, F.; Viranyi, Z.; Miklosi, A.; Kubinyi, E. Birds of a feather flock together? Perceived personality matching in owner-dog dyads. *Appl. Anim. Behav. Sci.* **2012**, *140*, 154–160. [CrossRef]
16. Bauer, A.; Woodward, L. People and Their Pets: A Relational Perspective on Interpersonal Complementarity and Attachment in Companion Animal Owners. *Soc. Anim.* **2007**, *15*, 169–189. [CrossRef]
17. Konok, V.; Kosztolanyi, A.; Rainer, W.; Mutschler, B.; Halsband, U.; Miklosi, A. Influence of Owners' Attachment Style and Personality on Their Dogs' (Canis familiaris) Separation-Related Disorder. *PLoS ONE* **2015**, *10*, e0118375. [CrossRef]
18. Deutscher Blinden-und Sehbehindertenverbandes e. V. Hilfsmittel und Alltagshilfen für Menschen mit Sehbeeinträchtigungen. 2006. Available online: https://blickpunkt-auge.de/files/dokumente/Broschueren/blickpunkt-auge-broschuere-hilfsmittel-2016.pdf (accessed on 18 August 2023).
19. Lloyd, J. Exploring the Match between People and Their Guide Dogs. Ph.D. Thesis, Massey University, Turitea, New Zealand, 2003.
20. Meyer, D.; Bhowmik, J.; Islam, F.M.A.; Deverell, L. Profiling guide dog handlers to support guide dog matching decisions. *Disabil. Rehabil.* **2018**, *42*, 137–146. [CrossRef]
21. Zapf, S.A.; Rough, R.B. The development of an instrument to match individuals with disabilities and service animals. *Disabil. Rehabil.* **2002**, *24*, 47–58. [CrossRef] [PubMed]
22. Lloyd, J.; Budge, C.; Stafford, K. Handlers' Expectations and Perceived Compatibility regarding the Partnership with Their First Guide Dogs. *Animals* **2021**, *11*, 2765. [CrossRef] [PubMed]
23. Gravrok, J.; Bendrups, D.; Howell, T.; Bennett, P. Beyond the Benefits of Assistance Dogs: Exploring Challenges Experienced by First-Time Handlers. *Animals* **2021**, *11*, 203. [CrossRef]
24. Lloyd, J.; Budge, C.; La Grow, S.; Stafford, K. The End of the Partnership With a Guide Dog: Emotional Responses, Effects on Quality of Life and Relationships With Subsequent Dogs. *Front. Vet. Sci.* **2021**, *8*, 543463. [CrossRef]
25. Whitmarsh, L. The Benefits of Guide Dog Ownership. *Vis. Impair. Res.* **2005**, *7*, 27–42. [CrossRef]

26. Kuckartz, U. *Qualitative Text Analysis: A Guide to Methods, Practice and Using Software*; SAGE: Thousand Oaks, CA, USA, 2014.
27. Hauser, S.; Wakkary, R.; Neustaedter, C. Understanding guide dog team interactions: Design opportunities to support work and play. In Proceedings of the DIS'14: Proceedings of the 2014 Conference on Design Interactive Systems, Vancouver, BC, Canada, 21–25 June 2014; pp. 295–304. [CrossRef]
28. Dr. Dresing & Pehl GmbH. F4transkript (7.0.6) [Software]. 2021. Available online: https://www.audiotranskription.de/f4transkript/ (accessed on 1 May 2022).
29. Mayring, P. Qualitative Content Analysis—Theoretical Foundation and basic Procedures. 2013. Available online: www.qualitative-content-analysis.aau.at (accessed on 1 May 2022).
30. Kuckartz, U. Qualitative Text Analysis: A Systematic Approach. In *Compendium for Early Career Researchers in Mathematics Education*; Kaiser, G., Presmeg, N., Eds.; Springer: Berlin, Germany, 2019. [CrossRef]
31. Dr. Dresing & Pehl GmbH. F4analyse (3.4.3) [Software]. 2022. Available online: https://www.audiotranskription.de/f4analyse/ (accessed on 1 May 2022).
32. Funder, D.C. On the accuracy of personality judgment: A realistic approach. *Psychol. Rev.* **1995**, *102*, 652–670. [CrossRef] [PubMed]
33. Leary, M.R. Motivational and emotional aspects of the self. *Annu. Rev. Psychol.* **2007**, *58*, 317–344. [CrossRef]
34. Paulhus, D.L.; John, O.P. Egoistic and moralistic biases in self-perception: The interplay of self-deceptive styles with basic traits and motives. *J. Pers.* **1998**, *66*, 1025–1060. [CrossRef]
35. Kurdek, L.A. Pet dogs as attachment figures. *J. Pers. Soc. Psychol.* **2008**, *25*, 247–266. [CrossRef]
36. Cavanaugh, L.A.; Leonard, H.A.; Scammon, D.L. A tail of two personalities: How canine companions shape relationships and well-being. *J. Bus. Res.* **2008**, *61*, 469–479. [CrossRef]
37. O'Haire, M. Companion animals and human health: Benefits, challenges, and the road ahead. *J. Vet. Behav.* **2010**, *5*, 226–234. [CrossRef]
38. Beck, A.M.; Katcher, A.H. Future Directions in Human-Animal Bond Research. *Am. Behav. Sci.* **2003**, *47*, 79–93. [CrossRef]
39. Chopik, W.J.; Weaver, J.R. Old dog, new tricks: Age differences in dog personality traits, associations with human personality traits, and links to important outcomes. *J. Res. Pers.* **2019**, *79*, 94–108. [CrossRef]
40. Van Herwijnen, I.R.; van der Borg, J.A.M.; Naguib, M.; Beerda, B. Dog ownership satisfaction determinants in the owner-dog relationship and the dog's behaviour. *PLoS ONE* **2018**, *13*, e0204592. [CrossRef]
41. Chur-Hansen, A.; Winefield, H.R.; Beckwith, M. Companion Animals for Elderly Women: The Importance of Attachment. *Qual. Res. Psychol.* **2009**, *6*, 281–293. [CrossRef]
42. Peacock, J.; Chur-Hansen, A.; Winefield, H. Mental Health Implications of Human Attachment to Companion Animals. *J. Clin. Psychol.* **2012**, *68*, 292–303. [CrossRef]
43. Bouma, E.M.C.; Vink, L.M.; Dijkstra, A. Expectations Versus Reality: Long-Term Research on the Dog–Owner Relationship. *Animals* **2020**, *10*, 772. [CrossRef]
44. Powell, L.; Chia, D.; McGreevy, P.; Podberscek, A.L.; Edwards, K.M.; Neilly, B.; Guastella, A.J.; Lee, V.; Stamatakis, E. Expectations for dog ownership: Perceived physical, mental and psychosocial health consequences among prospective adopters. *PLoS ONE* **2018**, *13*, e0200276. [CrossRef]
45. Barnard-Nguyen, S.; Breit, M.; Anderson, K.; Nielsen, J. Pet loss and grief: Identifying at-risk pet owners during the euthanasia process. *Anthrozoös* **2016**, *29*, 421–430. [CrossRef]
46. Thorn, P.; Howell, T.J.; Brown, C.; Bennett, P.C. The Canine Cuteness Effect: Owner-Perceived Cuteness as a Predictor of Human–Dog Relationship Quality. *Anthrozoös* **2015**, *28*, 569–585. [CrossRef]
47. Wood, L.; Martin, K.; Christian, H.; Nathan, A.; Lauritsen, C.; Houghton, S.; Kawachi, I.; McCune, S. The pet factor—Companion animals as a conduit for getting to know people, friendship formation and social support. *PLoS ONE* **2015**, *10*, e0122085. [CrossRef] [PubMed]
48. Zeigler-Hill, V.; Highfill, L. Applying the interpersonal circumplex to the behavioral styles of dogs and cats. *Appl. Anim. Behav. Sci.* **2010**, *124*, 104–112. [CrossRef]
49. Denzin, N.K.; Lincoln, Y.S. *The Sage Handbook of Qualitative Research*, 4th ed.; SAGE: Thousand Oaks, CA, USA, 2011.
50. Bergen, N.; Labonté, R. "Everything Is Perfect, and We Have No Problems": Detecting and Limiting Social Desirability Bias in Qualitative Research. *Qual. Health Res.* **2020**, *30*, 783–792. [CrossRef]
51. Krumpal, I. Determinants of social desirability bias in sensitive surveys: A literature review. *Qual. Quant.* **2013**, *47*, 2025–2047. [CrossRef]
52. Block, E.S.; Erskine, L. Interviewing by Telephone: Specific Considerations, Opportunities, and Challenges. *Int. J. Qual. Methods.* **2012**, *11*, 428–445. [CrossRef]
53. Groves, R.M.; Kahn, R.L. *Surveys by Telephone: A National Comparison with Personal Interviews*; Academic Press: Cambridge, MA, USA, 1979.
54. Rosenthal, R.; Rosnow, R.L. *The Volunteer Subject*; John Wiley & Sons: Hoboken, NJ, USA, 1975.
55. Herzog, H. Women dominate research on Human-Animal Bond. *Am. Psychol.* 24 May 2021. Available online: https://www.psychologytoday.com/us/blog/animals-and-us/202105/women-dominate-research-the-human-animal-bond (accessed on 18 August 2023).

animals

MDPI

Brief Report

Salivary Cortisol in Guide Dogs

Enrique De la Fuente-Moreno [1], Pedro Paredes-Ramos [2,*], Apolo Carrasco-García [2], Bertha Hernandez-Cruz [2], Mayvi Alvarado [3] and Claudia Edwards [4]

1 Doctorado en Neuroetología, Xalapa Universidad Veracruzana, Xalapa 91190, Mexico; efm121@yahoo.com.mx
2 Facultad de Medicina Veterinaria y Zootecnia, Universidad Veracruzana, Veracruz 91190, Mexico; acarrasco@uv.mx (A.C.-G.); behernandez@uv.mx (B.H.-C.)
3 Instituto de Neuroetología, Xalapa Universidad Veracruzana, Xalapa 91190, Mexico; malvarado@uv.mx
4 Facultad de Medicina Veterinaria y Zootecnia, Universidad Nacional Autónoma de México, Mexico City 04510, Mexico; dra.edwards@yahoo.com
* Correspondence: pparedes@uv.mx

Simple Summary: We compared cortisol levels in the saliva of guide dogs and dogs that were trained as such but became companion dogs during a period of social isolation and exposure to a gunshot sound. The results showed that cortisol levels were higher in guide dogs than in companion dogs throughout the test. No changes were observed as a consequence of social isolation or exposure to the gunshot. This suggests that guide dogs maintain higher levels of basal cortisol compared with companion dogs, which could be associated with cognitive processes derived from working as guide dogs.

Abstract: Guide dogs work for extended periods and are exposed to multiple environmental stimuli that could lead to higher stress compared with companion dogs. Cortisol is the main hormone associated with stress in most mammals. This study included seven guide dogs and seven same-breed dogs that were trained as guide dogs but became companion dogs to compare their salivary cortisol levels before, during, and after a period of social isolation and exposure to a 110-decibel gunshot sound. Each dog was left alone in an empty room for 60 min. After 15 min, the dogs were exposed to the sound. We collected four saliva samples from each dog. The first one was taken 5 min before starting the social isolation period, and the following ones at 15, 30, and 45 min after the test started. A two-way ANOVA was used to compare the group effect and the time effect during isolation and noise exposure. The results showed higher levels of cortisol in the guide dogs compared with the companion dogs throughout the test. No differences were found in time or in the interaction between time and group. This suggests that being a guide dog increases levels of basal cortisol when compared with dogs that live as companion animals and family members.

Keywords: cortisol; welfare; guide dogs; companion dogs

Citation: De la Fuente-Moreno, E.; Paredes-Ramos, P.; Carrasco-García, A.; Hernandez-Cruz, B.; Alvarado, M.; Edwards, C. Salivary Cortisol in Guide Dogs. *Animals* **2023**, *13*, 1981. https://doi.org/10.3390/ani13121981

Academic Editor: Betty McGuire

Received: 19 April 2023
Revised: 7 June 2023
Accepted: 11 June 2023
Published: 14 June 2023

1. Introduction

The use of dogs as guides for blind or visually impaired people began in Germany during World War I, when it is estimated that between 5 and 8% of the soldiers' injuries were to the eyes. In 1916, Dr. Gerhard Stalling founded the first guide dog training school (GDTS) using German Shepherd dogs [1]. Currently, there are many GDTSs around the world, where mainly Labrador Retriever and Golden Retriever dogs are raised and trained.

A guide dog must be able to carry out various cognitive processes such as maintaining concentration, paying no attention to distractions, remembering its training, making decisions autonomously, and even ignoring its handler if they put their own life at risk [2,3]. Guide dogs are known to have a strong attachment or social connection to their handlers, as well as bilateral non-verbal communication [4,5]. This suggests that being a guide dog is demanding and exhausting and might require maintaining higher blood cortisol levels since increases in cortisol are positively associated with improvements in cognitive and emotional processes [6].

On the other hand, increases in cortisol are associated with the stress response in most mammals [6–8]. In dogs, cortisol can be measured in blood, saliva, hair, feces, and urine. While cortisol in hair and feces reflects its levels for several days to months, the evaluation of blood and saliva reflects acute increases experienced within the last few minutes [9]. In healthy dogs, salivary cortisol levels are the result of the passive diffusion of cortisol through the acinar cells of the salivary lumen [10,11]. Since the method for saliva collection is noninvasive, the measurement of salivary cortisol is a practical test in dogs for evaluating stress [10–12].

Since high cortisol levels are associated with the success of working dogs [13], it seems important and useful to know the stress levels that guide dogs maintain compared with dogs that have a similar breed but live as companion animals. Herein, we compared the salivary cortisol concentrations of active guide dogs with those of dogs that failed to achieve specialized guide dog training and are currently living as companion animals, which, to the best of our knowledge, is unknown information and may be useful for selecting guide dogs in the future.

2. Materials and Methods

2.1. Subjects

We included Labrador Retrievers and Golden Retrievers that were bred and trained at one GDTS in Mexico City. At three months of age, pups were adopted by foster families and raised over the next ten months. When dogs returned to the GDTS, they were evaluated to identify degenerative diseases and behavioral problems. When dogs showed dysplasia, aggression, fear, and anxiety, they were discarded from the guide dog training program (GDTP) and adopted by their foster families, becoming companion animals. The dogs that continued into the program (n = 14) received basic training. During this stage, the dogs lived at the GDTS facilities in individual kennels 3 m long × 1.6 m wide that had a covered area and an open area. Dogs were fed specialized Royal Canin® (Gard, France) dry food twice daily with free access to water. After four months at the GDTP, 7 out of the 14 dogs successfully finished their training and began working as guide dogs, whereas the remaining dogs (n = 7) were discharged and adopted by their foster families to continue their lives as companion animals. The characteristics of the dogs included in the study are presented in Table 1.

Table 1. Name, age, breed, and time living as guide or companion dogs. Years, y; months, m.

Guide Dogs				Companion Dogs			
Name	**Age (Years)**	**Breed**	**Time as Guide Dog**	**Name**	**Age (Years)**	**Breed**	**Time as Companion Dog**
Nusa	2	Golden retriever	1 y	Jock	4	Labrador retriever	2 y 4 m
Meli	5	Labrador retriever	3 y 2 m	Camila	6	Labrador retriever	4 y 3 m
Ninfa	2	Golden retriever	1 y	Gupy	7	Labrador retriever	4 y 9 m
Einy	7	Labrador retriever	6 y 2 m	Heidi	7	Labrador retriever	4 y 7 m
Ita	6	Labrador retriever	4 y 6 m	Elmo	7	Labrador retriever	5 y 7 m
Joe	4	Labrador retriever	1 y 7 m	Hunter	7	Labrador retriever	5 y 10 m
Fiona	7	Labrador retriever	5 y 8 m	Lancelot	7	Labrador retriever	3 y 2 m

Prior to the start of the study, approval from the Bioethics and Animal Welfare Commission of the Facultad de Medicina Veterinaria y Zootecnia, Universidad Veracruzana, was obtained (No.008/22). Before the start of the study, all authors completed an education program on the care and use of animals.

2.2. Experimental Procedure

We obtained and compared 4 saliva samples from all 14 dogs before, during, and after a period of social isolation and exposure to a 110-decibel sound capable of producing surprise and fear in animals [13]. On different days, the dogs, accompanied by their tutors (blind handlers or family members), were received at the GDTS between 9 and 12 h. Thirty minutes after arriving at the GDTS, the dog and its tutor entered a 3 m × 4 m room. Five minutes later, the first saliva sample was taken; then, the tutor was asked to leave the room, and the dog stayed, alone. Fifteen minutes later, a second saliva sample was taken, and immediately after that, a blank gun was fired to produce a 110-decibel sound. Fifteen and thirty minutes after the gun was fired, the third and fourth saliva samples were collected, respectively. Disposable cotton cords 10 cm long × 4 mm wide were used to obtain the saliva samples. The cord was inserted into the dog's mouth while the handler held onto the other end. The animal was allowed to chew it for 60 s until moistened. The cord was cut into 3 cm pieces and inserted into a sterile 5 mL syringe, the plunger was pushed to extract the saliva, and it was collected in a 1.5 mL microtube. The samples were refrigerated (4 °C) for two hours and then frozen at −20 °C until processing [6]. To produce a 110-decibel sound, a Mendoza PK-62 sports gun was used. Salivary cortisol concentrations were assessed using a solid-phase immunoenzyme assay (ELISA) using the commercial kit Cortisol ELISA EIA-1887 (DRG® International, Inc., Springfield, NJ, USA). The sensitivity of the assay was 2.5 ng/mL. The range of the curve was 2.5–200 ng/mL. The intra- and inter-assay coefficients of variation were 5.6% and 6.9%, respectively. Concentrations are expressed as ng/mL.

2.3. Statistical Analysis

A two-way ANOVA was used to identify significant differences between groups and between the number of saliva samples (1 to 4), as well as the interactions among them. The Fisher post hoc test was used to evaluate mean differences. The significance value for all comparisons was $p < 0.05$.

3. Results

There were significant differences between the guide dog and the companion dog groups: F (1, 3) = 16.31 and $p < 0.001$; the post hoc test showed that the guide dogs had higher levels of salivary cortisol than dogs living as companion animals (Figure 1). On the other hand, no differences were found between the sample times concerning the gunshot noise, F (1, 3) = 0.170 and p = 0. 916, or in the interaction between the group and saliva sample number, F (1, 3) = 0.97 and p = 0.979 (Figure 2).

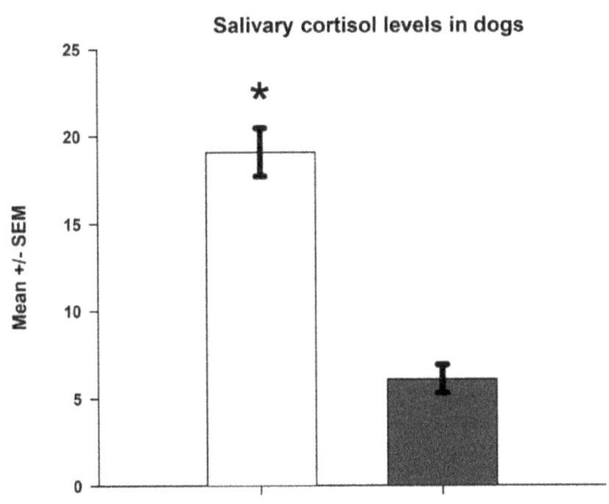

Figure 1. Guide dogs had higher salivary cortisol levels than companion dogs. * Indicates $p < 0.05$.

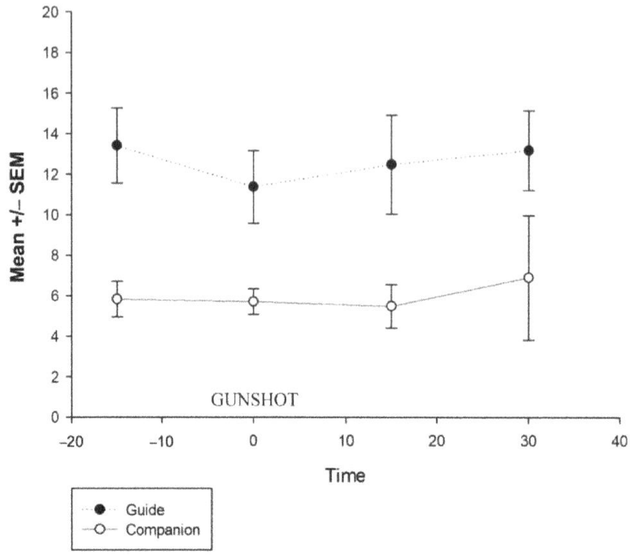

Figure 2. No differences were found between the sample times concerning the gunshot noise or in the interaction between the group and sample number. The first saliva sample was taken fifteen minutes before the gunshot (time 0). The second sample was collected 10 s after the gunshot, while the 3rd and 4th saliva samples were collected 15 and 30 min after the gun was fired.

4. Discussion

Basal cortisol levels are influenced by internal and external factors in an organism, such as age and sex and temperature and time of day, respectively [6,7]. However, stress, understood as a biological response of the organism caused by threats to its homeostasis, is the main cause of increased cortisol concentrations [6–8]. The hypothalamus–hypophysis–adrenal (HHA) axis is responsible for the release of cortisol as a consequence of stress [14,15]. In our study, guide dogs exhibited higher levels of cortisol (almost double) than dogs that live as companion animals. Research has shown that early training increases cortisol levels in dogs [16]. Furthermore, it has been observed that working dogs, such as those used in the military, increase cortisol levels when subjected to physical and sensory challenges [17]. While a companion dog lives without schedules, obligations, or social limits and with low demand for physical and mental activity, a guide dog is exposed daily to multiple and changing environmental stimuli and variable periods of activity and rest, requiring great concentration; sometimes, playing behavior is inhibited, and sometimes, it has to explore new territories, among other activities [1,4]. This suggests that guide dogs may have increased HHA axis activity as both a cause and a consequence of their daily activity. However, our results must be taken with caution since they represent only a small sample of animals and could be affected by factors such as genetics and environmental conditions in Mexico City, which is extremely noisy and crowded, and thus, they may not represent the reality for most guide dogs around the world.

Increases in cortisol levels are not necessarily bad but rather are necessary to better perform cognitive and emotional tasks such as those faced by a guide dog daily [18]. Studies have shown that individuals with hypoadrenocorticism, which produces low cortisol levels, show poor cognitive performance, poor sleep quality, low motivation, and mental fatigue [19]. On a typical day, a guide dog must maintain mental concentration for hours, ignore distractions, and make decisions, suggesting a high demand on its cognitive functions [3]. Based on this, we believe that the increase in cortisol in guide dogs compared with companion dogs is due to the high emotional and cognitive demands of adequately performing their work. Cortisol levels increase significantly in stressful periods compared with relaxation periods [20]. Social isolation is an important inducer of activity in the HHA axis and can increase cortisol release in animals, including humans [21,22]. In the dog, social isolation and separation from affection figures can lead to significant behavioral and physiological changes [23]. It is known that guide dogs form strong bonds with their handlers [24]; the guide dogs in our study had been interacting with their handlers for at least 1 year and did not separate from them at any time. Based on this assumption, it might be thought that greater levels of cortisol in guide dogs compared with companion dogs may be due to the separation period that the dogs had from their handlers during the test; however, the cortisol levels in both the guide and companion dogs were maintained throughout the test without increasing with the time the users were separated. Thus, we believe that social isolation and separation from their handlers had little to no effect on their cortisol levels. We must emphasize that, in our study, companion dogs do not represent the typical companion animal because, just like guide dogs, they belong to an elite group of dogs with suitable characteristics for work that received early stimulation from their very first days of life. Thus, we believe that a lack of cortisol increment in both groups of dogs, as a consequence of isolation or the gunshot, could be due to a remarkable resilience developed during their time at the GDTP.

Finally, neither the gunshot nor the isolation caused an observable effect on the cortisol levels of the dogs of either group. Given that all the dogs lived in Mexico City, where there is high noise pollution [25], we believe that the animals included in this study may be accustomed to noises of the same intensity as the gunshot and, therefore, showed no changes. Additionally, it is likely that the lack of response to noise in the dogs from both groups is because they received an auditory stimulation protocol when they were puppies when they belonged to the GDTP. We believe that increasing our understanding of guide dogs and working dogs may inspire further research into how the mind of this animal works and how we can improve selective breeding and training methods for the benefit of both dogs and humans.

5. Conclusions

In conclusion, the findings of this study suggest that the task of guiding individuals through various and unpredictable environments places significant demands on dogs. The results clearly demonstrate that guide dogs exhibited higher levels of cortisol compared with companion dogs. These results indicate that guide dogs consistently maintain elevated cortisol levels as a necessity or consequence of their working activity with blind individuals. The demanding nature of their role likely contributes to the heightened stress levels observed in these dogs. These findings emphasize the importance of recognizing and addressing the unique stressors faced by guide dogs to ensure their overall well-being and quality of life. Further research and investigation into effective stress management strategies for guide dogs are warranted to mitigate the potential negative impact of prolonged elevated cortisol levels. Ultimately, providing appropriate support and care for these remarkable animals is crucial to ensure their health, happiness, and ability to fulfill their important role as guides for the visually impaired.

Author Contributions: The idea for the study was conceived by P.P.-R. The data collection was designed by P.P.-R., B.H.-C., E.D.l.F.-M. and A.C.-G. The data were analyzed by P.P.-R., E.D.l.F.-M., M.A. and C.E. The article was written by P.P.-R. and E.D.l.F.-M. All authors have read and agreed to the published version of the manuscript.

Funding: This study was supported by a scholarship, UV-SNI CONAHCYT number 236463 from P.P.-R., and PhD CONAHCYT scholarship 565139.

Institutional Review Board Statement: Approval from the Bioethics and Animal Welfare Commission of the Facultad de Medicina Veterinaria y Zootecnia, Universidad Veracruzana, was obtained (No.008/22). Before the start of the study, all authors completed an education program on the care and use of animals.

Informed Consent Statement: Not applicable.

Data Availability Statement: The authors prefer not to make the data available for public.

Conflicts of Interest: The authors declare no conflict of interest. The funders had no role in the design of the study; in the collection, analyses, or interpretation of data; in the writing of the manuscript, or in the decision to publish the results.

References

1. Ostermeier, M. History of Guide Dog Use by Veterans. *Mil. Med.* **2010**, *175*, 587–593. [CrossRef]
2. Lloyd, J.K.F. Exploring the Match between People and Their Guide Dogs: A Thesis Presented in Fulfilment of the Requirements for the Degree of Doctor of Philosophy in Veterinary Science at Massey University, Turitea, Aotearoa/New Zealand. Doctoral Dissertation, Massey University, Massey, New Zealand, 2004.
3. Bensky, M.K.; Gosling, S.D.; Sinn, D.L. The world from a dog's point of view: A review and synthesis of dog cognition research. *Adv. Study Behav.* **2013**, *45*, 209–406.
4. Bray, E.E.; Sammel, M.D.; Seyfarth, R.M.; Serpell, J.A.; Cheney, D.L. Temperament and problem solving in a population of adolescent guide dogs. *Anim. Cogn.* **2017**, *20*, 923–939. [CrossRef] [PubMed]
5. Gaunet, F. How do guide dogs of blind owners and pet dogs of sighted owners (*Canis familiaris*) ask their owners for food? *Anim. Cogn.* **2008**, *11*, 475–483. [CrossRef]
6. Chmelíková, E.; Bolechová, P.; Chaloupková, H.; Svobodová, I.; Jovičić, M.; Sedmíková, M. Salivary cortisol as a marker of acute stress in dogs: A review. *Domest. Anim. Endocrinol.* **2020**, *72*, 106428. [CrossRef] [PubMed]
7. Koyama, T.; Omata, Y.; Saito, A. Changes in Salivary Cortisol Concentrations During a 24-Hour Period in Dogs. *Horm. Metab. Res.* **2003**, *35*, 355–357. [CrossRef] [PubMed]
8. Wojtaś, J.; Karpiński, M.; Czyzowski, P. Salivary Cortisol Interactions in Search and Rescue Dogs and Their Handlers. *Animals* **2020**, *10*, 595. [CrossRef] [PubMed]
9. Bergamasco, L.; Osella, M.C.; Savarino, P.; Larosa, G.; Ozella, L.; Manassero, M.; Badino, P.; Odore, R.; Barbero, R.; Re, G. Heart rate variability and saliva cortisol assessment in shelter dog: Human–animal interaction effects. *Appl. Anim. Behav. Sci.* **2010**, *125*, 56–68. [CrossRef]
10. Granger, D.A.; Kivlighan, K.T.; Fortunato, C.; Harmon, A.G.; Hibel, L.C.; Schwartz, E.B.; Whembolua, G.L. Integration of salivary biomarkers into developmental and behaviorally-oriented research: Problems and solutions for collecting specimens. *Physiol. Behav.* **2007**, *92*, 583–590. [CrossRef]
11. Dreschel, N.A.; Granger, D.A. Methods of collection for salivary cortisol measurement in dogs. *Horm. Behav.* **2009**, *55*, 163–168. [CrossRef]
12. Bennett, A.; Hayssen, V. Measuring cortisol in hair and saliva from dogs: Coat color and pigment differences. *Domest. Anim. Endocrinol.* **2010**, *39*, 171–180. [CrossRef] [PubMed]
13. Foyer, P.; Svedberg, A.-M.; Nilsson, E.; Wilsson, E.; Olsen Faresjö, Å.; Jensen, P. Behavior and cortisol responses of dogs evaluated in a standardized temperament test for military working dogs. *J. Vet. Behav.* **2016**, *11*, 7–12. [CrossRef]
14. García-Morato Fernández-Baíllo, C. Respuestas Comportamentales y Fisiológicas en Situaciones de Estrés en el Perro y el Gato. Tesis Doctoral, Universidad Autónoma de Barcelona, Barcelona, Spain, 2019.
15. Shirtcliff, E.A.; Vitacco, M.J.; Graf, A.; Gostisha, J.B.A.A.; Merz, L.B.A.J.; Zahn-Waxler, C. Neurobiology of empathy and callousness: Implications for the development of antisocial behavior. *Behav. Sci. Law* **2009**, *27*, 137–171. [CrossRef] [PubMed]
16. Ando, I.; Karasawa, K.; Shioya, T.; Matsuda, H.; Tanaka, A. Evaluation of stress status using the stress map for guide dog candidates in the training stage using variations in the serum cortisol with nerve growth factor and magnesium ions. *Vet. Anim. Sci.* **2020**, *10*, 100129. [CrossRef]
17. Haverbeke, A.; Diederich, C.; Depiereux, E.; Giffroy, J.M. Cortisol and behavioral responses of working dogs to environmental challenges. *Physiol. Behav.* **2008**, *93*, 59–67. [CrossRef]
18. Stauble, M.R.; Thompson, L.A.; Morgan, G. Increases in cortisol are positively associated with gains in encoding and maintenance working memory performance in young men. *Stress* **2013**, *16*, 402–410. [CrossRef]
19. Kalafatakis, K.; Russell, G.M.; Harmer, C.J.; Munafo, M.R.; Marchant, N.; Wilson, A.; Brooks, J.C.; Durant, C.; Thakrar, J.; Murphy, P.; et al. Ultradian rhythmicity of plasma cortisol is necessary for normal emotional and cognitive responses in man. *Proc. Natl. Acad. Sci. USA* **2018**, *115*, E4091–E4100. [CrossRef]
20. Cay, M.; Ucar, C.; Senol, D.; Çevirgen, F.; Ozbag, D.; Altay, Z.; Yildiz, S. Effect of increase in cortisol level due to stress in healthy young individuals on dynamic and static balance scores. *North. Clin. Istanb.* **2018**, *5*, 295.
21. Stafford, M.; Gardner, M.; Kumari, M.; Kuh, D.; Ben-Shlomo, Y. Social isolation and diurnal cortisol patterns in an ageing cohort. *Psychoneuroendocrinology* **2013**, *38*, 2737–2745. [CrossRef]
22. Cacioppo, J.T.; Cacioppo, S.; Capitanio, J.P.; Cole, S.W. The neuroendocrinology of social isolation. *Annu. Rev. Psychol.* **2015**, *66*, 733–767. [CrossRef]
23. Lensen, R.; Moons, C.; Diederich, C. Physiological stress reactivity and recovery related to behavioral traits in dogs (*Canis familiaris*). *PLoS ONE* **2019**, *14*, e0222581. [CrossRef] [PubMed]

24. Valsecchi, P.; Previde, E.P.; Accorsi, P.A.; Fallani, G. Development of the attachment bond in guide dogs. *Appl. Anim. Behav. Sci.* **2010**, *123*, 43–50. [CrossRef]
25. Rodríguez Manzo, F.E.; Juárez González, L. Exploración cualitativa sobre el ruido ambiental urbano en la Ciudad de México. *Estud. Demogr. Urbanos* **2020**, *35*, 803–838. [CrossRef]

animals

MDPI

Article

Defining the Characteristics of Successful Biosecurity Scent Detection Dogs

Ariella Y. Moser [1,*], Wendy Y. Brown [1], Pauleen Bennett [2], Peta S. Taylor [1], Bethany Wilson [3] and Paul McGreevy [4]

1 School of Environmental and Rural Science, University of New England, Armidale, NSW 2350, Australia
2 Anthrozoology Research Group, Department of Psychology, Counselling & Therapy, La Trobe University, Melbourne, VIC 3083, Australia
3 School of Life and Environmental Science, Faculty of Science, University of Sydney, Camperdown, NSW 2006, Australia
4 One Welfare Research Institute, Faculty of Science, Agriculture, Business and Law, University of New England, Armidale, NSW 2350, Australia
* Correspondence: ariellamoser0@gmail.com

Simple Summary: Dogs should display certain physical, behavioural, and cognitive characteristics to be suitable for training and deployment in a scent detection role. Identifying these characteristics is a vital first step to determine whether particular dogs will be suitable for selection and training. This study identified the behavioural traits that stakeholders consider relevant for dogs trained in biosecurity detection, and then assessed these traits in an active cohort of Australian biosecurity detector dogs. Our research revealed seven relevant traits: search motivation, emotional stability, search arousal, food motivation, play motivation, search independence, and search focus. Current biosecurity detector dogs had consistently high ratings from their handlers for search motivation, emotional stability, and food motivation, whereas other traits showed more variation. We found that dogs rated by handlers as high in search arousal and search motivation were more likely to also be rated highly for their overall detection performance. These findings will help to inform decisions about the selection and training of scent detection dogs.

Abstract: To perform their role effectively, scent detection dogs require certain characteristics. Identifying these characteristics will inform the selection of prospective dogs and preferred approaches to their training. The current study drew upon the perspectives of industry stakeholders to identify the behavioural traits considered relevant for detection dogs in biosecurity screening roles. Dog handlers, trainers, and supervisors (n = 25) in Australian biosecurity operations participated in focus group interviews to determine the perceived characteristics that, in their experience, influence detection performance. Their descriptions were used to create a questionnaire which was then administered to handlers to assess the working behaviours of current biosecurity dogs. Responses were collected for 88% of the operational dogs (n = 36). An exploratory factor analysis revealed seven tentative dimensions: search motivation, emotional stability, search arousal, food motivation, play motivation, search independence, and search focus. Search motivation and search arousal were both positively associated with handler ratings of detection performance ($p \leq 0.006$). In general, biosecurity dogs were scored consistently high in ratings of search motivation, emotional stability, and food motivation. Our approach has advanced our understanding of the working behaviours and characteristic profile of biosecurity detector dogs and will be used to inform candidate selection processes.

Keywords: detector dog; sniffer dog; working dog; selection; personality; traits; behaviours; focus group; survey; drive

Citation: Moser, A.Y.; Brown, W.Y.; Bennett, P.; Taylor, P.S.; Wilson, B.; McGreevy, P. Defining the Characteristics of Successful Biosecurity Scent Detection Dogs. *Animals* **2023**, *13*, 504. https://doi.org/10.3390/ani13030504

Academic Editors: David C. Dorman and Betty McGuire

Received: 14 December 2022
Revised: 19 January 2023
Accepted: 27 January 2023
Published: 31 January 2023

1. Introduction

Scent detection dogs are widely recognised as highly effective in screening and locating elusive targets across a number of fields. Their use in policing, rescue, and military [1,2] is well-established, but more recent applications have emerged in biosecurity, human health screening, and wildlife conservation [3–5].

Several behavioural and cognitive processes converge when a trained scent detection dog searches an area and signals the source of a target odour [6]. They must exhibit scenting behaviours appropriate to the task, disregard distractions, retain in their working memory the task and target odour, and evaluate odours against their trained prototype odours to discriminate or generalise accordingly. Meanwhile, they should often also be responsive to handler cues and directional input. Some dogs are better suited to this complex task than others and, as a reflection of their individual characteristics, their performance can vary accordingly [6,7]. As such, selecting dogs on the basis of appropriate characteristics can improve outcomes and optimise the performance of scent detection dogs [8,9]. Good selection processes reduce costs incurred for training and housing unsuitable dogs, and also potentially improve dog welfare by only recruiting dogs that will be behaviourally and cognitively equipped for the demands of the role. However, identifying these specific characteristics is not always straightforward due to the inherent complexity of defining and predicting animal behaviour.

The characteristics of successful scent detection dogs have been investigated with a variety of methods, including subjective ratings and direct behavioural assessments [10]. Methods that draw upon the perspectives and practical experiences of industry professionals can provide real-world insight into the tendencies and challenges that professionals experience in a given role [9,11]. Additionally, this approach can help to bridge gaps between industry needs and empirical research, improving the applicability of research outcomes [9,11]. Meanwhile, objective behavioural assessment allows handlers' subjective expectations to be tested. Together, both approaches underpin a comprehensive understanding of dogs' characteristics and working performance in scent detection roles.

To date, research into the selection of scent detection dogs has revealed an industry preference for dogs with a strong intrinsic motivation to search, high levels of playfulness, boldness, and environmental stability [6,7,11,12]. Furthermore, behavioural research has revealed associations between various constructs and performance outcomes. These include positive associations among performance and search desire and ability [2,12], activity [13], trainability [2], environmental stability [2,14], inhibitory control [15], short-term memory [8,16], and search thoroughness [17]. These traits seem to underpin many specific working behaviours.

This body of literature offers a robust basis to predict desirable traits in scent detection dogs, but some important knowledge gaps remain. Importantly, there are discrepancies in the way certain characteristics are labelled, defined, or measured, which may help to explain apparently contradictory findings in the literature. For example, the association of inhibitory control with detection performance has been reported to be positive [15], negative [18], and inconsistent [16]. Additionally, the overall literature is still somewhat limited in scope. Certain applications, such as the detection of explosives and narcotics, are predominantly represented [10]. Although relevant, such findings may not be directly applicable to all scent detection roles, which present different environmental challenges, working conditions, and target odours. Furthermore, the most commonly used performance measure is whether or not a dog completes training (i.e., reaches qualification or certification) [2,12,14,19,20], but this outcome does not necessarily predict the quality of ongoing performance in an operational context. These gaps necessitate further research using various methods to investigate, validate, and confirm findings.

The current research is a first step in determining the canine behaviours and characteristics that contribute to scent detection dog success within Australian biosecurity detector dog (BDD) operations. Australian BDDs are used to detect organic material that may pose a biological threat to native species or domestic agricultural industries. To do so, they are trained using positive reinforcement to respond to more than 200 different commodities. They predominantly work for extended periods in busy, relatively target-rich environments, such as airports, mail centres, and wharves, performing fairly repetitive search tasks. Additionally, any BDD is expected to be able to be handled by any handler in the agency. Some of these tasks and conditions are comparable with those of other scent detection dogs, such as those used for explosives detection. We therefore expected overlap of desirable traits for BDDs and other types of scent detection dogs. On the other hand, some of these conditions are less common among scent detection dogs and may involve specific attributes. For example, working for extended periods with frequent detections may be aided by a particularly insatiable motivation for rewards. Working for any handler may require a very independent approach to detection and relatively low attachment to individual handlers.

In this initial study, we sought to determine stakeholders' perspectives on the behavioural characteristics and performance of their current cohort of BDDs. This approach will inform ongoing research to design a standardised behavioural test battery for the evaluation and selection of candidate dogs for this agency. Specifically, we sought to probe the relevant agency staff's perspectives of canine working behaviours and their associations with performance within the agency. This was an important first step to strengthen the relevance of the ongoing research by drawing upon the authentic experiences of stakeholders and dogs in this role. Furthermore, since definitions for canine traits are rarely unequivocal, this approach sought to determine a shared vocabulary of traits with behavioural descriptors in order to collect data about individual dogs' working behaviours and performance within this agency.

To serve this purpose, we aimed to (1) develop a framework that captures scent detection traits relevant to BDDs; (2) using this framework, ascertain the behavioural profile of currently operational BDDs; and (3) assess the utility of a questionnaire for investigating work-related behaviours for BDDs. The research was conducted in two parts. Firstly, Australian government BDD stakeholders were interviewed in focus groups. Subsequently, a questionnaire based on findings from the focus group discussions was administered to the agency's dog handlers to assess the working characteristics of their currently operational dogs.

2. Focus Group Materials and Methods

2.1. Recruitment, Participants, and Procedures

All eligible employees (n = 75) were invited via email to participate on a voluntary basis. An information sheet that provided a brief overview of the study was distributed at this time. A total of 8 mini focus group interviews, with a mean of 3 participants per group (min = 2; max = 5), were conducted with employees (n = 25) of the Australian Government Department of Agriculture, Water, and Environment (DAWE) whose roles were directly related to BDD operations. They were grouped according to their roles to facilitate open discussion without potential impediment from the presence of line managers. Small group sizes with frequent time availabilities sought to encourage participation based on availability and allow individuals to speak at length. Staff participated during their usual working hours and each interview lasted approximately 60–90 min. The participants included dog handlers and kennel staff (n = 16), dog trainers (n = 3), supervisors (n = 3), and management (n = 3). Encouraging the participation of staff in various roles sought to invite different perspectives about various facets of BDD behaviour that might otherwise be overlooked.

The focus groups were conducted over Zoom and moderated by two members of the research team. Each participant joined the teleconference meeting from a separate device at a work location and each voluntarily consented to their participation. Participants were reminded that sharing personal opinions or experiences was encouraged.

A structured guide was used for the discussions, and the following three questions were asked in each group discussion:

1. What qualities do you think are important for success in a biosecurity detector dog?

2. What characteristics do you think make some dogs less successful?
3. If you were given the task of selecting dogs for this organisation, what characteristics would you prioritise in new candidates?

Finally, participants were asked to reiterate an important point they thought had been made in the discussion or one that had not yet been mentioned. These questions and structure were chosen to encourage participants to respond to the central question—what qualities influence the performance of BDDs?—from different perspectives.

The moderators asked additional questions throughout to clarify and further explore participants' views. These questions sought to encourage participants to describe specific behaviours, rather than traits or characteristic adjectives. Examples of such prompts included "what does that look like to you?" and "what do you mean by [statement]?"

2.2. Data Coding and Analysis

Focus group interviews were audio recorded, with the formal consent of all participants. Transcription and coding were conducted using NVivo (Version 12, 2018). The answers to the three questions were analysed together. The transcriptions were coded, and the coding rubric was refined iteratively using grounded theory analysis [21]. Firstly, transcripts were open coded using labels drawn from participants' descriptions. Following this, axial coding was used to draw connections between these nodes and group them into categories drawn from participants' explanations of the behaviours. Finally, these categories were selectively coded to arrive at the overall themes of the responses.

2.3. Ethics Statement

Approval to conduct this research was granted by the University of New England Human Research Ethics Committee (HE21–255).

3. Focus Group Results

Overall, participants described successful BDDs as those which were willing and capable of performing the role. Dogs' willingness was reflected in their perceived motivation and independence, which were qualities mentioned most frequently in discussions. Meanwhile, their capability was reported to depend upon their emotional stability, cognitive ability, and physical suitability.

3.1. Motivation

Dogs' willingness to work was the theme most frequently discussed by participants. This was generally described as "drive", a term commonly used in industry that refers to different aspects of canine motivation. In the context of detection work, participants referred to reward drive (food, play/prey, toy/dummy drives) and hunt drive.

In every group, participants described a dog with high drive as most suited for detection work. They were asked to expand upon this in each instance, and the resulting behavioural descriptors were used in the analysis. Reward drives were described as dogs' motivation or desire to access a primary reinforcer. Hunt drive tended to refer to a more complex set of behaviours that, taken as a whole, indicated a dog with a strong intrinsic desire to search using odour. Figure 1 presents participants' descriptions of these traits in terms of the dogs' primary motivators, their conditioned motivation to search for target odours, and the behavioural indicators of high motivation in a deployment context. These are discussed in the following sections.

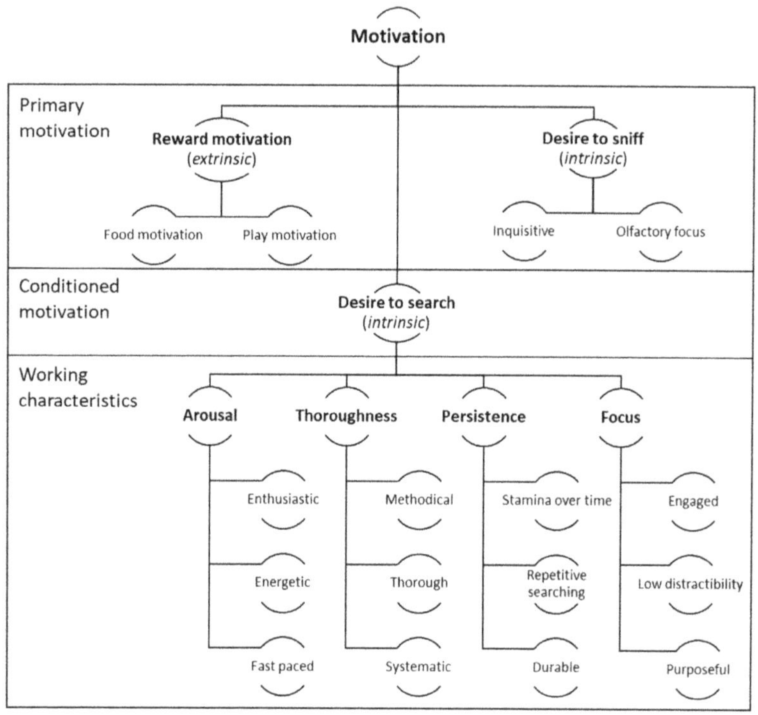

Figure 1. Characteristics and descriptors associated with work motivation among biosecurity detector dogs, as described by focus group participants.

3.1.1. Reward Motivation

Most participants (18/25) said that BDDs need to have a strong desire for a given reward. All eight groups raised this and no participants disagreed. Indeed, some described this level of desire as an "obsession". The behavioural indicators of this attribute were described as a consistent eagerness to eat (for food rewards) or to grab or chase an item (for play rewards).

> *"The higher the drive for those foods or play reward, I think the more successful you'll be with your training because if the dog values the reward, then they're more likely to put the effort into the actual task that we're asking them to do, especially when we're doing large volume screening with quite a low target rate."*
>
> (Speaker 11)

Participants indicated that this motivation should persist over time and/or multiple reinforcing events.

> *"A dog that doesn't reach satiation levels, so a dog that will constantly want more rewards … it constantly wants more food, more play, whatever the reward is. Otherwise they get to that level where they're content and have no interest in continuing to work."*
>
> (Speaker 12)

A dog's desire for rewards was considered an important factor for success as a scent detection dog. It was proposed that this motivation can be increased through training if required, but it was preferable that the dog naturally had a strong motivation for a reward. At the time the interviews were conducted, a desire for both food and dummy play rewards were perceived to be important for dogs to be effectively deployed in different locations where different types of rewards were used.

3.1.2. Desire to Sniff

Dogs' innate desire to seek and investigate odours was discussed in five of the eight groups, by 11 of the 25 participants. These participants described this as an underlying trait for a strong hunt drive or motivation to search. Several participants described dogs with this quality as curious or inquisitive, and most suggested that this is reflected in their tendency to explore their environment, especially through olfaction.

> *"They're nose to the ground straight away, they're looking for something ... actively engaging with things in their environment .... and investigating smells... So, for me, that's what the hunt drive looks like initially."*
>
> (Speaker 23)

3.1.3. Search Motivation

Finally, most participants in all eight groups (18/25), with no disagreement, stated that the dogs' desire to search for their trained target odours was one of the most important qualities for success. They explained that this quality is underpinned by the dogs' desire for a reward and their innate desire to sniff, resulting in the working task being highly reinforcing for the dogs.

> *"We don't want to make them work. We want them to love doing it. So, you want it to be just in their nature—they love it. Work should be the funnest [sic] thing ever."*
>
> (Speaker 8)
>
> *"Hunting and finding what they're after is almost like its own reward and they're keen to get back into it."*
>
> (Speaker 11)

The observed behaviours of dogs with this quality were described in several ways, as outlined in Figure 1.

According to participant descriptions, one behavioural indicator of dogs' motivation to search is their arousal level, as reflected in a dog that shows excited or energetic behaviours.

> *"They're showing that intensity."*
>
> (Speaker 10)
>
> *"A high drive dog that really wants to get out there, work at a fast pace ... go in, work hard, work fast"*
>
> (Speaker 17)
>
> *"A dog that has good energy and responds well to the handlers' energy as well, you know, eager to come out."*
>
> (Speaker 22)

However, importantly, participants clarified that over-arousal was not desirable, and could lead to tiring out quickly or an inability to focus, sometimes resulting in missed targets.

> *"When we talk about dogs that are highly motivated, often that gets confused with highly aroused ... [but] dogs with high levels of arousal [are] generally pretty poor searchers. The search suffers as a result; it's not methodical, it's not planned, it's not detailed and ... it's not sustainable over an extended period of time. So those dogs, regardless of how fit they are, generally have much shorter effective deployment times than a dog that has a lower state of arousal with a similar level of motivation."*
>
> (Speaker 24)

Dogs with a strong desire to search were also described as being focused and purposeful in their searching behaviour. This may manifest in their searching with more effort.

> *"Coming out with a purpose to work ... [they] work with intent"*
>
> (Speaker 17)

"[Dogs with low motivation] go through the motions, but it takes a lot of energy and work from the handlers to keep them on track and to keep them actively smelling . . . they might be walking where you need them to [but] they're not actively doing their job"

(Speaker 5)

It was also stated that this motivation should be persistent. If dogs are highly motivated, they should persist even when the search is difficult and should maintain motivation despite long durations, changing search conditions, or varying reinforcement schedules.

"[The] natural willingness to engage and hunt for it and maintain that hunt for a period of time rather than disengaging and losing interest."

(Speaker 22)

Thoroughness in their search was also proposed to characterise dogs that are highly motivated to find a target.

"There's quite often an emphasis on speed with the detection dogs, but I think that there is a lot of value in those dogs that are really methodical in the way that they search, whether it be luggage or parcels on a mail belt. I find that impressive when you see a dog that basically doesn't want to let anything go unsniffed [sic]."

(Speaker 11)

Overall, if dogs are highly motivated and are otherwise behaviourally, cognitively, and physically capable of the work, they would be expected to demonstrate arousal, focus, persistence, and thoroughness while searching.

3.2. Emotional Stability

BDDs in this agency are exposed to many potential stressors throughout their working life. Such stressors include busy and noisy work environments, frequent crating and transport, change of handlers, and kennel environments shared with other dogs. To be capable of working in these conditions, participants said that dogs should be free from excessive nervousness or anxiety while working and off duty (see Figure 2). Poor emotional stability and inability to cope with the stressors of the working role were identified as the most common causes of early retirement for the agency's scent detection dogs.

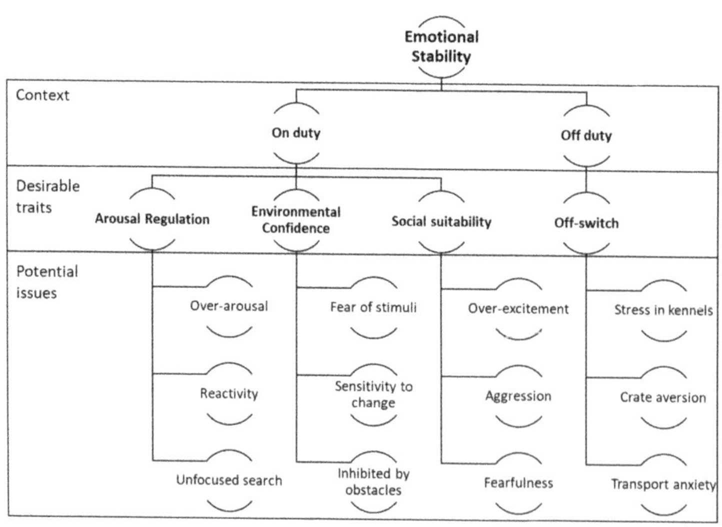

Figure 2. Emotional stability traits and descriptors in biosecurity detector dogs as described by focus group participants.

3.2.1. On Duty

Appropriate arousal, in terms of excitement and stress responses, was discussed in all eight groups. Arousal can be an indicator of a dog's willingness to work and dogs typically tend to be selected on the basis of showing high arousal behaviours, such as excitement and energy, in a search context. However, several participants also posited that over-arousal could contribute to chronic and acute stress responses, inappropriate reactivity to stimuli, slow recovery, and poor-quality searching.

According to participant responses, some dogs are less capable of coping with the stressors inherent in their working role. These dogs were often described as anxious or "high-strung". Participants commented that arousal behaviours stemming from anxiety, such as panting, tail-wagging, and restlessness, could be misinterpreted as excitement or drive.

> *"I think for some of our nervous dogs that that's been confused as being drive. So, they might look drivey [sic], but actually they're nervous and they're anxious."*
>
> (Speaker 16)

It was expressed in discussions that this anxiety and reactivity tended to detract from dogs' detection performance.

> *"You only get 50 percent as opposed to 95 percent of what the dog is capable of . . . A lot of dogs will look like they're actually going through the motions . . . but they're really not engaged in the whole game and 100 percent committed to what they're doing because their mind is half full about stuff that's happening around them."*
>
> (Speaker 1)

Similarly, dogs that work in a high state of arousal with little behavioural regulation may tend towards undesirable reactivity behaviours, including fixating on or barking at stimuli such as people or other dogs. This form of distractibility was identified as an issue for some dogs.

> *"We have had one or two dogs that I can recall that once they even see another dog walking even 50 meters away, they become fixated on that and that detracts from the focus of their work."*
>
> (Speaker 20)

In terms of social suitability, dogs are expected to behave neutrally towards strangers and other dogs while being relaxed and happy to be approached and handled. Participants agreed that, when working, dogs should not show emotional responses such as fearfulness, aggression, or over-excitability.

> *"[They have been] socialised with people. Not that they're wanting to run up and say hello, but that they understand that [for example] kids are there and they don't mind. They're not showing any aggression or fearfulness around different types of people."*
>
> (Speaker 12)

Similarly, participants expressed that dogs should not be fearful or stressed in response to environmental stimuli but instead should be able to confidently perform their working task regardless of the environmental conditions or obstacles.

> *"[Successful dogs are] able to work through pressure, they're not sensitive to changes in the environment. They're observant and acknowledge changes instead of being fearful of things that change and a different environment."*
>
> (Speaker 1)

The term "resilience" was mentioned in six of eight groups, with reference to being impervious to environmental stressors, sub-optimal handling, or changes in routine. This trait also embraced the dogs' ability to recover or cope with potentially stressful events and reflects an overarching quality of emotional regulation.

3.2.2. Off Duty

Participants agreed that it was important that dogs show appropriate emotional stability when off-duty, particularly in being able to rest and adapt to changes in routine. In the course of deployment, dogs are regularly crated to rest in between tasks and are transported to different areas throughout the day. Some can "switch off" and rest during these times, whereas other dogs show significant stress behaviours such as panting when confined.

> *"[A retired dog] was extremely anxious in the vehicle ... and actually got so bad that she would then not be able to be used for deployment because she was panting too much."*
>
> (Speaker 13)

Similarly, participants discussed chronic anxious behaviours at kennels. This not only represents a risk to the dog's welfare but may also detract from their working performance.

> *"Because they spend so much time at kennels, we have to have dogs that are able to have an off switch and able to turn off, chill out. We don't want dogs that are here for the time that they're not working to be anxious, pacing, barking, fence fighting [and showing] aggression to other dogs."*
>
> (Speaker 9)

Overall, dogs with poor emotional stability would be expected to show poor performance and be predisposed to negative welfare outcomes in the long term.

3.3. *Independence*

Dogs' independence and self-assuredness in their working role were discussed in seven of the eight groups and were identified as important for success in a detection role. Participants expressed that successful scent detection dogs tend to be only moderately obedient. An appropriate balance between obedience and independence allows a dog to respond to the handler, while being primarily interested in interacting with the environment rather than with people.

> *"Dogs that ended up being good dogs, dogs with drive, those dogs are hard to get back. Dogs that aren't that interested in a handler ... they're more engaged with the environment."*
>
> (Speaker 23)

Ideally, while working, preferred dogs were said to be minimally aware of the handler. Instead, the dogs are expected to make independent decisions about following or responding to an odour. Dogs lacking this quality were described as handler-dependent.

> *"[It is] an issue if they become handler dependent and they're just looking at the handler for the cues to find the target rather than actually just finding it themselves."*
>
> (Speaker 11)

In the agency, dogs are handled by several different handlers and, once fully trained, are expected to work with any handler. As such, they should work consistently regardless of the handler. This was described as being a particular challenge for some dogs.

> *"They're so dependent on my input and me being there and developing that bond is so much more important to them before they were willing to work."*
>
> (Speaker 7)

The extent to which dogs will search independently is likely impacted by their level of motivation. However, discussions revealed that even a highly motivated dog should also have a high level of self-assuredness and confidence to search independently. According to examples, dogs lacking this quality could become overly sensitive to handlers' behaviour while searching, thereby detracting from their detection performance.

3.4. Cognitive Ability

Participants in half of the groups also mentioned characteristics related to dogs' intelligence or cognitive aptitude. Some suggested that the dogs should be "trainable". Although this depends at least partly on the dogs' motivation to learn, participants expressed that some dogs appeared to be less capable of learning new tasks, despite their apparent motivation. Furthermore, one participant identified that some dogs were better than others at generalising between odours, which is an important part of the training process.

Additionally, four participants reported a preference for dogs that were good at problem solving, and notably at attempting new behaviours when trying to access a reward or reach a target odour. On the other hand, at least one handler mentioned that more intelligent dogs tended to seek potential shortcuts, such as handler cues in the training environment or environmental cues. That said, a trainer identified that this apparent reliance on handler cues could be overcome with training.

> *"[A trainable dog] looks like a dog that's willing to learn and is able to problem solve. There [are] some dogs that you can teach the basics to, and they're really good at "sitting" but . . . have difficulty . . . working more independently and problem-solving different situations."*
>
> (Speaker 2)

This theme was not frequently identified in discussions with handlers but was discussed in the trainer and technical supervisor groups.

3.5. Physical Capability

Finally, many of the participants mentioned or agreed (in six of eight groups) with the requirement for dogs to be physically capable of performing the role. This included their health and physical wellness, good physical condition and fitness, a structural conformation appropriate for efficient and uninhibited movement, and a physical size and type conducive to navigating obstacles.

This tended to be mentioned as a basic requirement that should be considered at the time of breeding and initial assessment. The discussions focused on behavioural attributes as the most important characteristics determining success among dogs that were physically suited for the task.

4. Survey Materials and Methods

4.1. Survey Development

Subsequent to the mini focus group interviews described above, a survey was created to collect information about the characteristics and performance of Australia's current operational BDDs. The survey had three components: (1) a validated canine personality questionnaire, (2) a questionnaire of working traits extracted from focus group discussions, and (3) ratings of working performance.

4.1.1. Component One: Monash Canine Personality Questionnaire—Revised

The first section of the survey was a previously validated measure of dog personality/behavioural traits, the Monash Canine Personality Questionnaire—Revised (MCPQ-R) [22,23]. It comprises a list of descriptive words (Table 1) with a 6-point Likert response scale from "Really does not describe [Dog Name]" to "Really describes [Dog Name]".

4.1.2. Component Two: Work Behaviour Questionnaire

The themes and constructs extracted from the focus group sessions were used to formulate questionnaire items about work-related traits that were expected to be important. Only the behaviours that all handlers could reasonably have been expected to witness were included, whereas questions about dogs' kennel behaviours, training, and cognitive abilities were excluded. A selection of participants' statements and examples that represented the prevalent themes were phrased as possible questionnaire statements. Based on clarity and anticipated ease of responding, statements were chosen by consensus among the experimenters and a DAWE representative. The statements were separated into overarching themes as determined in focus groups, and were then separated into the coded categories that were expected to be distinct, and condensed with a maximum of three statements from each category (presented in Table 2). These categories were then refined based on the outcomes of internal consistency testing using Cronbach's alpha (Table 2).

Table 1. Descriptive words used in the MCPQ-R and the label of their underlying dimension. Scores for the items under each label were averaged to produce an overall score.

Extraversion	Motivation [1]	Training Focus	Amicability	Neuroticism
Active Energetic Excitable Hyperactive Lively Restless	Assertive Determined Independent Persevering Tenacious	Attentive Biddable Intelligent Obedient Reliable Trainable	Easy-going Friendly Non-aggressive Relaxed Sociable	Fearful Nervous Submissive Timid

[1] Alternatively labelled as self-assurance.

For this component of the survey, participants were asked to rate their agreement with each statement on a 7-point Likert scale from "Strongly disagree" to "Strongly agree". The end of this section included a text area for optional comments, asking, "Do you have any other comments to add?" Responses to this were not analysed, but this question was included for clarification or feedback opportunity.

Table 2. Statements included for each construct in the work behaviour questionnaire with the overarching themes, coded categories, and subsequently refined categories. [R] indicates the items that were reverse-scored.

Theme	Original Category	Refined Category	Statement	Internal Consistency [1]
Motivation	Reward Motivation	Food Motivation	[Dog Name] has a strong desire for food rewards [Dog Name] will work really hard for a food reward [Dog Name] loses interest in a food reward after a few repetitions [R]	0.845
		Play Motivation	[Dog Name] has a strong desire for play rewards [Dog Name] will work really hard for a toy reward [Dog Name] loses interest in a toy reward after a few repetitions [R]	0.852
	Search Motivation	Search Motivation	[Dog Name] is always eager to start searching [Dog Name] will not start searching of their own accord [R]	0.731
		Desire to sniff	[Dog Name] always tends to sniff and investigate their surroundings	-
	Search Arousal	Speed and Intensity	[Dog Name] works at a fast pace [Dog Name] looks highly stimulated while searching	0.882
		Calmness	[Dog Name] appears calm while searching	-
	Focus	Engagement	[Dog Name] sometimes goes through the motions without actively searching [R]	-
		Distractibility	[Dog Name] ignores distractions while searching	-
	Persistence	Persistence	[Dog Name] wants to keep searching, even when the task is finished [Dog Name] sometimes gives up while searching [R] [Dog Name]'s search behaviour is consistent and durable	0.814
	Thoroughness	Thoroughness	[Dog Name] is methodical in their search pattern [Dog Name] searches areas superficially [R] [Dog Name] searches items thoroughly	0.721

Table 2. *Cont.*

Theme	Original Category	Refined Category	Statement	Internal Consistency [1]
Independence	Search Independence	Handler Consistency	[Dog Name] would work reliably for any handler	-
		Search independence	[Dog Name] relies heavily on direction from their handler [R] [Dog Name] often looks to their handler before indicating on an item [R]	0.863
Emotional Stability	Environmental Confidence	Environmental Confidence	[Dog Name] works well regardless of their surroundings [Dog Name] is sensitive to change in their environment [R] [Dog Name] recovers quickly after a stressful event	0.802
	Off-duty coping	Off-switch	[Dog Name] can "switch off" when not working	-
		Energy use	[Dog Name] tires themselves out while off-duty [R]	-
		Crate behaviour	[Dog Name] settles calmly in their crate	-

[1] Cronbach's alpha.

4.1.3. Component Three: Handler-Rated Working Performance

The final section asked handlers to rate the dog's detection performance on a sliding scale of 1–10 with 0.1 decimal steps, with 1 representing poor and 10 representing excellent. They were asked, based on their own perceptions drawn from working with the dog, to rate the dog's sensitivity ("How well you think they find all the targets that are present"), specificity ("How well they avoid making false responses"), and their general detection performance in the airport, mail centre, and overall. The end of this section included another text area for optional comments.

4.2. *Survey Distribution*

The survey was distributed electronically, via Qualtrics software (Qualtrics, Provo, UT), to all DAWE dog handlers. They were invited to report and rate the behaviours and performance of the individual dog(s) that they handled. The response rate was high; 88% of the currently operational dogs were reported on. A separate survey was completed for each individual dog (n = 36). The dogs were Labrador retrievers, 17 male and 19 female, aged 2–8 years (M = 67.8 months). Some handlers completed the survey for more than one dog. Because all responses were anonymous, we were unable to identify when this occurred and so made the decision to include all available data. Where there was more than one survey response for an individual dog, the responses were averaged to produce a single set of variables for that dog. This approach was taken to retain the maximum amount of information for greater accuracy. Descriptive statistics were then calculated for each new variable.

4.3. *Data Analysis*

Analyses were carried out using IBM SPSS Statistics Version 28.0.1.0 and RStudio Version 22.02.2.485 [24]. To develop a framework that captures scent detection traits relevant to BDDs, we first determined the underlying dimensions of working behaviours from the working behaviour questionnaire (Aim 1). Following this, we sought to ascertain the behavioural profile of currently operational BDDs (Aim 2). Finally, the utility of the questionnaire for investigating work-related behaviours for BDDs was assessed by its associations with the previously validated MCPQ-R (Aim 3) and whether dimensions of working behaviours were associated with performance ratings (Aim 4).

4.3.1. Aim 1: Determine Underlying Dimensions of Working Behaviours

The items from the work behaviour questionnaire were analysed using both a theory-grounded method and data-driven method to investigate the underlying dimensions. First, the framework derived from focus group coding was used as the basis for the construct subscales. The internal consistencies of these constructs were calculated and, in cases where Cronbach's α could be improved or was unacceptably low (< 0.5), were refined by separating items (Table 2). Items were averaged into a single score for each resulting category, with items reverse-coded, as marked in Table 2. Following this, an exploratory factor analysis (EFA) was conducted to determine whether the variance could be better explained with fewer dimensions. Bartlett's test of sphericity and KMO's test of sampling adequacy were used to determine the appropriateness of this method. An exploratory factor analysis using unweighted least squares, determined to be the most appropriate method for a small sample size, was conducted [25]. The number of factors was confirmed based on examination of the scree plot, eigenvalues > 1, and their interpretability. Orthogonal and oblique-rotated solution matrices were examined, and item loadings were found to be generally consistent between methods. An oblique rotation, Promax, was chosen due to the likelihood of correlations between factors. For each factor, a composite score was calculated from the mean of the items that had their primary loading on that factor.

4.3.2. Aim 2: Determine General Behavioural Profile of Operational Dogs

Five personality variables were calculated from the MCPQ-R items as in Ley et al. [23] (see Table 1). Cronbach's alpha was computed for each of these construct subscales to determine their reliability in this participant population. The novel work behaviour variables revealed from the EFA were calculated and assessed in the same way.

Descriptive statistics were calculated and plotted for each personality and work behaviour variable.

4.3.3. Aim 3: Assess Expected Associations between Work Behaviour Questionnaire and MCPQ-R

Associations that were theoretically expected between personality variables and working behaviour variables were tested as a measure of criterion validity of the work behaviour questionnaire. Many of the variables did not have normal distributions, and therefore, Spearman's rank correlation coefficients were used to analyse associations between them. The family-wise error rate was controlled for using Holm's sequential procedure.

4.3.4. Aim 4: Identify Dimensions Associated with Detection Performance

The personality and work behaviour dimensions were tested for associations with detection performance using the same method as in Aim 3. Additionally, groups were determined from the dogs' overall performance rating scores, allocated into three equal groups of the lowest, middle, and highest ratings. These groups were used to illustrate trends between performance and other dimensions.

4.4. Ethics Statement

The survey and methods were approved by the University of New England Human Ethics Research Committee (approval number HE22–018).

5. Survey Results

5.1. Aim 1: Determine Underlying Dimensions of Working Behaviours

Five of the initial theory-based constructs in the work behaviour questionnaire had acceptable (>0.7) or good (>0.8) internal consistencies. These were food motivation, play motivation, persistence, thoroughness, and environmental confidence. Others were refined into separate constructs, and are presented in Table 2. This process yielded 16 constructs in total.

A Bartlett's test of sphericity was significant for the work behaviour items (Chi-square = 945.98, df = 406, $p < 0.001$). A Kaiser–Meyer–Olkin measure of sampling adequacy was poor, at 0.592, likely owing to the small population size, and suggesting that results from the exploratory factor analysis (EFA) should be tentatively regarded. The EFA yielded seven factors (Table 3), accounting for 73% of the variance (Table 4).

Table 3. Factor loadings for each work behaviour questionnaire item. Loadings of <0.4 are not reported in this table.

Item	1	2	3	4	5	6	7
has a strong desire for food rewards				0.852			
will work really hard for a food reward				0.857			
loses interest in a food reward after a few repetitions				−0.772			
has a strong desire for play rewards					0.763		
will work really hard for a toy reward					0.838		
loses interest in a toy reward after a few repetitions					−0.912		
is always eager to start searching	0.566						
will not start searching of their own accord	−0.756						
always tends to sniff and investigate their surroundings	0.401						
works at a fast pace			1.015				
appears calm while searching		0.714					
looks highly stimulated while searching			0.973				
sometimes goes through the motions without actively searching	−0.767						
ignores distractions while searching							0.933
wants to keep searching, even when the task is finished			0.470				
sometimes gives up while searching	−0.561						
search behaviour is consistent and durable							
is methodical in their search pattern	0.467						
searches areas superficially	−0.871						
searches items thoroughly	0.947						
would work reliably for any handler		0.652					
relies heavily on direction from their handler						0.689	
often looks to their handler before indicating on an item						0.923	
works well regardless of their surroundings		0.682					
is sensitive to change in their environment		−0.673					
recovers quickly after a stressful event		0.576					
can "switch off" when not working		0.540					
tires themselves out while off-duty		−0.505					
settles calmly in their crate		0.618					

Table 4. Factor labels and internal consistencies of items with the highest loading onto factor.

Factor	Factor Label	No. of Items	% Variance Explained	Cronbach's Alpha
1	Search Motivation	8	27.67	0.866
2	Emotional Stability	8	11.59	0.823
3	Search Arousal	3	9.78	0.819
4	Food Motivation	3	8.81	0.845
5	Play Motivation	3	6.14	0.852
6	Search Independence	2	1.08	0.863
7	Focus	1	4.47	-

The dimensions yielded from factor analysis were meaningful, consistent with the original theoretical framework, had high internal consistencies, and were able to describe variation with fewer variables than the original constructs. As such, these factors were scored as new variables using an average composite score and used for the following analyses. Items loaded to factor six were reverse-scored and the factor labelled "search independence" for consistency and clarity.

5.2. Aim 2: Determine General Behavioural Profile of Operational Dogs

The descriptive statistics of each survey section are provided here as a baseline of the overall profile of the performance, behaviours, and personalities of this cohort of operational dogs.

5.2.1. Performance Ratings

Overall, dogs were rated highly ($M \geq 7.5$) for sensitivity, performance in the airport and mail centre, and their performance overall (Figure 3).

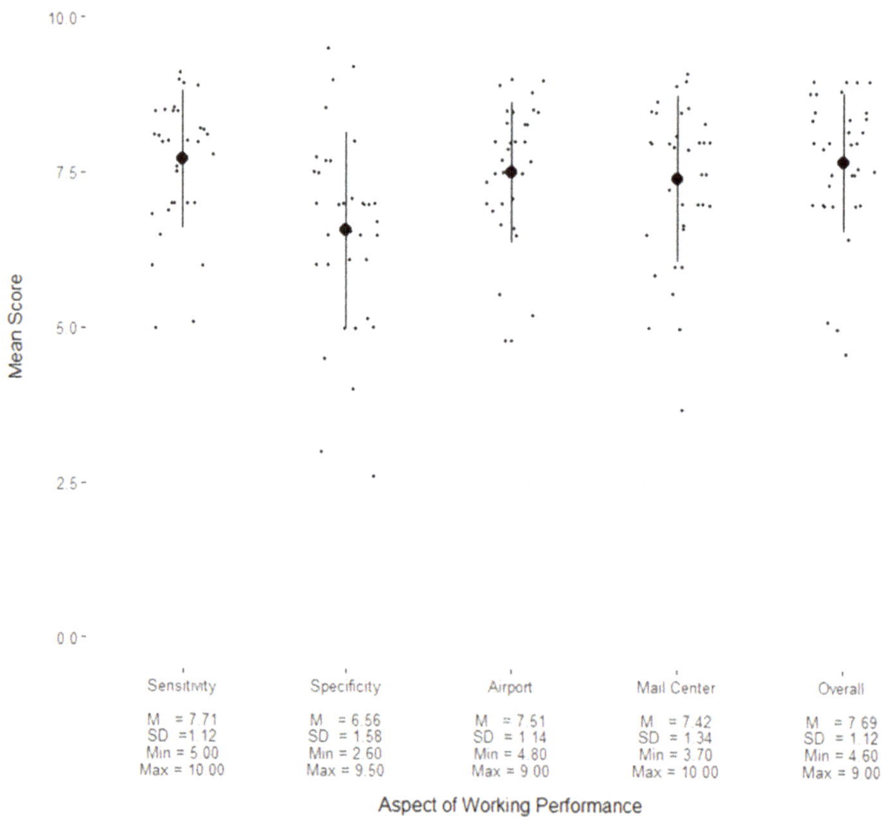

Figure 3. Working performance scores, where 0 indicates poor and 10 indicates excellent. The circles indicate the mean and lines indicate the standard deviation. Each point is an individual dog's score. "Sensitivity" is the label for the item, "How well you think they find every target that is present?" "Specificity" is the label for the item, "How well they avoid making false responses?" "Airport" and "Mail Center" are the labels for items asking how well dogs perform at each of these locations. "Overall" is the label for the item asking for a rating of the dog's overall performance.

5.2.2. Work Behaviours

Handlers tended to agree that the dogs exhibited behaviours related to search motivation, emotional stability, search arousal, food motivation, play motivation, search independence, and focus, with mean ratings of >4 on these variables (Figure 4).

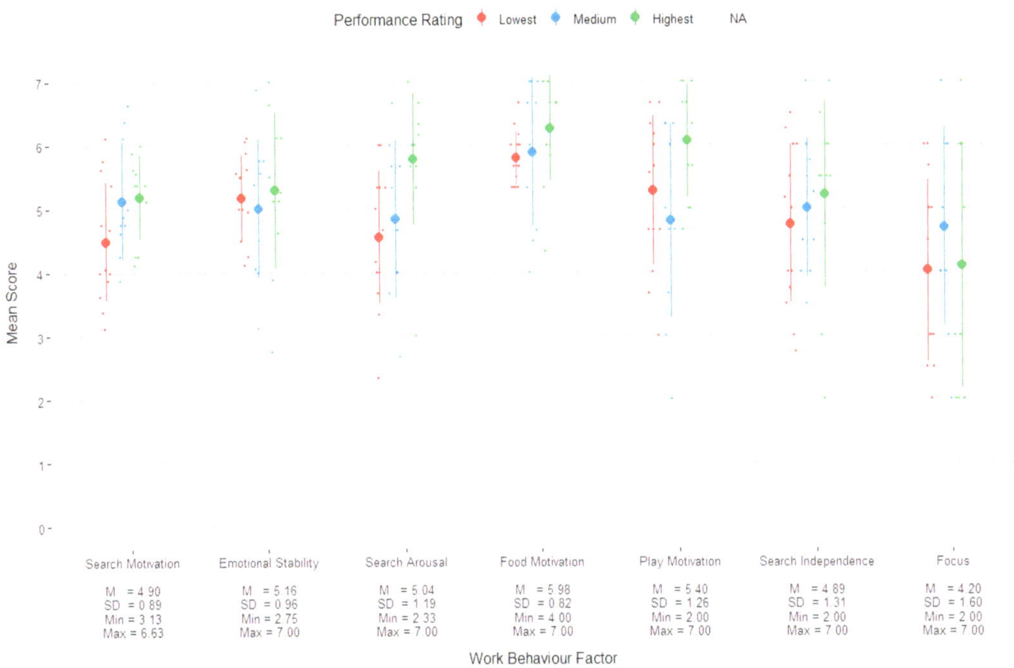

Figure 4. Scores of work behaviour dimensions for each group of performance rating. The circles indicate the mean and lines indicate the standard deviation. Each point is an individual dog's score. For illustrative purposes, performance ratings groups were determined from the dogs' overall performance rating scores allocated into three equal groups of the lowest, middle, and highest ratings. Descriptive statistics are presented below each factor label.

5.2.3. MCPQ-R

Most of the MCPQ-R subscales had satisfactory internal consistencies. The calculated Cronbach alphas were moderate to high: 0.87 for extraversion, 0.78 for amicability, 0.81 for motivation, and 0.69 for neuroticism. However, the items contributing to training focus yielded poor internal consistency in this sample, with an alpha of 0.48. As such, results related to this variable should be considered with caution. The scores for each of these subscales are presented in Figure 5.

5.3. Aim 3: Assess Expected Associations between Work Behaviour Questionnaire and MCPQ-R

To assess the criterion validity of the work behaviour questionnaire, hypothesis testing was carried out to determine whether associations existed between this questionnaire and personality traits measured with the MCPQ-R. Reported here are tests of the expected associations based on theoretical relatedness. Most of the expected correlations were significant, and of those that were, all were in the expected direction (Table 5). All other correlation values can be found in Table S1.

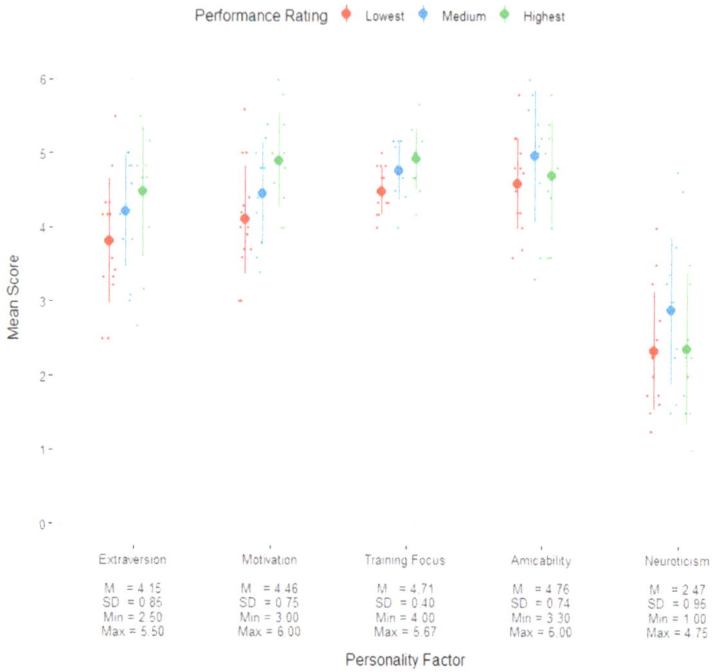

Figure 5. Scores of MCPQ-R personality factors for each group of performance rating. The circles indicate the mean and lines indicate the standard deviation. Each point is an individual dog's score. For illustrative purposes, performance ratings groups were determined from the dogs' overall performance rating scores allocated into three equal groups of the lowest, middle, and highest ratings.

Table 5. Expected correlations between work behaviour factors and personality factors with calculated correlations and significance with Holm–Bonferroni correction. The Holm alpha is the alpha needed for significance to correct for family-wise error rates.

Correlation	Spearman's *rho*	*p*-Value	Holm Alpha
Search motivation with Motivation (Self-assuredness)	0.526 *	< 0.001 *	0.007
Emotional stability with			
Neuroticism	−0.617 *	< 0.001 *	0.008
Amicability	0.537 *	< 0.001 *	0.01
Search arousal with			
Extraversion	0.444 *	0.007 *	0.025
Motivation	0.695 *	<0.001 *	0.013
Play motivation with			
Extraversion	0.111	0.519	0.05
Search independence with Motivation (Self-assuredness)	0.549 *	<0.001 *	0.017

* denotes significance following Holm–Bonferroni correction.

5.4. Aim 4: Identify Dimensions Associated with Detection Performance

5.4.1. Personality

Trait motivation, alternatively labelled self-assurance in the original study [22], was associated with detection performance outcomes (Table 6). Training focus, despite having low internal consistency in the current population, was also associated with overall detection performance.

Table 6. Correlations between overall working performance score and work behaviour factors with calculated correlations and significance with Holm–Bonferroni correction.

Factor	Spearman's *rho*	*p*-Value	Holm Alpha
Extraversion	0.204	0.239	0.017
Motivation	0.473 *	0.004 *	0.0125
Training Focus	0.528 *	0.001 *	0.01
Amicability	0.125	0.475	0.025
Neuroticism	−0.053	0.762	0.05

* denotes significance following Holm–Bonferroni correction.

5.4.2. Work Behaviour

Only two of the work behaviour factors, search motivation and search arousal, were associated with ratings of overall performance (Table 7).

Table 7. Correlations between overall working performance score and work behaviour factors with calculated correlations and significance after Holm–Bonferroni correction.

Factor	Spearman's *rho*	*p*-Value	Holm Alpha
Search motivation	0.458 *	0.006	0.008
Emotional stability	0.218	0.209	0.025
Search arousal	0.491 *	0.003	0.007
Food motivation	0.317	0.064	0.01
Play motivation	0.230	0.184	0.0167
Search independence	0.256	0.138	0.0125
Focus	0.173	0.319	0.05

* denotes significance following Holm–Bonferroni correction.

6. Discussion

The current study achieved its aim of identifying traits that are considered important for Australia's biosecurity detector dogs (BDDs) and evaluating how these traits are expressed in the current population of operational dogs. First, a collection of traits that are relevant to the performance and welfare of BDDs was determined from focus group interviews. This process sought a more nuanced understanding of specific working behaviours and their meaning in the context of this agency.

Subsequently, a questionnaire, designed to quantitatively assess working traits relevant for dogs in this agency, revealed several underlying dimensions of work behaviours. These dimensions were labelled search motivation, emotional stability, search arousal, food motivation, play motivation, search independence, and focus. Using these labels, we investigated the trait expressions of currently operational scent detection dogs to provide a reference for the typical behaviours of BDDs.

This methodical approach to consulting with key stakeholders will inform an ongoing project. It bolsters content validity for a new selection testing procedure and a metric by which to collect information about their day-to-day working behaviour and performance.

6.1. Work Behaviour Framework

A series of behaviours and descriptions were categorised into a smaller number of variables, reflecting the dimensions that underlie these work-related behaviours [26]. Initially, interview transcripts were coded and behaviours were categorised into ostensible constructs. Subsequently, a data-driven analysis of questionnaire results revealed factors that largely aligned with the original constructs but were more parsimonious, grouping the behaviours into broader but meaningful dimensions. A framework using these dimension labels was adopted for analyses and ongoing research within this population of dogs.

The dimension of "search motivation" incorporated the largest number of questionnaire items. The items contributing to this dimension described the dogs' eagerness, engagement, and thoroughness while searching for a target. Meanwhile, a separate factor also emerged, labelled 'search arousal', which described dogs' energy and pace while working. According to focus group findings, search arousal may relate to one aspect of search motivation but may not necessarily reflect search effort. On the other hand, the behaviours clustered in the dimension of search motivation suggest effort and engagement in the task of searching. Both search motivation and search arousal reflect key descriptions of working behaviour that many focus group participants and industry professionals tend to describe as "drive" or "hunt drive" [2,7,11,27].

Behaviours indicative of "emotional stability" accounted for approximately 12% of the variation among dogs in the current population. They related to the dogs' environmental sensitivity, stress coping, and "off-switch" behaviours. In addition to the questionnaire items intended to measure this construct, two other items, one rating dogs' perceived calmness while searching and the other rating their ability to work with any handler, also loaded onto this factor. Connections to this dimension are logical in that calm search behaviour suggests low apparent stress, whereas especially sensitive or reactive dogs may require particular handling or the support of a known handler. Emotional stability, or aspects of it, have been investigated in other canine scent-detection research, and behaviours relating to this dimension have been considered an important indicator of dogs' overall success in the working environment [2,11,14].

Dogs' desire for primary reinforcers was reflected by two factors, labelled "food motivation" and "play motivation". These constructs may underpin dogs' initial trainability [8] and the strength of their conditioned motivation to search.

"Search independence" emerged as a separate factor and reflected dogs' tendency not to rely on handler input while working. According to focus group interviews, this working trait likely relates to a dog's self-assurance and confidence to make decisions. This may relate to trait "boldness", which has been investigated in previous working dog research and has been found to positively predict performance [28]. However, it is also possible that search independence behaviours are influenced by the dog's training and experience in the role.

Finally, one item—"ignores distractions while searching"—loaded separately to other items. This may be because dogs' distractibility is moderated by more than one underlying dimension, such as motivation and emotional stability, or it may be a separate dimension that no other items measured. The inclusion of this dimension requires further consideration in a future iteration of the questionnaire, as it would require more relevant items for it to be a reliable measure of this trait.

6.2. Traits of Operational Dogs

All of the individuals in this study are operational dogs that have completed training, and so all are considered examples of successful scent detection dogs. Indeed, when rated for overall detection performance, their mean score was above 7.5 on a scale of 10. Accordingly, we expected that trends in this population as a whole would be useful to inform our understanding of the characteristics generally required to perform the role.

Additionally, although this narrow range of variability limits our ability to detect all associations between performance and individual traits, there was some meaningful variation in performance ratings among the dogs. As such, the observed associations and trends between behavioural characteristics and working performance were also explored. This process sought to investigate the behaviours that may predict performance potential and work towards identifying and selecting the highest performing dogs.

6.2.1. Work Behaviour Traits

Overall, the average ratings of operational dogs for work-related behavioural traits were predominantly as expected (Figure 4). This cohort of dogs was rated highly for "search motivation", "emotional stability", and "food motivation", with mean scores between 4.9 and 5.9 on a scale of 7, and little variation, with standard deviations below 1. This suggests that these are key traits that operational BDDs consistently demonstrate. On the other hand, more inter-individual variation was observed for "search arousal", "play motivation", "search independence", and "focus", with standard deviations between 1.2 and 1.6. This may suggest that these traits are not essential indicators of a dog's ability to complete training and become operational, although they may still contribute to performance outcomes. Comparing successful working dogs against unsuccessful dogs will be helpful in future to determine to what extent variability in specific traits can be tolerated.

Search motivation and search arousal were both significantly positively associated with ratings of overall detection performance. It was expected that more effort and eagerness while searching would translate to better performance outcomes. Similarly, it is feasible that search arousal, which, to some extent, can reflect a dog's enthusiasm to perform the task, would predict detection performance. However, according to interview discussions, it is likely that this relationship is not always linear and may instead have an inverted U-shape trend. While some degree of search arousal is desirable, excessive arousal could interfere with the dogs' ability to search effectively, due to its effect on cognitive and attentional factors [6,8,29]. In this population of successful operational dogs, search arousal likely manifests at an appropriate and adaptive level, and therefore, inappropriately high levels of arousal may not be present in this sample. Overall, our findings suggest that search motivation and arousal contribute positively to perceived detection performance in an operational context. It is possible that other work-related traits may contribute to performance outcomes, but there was insufficient variability within this relatively small population to observe a statistical association.

Emotional stability was frequently mentioned in focus group interviews but did not appear to predict detection outcomes in this population. It was hypothesised that dogs low in emotional stability would perform more poorly in a detection role than dogs high in emotional stability. Stress responses, such as fear and inability to rest, have been found to affect cognitive processes [6,30] which are believed to contribute to detection ability [10]. It is possible that the current population did not include any dogs with emotional stability so low as to compromise their performance. In the focal agency, dogs are required to work in public and potentially stressful locations, be kennelled at central facilities, and be handled by different handlers, all of which may require above-average resilience [6,31]. Therefore, there was likely a minimum threshold of emotional stability to be included in this sample, and this sampling bias may have obscured an association. Although in this specific population we did not observe an association between emotional stability and performance ratings, this dimension should not be discounted due to its likely impact on dogs' welfare, safety, and ease of handling.

6.2.2. Personality

Operational dogs tended to be rated highly in amicability, extraversion, and motivation (alternatively labelled as self-assuredness) [22], whereas they were rated fairly low in neuroticism, as measured by the MCPQ-R (Figure 5). Trait motivation (or self-assuredness) was positively associated with performance ratings. This aligns with the perspectives of focus group participants of the descriptions of dogs suited for this role, particularly that they are non-aggressive, energetic, confident, and have high emotional stability, all of which are theoretically related to the above personality traits.

Although dogs had high scores for training focus and this trait was associated with performance outcomes, this construct had low internal consistency in this population, and so may not have offered an accurate representation of the construct. This is possibly due to interpretations of the adjectives (e.g., intelligent, obedient, reliable) by professional participants that diverge from common expectations. This highlights one of the key difficulties of identifying and measuring traits using subjective measures and reinforces the importance of considering different populations' understandings of adjectives depending on context when using such measures.

6.3. Work Behaviour Questionnaire Utility

Survey data can have the advantage of providing information based on an extended period of observation that is not always accessible or feasible to collect using objective measures [26]. This information can therefore offer a more granular measure of ongoing detection performance and variation than can be gleaned from measures such as pass-fail training outcomes, artificial detection tasks, or overall detection statistics. These objective measures are certainly an important component in validating new testing procedures; however, they can be limited in their information and in some cases can be misleading. For example, artificial tasks rarely present all of the same challenges as faced in deployment, and operational detection statistics can be impacted by differences in opportunity to make detections. As such, survey data can be a useful additional tool for the validation of behavioural tests, particularly to develop in-house assessment methods suited for a particular application or context.

We therefore assessed the utility of this work behaviour questionnaire to measure accurately the working traits in scent detection dogs and to warrant its ongoing use in this context. The face validity of the work behaviour questionnaire was supported by its reliance upon statements used by various focus group participants to describe each construct. We further assessed the validity of this questionnaire by its associations with other measures.

The majority (6/7) of hypothesised correlations between work behaviour factors and personality factors were significant and in the expected direction (Table 5). For example, emotional stability in the workplace was related to trait neuroticism and amicability, as described by the MCPQ-R. Trait motivation (alternatively labelled as self-assurance) was positively associated with search motivation, search arousal, and search independence. Extraversion, which in the MCPQ-R suggests high energy, predicted search arousal, although not play motivation. Overall, these associations provide some evidence that the questionnaire items reflected the intended domains.

A key aim of the work behaviour questionnaire was to probe dogs' detection behaviours and performance on a granular level. As intended, the questionnaire appeared to glean information about specific dimensions of work behaviour that are relevant to performance. The survey revealed two domains positively associated with performance ratings in this population of dogs which align with two different descriptions of "drive" cited in focus group discussions. "Drive" was the most commonly cited trait of successful scent detection dogs in the focus group discussions, and the questionnaire findings support the importance of this over-arching trait.

A recognised limitation of the current study design is that detection performance could not be measured objectively in such a way that reliably reflected BDD general performance. As such, we relied on subjective perceptions of performance, which may not be entirely accurate or may capture only one part of the picture. However, handlers in this agency handle a variety of different dogs and so likely would be well-versed in assessing dog behaviour. Furthermore, they do not own the dogs they handle and do not carry out the initial training for the dogs, which removes much of the motivation to purposefully under- or over-estimate the dogs' performance. Nevertheless, since unconscious biases can persist in any subjective measure, future comparisons with an objective measure of performance and work behaviour may provide a valuable indication of convergent validity.

Furthermore, the work behaviour questionnaire did not collect information about other aspects that were identified as important during focus group discussions, including their kennel behaviour, training, cognition, and physical capability. For future selection processes, these aspects will be considered and measured using other methods.

6.4. Future Directions

These findings will inform the development of an in-house selection testing procedure that addresses the needs of this detection role while also considering the current scientific knowledge base. This first step sought to consult with stakeholders about their experiences with BDDs, and thus is limited to a group of detection dogs that are performing the role successfully with a generally high standard of performance. As such, only general associations with performance were investigated in this instance. The findings warrant further scrutinization and predictive modelling applied to a larger group of candidate dogs with greater variation in their working suitability. This will be achieved by administering to a subsequent cohort of candidate dogs a behaviour testing procedure designed to measure relevant traits, and then comparing those behaviours to training outcomes and survey ratings of working behaviour for those dogs which become operational.

7. Conclusions

Our findings suggest that this methodology can identify and assess the important characteristics of a specific working dog population and role. As each working dog role is different, it is logical that different trait frameworks and behavioural examples will apply depending on the context. Consultations with stakeholders and assessment of experienced dogs in a particular role are valuable contributions to the design and advancement of behavioural testing and selection procedures.

This research revealed a collection of work-related attributes in a population of scent detection dogs used for biosecurity. These were food motivation, play motivation, search motivation, search arousal, emotional stability, search independence, and search focus. In particular, search motivation and search arousal were positively associated with detection performance ratings. These domains mirror two different examples of "drive", as described by industry professionals in the focus group interviews. Emotional stability was another broad dimension that encompassed many important behavioural traits. Although this construct was not directly associated with detection performance, focus group discussions strongly emphasised its importance for positive welfare and handling of the dogs. Overall, there was concordance between the qualitative and quantitative methods to describe the important domains of working detector dog behaviour.

In addition, the current questionnaire, developed to assess these work behaviour factors among detector dogs, had preliminary validity, as evidenced by correlations with validated measures of personality and associations with detection performance ratings. As such, it may be a useful tool to assess the predictive validity of other indirect measures, such as standardised behavioural testing. The ultimate aim of this will be to predict the future working behaviours and performance of unfamiliar candidate dogs based on a behavioural measure taken at a single time point.

Measuring and understanding the individual characteristics of dogs and the association of these traits with working behaviours will pave the way to advance the selection, training, and handling of scent detection dogs. Improvements in these processes could reduce the economic and time investments of purchasing and training dogs and enhance detection performance in operational contexts. Furthermore, we could expect improved welfare outcomes by selecting dogs that are capable and motivated to perform scent detection work, resulting in enriching experiences for both the dogs and their handlers.

Supplementary Materials: The following supporting information can be downloaded at: https://www.mdpi.com/article/10.3390/ani13030504/s1. Table S1. A correlation table depicting Spearman's rho correlation coefficients and significance values between MCPQ-R variables and work behaviour questionnaire variables. Highlighted are the expected correlations based on theoretical relatedness.

Author Contributions: Conceptualization, A.Y.M., W.Y.B., P.B., P.S.T. and P.M.; methodology, A.Y.M., W.Y.B., P.B., P.S.T. and P.M.; formal analysis, A.Y.M. and B.W.; investigation, A.Y.M. and W.Y.B.; data curation, A.Y.M.; writing—original draft preparation, A.Y.M.; writing—review and editing, W.Y.B., P.B., P.S.T. and P.M.; visualization, A.Y.M.; supervision, W.Y.B., P.B., P.S.T. and P.M.; project administration, A.Y.M. and W.Y.B.; funding acquisition, A.Y.M. and W.Y.B. All authors have read and agreed to the published version of the manuscript.

Funding: This research was funded by the Biosecurity Innovation Program through the Australian Government Department of Agriculture, Water, and Environment.

Institutional Review Board Statement: The study was conducted in accordance with the Declaration of Helsinki, and approved by the Ethics Committee of the University of New England (HE21-255 and HE22-018).

Informed Consent Statement: Informed consent was obtained from all subjects involved in the study.

Data Availability Statement: The data presented in this study are available on request from the corresponding author. The data are not publicly available due to confidentiality requirements.

Acknowledgments: The authors acknowledge the considerable contribution from the Biosecurity Detector Dog team of the Australian Government Department of Agriculture, Water, and Environment. Jeff Smith collaborated with us to create the project proposal and research plan, and continued to advise, support, and provide technical expertise throughout. We wish to thank all of the staff who participated in the focus group interviews and/or completed surveys who provided their experience and expertise.

Conflicts of Interest: The authors declare no conflict of interest.

References

1. Lorenzo, N.; Wan, T.; Harper, R.J.; Hsu, Y.-L.; Chow, M.; Rose, S.; Furton, K.G. Laboratory and Field Experiments Used to Identify *Canis Lupus* Var. *Familiaris* Active Odor Signature Chemicals from Drugs, Explosives, and Humans. *Anal. Bioanal. Chem.* **2003**, *376*, 1212–1224. [CrossRef] [PubMed]
2. Lazarowski, L.; Haney, P.S.; Brock, J.; Fischer, T.; Rogers, B.; Angle, C.; Katz, J.S.; Waggoner, L.P. Investigation of the Behavioral Characteristics of Dogs Purpose-Bred and Prepared to Perform Vapor Wake® Detection of Person-Borne Explosives. *Front. Vet. Sci.* **2018**, *5*, 50. [CrossRef] [PubMed]
3. Beebe, S.C.; Howell, T.J.; Bennett, P.C. Using Scent Detection Dogs in Conservation Settings: A Review of Scientific Literature Regarding Their Selection. *Front. Vet. Sci.* **2016**, *3*, 96. [CrossRef] [PubMed]
4. Needs, S.; Bennett, E.; Mao, B.; Hauser, C.E. Do Detection Dogs Respond Differently to Dried, Frozen and Live Plant Targets? *Appl. Anim. Behav. Sci.* **2021**, *236*, 105276. [CrossRef]
5. Jendrny, P.; Twele, F.; Meller, S.; Osterhaus, A.D.M.E.; Schalke, E.; Volk, H.A. Canine Olfactory Detection and Its Relevance to Medical Detection. *BMC Infect. Dis.* **2021**, *21*, 838. [CrossRef] [PubMed]
6. Troisi, C.A.; Mills, D.S.; Wilkinson, A.; Zulch, H.E. Behavioral and Cognitive Factors That Affect the Success of Scent Detection Dogs. *Comp. Cogn. Behav. Rev.* **2019**, *14*, 51–76. [CrossRef]
7. Jamieson, L.T.J.; Baxter, G.S.; Murray, P.J. Identifying Suitable Detection Dogs. *Appl. Anim. Behav. Sci.* **2017**, *195*, 1–7. [CrossRef]
8. Lazarowski, L.; Waggoner, L.P.; Krichbaum, S.; Singletary, M.; Haney, P.; Rogers, B.; Angle, C. Selecting Dogs for Explosives Detection: Behavioral Characteristics. *Front. Vet. Sci.* **2020**, *7*, 597. [CrossRef]
9. Rooney, N.J.; Bradshaw, J.W.S.; Almey, H. Attributes of Specialist Search Dogs—A Questionnaire Survey of UK Dog Handlers and Trainers. *J. Forensic Sci.* **2004**, *49*, 300–306. [CrossRef]
10. Bray, E.E.; Otto, C.M.; Udell, M.A.R.; Hall, N.J.; Johnston, A.M.; MacLean, E.L. Enhancing the Selection and Performance of Working Dogs. *Front. Vet. Sci.* **2021**, *8*, 644431. [CrossRef] [PubMed]
11. Rocznik, D.; Sinn, D.L.; Thomas, S.; Gosling, S.D. Criterion Analysis and Content Validity for Standardized Behavioral Tests in a Detector-Dog Breeding Program. *J. Forensic Sci.* **2015**, *60* (Suppl. S1), S213–S221. [CrossRef] [PubMed]
12. Maejima, M.; Inoue-Murayama, M.; Tonosaki, K.; Matsuura, N.; Kato, S.; Saito, Y.; Weiss, A.; Murayama, Y.; Ito, S. Traits and Genotypes May Predict the Successful Training of Drug Detection Dogs. *Appl. Anim. Behav. Sci.* **2007**, *107*, 287–298. [CrossRef]
13. Ganitskaya, Y.V.; Feoktistova, N.Y.; Vasukov, D.D.; Surov, A.V. Some Behavioral Features Required for the Selection of Detection Dogs. *Biol. Bull.* **2020**, *47*, 501–505. [CrossRef]

14. McGarrity, M.E.; Sinn, D.L.; Thomas, S.G.; Marti, C.N.; Gosling, S.D. Comparing the Predictive Validity of Behavioral Codings and Behavioral Ratings in a Working-Dog Breeding Program. *Appl. Anim. Behav. Sci.* **2016**, *179*, 82–94. [CrossRef]
15. Tiira, K.; Tikkanen, A.; Vainio, O. Inhibitory Control—Important Trait for Explosive Detection Performance in Police Dogs? *Appl. Anim. Behav. Sci.* **2020**, *224*, 104942. [CrossRef]
16. MacLean, E.L.; Hare, B. Enhanced Selection of Assistance and Explosive Detection Dogs Using Cognitive Measures. *Front. Vet. Sci.* **2018**, *5*, 236. [CrossRef] [PubMed]
17. Rooney, N.J.; Gaines, S.A.; Bradshaw, J.W.S.; Penman, S. Validation of a Method for Assessing the Ability of Trainee Specialist Search Dogs. *Appl. Anim. Behav. Sci.* **2007**, *103*, 90–104. [CrossRef]
18. Lazarowski, L.; Krichbaum, S.; Waggoner, L.P.; Katz, J.S. The Development of Problem-Solving Abilities in a Population of Candidate Detection Dogs (*Canis Familiaris*). *Anim. Cogn.* **2020**, *23*, 755–768. [CrossRef] [PubMed]
19. Hare, E.; Kelsey, K.M.; Serpell, J.A.; Otto, C.M. Behavior Differences Between Search-and-Rescue and Pet Dogs. *Front. Vet. Sci.* **2018**, *5*, 118. [CrossRef]
20. Lazarowski, L.; Strassberg, L.R.; Waggoner, L.P.; Katz, J.S. Persistence and Human-Directed Behavior in Detection Dogs: Ontogenetic Development and Relationships to Working Dog Success. *Appl. Anim. Behav. Sci.* **2019**, *220*, 104860. [CrossRef]
21. Charmaz, K. *Constructing Grounded Theory*; SAGE: Thousand Oaks, CA, USA, 2014; ISBN 9781446297223.
22. Ley, J.; Bennett, P.C.; Coleman, G.J. A Refinement and Validation of the Monash Canine Personality Questionnaire (MCPQ). *Appl. Anim. Behav. Sci.* **2009**, *116*, 220–227. [CrossRef]
23. Ley, J.; McGreevy, P.; Bennett, P.C. Inter-Rater and Test–Retest Reliability of the Monash Canine Personality Questionnaire-Revised (MCPQ-R). *Appl. Anim. Behav. Sci.* **2009**, *119*, 85–90. [CrossRef]
24. R Core Team R. A Language and Environment for Statistical Computing 2022. Available online: http://www.r-project.org (accessed on 1 December 2022).
25. Jung, S. Exploratory Factor Analysis with Small Sample Sizes: A Comparison of Three Approaches. *Behav. Process.* **2013**, *97*, 90–95. [CrossRef] [PubMed]
26. Wiener, P.; Haskell, M.J. Use of Questionnaire-Based Data to Assess Dog Personality. *J. Vet. Behav.* **2016**, *16*, 81–85. [CrossRef]
27. Cobb, M.; Branson, N.; McGreevy, P.; Lill, A.; Bennett, P. The Advent of Canine Performance Science: Offering a Sustainable Future for Working Dogs. *Behav. Process.* **2015**, *110*, 96–104.
28. Svartberg, K. Shyness–Boldness Predicts Performance in Working Dogs. *Appl. Anim. Behav. Sci.* **2002**, *79*, 157–174. [CrossRef]
29. Bray, E.E.; MacLean, E.L.; Hare, B.A. Increasing Arousal Enhances Inhibitory Control in Calm but Not Excitable Dogs. *Anim. Cogn.* **2015**, *18*, 1317–1329. [CrossRef]
30. Foraita, M.; Howell, T.; Bennett, P. Environmental Influences on Development of Executive Functions in Dogs. *Anim. Cogn.* **2021**, *24*, 655–675. [CrossRef]
31. Lazarowski, L.; Waggoner, P.; Katz, J.S. The Future of Detector Dog Research. *Comp. Cogn. Behav. Rev.* **2019**, *14*, 77–80. [CrossRef]

animals MDPI

Systematic Review

Digital Technology Supporting the Remote Human-Dog Interaction: Scoping Review

Liliana Rodríguez-Vizzuett, Ismael E. Espinosa-Curiel and Humberto Pérez-Espinosa *

CICESE-UT3, Tepic 63155, Mexico
* Correspondence: hperez@cicese.mx

Simple Summary: Due to the close affective and collaborative relationship between dogs and humans, in several situations there is a need to maintain communication when it is not possible to do it face to face. The objective of this review is to analyze the main aspects of current technologies that support remote communication between dogs and humans. Fifteen articles were selected which were conscientiously analyzed. The most widely used technologies to allow dogs to generate messages are wearable devices equipped with sensors. The most used technologies for dogs to receive messages are wearable devices equipped with vibrotactile actuators. Most of the proposals developed only include one-way communication, and those that include bidirectional communication uses videochats. All reported evaluations were pilot studies with positive feasibility results. The use of technology to support remote human-dog interaction is generating a lot of anticipation and excitement. However, there is still a long way to go in terms of technological developments, integration into the activities and context of dogs, support for new modalities of dog interaction, adaptation of technology to the canine traits and the determination of its efficacy.

Abstract: For thousands of years, dogs have coexisted with humans and have been adopted as companion pets and working animals. The communication between humans and dogs has improved their coexistence and socialization; however, due to the nature of their activities, dogs and humans occasionally lose face-to-face contact. The purpose of this scoping review is to examine five essential aspects of current technology designed to support intentional communication between humans and dogs in scenarios where there is no face-to-face contact: (1) the technologies used, (2) the activity supported, (3) the interaction modality, (4) the evaluation procedures, and the results obtained, and (5) the main limitations. In addition, this article explores future directions for research and practice. The PRISMA-ScR (Preferred Reporting Items for Systematic Reviews and Meta-Analyses extension for Scoping Reviews) guidelines were followed when conducting the review. Scopus (Elsevier), Springer-Link, IEEE Xplorer, ACM Digital Library, and Science Direct were used as data sources to retrieve information from January 2010 to March 2022. The titles and abstracts were individually reviewed by the authors (L.R.-V., I.E.E.-C., and H.P.-E.), and the full articles were then examined before a final inclusion determination. 15 (3%) out of the 571 records that were obtained met the requirements for inclusion. The most used technologies for dogs are: (1) 71% of technologies focused on generating messages are wearable devices equipped with sensors (bite, tug, or gesture), (2) 60% of technologies focused on receiving messages are wearable devices equipped with vibrotactile actuators, and (3) 100% of technologies focused on bidirectional communication are videochats. 67% of the works are oriented to support search and assistance tasks. 80% of the works developed technology for one-way communication. 53% of the technologies have a haptic dog interaction modality, that is, there is an object that the dog must wear or manipulate in a certain way. All of the reported evaluations were pilot studies with positive feasibility results. Remote human-dog interaction technology holds significant promise and potential; however, more research is required to assess their usability and efficacy and to incorporate new technological developments.

Citation: Rodriguez-Vizzuett, L.; Espinosa-Curiel, I.E.; Pérez-Espinosa, H. Digital Technology Supporting the Remote Human-Dog Interaction: Scoping Review. *Animals* **2023**, *13*, 699. https://doi.org/10.3390/ani13040699

Academic Editor: Betty McGuire

Received: 12 January 2023
Revised: 11 February 2023
Accepted: 12 February 2023
Published: 16 February 2023

Keywords: animal computer interaction; dogs; human-dog interaction; interactive technology; remote communication

1. Introduction

1.1. Background

Humans have the intuition that animals are intentional beings who know and feel and can communicate their intentions, knowledge, and feelings among themselves and humans [1]. Animals communicate through various modalities, such as visual, acoustic, semiochemical, or gestural behaviors [2]; therefore, the study of human-animal communication expands what can be considered language beyond grammar, words, and human language. In addition, a better understanding of how animals naturally communicate and the structure of their messages will significantly help develop more effective technological tools to assist human-animal communication. There are relevant precedents in the communication analysis between humans and different species of animals, for example, parrots [3], dolphins [4], and apes [5].

For thousands of years, humans and domestic dogs, also known as Canis familiaris, have formed close friendships and strong socialization bonds [6–8]. At the beginning of the domestication process (over 15,000 years), dogs were associated with human groups, and later the interaction between the two species intensified. Dogs began collaborating with humans in various activities such as hunting, herding, guarding, and pulling sleds [9–11]. Given the long-standing relationship between both species, dogs developed social-cognitive skills and abilities. Dogs can identify human social gestures and understand human communicative signals, especially, social signs [12], and human vocalizations [13]. Given those valuable capabilities, dogs were widely adopted as working animals to perform a wide range of support and assistance tasks [14,15]. Furthermore, as companion animals, dogs are able to positively affect psychologically and physiologically humans [16–18].

Humans and dogs occasionally lose face-to-face contact, for example, when a dog searches for a person in rural areas or when rescue dogs pass through extraordinarily narrow or difficult-to-reach locations for humans. In the case of companion animals, humans frequently leave their beloved dogs alone for long periods at home, which can lead to separation anxiety [19]. Supporting remote communication and interaction through current digital technologies opens up an exciting range of applications. These new technologies can enhance dogs' abilities to perform companion tasks, search and assist, and improve their well-being [20]. However, the development of digital technology for animals poses many challenges. It is essential to discover the best communication and interaction technology that allows dogs to readily transmit messages to their owners or handlers, ideally with no or minimal training required, and that delivers messages from the human to the dog in a form that it can correctly understand.

Animal Computer Interaction (ACI) is a new branch of computer science that seeks to understand the interaction between animals and computer technology in contexts where animals live, are active, and socialize with members of their own or other species, including humans [21]. Relevant advances have been made in ACI to understand the aspects of usability and user experience critical in the design of animal-oriented interactive systems [22] and to develop interactive interface technologies for various species [23]. The design, development, and evaluation of interactive technologies that enable intentional communication between humans and dogs who do not have face-to-face contact is an exciting aspect of ACI. Various research efforts are being made to develop and evaluate technology in order to gain knowledge and facilitate remote human-dog interaction. Therefore, it's crucial to comprehend how these solutions are created, put into practice, and assessed in terms of the following inquiries:

1. What digital technologies are employed to facilitate remote human-dog interaction?
2. What activities have been supported by remote human-dog interaction technology?

3. What interaction modalities have been used for remote human-dog interaction systems?
4. What are the types of evaluations applied to validate the technologies, and what are the primary outcomes assessed when validating remote human-dog interaction technology?
5. What are the reported limitations of technology employed for remote human-dog interaction?

1.2. Objective

To our knowledge, only one previous review addresses the use of technology for animal welfare. However, it is exclusively focused on smart computing and sensing technologies and for a broad range of species [24]. As a result, the goal of this study was to conduct a scoping review of scientific and technological advances in interactive technology for remote human-dog interaction. This review will help us better understand how this type of technology is designed, used, and evaluated by answering the five questions mentioned above. This article also analyzes the impact of digital interventions for remote human-dog interaction in various contexts and activities and explores future directions for research and practice.

2. Methods

To ensure that our review was conducted systematically and without bias, we conducted a scoping review using the PRISMA-ScR (Preferred Reporting Items for Systematic Reviews and Meta-Analyses extension for Scoping Reviews) methodology [25]. The study has not been registered in PROSPERO since it is not for human health.

2.1. Eligibility Criteria

The studies included in this scoping review were English-language research articles published in journals and conference proceedings between January 2010 and March 2022 that described (1) interactive digital technology with the goal of (2) supporting remote human-dog communication and interaction and included (3) an evaluation procedure. Thus, studies that (1) were not research articles, (2) were not written in English, (3) did not describe a digital interaction technology, (4) did not support remote human-dog or dog-human interaction, (5) did not include an evaluation procedure, (6) were literature reviews, and (7) were repeated were excluded.

2.1.1. Information Sources

The databases used for this review were: Scopus, SpringerLink, IEEE Xplorer, ACM Digital Library, and Science Direct. These five databases were chosen because they are recognized as reliable sources of high-quality publications from computer science, technology, and engineering. The search also took into account some hand-searched papers that were cited in the articles that were retrieved.

2.1.2. Search

The specific syntax of the queries varied depending on the database. However, the concepts of (1) digital interactive technology and (2) human-dog communication and interaction were always expressed using the same words. The following words were included in the query: ("dog") AND ("assistance" OR "service" OR "search and rescue" OR "working" OR "companion") AND ("technology" OR "wearable" OR "computer" OR "system" OR "platform") AND ("interaction" OR "communication").

2.1.3. Study Selection

The screening process was carried out in stages. The titles and abstracts were initially screened by the three authors (LRV, IEEC, and HPE). The full texts of the selected articles by the three researchers were examined in a subsequent stage before final inclusion. When numerous publications were published for the same study or application, it was reviewed to see if there were any major differences in the evaluation, such as if it was evaluated with a different population or other variables. The data were independently examined and extracted by the reviewers, and any discrepancies were settled through discussion until an agreement was reached.

2.1.4. Data Charting and Result Synthesis

The review included all studies that met the inclusion criteria, and the data extracted were those that allowed for the answers to the five questions listed in Section 1: (1) the type of technology (sensing gesture sensors, touchscreens, objects with capacitive sensors) aimed to support the remote human-dog interaction, (2) the type of activity supported (search-and-rescue, assistance, companion, or general purpose), (3) the dog interaction modality (haptic, sound, video, vibrotactile), (4) the reported findings in terms of the various outcomes related to remote human-dog interaction, and (5) the reported limitation of the technology. To extract and summarize the above data, the authors created, calibrated, and used a template with various sections. We proceed to describe the major findings that emerged from the studies.

3. Results

3.1. Overview

The search resulted in the identification of 571 records. 99.1% (566/571) of the records were obtained from the five digital libraries, with an additional 0.9 percent (5/571) obtained through hand searching. After removing all duplicated records, 535 papers were screened for eligibility in the first stage. Based on the exclusion criteria, 75.9% (406/535) of the records were discarded after reading the titles and abstracts. After reviewing the full text of 129 articles, 94.1 % (112/129) were excluded. As a result, 15 studies were chosen for further examination. Figure 1 shows the flow diagram for the several stages of the review. Table 1 summarizes the overall findings. Table 2 presents, in chronological order, all the articles that were selected, analyzed, and summarized. The information in this table is the paper author, year, technology, main functionalities, addressed activities, communication direction, dog interaction modality, and evaluation. Next, we summarize the main findings to respond to the research questions.

Table 1. Overview of the main characteristics of the reviewed technology (n = 15).

Characteristic	**Studies, n (%)**
Interfaces for dogs	
To generate messages	
Touchscreen	2(13)
Bite sensor	1(7)
Tug sensor	1(7)
Gesture sensor, bite sensor, tug sensor	1(10)
Gesture sensor	2(13)
To receive messages	
Vibrotactile	3(20)
Audio	1(7)
Audio and vibrotactile	1(7)
To generate and receive messages	
Videochat	2(13)
Videochat and bite sensor	1(7)
Interfaces for humans	
To generate messages	
Mobile application	3(20)
Handheld transmitter	1(7)
To receive messages	
Mobile applications	3(20)
To generate and receive messages	
Videochat	3(20)
Do not show human interface	5(15)

Table 1. *Cont.*

Characteristic	Studies, n (%)
Activities addressed	
Search-and-rescue	5(33)
Assistance	5(33)
Companion	3(20)
General purpose	2(14)
Interaction modalities for dogs	
Haptic	8(53)
Haptic and sound	1(7)
Video and sound	2(13)
Sound	1(7)
Wearable	2(13)
Video, sound and haptic	1(7)
Evaluation protocol	
Pilot study with dogs	14(93)
Pilot study with trainers	1(7)

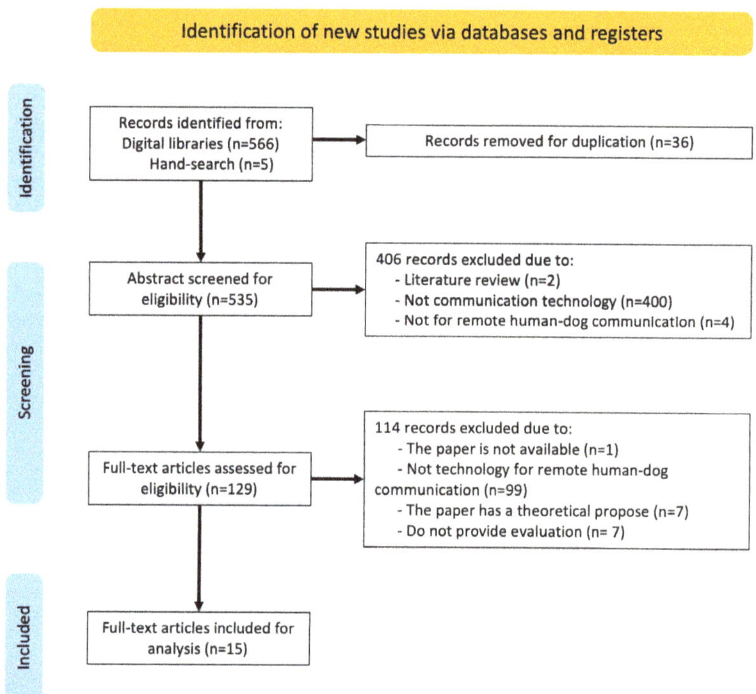

Figure 1. Flow diagram of the study selection.

Table 2. Summary of the main characteristics of the reviewed technologies (n = 15).

Technology	Year	Main Functionalities	Dog Activities Addressed	Message Direction	Dog Interaction Modality	Evaluation Protocol	Dog Training Sessions	Main Reported Results
1. Harness with sensors and actuators [26]	2011	Vest that allows handlers to command and track a trained canine in real-time using a sensor suite and a tone generator.	Search-and-rescue	Human to Dog	Haptic and sound-Vibrations and audio commands	Pilot test-1 dog	The dog had previous field/hunt trials and explosive detection training. The dog was trained to respond to tones and vibrations, but no training session details are provided.	The system can use data from the canine's sensors to provide audio and vibration commands and control signals to autonomously guide the dog to destinations and send information to the handler.
2. PC with Skype [27]	2012	Video chat system that augments a Skype audio-video connection with remote interaction features.	Companion	Bidirectional	Video and sound	Pilot test-10 dogs	Authors do not mention the dog's previous training. The dogs were trained to interact with the screen but no training session details are provided.	They demonstrated the potential of pet-based videochat systems that enable owners to watch their pets while not at home and communicate with them via audio and visuals, promoting animal engagement.
3. Harness with speakers [28]	2013	Vest allowing to command the dog through an embedded voice and recognizes some activities of the dog: walk, seat, run, lying.	General	Human to Dog	Hearing commands	Pilot test-1 dog	The participant dog was already familiar with the basic vocal commands. The authors do not specify a training period.	They demonstrated that dogs obey recorded vocal commands. Furthermore, using their system, the handler can remotely monitor the dog's activity.

Table 2. *Cont.*

Technology	Year	Main Functionalities	Dog Activities Addressed	Message Direction	Dog Interaction Modality	Evaluation Protocol	Dog Training Sessions	Main Reported Results
4. Harness with sensors [29]	2015	Five different haptic sensors that dogs could activate based on natural dog behaviors such as biting, tugging, and nose touches.	General	Dog to Human	Haptic-Bit, tug, and nose gestures	Pilot test-3 dogs	The dogs had previous assistance training. No more than four sessions per day, with a 30-min break in between, were held for sensor-specific training, and each session lasted no longer than 15 min.	It is feasible to design wearables that canine handlers can dependably activate. 100% of the commands transmitted through capacitive and pneumatic sensors resulted in successful activations for all eight dogs.
5. Collar with sensors [30]	2015	Collar that detects gestures using a motion sensor. Each gesture is paired with a predetermined message that is voiced to the handler by a smartphone.	Search-and-rescue	Dog to Human	Gestures recognition	Pilot test-3 dogs	The dogs had previous training in alert, assistance, and police tasks. The gesture training occurred in at most four 30-min long sessions for each dog.	The authors demonstrated that working dogs could use gestures (spin, twirl, right sequence, left sequence) to communicate with humans. The results when evaluating the gestures recognition collar were not totally satisfactory since important aspects to improve in the design were detected.

Table 2. *Cont.*

Technology	Year	Main Functionalities	Dog Activities Addressed	Message Direction	Dog Interaction Modality	Evaluation Protocol	Dog Training Sessions	Main Reported Results
6. Trigger that activates an alarm upon detachment [31]	2015	Alarm system that enables assistance dogs to call for help using a detaching component that the dog could pull off to trigger the alarm.	Assistance	Dog to Human	Haptic-Pulling rope	Pilot test-4 dogs	All dogs in this study already knew how to 'tug' and retrieve on command. The dogs were trained to use the mechanism but no training session details are provided.	The main contribution was the set of lessons learned from the particular design application and design process that included the dogs, their owners, and trainers in the process.
7. Harness with movement sensors [32]	2016	Collar that senses gestures using an inertial measurement unit and relays specific alerts to a smartphone application.	Assistance	Dog to Human	Gestures recognition	Pilot test-2 dogs	Only one dog had previous experience with gestures. Each training session lasted at most 10 min. The learning time for each gesture varied depending on the dog's prior training experience but did not exceed 15 training sessions per gesture.	Dogs were successfully trained to perform gestures. However, the sensing harness was not evaluated. After training, dogs could continue to make the signals without the harness, but less precisely. Dogs still recall the gesture when prompted with a verbal and physical cue over three months.

Table 2. *Cont.*

Technology	Year	Main Functionalities	Dog Activities Addressed	Message Direction	Dog Interaction Modality	Evaluation Protocol	Dog Training Sessions	Main Reported Results
8. Harness with bite sensor [33]	2016	Wearable bite sensor for search-and-rescue dogs that communicates with their handler via a mobile application.	Search-and-rescue	Dog to Human	Haptic- bite	Pilot test-3 trainers	The dogs had previous training in search and rescue. A dog was trained to use the vest, but no training session details are provided.	Three K9-Search and Rescue experts evaluated the system. They recommended improvements to the vest for durability, visibility, and connectivity to the handler; and improvements to the app regarding mapping, iconography, and annotation.
9. PC with Skype [34]	2016	Video call interactions with the dog and a treat dispenser triggered to release food from a distant location.	Companion	Bidirectional	Video and sound	Pilot test-1 dog	The dog was trained to correctly respond to verbal cues requested through Skype, but no training session details are provided.	The video capability not only gives the owner the option to check on their pet frequently to ensure that it is safe, but it also allows them to engage in meaningful communication with the pet.

Table 2. *Cont.*

Technology	Year	Main Functionalities	Dog Activities Addressed	Message Direction	Dog Interaction Modality	Evaluation Protocol	Dog Training Sessions	Main Reported Results
10. Touch-screen [35]	2016	Touchscreen interfaces usable for assistance dogs in the home. Validation of interaction techniques such as lift-off selection and sliding gestural motions.	Assistance	Dog to Human	Haptic-Nose touching	Pilot test-5 dogs	The dogs were trained in a 15–20 min sessions with at least 30 min rest between each training session.	The most effective technology for canine interaction is infrared touchscreens with backing projection monitors. The most efficient training technique for touchscreen interactions involving tapping is shaping. Luring can be used successfully to train sliding/gestural interactions at first, but it should be swiftly replaced with shaping.
11. Harness with vibrator [36]	2016	Harness to provide vibrotactile commands to dogs, working with variable-intensity vibrating motors mounted to a modified hug shirt.	Search-and-rescue	Human to Dog	Haptic-vibrations	Pilot test-4 dogs	The dogs had previous hunting, track, and obedience training. A dog was trained to correctly respond to vibrotactile cues, but no training session details are provided.	The authors tested the design of the vest and vibrating actuators. They concluded that it is crucial to correctly identify previous training methods and prepare modified experimental settings that consider each dog's learning experience.

Table 2. *Cont.*

Technology	Year	Main Functionalities	Dog Activities Addressed	Message Direction	Dog Interaction Modality	Evaluation Protocol	Dog Training Sessions	Main Reported Results
12. Harness with vibrator [37]	2017	Vest with vibration actuators at different points on the dog body that is evaluated measuring the working dog's ability to perform distinct tasks.	Assistance	Dog to Human	Haptic-vibrations in shoulders and bite	Pilot test-11 dogs	The dogs had diverse previous training. Training sessions were no more than fifteen minutes long. Each dog had no more than four training sessions per day, with at least thirty minutes between them. Training sessions were conducted until the dog mastered the haptic cue.	They demonstrated that canines can be taught to react to haptic stimuli. Over 93% of haptic cues resulted in the dog reporting perceiving the cue, for the highest power level of vibration. Not surprisingly, the lower power levels resulted in lower Dog Response Rates, with the lowest power level under 57%.
13. Touch-screen [38]	2018	Touchscreens mounted in the home triggered by the dog interaction to alert in emergencies.	Assistance	Dog to Human	Haptic-Nose touch	Pilot test-3 dogs	The dogs had diverse previous training and diverse experience with touchscreens. Dogs require less than 40 min of total training time, spread out over less than a week.	Dogs can be taught to use their noses to press a sequence of touchscreen icons to signal a medical emergency in less than a week with just five-minute training sessions each day. Dogs can locate the touchscreen from different rooms and only activate the touchscreen only when given the training cue.

Table 2. *Cont.*

Technology	Year	Main Functionalities	Dog Activities Addressed	Message Direction	Dog Interaction Modality	Evaluation Protocol	Dog Training Sessions	Main Reported Results
14. Harness with vibrator [39]	2019	Harness embedded with vibration motors associating four different types of vibrations with different commands.	Search-and-rescue	Human to Dog	Haptic-vibrations in shoulders	Pilot test-1 dog	The dogs had never received any formal training. The dog was trained to correctly respond to vibrotactile cues, but no training session details are provided.	Vibrotactile indications successfully directed dogs to carry out several tasks (turn around, lie down, approach handler, walk backward). Dogs responded well to a single haptic command, matching their vocal command sensitivity.
15. PC with Skype and bite sensor [40]	2021	Video call device to allow a dog to remotely call their human, giving the animal control and agency over technology in their home.	Companion	Bidiectional	Haptic, video and sound	Pilot test-1 dog	The dog had no previous professional training but had previous technology experience with screen devices and motion and facial trackers. The trainer performs five use demonstration actions to teach the dog how to use the ball to call.	Thanks to the system, dogs could video call their human whenever and wherever they wanted. The experimental design provided knowledge on how to create Internet of Things systems using canines. Through prototyping, dogs were incorporated into the early stages of design.

3.2. Digital Technologies Implemented

According to our analysis, the digital technology implemented can be divided into three groups: technology for dogs, technology for humans, and technology for processing and interconnection.

3.2.1. Technologies for Dogs

Dog technology can be divided into three categories: to send messages to humans, to receive messages from humans, or to send and receive messages. The technology to send messages to humans is mainly based on wearable devices with sensors [29,30,32,33], pulling sensor installed on a wall [31], and touchscreens [35,38]. To receive messages from humans, the technology is mainly based on vibrotactile actuators [36,37,39] and audio playback [28], or both, vibration and sound [26]. The technology to generate and receive messages is based on videochats [27,34,40]. Concerning wearable devices to generate messages by the dogs, some works [29,33] used harnesses with different sensors activated by bite, tug, and nose gestures. They used force-sensitive resistors to implement bite sensors, an ultrasonic range finder that detects nose movement at 3 cm. The tug sensor was made into an elastic band with a stretchable rubber variable resistor. Other wearable devices were developed [30,32] in order to identify head and body gestures, respectively. These systems obtain data from a collar that includes inertial sensors to detect gestures paired with predetermined behaviors. Other works [31] developed a detaching component that the dog could pull off to trigger a medical alarm. A couple of works [35,38] explored how to obtain dogs' input with a touchscreen interface and the difficulties they have when interacting with this kind of device. Videochats allow bidirectional communication where dogs and humans can generate and receive messages synchronously. The DogPhone hardware prototype [40] includes an orientation sensor that combines an accelerometer, magnetometer, and gyroscope to detect movement, interpret the dog input, and start a phone call. In [27,34], a pet video chat system was designed using Skype's audio-video connection with remote interaction features.

3.2.2. Technologies for Humans

Mobile applications are the most common device to send or receive messages from dogs. In [28] used a mobile application to send spoken commands to the wearable device. In [36,37] used a mobile application to send vibrotactile commands to the wearable device of search-and-rescue dogs. In [30] used a mobile application that receives sensor readings via a Bluetooth connection and plays synthesized speech messages of the gesture being performed. A corresponding message is communicated if the collar is out of range or more than five samples were skipped. In another work [33], the authors developed an application that receives notifications when the dog bites the sensor. The application also shows the dog's location concerning the handler, a compass, and general wind direction. In the system developed by Golan et al. [39], a vibrator in the dog's harness was activated by a handheld transmitter (remote control).

3.2.3. Technologies for Processing and Interconnection

For processing and interconnection, most studies used an all-in-one development board. In [33] they used a central hub with a Bluetooth radio to broadcast sensor activation feedback tones to the dog and send alerts to a cell phone. An Arduino board that activates the appropriate vibrators and interconnects via WiFi to a mobile application was used in Morrison et al. [36]. The Adafruit Feather Huzzah ESP8266 board, which included a Built-in WiFi 802.11 b/g/n was used in [40]. In [37], the authors used the Intel Next Unit of Computing (NUC) KIT NUC5i3RYH to control the prompts displayed on the screen, manage the interactions, and upload the data to the server. In the case of human technology, all the processing and interconnection were made in the smartphone. In [28], two prototypes were developed; the prototype is a harness equipped with speakers connected to a smartphone that is attached to the harness. The smartphone is connected to two amplified speakers attached to the harness under the dog's ears. Voice commands can be activated remotely with a second smartphone. The two smartphones communicate using the Direct WiFi standard that allows a distance of 50 m between the two smartphones without much delay to reproduce the sounds. The second prototype was for detecting dog activity. Two smartphones were used for this prototype, adding the 3-axis accelerometer and 3-axis gyroscope. The smartphone is placed on the dog's back, and the sensors with a specific location so that the sound can be reproduced correctly.

3.3. *Activities Intended to Support*

According to our analysis, the main activities supported by remote interaction technology can be grouped into assistance, search-and-rescue, companion, and general-purpose activities.

3.3.1. Assistance and Service Activities

Assistance dogs have become part of the daily life of many people with conditions that limit them from carrying out their daily activities. In terms of assistance activities, the works have addressed technology allowing assistance dogs to alert in case of events that require attention or in an emergency and ask for help on behalf of their owners [31,32,35], and generate medical alerts by operating emergency notification systems [37,38].

3.3.2. Seeking, Locating, and Rescuing Activities

Dogs that assist in seeking, locating, and rescuing activities are called search-and-rescue dogs [41]. We found that remote interaction technology has been designed for search and rescue and hunting dogs. Both tasks have in common the use of their powerful sensory abilities, mainly olfactory, to locate a target. Search and rescue (SAR) dogs are trained to locate people in extreme situations, in terrain that is often difficult for humans to access: in the snow, in the open air, in the mountains or at sea, and after earthquakes and other catastrophes that can generate large amounts of rubble. Concerning SAR dogs, we identified two studies. The first study is to alert the handler when a SAR dog finds something interesting [33]. This system sends a signal via cell phone to the handler's smartphone, including GPS data and activation information. As the SAR dog moves, a trajectory is drawn on the map showing where the dog has searched. In [26], the authors developed a system that tracks a canine's position, motion behavior, and orientation. It also supports the remote actuation of tone and vibration commands and reports commands in real-time alongside sensor data. For the case of hunting dogs, in [36] created a vibrotactile vest (VTV) to give commands to execute the tasks for which they were trained. Valentin et al. [30] proposed a system including a collar and an app for dangerous tasks such as search and rescue or explosive detection. The collar identifies specific movements of the dog's head, and the app receives messages about the movements detected. Golan et al. [39] also tackled complicated scenarios humans cannot do alone, such as detecting explosives or searching narrow spaces. They implemented a vest with four embedded vibration motors. The vest applies vibrotactile cues to the dog that wears it, and the dog is trained to associate the cues with useful commands.

3.3.3. Companion Activities

Companion dogs live in their owners' homes and may learn to perform specific tasks. The research in this category is motivated by the bond between humans and domesticated dogs and the need to stay connected from distant places. The following three studies were identified in this category. In [27], investigate the potential of interactive cameras for dogs throughout a pet video chat system to augment Skype's audio and video connection with remote interaction features. The work by Rossi et. al. [34] aims to show the ability of a canine to provide verbal cues given through a video chat. The authors used the software Skype and an automatic kibble dispenser to improve the bond between domestic dogs and their owners through remote audiovisual interaction. Communication was bidirectional from the human to the dog and vice-versa. The owner communicated with the dog to give directions, and the dog could emit a vocalization or another signal in response. A more recent studio explores the creation of a video call device to allow a dog to initiate a video call with its owner [40]. In this case, the dog has control over this home communication device. The authors chose a tennis ball as the interface to initiate the call because the behavior of biting such an object already had a prior meaning and use for the dog-human relationship.

3.3.4. General-Purpose Activities

These works have developed remote interaction technology without focusing on a specific activity. In this category, one study focused on a technology that enables dogs to communicate events through several interfaces that detects head movements, bites and tugs [29]. In addition, other study focused on technology that allows humans to provide audio cues and commands [28] to dogs in several scenarios or activities. The authors argue that this technology may benefit activities such as training and communication with deaf dogs and training by handlers with speech impairments.

3.4. Interaction Modalities

The interaction modalities are the means that allow the user (dog or human) to communicate with the computer system, that is, they allow it to generate the input or receive the output. According to our analysis, the interaction modalities can be divided between those that are for dogs and humans.

3.4.1. Interaction Modalities for Dogs

The primary interaction modalities for dogs are haptic, sound, and audio/video. Dogs can bit, tug, and make nose gestures [29,33], nose touching [35,38], head and body movements [30,32], and pulling a rope to generate messages [31]. Three works explore the interaction through audio/video [27,34,40]. Most works that send messages to the dog use vibrotactile devices [36,37,39]. In addition, two works used sound to provide a message to dogs. The first is speech prompts [28], and the second uses different tones that correspond to the commands forward, stop, and recall [26].

3.4.2. Interaction Modalities for Humans

Several of the identified works only focus on the design of technology for dogs and do not provide an interface for humans [26,29,31,35,38]. In the works that include interaction technology for humans, most are touch interfaces into mobile apps to send [28,37] or receive messages [30,32,33,36], and wireless remote devices to send commands [39]. Finally, three works use an audio/video interface to support that humans interact with dogs [27,34,40].

3.5. Evaluation Procedure and Reported Results

All the analyzed studies conducted exploratory studies to evaluate the feasibility of the proposed technology. In particular, most studies focus on assessing dogs' capacities to use the technology and the time required to train them to use it. The evaluations were conducted in pilot studies with 1 to 12 dogs; nevertheless, most were conducted with four dogs. Next, we analyze the evaluation studies and the reported results.

3.5.1. Evaluation of Technologies to Support Dogs Sending Messages

We identified that most of the studies that propose technology that enables dogs to send messages focused on evaluating the capacities of dogs to use the technology and the time required to train them to use it. The evaluation of vests and collars focused on validating if dogs can reliably activate the sensor mounted on them to interact with their handlers.

Variables such as training time, dog accuracy, sensor accuracy, sensor range, and overall success were measured [29,32]. The authors were able to verify that wearable electronics mounted on harness can be reliably activated by dogs to interact with their handlers.

However, the sensors must be more compact, durable, and power-efficient. In addition, requirements related to durability, visibility, and connectivity of the vest and the mobile application's mapping, iconography, and annotation were identified [33]. They showed the viability of the proposed technology by evaluating it with feedback from expert trainers. Also, it was evaluated if dogs can use a wearable device such as a collar or vest that detects a set of gestures for dogs to communicate with handlers [30]. Their findings demonstrated the kinds of gestures that can help working dogs communicate important information to trainers and the significance of taking into account the devices the dog is currently wearing, such as a leash, harness, or existing collar, when choosing gestures.

Similarly, the evaluation of touchscreen interfaces focused on validating their usability and precision. In [35], the authors evaluated the usability of a touchscreen interface by training five dogs on the take-off tapping task. In previous work, the same author [42] had evaluated their touchscreen-based system by counting the number of touches of the dogs vs. time, the time of the dogs vs. distance from the edge of the circle, and the time of the dogs vs. difficulty index. They highlighted a number of best practices, including the use of infrared touchscreens with non-projection monitors as the background, the need that tapping targets be at least 3.5" long, and the fact that shape is the most efficient mode of instruction for touchscreen interactions. Byrne et al. [38] assessed the feasibility of using a touch screen as a real-time medical alert system. They showed how dogs may be taught to use their noses to press a sequence of touchscreen symbols to transmit a medical alert. Even dogs with no prior touchscreen training can learn a complicated alert behavior chain in less than a week with just daily training sessions of five minutes.

Similarly, a medical alert system activated by pulling a rope was evaluated by Robinson et al. [31]. The results of this evaluation were to provide a series of recommendations on user-centered design for assistance dogs and humans to develop a system that would allow assistance dogs to call for help remotely.

3.5.2. Evaluation of Technologies to Support Dogs Receiving Messages

The studies that propose technology for supporting dogs receiving remote messages focused on evaluating the capacities of dogs to interpret these messages and the time required to train them to use them. In [28], the authors verified that dogs could obey a recorded vocal command of the owner's voice when it is not in visual contact with their owner. In addition, several works evaluated how well dogs interpret orders remotely sent and played by vibrotactile actuators. In [36] the authors measured interaction variables; whether the dog was treated or praised, the lack or type of mark the dog made, whether the dog looked away, looked at the facilitator, followed the facilitator's hand with its head, or walked towards the facilitator, and the number of times it responded correctly to the command. They emphasize the significance of accurately identifying previous training methods and planning modified trial settings in advance to accommodate each dog's unique learning experience. In the evaluation reported in [37], the authors measured the accuracy of the dog's response to a series of stimuli; these responses are divided into three variables: Deletions (D), Substitutions (S), and Insertions (I). They demonstrated that dogs could be trained to respond to vibrotactile cues. Moreover, in [39], the evaluation was carried out with a dog trained to associate four different types of vibrations with other commands to assess the number of successfully performed orders. They proved that instructing dogs to carry out numerous activities using vibrotactile cues was quite effective. The test subjects responded well to a single haptic command, coming close to matching the vocal command sensitivity. In [26], the authors measured the success rate for simple and more complicated multi-point paths where dogs had to leave the point of origin, go to a waypoint, stop, and then return to the end of the head. A "success" means that the canine came close enough to the destination waypoint and stopped when commanded. A "failure" indicates that the dog could not be commanded to arrive at the waypoint. They demonstrated how the sensor data may be used to recognize when a dog assumes a different stance in addition to guiding the dog to a predefined place.

3.5.3. Technologies for Bidirectional Communication

The studies that propose bidirectional communication technology are mainly audio-visual systems used by pet dogs. In [34], it is shown that the dog responds to commands given from a distance and that video call interactions can benefit the dog. To evaluate the use of DogPhone [40], the authors employed HCI's established mixed-method approach of combining a diary study and recommended interpretations from the human side with quantitative interaction data from the DogPhone interactions. They examine how interactions should be managed, how to measure interactions, how dog devices are created through prototyping, and what these things entail for dogs. In [27], the effectiveness of the pet video chat was tested. The results are encouraging for pet video chat systems that allow owners to see and interact with their pets while away. They demonstrated how it is essential to be able to see the animal in order to properly promote interaction.

3.6. Limitations of Remote Human-Dog Communication Technology

Although significant advances have been made with current technologies, the studies identified the following limitations that should be considered. Jackson et al. [29] identified that sensors need to be smaller, robust, and less power-consuming to adapt to the characteristic of dogs. In addition, the breed and dog body types affect the effectiveness of the technology. Coat density, body shape, and fat/muscle distribution could affect the fidelity of the message conveyed. In addition, differences in cognition and experience can also be a problem. These issues also affect the design and positioning of the sensing devices [30]. In addition, other limitations are related to the balance between canine and human requirements, an issue that must be considered during the design process [31]. Finally, the evaluation is a limitation since more research is necessary to test their results' validity, reliability, reproducibility, and generalization. Future efforts should focus on trying the technologies in a more significant number of dogs of different breeds, ages, and training histories [39].

4. Discussion

4.1. Principal Findings

Considering the rising interest in building digital technology to enable remote human-dog communication, future research should highlight critical elements that remote human-dog communication designers should consider. This scoping study helps by identifying and summarizing the description of five key features that characterize how these devices are currently constructed and the intervention results provided. The results discussion is organized around the primary objectives addressed in this scoping review.

4.1.1. Digital Technologies Implemented

The results obtained from this literature review indicate that the development of technology to support communication between dogs and humans is still incipient. It has been possible to validate suitable interfaces for dogs to send messages (touch screens, devices activated by biting, pulling, and gestures) and receive messages (vibrating vests, speakers). Wearable devices like harnesses, collars, and modified toys (balls, rope toys) have been studied extensively. The prototypes include a variety of sensors and actuators for two-way communication.

Undoubtedly, the results achieved are valuable and relevant for the design of dog-computer interfaces. However, it is clear that these technological proposals still need to mature and be evaluated more extensively, as well as testing other means of communication, taking advantage of dogs differentiating characteristics. Therefore, there is a need to investigate new sensors and actuators, for vocal or olfactory interaction, for example. Most of the revised works focused only on one-way communication technology (dog-human or human-dog), and few works focused on bidirectional communication, all of them videochats. Further research is needed to integrate the advance of one-way communication technology to develop bidirectional communication systems.

Information technologies such as the Internet of Things, augmented and virtual reality, big data, 3D printing, and artificial intelligence have not been exploited when implementing prototypes for remote human-dog interaction. In the coming years, these technologies have the potential to enable significant advances in remote human-dog interaction. In order to scale current technological solutions, it is necessary to integrate these trending technologies into more robust communication platforms taking advantage of their benefits. Surprisingly, advances in technology for interaction with dogs are not as outstanding as expected, given the very close human-dog relationship. It is perceived that the area of animal-computer interaction is in the early stages of growth, specifically in dog-computer interaction, in which it is beginning to take inertia thanks to the push of a few research groups and the financing of projects around the world. In later stages, it is expected that synergy will be generated between the different groups to create shared resources and tools, leading to a more accelerated advancement of the area.

4.1.2. Activities Supported by Remote Human-Dog Interaction Technology

Significant efforts have been made to implement technology to support remote interaction in assistance, search and rescue, hunting, companionship, or general-purpose dog activities. However, according to the taxonomy of assistance animals proposed by Parenti et al. [41], many activities could benefit from human-dog remote interaction technology. For example, guide, autism, herding, emotional support, mobility assistance, and patrol. Applying this technology to these activities could even revolutionize how dogs currently perform these activities. Additionally, while the identified technologies were created to support specific activities, they can be adapted to new situations and activities with minor changes. However, more research is needed to determine the viability and implications of these actions.

4.1.3. Interaction Modalities

Dogs' key interaction modalities to create signals include nose touch, bite, pull, tug, body, head and nose movements, and audio/video. Furthermore, the key interaction modes for receiving signals are vibrotactile and sound.

There are interaction modalities that have not yet been investigated and used in both circumstances, taking into consideration the order of significance of dogs' senses (smell, hearing, vision, touch, and taste) [28] and dogs' communication methods (e.g., touch, vocalizations, and movements). For instance, a vocal interface for dogs to create messages, assuming that the dog expresses itself through vocalizations. Another example is a system that reads dog body motions and converts them into messages. Another option is to use the dog's sense of smell to create an odor-based interface for communicating with them. Concerning interaction technology for humans, most prototypes generate messages through touch interfaces implemented into mobile devices. Mobile apps are the most common interface for receiving visual or audible messages. Similarly, human vocal interfaces could send direct audio messages to dogs.

It is important to mention that in all the studies safe technologies were proposed in terms of canine well-being. The devices that could be less comfortable for dogs are harnesses and vests since due to the electronics with which they are equipped they can be heavy and generate a little heat. However, the signals sent by these devices to dogs are harmless as they are mainly mild vibrations.

4.1.4. Evaluation and Results

Some human-oriented interface design techniques have been used in research efforts with dogs. However, a methodological adaptation and specialization phase are still required to create effective human-dog interaction interfaces. On the other hand, most of the current validations of human-dog interaction interfaces are inconclusive because they were conducted with a small number of test subjects or prototypes in the early development stages.

4.1.5. Technology Limitations

The constraints noted in the papers under consideration are connected to the difficulty of generalizing system design advances. Depending on the breed, dogs have a variety of physical and behavioral characteristics. It is worth noting that the validations were only done with a few dogs (between one and four). As a result, there is insufficient data to draw definite judgments. Several challenges developing interspecies communication technologies can be addressed. Canines that function as therapy, assistance, skilled companion, and service among others [43], have piqued the curiosity of the scientific community.

4.2. Gaps in the Research

The following areas of opportunity and research requirements to support the development of this type of technology came from the findings of this review.

- Creation of cutting-edge new technology. Wearable technologies, touchscreens, video and audio interfaces, specialized network systems, and artificial intelligence advances must be integrated to enable future developments that allow humans and canines to execute sophisticated remote socializing and collaboration tasks more naturally.
- Integrate new cross-application research. There is a need to build technology that can be utilized easily and effectively in diverse environments or for different activities.
- Create new interaction modalities. It is necessary to build and create new dog-computer interfaces and multimodal communication systems that consider the whole range of a dog's senses and interaction methods.
- Develop dog-centered technology. There is a need to shift the technology design paradigm in favor of one focused on the characteristics of dogs. For instance, create small and low-power devices considering the dogs' breed, size, and body type.

5. Limitations

This review raises critical issues that should be taken into consideration when interpreting the results of this review. One of these disadvantages is that only studies that provided information on an evaluation process were considered. Some advanced technology (electronic devices with machine learning or other artificial intelligence methods) was not considered due to a lack of testing. Furthermore, most studies do not include a medium-term evaluation of intervention efficacy in their research design to confirm the generalizability of the developed technology. Another limitation is the small number of included studies that were examined. Due to the possibility that some other pertinent studies were missed during the search, this study's database consideration is also constrained. If too many databases are used, the search may be predisposed to excessive, unjustified duplicates of the searched results, even though the five databases that were searched may overlap with other databases. However, we considered that this study makes a significant contribution because it shows the state of the art in this area of technological research and development. In addition, this study also highlights the need for more research in this area because there have only been a few publications in this particular field.

6. Conclusions

The results highlight digital technology's significant promise and potential to support remote human-dog interaction. Wearable technology for dogs and mobile apps for humans was the most extensively studied technologies for remote human-dog interaction. Most technologies were created to aid dogs in their assistance activities, and the most commonly reported interaction mode for dogs was haptic. Most of the reported evaluation protocols are pilot studies with fewer dogs that reported positive results regarding the feasibility of the technology focusing on assessing dogs' capacities to use the technology and the time required to train them. The use of technology to support remote human-dog interaction generates much expectation and excitement. However, there remains a long way to go regarding technological developments, integration into the activities and context of dogs, supporting new dogs' interaction modalities, adapting the technology to the dog's characteristics, and establishing effectiveness.

Author Contributions: Conceptualization, L.R.-V., I.E.E.-C., H.P.-E.; methodology, L.R.-V., I.E.E.-C.; formal analysis, L.R.-V., I.E.E.-C., H.P.-E.; writing—original draft preparation, L.R.-V.; writing—review and edit-ing I.E.E.-C., H.P.-E. All authors have read and agreed to the published version of the manuscript.

Funding: This research was funded by CONACyT Mexico under the grant CF-2019/2275. We are grateful to the postdoctoral fellowship program with application number 1143103 corresponding to the call for applications "Estancias Posdoctorales por México Modalidades 1 y 2" in the "Modalidad 2: Estancia Posdoctoral de Incidencia" for the realization of the present research project.

Institutional Review Board Statement: Not applicable.

Informed Consent Statement: Not applicable.

Data Availability Statement: Not applicable.

Conflicts of Interest: The authors declare no conflict of interest. The funders had no role in the design of the study; in the collection, analyses, or interpretation of data; in the writing of the manuscript; or in the decision to publish the results.

References

1. Bekoff, M.; Allen, C. Intentional communication and social play. In *Animal Play: Evolutionary, Comparative, and Ecological. Perspectives*; Cambridge University Press: Cambridge, UK, 1998; pp. 97–114.
2. Ma, Z.S. Towards computational models of animal communications, an introduction for computer scientists. *Cogn. Syst. Res.* **2015**, *33*, 70–99. [CrossRef]
3. Pepperberg, I.M.; Pepperberg, I.M. *The Alex Studies: Cognitive and Communicative Abilities of Grey Parrots*; Harvard University Press: Cambridge, CA, USA, 2009.
4. Herman, L.M. Can dolphins understand language? In *Proceedings of the LACUS forum. Linguistic Association of Canada and the United States*; Eastern Kentucky University: Richmond, KY, USA, 2007; Volume 34, pp. 3–21.
5. Patterson, F.G.; Gordon, W. Twenty-seven years of Project Koko and Michael. In *All Apes Great and Small*; Springer: Berlin/Heidelberg, Germany, 2002; pp. 165–176.
6. Wang, G.D.; Zhai, W.; Yang, H.C.; Fan, R.X.; Cao, X.; Zhong, L.; Wang, L.; Liu, F.; Wu, H.; Cheng, L.G.; et al. The genomics of selection in dogs and the parallel evolution between dogs and humans. *Nat. Commun.* **2013**, *4*, 1–9. [CrossRef]
7. Kis, A.; Ciobica, A.; Topál, J. The effect of oxytocin on human-directed social behaviour in dogs (Canis familiaris). *Horm. Behav.* **2017**, *94*, 40–52. [CrossRef]
8. Payne, E.; Bennett, P.C.; McGreevy, P.D. Current perspectives on attachment and bonding in the dog–human dyad. *Psychol. Res. Behav. Manag.* **2015**, *8*, 71. [CrossRef]
9. Losey, R.J.; Nomokonova, T.; Fleming, L.; Latham, K.; Harrington, L. Domestication and the embodied human–dog relationship: Archaeological perspectives from Siberia. In *Dogs in the North*; Routledge: London, UK, 2018; pp. 8–27.
10. Frantz, L.A.; Bradley, D.G.; Larson, G.; Orlando, L. Animal domestication in the era of ancient genomics. *Nat. Rev. Genet.* **2020**, *21*, 449–460. [CrossRef]
11. Morey, D. *Dogs: Domestication and the Development of a Social Bond*; Cambridge University Press: Cambridge, UK, 2010.
12. Hare, B.; Tomasello, M. Human-like social skills in dogs? *Trends Cogn. Sci.* **2005**, *9*, 439–444. [CrossRef]
13. Benjamin, A.; Slocombe, K. 'Who'sa good boy?!'Dogs prefer naturalistic dog-directed speech. *Anim. Cogn.* **2018**, *21*, 353–364. [CrossRef]
14. Olson, P.N. The modern working dog—A call for interdisciplinary collaboration. *J. Am. Vet. Med Assoc.* **2002**, *221*, 352–355. [CrossRef]
15. Elgier, A.M.; Jakovcevic, A.; Barrera, G.; Mustaca, A.E.; Bentosela, M. Communication between domestic dogs (Canis familiaris) and humans: Dogs are good learners. *Behav. Process.* **2009**, *81*, 402–408. [CrossRef]
16. Gee, N.R.; Rodriguez, K.E.; Fine, A.H.; Trammell, J.P. Dogs Supporting Human Health and Well-Being: A Biopsychosocial Approach. *Front. Vet. Sci.* **2021**, *8*. [CrossRef]
17. Bussolari, C.; Currin-McCulloch, J.; Packman, W.; Kogan, L.; Erdman, P. I Couldn't Have Asked for a Better Quarantine Partner: Experiences with Companion Dogs during Covid-19. *Animals* **2021**, *11*, 330. [CrossRef]
18. Friedmann, E.; Thomas, S.; Eddy, T.J. Companion animals and human health. In *Companion Animals and Us*; Cambridge University Press: Cambridge, UK, 2000; pp. 125–142.

19. Sherman, B.L.; Mills, D.S. Canine anxieties and phobias: An update on separation anxiety and noise aversions. *Vet. Clin. N. Am. Small Anim. Pract.* **2008**, *38*, 1081–1106. [CrossRef]
20. Paldanius, M.; Kärkkäinen, T.; Väänänen-Vainio-Mattila, K.; Juhlin, O.; Häkkilä, J. Communication technology for human-dog interaction: Exploration of dog owners' experiences and expectations. In Proceedings of the SIGCHI Conference on Human Factors in Computing Systems, Vancouver, BC, Canada, 7–12 May 2011; pp. 2641–2650.
21. Mancini, C. Animal-computer interaction: A manifesto. *Interactions* **2011**, *18*, 69–73. [CrossRef]
22. Robinson, C.L. *Animal-Computer Interaction: Designing Specialised Technology with Canine Workers*; Open University: Milton Keynes, UK, 2017.
23. Hirskyj-Douglas, I.; Pons, P.; Read, J.C.; Jaen, J. Seven years after the manifesto: Literature review and research directions for technologies in animal computer interaction. *Multimodal Technol. Interact.* **2018**, *2*, 30. [CrossRef]
24. Jukan, A.; Masip-Bruin, X.; Amla, N. Smart Computing and Sensing Technologies for Animal Welfare: A Systematic Review. *ACM Comput. Surv.* **2017**, *50*. [CrossRef]
25. Tricco, A.C.; Lillie, E.; Zarin, W.; O'Brien, K.K.; Colquhoun, H.; Levac, D.; Moher, D.; Peters, M.D.; Horsley, T.; Weeks, L.; et al. PRISMA extension for scoping reviews (PRISMA-ScR): Checklist and explanation. *Ann. Intern. Med.* **2018**, *169*, 467–473. [CrossRef]
26. Britt, W.R.; Miller, J.; Waggoner, P.; Bevly, D.M.; Hamilton, J.A. An embedded system for real-time navigation and remote command of a trained canine. *Pers. Ubiquitous Comput.* **2011**, *15*, 61–74. [CrossRef]
27. Golbeck, J.; Neustaedter, C. Pet video chat: Monitoring and interacting with dogs over distance. In Proceedings of the ACM CHI'12 Extended Abstracts on Human Factors in Computing Systems, Austin, TX, USA, 5–10 May 2012; pp. 211–220.
28. Lemasson, G.; Pesty, S.; Duhaut, D. Increasing communication between a man and a dog. In Proceedings of the 2013 IEEE 4th International Conference on Cognitive Infocommunications (CogInfoCom), Budapest, Hungary, 2–5 December 2013; pp. 145–148.
29. Jackson, M.M.; Valentin, G.; Freil, L.; Burkeen, L.; Zeagler, C.; Gilliland, S.; Currier, B.; Starner, T. FIDO—Facilitating interactions for dogs with occupations: Wearable communication interfaces for working dogs. *Pers. Ubiquitous Comput.* **2015**, *19*, 155–173. [CrossRef]
30. Valentin, G.; Alcaidinho, J.; Howard, A.; Jackson, M.M.; Starner, T. Towards a canine-human communication system based on head gestures. In Proceedings of the 12th International Conference on Advances in Computer Entertainment Technology, Iskandar, Malaysia, 16–19 November 2015; pp. 1–9.
31. Robinson, C.; Mancini, C.; van der Linden, J.; Guest, C.; Swanson, L.; Marsden, H.; Valencia, J.; Aengenheister, B. Designing an emergency communication system for human and assistance dog partnerships. In Proceedings of the 2015 ACM International Joint Conference on Pervasive and Ubiquitous Computing, Osaka, Japan, 7–11 September 2015; pp. 337–347.
32. Alcaidinho, J.; Valentin, G.; Abowd, G.D.; Jackson, M. Training collar-sensed gestures for canine communication. In Proceedings of the Third International Conference on Animal-Computer Interaction, Milton Keynes, UK, 15–17 November 2016; pp. 1–4.
33. Zeagler, C.; Byrne, C.; Valentin, G.; Freil, L.; Kidder, E.; Crouch, J.; Starner, T.; Jackson, M.M. Search and rescue: Dog and handler collaboration through wearable and mobile interfaces. In Proceedings of the Third International Conference on Animal-Computer Interaction, Milton, Keynes, 15–17 November 2016; pp. 1–9.
34. Rossi, A.P.; Rodriguez, S.; Cardoso dos Santos, C.R. A dog using skype. In Proceedings of the Third International Conference on Animal-Computer Interaction, Milton, Keynes, 15–17 November 2016; pp. 1–4.
35. Zeagler, C.; Zuerndorfer, J.; Lau, A.; Freil, L.; Gilliland, S.; Starner, T.; Jackson, M.M. Canine computer interaction: Towards designing a touchscreen interface for working dogs. In Proceedings of the Third International Conference on Animal-Computer Interaction, Milton, Keynes, 15–17 November 2016; pp. 1–5.
36. Morrison, A.; Møller, R.H.; Manresa-Yee, C.; Eshraghi, N. The impact of training approaches on experimental setup and design of wearable vibrotactiles for hunting dogs. In Proceedings of the Third International Conference on Animal-Computer Interaction, Milton, Keynes, 15–17 November 2016; pp. 1–10.
37. Byrne, C.; Freil, L.; Starner, T.; Jackson, M.M. A method to evaluate haptic interfaces for working dogs. *Int. J. Hum.-Comput. Stud.* **2017**, *98*, 196–207. [CrossRef]
38. Byrne, C.; Zeagler, C.; Freil, L.; Rapoport, A.; Jackson, M.M. Dogs using touchscreens in the home: A case study for assistance dogs operating emergency notification systems. In Proceedings of the Fifth International Conference on Animal-Computer Interaction, Atlanta, GA, USA, 4–6 December 2018; pp. 1–10.
39. Golan, Y.; Serota, B.; Shapiro, A.; Shriki, O.; Nisky, I. A vibrotactile vest for remote human-dog communication. In Proceedings of the 2019 IEEE World Haptics Conference (WHC), Tokyo, Japan, 9–12 July 2019; pp. 556–561.
40. Hirskyj-Douglas, I.; Piitulainen, R.; Lucero, A. Forming the Dog Internet: Prototyping a Dog-to-Human Video Call Device. In Proceedings of the ACM International Conference on Interactive Surfaces and Spaces, Lodz, Poland, 14–17 November 2021; pp. 1–20.
41. Parenti, L.; Foreman, A.; Meade, B.J.; Wirth, O. A revised taxonomy of assistance animals. *J. Rehabil. Res. Dev.* **2013**, *50*, 745. [CrossRef]

42. Zeagler, C.; Gilliland, S.; Freil, L.; Starner, T.; Jackson, M. Going to the dogs: Towards an interactive touchscreen interface for working dogs. In Proceedings of the 27th Annual ACM Symposium on User Interface Software and Technology, Honolulu, HI, USA, 5–8 October 2014; pp. 497–507.
43. Howell, T.J.; Nieforth, L.; Thomas-Pino, C.; Samet, L.; Agbonika, S.; Cuevas-Pavincich, F.; Fry, N.E.; Hill, K.; Jegatheesan, B.; Kakinuma, M.; et al. Defining terms used for animals working in support roles for people with support needs. *Animals* **2022**, *12*, 1975. [CrossRef]

animals

MDPI

Review

The Welfare of Dogs as an Aspect of the Human–Dog Bond: A Scoping Review

Peter Verbeek *, Chase Alan Majure, Laura Quattrochi and Stephen James Turner

Department of Anthropology, University of Alabama at Birmingham, 402 10th Avenue South, Birmingham, AL 35294-1241, USA; cmajure@uab.edu (C.A.M.); laura.quattrochi@gmail.com (L.Q.); sturner0@uab.edu (S.J.T.)
* Correspondence: pverbeek@uab.edu; Tel.: +1-205-975-9276

Simple Summary: The popular notion that dogs are our best friends suggests a positive bond between ourselves and our dogs. When people think and talk about their connection with dogs, they often do this from a human perspective, and they are less likely to try to think about this from the dog's perspective. We are interested in what the bond between humans and dogs means for dogs, and we are particularly interested in how a bond with a human partner affects the welfare of the dog. We decided to investigate to what degree and how research on the human–dog bond considers the welfare of dogs. We used a large database to select research publications on the human–dog bond published during the 2012–2023 period. We found 706 publications on the human–dog bond from around the globe, of which 246 had a research focus on dog welfare. We studied the publications with a focus on dog welfare, and we found that the characteristics and backgrounds of the dog and the owner affected both the nature of their bond and the welfare of the dog in both positive and negative ways. Most the publications that we studied were on pet dogs in Western industrialized societies.

Abstract: The close bond that can exist between humans and their dogs is an important aspect of the evolutionary, economic, and social connections between the two species. There is a need for a better understanding of the place of the dog within the human–dog bond and on ways the human–dog bond affects dog welfare. We conducted a scoping review to investigate to what extent and in what ways dog welfare is addressed in the research literature on the human–dog bond. We identified 706 publications on the human–dog bond from across the globe that were published from 2012 to 2023. We found that 246 of these 706 publications had a focus on dog welfare. Our review showed that the interplay of characteristics and backgrounds of owners/handlers and their dogs was linked to dog welfare in multiple, both positive and negative, ways. Our review is limited by the fact that most of the research that we reviewed involved pet dogs and in majority came from Western, Educated, Industrialized, Rich, Democratic (WEIRD) societies. There is a need for a better understanding of how the human–dog bond affects the welfare of working, assistance, and service dogs.

Keywords: human–dog bond; dog welfare; behavior

Citation: Verbeek, P.; Majure, C.A.; Quattrochi, L.; Turner, S.J. The Welfare of Dogs as an Aspect of the Human–Dog Bond: A Scoping Review. *Animals* **2024**, *14*, 1985. https://doi.org/10.3390/ani14131985

Academic Editor: Betty McGuire

Received: 24 May 2024
Revised: 26 June 2024
Accepted: 3 July 2024
Published: 5 July 2024

1. Introduction

1.1. A Tale of Two Dogs

Years ago, when the first author lived in Bretagne, France, he heard from a friend that a local hunter was planning to shoot and kill his male Gordon Setter for being unsatisfactory in the hunt. The dog would not point toward the game that the hunters were chasing, but rather would drive the game away from the hunters before they had a chance to aim their guns at it and shoot it. As an alternative to killing his dog, the hunter let it be known that he would give away the dog to anyone who wanted it, and so I (PV) got in touch with him and told him that I would take his dog. For the next several weeks, I collected the dog, named Sulky, every afternoon from the garage in which he spent his days, and then took him for a long walk along the Bretagne shoreline. I thought that this might help Sulky to become accustomed to me and help him with his transition from his current life to life with me. Sulky never showed fear or aggression toward me when I arrived at the garage to pick him up, and he would let me leash him with no problem. During our walks together, he mostly ignored me, however, and he seemed not interested in being petted by me, nor did he pay much attention to me when I talked to him. One day when we had arrived at a remote spot along the coast with no people or other dogs in sight, I decided to take a chance and I unhooked Sulky's leash from his collar. Sulky took off like a greyhound at a dog racetrack. I sat down on a rock overlooking the ocean and contemplated that this was perhaps not the smartest move that I had ever made in my life with dogs, releasing a powerful hunting dog in a remote area teeming with small mammals, quail, and even the occasional pheasant. Scent heaven! Then, just when I was trying to think of what to do next, a small black dot appeared on the rugged horizon, from behind some mighty granite rocks, gradually morphing into a large Gordon Setter, black feathery fur flowing in the ocean breeze, running toward me. When he finally made it to my side, Sulky sat down close to me—panting heavily and drooling profusely—and then put his head on my knee. He let me pet him and appeared to enjoy it, and he listened intently to my repeated "Good boy!" The next day, I took Sulky in for good. During the following years, my bond with Sulky was as close as with any of my other canine companions that I have been fortunate to share my life with thus far.

We were reminded of Sulky's entry in the first author's life by a recent news report about a politician in the US who shot and killed her 14-month-old German Wirehaired Pointer, Cricket, for reportedly ruining a pheasant hunt, being aggressive toward her owner, and for getting into a chicken coop with fatal consequences for the chicken. The news generated many comments from across the political spectrum [1], few of them positive.

We can hypothesize that both Cricket and Sulky behaved in part as a function of strong prey drives, fueled by their evolutionary heritage as a predator species, and shaped by generations of human-controlled selective breeding. We can also hypothesize that being well—feeling well—for Sulky may have meant running free, tracking scents and the animals that produced them in the Bretagne countryside, while for Cricket, it may have meant having a go at those cooped up chickens. As for Cricket's purported aggressiveness toward her owner, we cannot say much about this, other than that it would be relevant to know more about Cricket's owner's behavior toward her, and about the way Cricket was raised from puppyhood.

For both Sulky and Cricket, we can ask whether they should have been judged that harshly by their owners, or even judged at all, not to mention punished, for ruining hunts in which they were recruited to participate by their owners, and for which they were expected by their human minders to behave in certain ways and not in others. While a prey drive is inherent to a dog, human control over it is a training issue. Training creates a bond between the human and the dog, for better or for worse, and in the case of Sulky and Cricket, this was associated with an ultimate threat to the dog's welfare; Sulky escaped with his life, Cricket did not. Considering this tale of two dogs implies what concepts such as welfare and wellbeing of a dog may mean to the dogs themselves, and how that connects to the human–dog bond. As a step toward a better understanding of this issue, we set out and report here on a scoping review that explores to what extent and in which ways dog welfare is considered in research on the human–dog bond.

1.2. The Human–Dog Connection

There is a growing understanding that dealing with global environmental challenges requires a systems perspective that places humans squarely among other animals and considers mutualistic as well as antagonistic interactions between humans and other than human animals (hereafter animals) [2]. An integral aspect of such an approach needs to be a focus on interactions between humans and domestic animals, some of which can be described as mutually beneficial and others as exploitative when most benefits accrue to humans. A key connection that combines mutually beneficial as well as exploitative elements consists of the evolutionary, economic, and social links between humans and dogs, which is the oldest established connection between humans and another animal species (Ibid.). Leaving aside for a moment the importance of including this connection in a systems approach to dealing with the challenging environmental conditions that we have created for ourselves, it can be argued that if we truly want to understand ourselves as a species, we need to consider our evolved and developing connections with dogs, considering that dogs have been with us almost every step of the way.

Science is heeding the call, as research on the connections between humans and dogs is rapidly becoming a major interdisciplinary field [3]. Many questions about the complex connections between humans and dogs remain to be answered, including about the processes of domestication of dogs and whether the Eurasian grey wolf, *Canis lupus*, is indeed the direct ancestor of the domestic dog, *Canis lupus familiaris* [2,4]. A better understanding of the evolution and domestication of dogs is important for guiding research on the human–dog bond. The human–dog bond is integral to the comprehensive connection between humans and dogs, and, considering the diversity in cultural opportunities and constraints on the human–dog connection across the globe, the human–dog bond can be expected to be highly varied in its expression and needs a multifaceted approach.

1.2.1. The Human–Dog Bond

It is important at this point to distinguish between the comprehensive human–dog connection and the human–dog bond. We define the human–dog connection as a complex network of human associations with dogs, ranging from puppy mills to search and rescue dogs, and from pampered pets to free-ranging street dogs, and much more. We follow Oxford Languages in defining a bond in the social realm as "a relationship between people or groups based on shared feelings, interests, or experiences", and apply this definition to the close relationship between humans and dogs. As the sharing of feelings, interests, or experiences between humans and dogs is culture- and context-dependent, the human–dog bond can be expected to be highly varied in its expression. At the most basic level, research on the human–dog bond focuses on behavioral and cognitive interactions between humans and dogs that enable individuals or groups of the two species to coordinate their lives. Important work in this area is conducted on what dogs bring to the table for this, including ways in which dogs read and anticipate human behavioral cues, and how human language can function as a conduit for interactions between humans and dogs [5,6].

The ways in which dogs rely on and express their species-specific ability to attune to human behavior and language can also be expected to vary significantly across social and cultural contexts, and this has inspired ethnographic and culturally informed research on the form and function of the human–dog bond across the globe [7]. The questions here are how and why and what kinds of feelings, interests, or experiences are shared between humans and dogs, and in what contexts? The complexity of the task of understanding these contextually defined close connections between humans and dogs is apparent from the fact that the world population of dogs is estimated at between 700 million and 1 billion dogs, 70% of which are free-ranging dogs [8]. To date, much of the research on the human–dog bond as an aspect of the comprehensive human–dog connection has been on owned and non-free-ranging dogs, primarily in Western, Educated, Industrialized, Rich, Democratic (WEIRD) societies, and significantly less research on the human–dog bond has been conducted in other than WEIRD societies [7].

The need for culturally relevant research on the human–dog bond that focuses on when, how, why, and where, feelings, interests, and experiences are shared between humans and dogs, comes at a time when concepts traditionally eschewed in anthropological, psychological, and biological approaches to animal behavior, such as feelings (i.e., felt emotions [9]) and sentience [10], are increasingly seen as valid and important concepts to pursue in the study of how animals make their way through life. Theoretically and practically, the conceptual interplay between classical behaviorism (classical and operant conditioning) and classical ethology (instinct; fixed action patterns) that portrayed other animals mostly as automata, has largely run its course, clearing the way for newly realigned perspectives on emotion [11,12] and sentience [9,13] as important drivers of animal behavior. These developments in animal behavior science map onto contemporary perspectives on animal welfare (e.g., The Universal Declaration on Animal Welfare 2000–2014), and they inspired our review.

1.2.2. Function and the Human–Dog Bond

For thousands of years, humans have selectively bred dogs for services to be rendered to them, including hunting, guarding, tracking, and herding for humans. Starting about 150 years ago, selective breeding of dogs became formalized through the foundation of breed associations and national kennel clubs, including the American Kennel Club (AKC) and similar kennel clubs in WEIRD societies organized under the umbrella of the Federation Cynologique Internationale (FCI). Modern representatives of the more than 200 recognized dog breeds commonly carry pedigrees attesting to multiple generations of breeding according to an accepted breed standard that outlines a breed's desired form and function. Humans also train and breed dogs of differing ancestry to assist them with modern day life issues and to provide them with services and emotional comfort in times of need.

While most current dog breeds were created and selectively bred with a specific function in mind, there is mixed evidence about whether and how behaviors associated with function are genetically encoded and preserved in current breeding populations [14,15]. Moreover, in WEIRD societies, often only a minority of fanciers of a given dog breed still breed, train, and care for their dogs guided by adequate knowledge of—and dedication to—the original function of the breed.

All too often when it comes to acquiring and raising a pup in WEIRD societies, preferences for 'looks' trump informed knowledge of the function and character of the breed, or, in case of a mixed-breed animal, relevant information about the behavior of parents or family history of the pup [16,17]. Such uninformed acquisition of a canine companion can result in a mismatch between owner characteristics and owner expectations and a dog's developing temperament and behavior. As the stories of Sulky and Cricket at the beginning of this introduction suggest, a mismatch between owner characteristics and owner expectations of the dog can negatively affect the human–dog bond and a dog's welfare.

Owner characteristics and expectations and a dog's inherent or desired behavioral function do not only correlate with the quality of the human–dog bond and dog welfare in WEIRD societies, but also in non-WEIRD societies. A recent comprehensive review of ethnographic studies conducted in 124 globally distributed societies showed that human–dog bonds were closer, and positive care of the dogs increased, in the case of herding dogs, while the opposite was found for hunting dogs [7]. Form and function of the dog can thus be hypothesized to be important variables in the study of dog welfare in the context of the human–dog bond.

1.3. *Dog Welfare Defined*

Psychologist Alexandra Horowitz suggests that while keeping pets in the home can be seen as reflective of an interest in animals, it is worth remembering that this is a model of animal captivity that also produces millions of homeless or unmanageable animals who are killed annually in the United States alone [18]. Focusing on dogs as pets, and while commenting on the field of human–animal interaction (HAI) research, Horowitz comments that in HAI studies, the dog is usually the silent partner, with little or no attention to the dog in and of itself, or to its welfare (Ibid.). In part inspired by the work of Horowitz and others, including the work of the evolutionary biologist and canine expert Marc Bekoff [19], we set out to explore to what extent and in which ways dog welfare (a dog's state of being well) is considered in research on the human–dog bond. From the opening 'tale of two dogs' throughout the subsequent sections of this introduction, various aspects of dog welfare have come to the fore. However, for this scoping review, we needed a more systematic and established approach to the welfare of animals, and we decided to let our work be guided by the approach developed over the past 25 years by the bioethicist David Mellor and his colleagues [20].

As detailed in Section 2.4.3., the 'Five Domains Model' of human–animal interactions in assessments of animal welfare developed by Mellor et al. [20] focuses on (1) nutrition, (2) physical environment, (3) health, (4) behavioral interactions, and (5) mental state. From the initial assessment of our selected literature (see Section 2.3), we decided that the first four domains of the model would adequately cover aspects of dog welfare addressed in our sample. As Mellor and colleagues explain, the model is a guide to the assessment of the positive and negative impacts of human behavior on animal welfare, including the behavior of such persons as companion animal owners, owners of sport/recreational animals, animal trainers, service animal handlers, hunters, researchers, veterinary care staff, and pound/shelter staff, who are all persons serving as "the human" in the human–dog bond, illustrating the good fit of Mellor et al.'s model for the purpose of this scoping review.

1.4. Aims

The aim of this scoping review is to obtain insight into the extent to which dog welfare is considered in the research and literature on the human–dog bond. We were inspired to conduct this review in part by our work on a university course on the anthropology of peoples and their dogs, as well as by our own experiences and activities with dogs, which, for some of us, span many decades. From our earlier work, we had the impression that research on the human–dog bond has been on the rise during the past decade, and we wanted to test that by selecting the literature for our review from the period 2012–2022 We also included literature that was published up until the date of our initial search in May 2023. We hope that this review will inspire further research on dog welfare as an aspect of the human–dog bond. We also hope that this review will have some practical significance for the lives of pet dogs and working dogs of all kinds, and their human partners, as well as for the lives of service and assistance dogs and their human beneficiaries.

2. Materials and Methods

Given that the research on the human–dog relationship includes work in multiple disciplines, we decided on using the Scopus® database (Elsevier B.V., Amsterdam, The Netherlands), which is the largest database of peer-reviewed literature and covers nearly 36,377 titles from approximately 11,678 publishers, of which 34,346 are peer-reviewed journals in top-level subject fields: life sciences, social sciences, physical sciences, and health sciences. We utilized the PRISMA checklist for scoping reviews [21]. As registration is not deemed necessary for scoping reviews, the review was not pre-registered.

2.1. Specifying the Research Question

The research question for this scoping review developed from what we see as a need for a greater emphasis on dogs in research on the human–dog bond [22]. Even research that proclaims to take a 'dogcentric' perspective often focuses on the impact of dogs on the lives of humans and not the other way around. For our purpose, a 'dogcentric' approach starts with the dog and focuses on what the dog contributes to the human–dog bond by means of the dog's species-specific and domesticated nature, and on how interacting with humans affects the dog's life. We predicted that there is a growing segment within research on humans and their dogs that focuses on dog welfare, and we wanted to know how much of this trend is reflected in research on the human–dog bond. Our research question for this scoping review is therefore stated as, 'To what extent and in what ways is dog welfare addressed in the research literature on the human–dog bond?'.

2.2. Identifying Relevant Literature

As illustrated in Figure 1, we developed a search strategy to identify literature relevant to answering the research question. To cast a wide net, the following three search terms were used, separated by the Boolean operator 'AND': human, dog, bond. On 29 May 2023, the first author used this strategy to search the abstract and citation database Scopus®. The search was limited to publications in the English language.

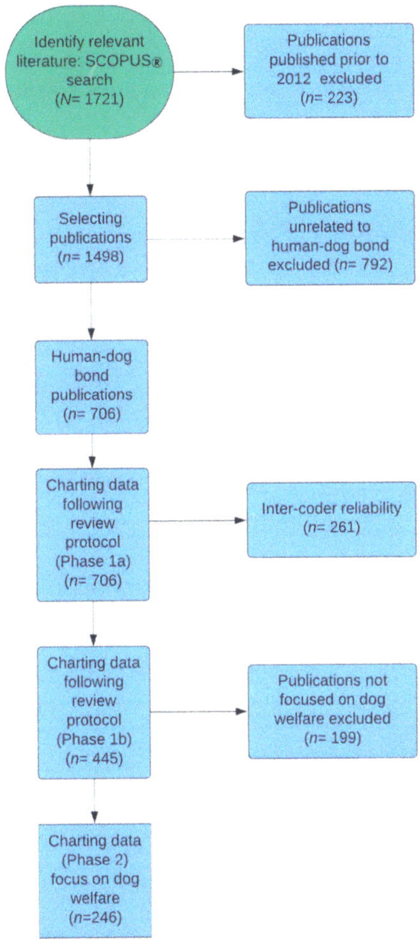

Figure 1. Selection of sources of evidence.

The initial search resulted in 1721 records, with publication dates spanning the period of 1939–2023. The search was downloaded and transformed into an MS Excel file (Version 16.84), and the 1721 publications identified by the search were downloaded from the publisher site or requested through inter-library loan.

From earlier work, we had noted that the literature on the human–dog bond appeared to have significantly increased in volume over the past decade [23], and to test this, we selected records published during the years 2012–2022, and added the publications published in 2023 up to the date of the search. This resulted in 1498 publications.

2.3. Selecting Publications

Next, we inspected abstracts of the 1498 preselected publications and eliminated those not related to the human–dog bond (e.g., records on strictly veterinary issues; chemical bonds, etc.). This data cleaning resulted in 706 publications, of which 680 were published during the years 2012–2022, and 26 in 2023 up to the date of our Scopus® database search. Of the 706 selected publications, 689 are journal articles and 17 are book chapters.

2.4. Charting the Data—Phase 1a

We developed a protocol to chart the data relevant to our research question. We prepared a data extraction form in Microsoft Excel to include each of the 706 publications selected from the original Scopus® master file that were charted following our protocol. Each article was read in its entirety and re-read if needed for clarity and proper understanding. We scored presence (1) or absence (0) for each level of each of the 3 criteria listed in the protocol.

Protocol criterion 1: Consideration of dog welfare

(a) Primary focus of research/discussion on dog welfare.
(b) Secondary/tertiary focus of research/discussion on dog welfare.
(c) Dog welfare mentioned but not a focus of the research/discussion.
(d) Dog welfare not mentioned in publication.

We coded a primary focus on dog welfare in case the research or possible application thereof was centered entirely on (the) dogs' state of being well (cf. 1.3). We coded a secondary/tertiary focus on dog welfare in cases where the interest or benefits for the human in the human–dog bond came first, while the welfare of the dog was also seriously or actively considered, for example, as in the case of ONE-HEALTH articles. We discussed articles that proved difficult for one of us to classify until we reached a consensus for how to code it.

We also scored (0 or 1) whether Institutional Animal Care and Use Committee (IACUC) approval of the research was mentioned in the publication, and whether the publication either referred to—or was specifically focused on—ONE-HEALTH.

Protocol criterion 2: Dogs studied/discussed

(a) Pet dogs
(b) Working dogs
(c) Assistance/Service dogs

Protocol criterion 3: Human partner

(a) Owner (handler) [non-assistance/service dog]
(b) Professional (e.g., K9 police officer; researcher, etc.)
(c) Beneficiary (assistance/service dog)

2.4.1. Inter-Coder Reliability

Charting of the 706 publications selected for our scoping review was performed by the first three authors. To assess inter-coder reliability upfront, the first three authors each worked on the same randomly selected subset [n = 261 (36.8%)] of the 706 publications selected for charting. Full agreement on the three protocol criteria (see above) was shown for 205 of the selected publications (78.5%; Fleiss' kappa = 0.571). The 56 records for which there was disagreement were discussed and updated after consensus was reached.

2.4.2. Charting the Data—Phase 1b

The task of charting the remaining 445 publications was approximately equally divided among the first three authors. Several meetings were held during this phase to review and discuss any issues that arose. Following completion of this task, the first author verified the coding of the 3 protocol criteria for each of the 445 publications, and any remaining issues were discussed and resolved with the respective coder prior to releasing the full dataset of 706 records for the second phase of data charting.

2.4.3. Charting the Data—Phase 2

For each of the 706 publications on the human–dog bond, the first author identified the country and greater geographical region of origin based on the country of residence of the main author of the publication (as determined by institutional affiliation). The following geographical regions were defined: Africa (publications from 5 countries), Asia (11 countries), Europe (23 countries), Middle East (1 country), North America (2 countries), Oceania (2 countries), and South America (5 countries).

Next, each of the 246 publications charted as having either a primary or secondary focus on dog welfare were inspected and scored by dog welfare domain, based on the 2020 five domains model developed by David Mellor and colleagues (Mellor et al. 2020 [20]). As mentioned previously, these welfare domains include Nutrition (nutritional conditions and their associated effects), Environment (physical environmental conditions and their associated effects), Health (health conditions and their associated effects), Behavior (behavioral interactions and their associated effects), to which was added the domain Other for any other aspect of welfare not specified in the first four domains adapted from Mellor et al.

Finally, for each of the 706 publications, the scientific domain best describing the research/discussion was determined and scored by the first author as one of the following: (Animal) Behavior, (Animal) Welfare, Medical Science, Social Science, and Veterinary Science. The determination by scientific domain was based on journal/article title as well as by the publication's main author's position and/or affiliation.

2.5. Collating, Summarizing, and Reporting the Data

In the results sections that follow, we first report on summaries derived from the full data set of 706 publications on the human–dog bond. We follow this with summaries and analyses derived from the publications on the human–dog bond with either a primary or secondary focus on dog welfare ($n = 246$). We conclude with a more detailed discussion of publications with a focus on dog welfare that represent the behavioral interaction domain (Behavior) of the five-domain model of animal welfare, i.e., behavioral interactions and their associated effects ($n = 114$). These latter publications are attached to this paper within the reference list, and the remaining publications charted in this scoping review can be accessed as Supplementary Material.

3. Results

3.1. Consideration of Dog Welfare in Human–Dog Bond Research Publications

As we predicted, and as illustrated in Table 1, our summative analysis showed that the relative number of publications on the human–dog bond increased notably from an average of 40.2 (SD = 27.9) per year during the years 2012–2017 to an average of 87.8 (SD = 26.2) per year during the years 2018–2022.

During the years 2012–2017, an average of 11.8% of the publications on the human–dog bond featured a primary focus on dog welfare, while for the following period 2018–2022, this percentage was 24.2%, which is also a notable increase. Publications with a secondary focus on dog welfare accounted for 10.2% during 2012–2017 and 14.2% during 2018–2022, mirroring, albeit less substantially so, the upward trend shown for publications with a primary focus on dog welfare. The records for the first months of 2023 suggest a continuation of the rise in human–dog bond publications with a primary focus on dog welfare.

Table 1. Consideration of dog welfare in human–dog bond research publications (2012–2023; *N* = 706).

Year	Primary Focus	Secondary Focus	Dog Welfare Mentioned	Dog Welfare Not Mentioned	Total
2012	1	1	0	8	10
2013	0	0	4	10	14
2014	9	4	5	16	34
2015	4	6	8	28	46
2016	6	5	11	28	51
2017	13	15	21	37	86
2018	14	2	10	35	61
2019	19	13	23	46	101
2020	21	21	17	47	106
2021	28	28	24	33	113
2022	20	6	15	17	58
Total	135	102	138	305	680
%	19.9	15	20.3	44.8	100
2023	7	2	4	13	26
%	26.9	7.7	15.4	50	100

Taken together, over the period 2012–2022, 34.8% of publications on the human–dog bond had a focus on dog welfare, with 20.2% of the 680 publications at least mentioning dog welfare. Next, we compared how the focus on dog welfare was distributed among publications on the types of dogs that we defined for our review: pet dogs, working dogs and service/assistance dogs.

3.1.1. Consideration of Dog Welfare by Category of Dog

As shown in Table 2, pet dogs were by far the most studied type of dog, representing 80.7% of the 706 publications on the human–dog bond, compared to working dogs, 3.4% of the total, and service dogs, 12.3% of the total. Mixed samples, mostly made up of pet dogs and service dogs, accounted for 3.5% of the total.

Table 2. Focus on dog welfare in human–dog bond articles by category of dog (2012–2023; *N* = 706).

Category	Pet Dogs	Working Dogs	Service Dogs	Mixed Sample	Total
Primary Focus	129	6	3	4	142
Secondary Focus	95	3	6	0	104
Dog Welfare Mentioned	105	4	27	6	142
Dog Welfare Not Mentioned.	241	11	51	15	318
Total	570	24	87	25	706

Looking within each type of dog, publications with a primary focus on dog welfare accounted for 22.6% of all publications on pet dogs, with 16.6% having a secondary focus on dog welfare, and 18.4% mentioning—and 42.3% not mentioning—dog welfare. The percentages for working dogs, presented here in the same order as for pet dogs, were 25.0%, 12.5%, 16.7%, and 45.8%, respectively, and for service dogs, 16.0%, 0%, 24.0%, and 60.0%.

IACUC compliance was reported in 66 research publications, of which 19.7% had a primary focus and 12.2% a secondary focus on dog welfare, with 9.1% mentioning, and 59.0% not mentioning dog welfare.

Our next analysis looked at the country and geographical region of origin of the 706 publications on the human–dog bond, and we noted the focus or lack thereof on dog welfare for each country and region.

3.1.2. Consideration of Dog Welfare by Country and Geographical Region

Our analysis showed that Europe topped the number of publications on the human–dog bond with 301 publications (42.6% of total), followed by North America, 277 (39.2%), Oceania, 52 (7.4%), Asia, 37 (5.2%), South America, 30 (4.2%), Africa, 6 (0.8%), and the Middle East, 3 (0.4%).

As detailed in Table 3, for regions with 10 or more human–dog bond publications, Europe showed the highest percentage of publications with a primary focus on dog welfare, 24.3%, compared to North America, 18.4%, Asia, 16.2%, Oceania, 13.5%, and South America, 13.3%. Taken together, the findings of this geographical analysis show that the human–dog bond is a topic of scientific inquiry across the major regions of the world, with relatively high numbers of publications concentrated in Europe and the Americas.

Table 3. Consideration of dog welfare in human–dog bond research publications (2012–2023; $N = 706$) by country and geographic region.

Africa	Primary	Secondary	Mentioned	Not Mentioned	Total
Burkina Faso		1			1
Egypt				1	1
Ethiopia			1		1
Nigeria			1		1
Rwanda			1		1
South Africa				1	1
Subtotal (%)	0 (0)	1 (16.6)	3 (50.0)	2 (33.3)	6 (100)
Asia	**Primary**	**Secondary**	**Mentioned**	**Not Mentioned**	**Total**
China				2	2
Hong Kong		1	1		2
India	1		2	3	6
Japan	2	2	4	11	19
Malaysia		1			1
Pakistan	1				1
Philippines		1			1
South Korea		1	1		2
Sri Lanka	1				1
Taiwan				1	1
Thailand	1				1
Subtotal (%)	6 (16.2)	6 (16.2)	8 (21.6)	17 (45.9)	37 (100)
Europe	**Primary**	**Secondary**	**Mentioned**	**Not Mentioned**	**Total**
Austria	1	2	4	15	22
Belgium		2		5	7
Czech Republic				4	4
Denmark	1		2	1	4
Finland	4	1			5
France	8	2	3	6	19
Germany	5	1	2	7	15
Greece		1			1
Hungary	3	2	2	11	18
Ireland				2	2
Italy	13	4	5	26	48
The Netherlands	2	2	5	10	19
Northern Ireland	1				1
Norway			1	1	2
Poland	2			2	4
Romania		1	1	1	3
Russia			1	1	2
Serbia			1		1
Slovenia	1				1
Spain	2	2	2	4	10

Table 3. *Cont.*

Europe	Primary	Secondary	Mentioned	Not Mentioned	Total
Sweden	4	1	1	8	14
Switzerland	1	1		5	7
UK	25	16	25	22	88
Subtotal (%)	73 (24.3)	41 (13.6)	55 (18.3)	132 (43.8)	301 (100)
Middle East	**Primary**	**Secondary**	**Mentioned**	**Not Mentioned**	**Total**
Iran	1			2	3
Subtotal (%)	1 (33.3)			2 (66.7)	3 (100)
North America	**Primary**	**Secondary**	**Mentioned**	**Not Mentioned**	**Total**
Canada	4	10	9	14	37
USA	47	29	50	114	240
Subtotal (%)	51 (18.4)	39 (14.1)	59 (21.3)	128 (46.2)	277 (100)
Oceania	**Primary**	**Secondary**	**Mentioned**	**Not Mentioned**	**Total**
Australia	6	8	8	26	48
New Zealand	1	1	1	1	4
Subtotal (%)	7 (13.5)	9 (17.3)	9 (17.3)	27 (51.9)	52 (100)
South America	**Primary**	**Secondary**	**Mentioned**	**Not Mentioned**	**Total**
Argentina		1	3	6	10
Brazil	1	4	1	3	9
Chile	1	1	2		4
Colombia	1				1
Mexico	1	2	2	1	6
Subtotal (%)	4 (13.3)	8 (26.7)	8 (26.7)	10 (33.3)	30 (100)

3.1.3. Top Journals Publishing Articles on the Human–Dog Bond with a Focus on Dog Welfare

As shown in Table 4, the journal *Animals* stands out as the leading journal in our selection of journals with three or more articles on the human–dog bond with a focus on dog welfare, publishing 14.3% of the 108 articles. Eight (42.1%) of the 19 journals in this selection are from veterinary science, and together, they represent 47.2% of the 108 articles with a primary focus on dog welfare and 41.7% of the 60 articles with a secondary focus on dog welfare, illustrating the importance of veterinary science for the study of welfare as an aspect of the human–dog bond.

Table 4. Journals publishing three or more articles on the human–dog bond with a focus on dog welfare ranked by primary focus (2012–2023).

Journal	Primary Focus	Secondary Focus	Total
Animals	13	11	24
Scientific Reports	12	1	13
Frontiers in Veterinary Science	11	8	19
Journal of Veterinary Behavior	11	8	19
Journal of Applied Animal Welfare Science	9	1	10
Veterinary Clinics of N. America—Small Animal Practice	9	1	10
PLoS ONE	8	6	14
Veterinary Record	8	4	12
Veterinary Sciences	5	0	5
Applied Animal Behavior Science	4	2	6
Physiology and Behavior	4	0	4
Preventive Veterinary Medicine	3	1	4
Society and Animals	3	1	4
Animal Cognition	2	1	3

Table 4. *Cont.*

Journal	Primary Focus	Secondary Focus	Total
International Journal of Comparative Psychology	2	6	8
Topics in Companion Animal Medicine	2	2	4
Veterinary Journal	2	1	3
Anthrozoös	0	3	3
Journal of Comparative Pathology	0	3	3

3.1.4. Focus on Dog Welfare by Scientific Domain

As mentioned under Section 2.4.3, the scientific domain for each of the 706 publications on the human–dog bond was categorized as either (Animal) Behavior, (Animal) Welfare, Medical Science, Social Science, or Veterinary Science, with Behavior broadly defined and including work on animal cognition. As shown in Table 5, the greatest number of publications on the human–dog bond concerned work on Behavior (26.3% of the total), closely followed by Social Science (24.4%), Medical Science (23.2%), and Veterinary Science (22.7%). Publications from the domain of Welfare accounted for 3.4% of the 706 publications on the human–dog bond.

Table 5. Focus on dog welfare in human–dog bond publications (2012–2023; $N = 706$) by scientific domain.

Science Domain:	Behavior	Welfare	Medical	Social	Veterinary
Primary Focus	37	13	10	16	66
Secondary Focus	21	1	26	23	33
Dog Welfare Mentioned	35	5	40	34	28
Dog Welfare Not Mentioned	93	5	88	99	33
Total	186	24	164	172	160

Underscoring the significance of Veterinary Science for the study of dog welfare in the context of the human–dog bond, 61.9% of the publications in Veterinary Science had either a primary or secondary focus on dog welfare, compared to 58.3% in Welfare, 48.3% in Behavior, 22.7% in Social Science, and 22.0% in Medical Science (Table 5).

3.2. *Aspects of Dog Welfare*

Following up on our analyses on the prevalence and origin of work on dog welfare in the context of the human–dog bond, we investigated the distribution of specific aspects of dog welfare among publications addressing the human–dog bond.

3.2.1. Aspects of Dog Welfare by Category of Dog

Behavioral interactions and their associated effects (Behavior; see Sections 1.3 and 2.4.3) were the most common welfare domain addressed in publications on the human–dog bond with pet dogs as subjects (46.0% of 224), followed by Health (36.6%) [Table 6]. Studies on working dogs mirrored this pattern, with 55.6% of the publications dealing with Behavior, and 22.2% dealing with Health. Studies on service dogs showed the reverse of this pattern: Behavior 33.3% and Health 55.6%. However, considering the small sample size of publications on working dogs and service dogs, it is hard to draw conclusions from these findings.

Table 6. Welfare domains in human–dog bond publications with a primary or secondary focus on dog welfare by category of dog (2012–2023; n = 246).

	Pet Dogs	Working Dogs	Service Dogs	Mixed Sample
Nutrition	14	1	0	0
Environment	17	0	0	0
Health	82	2	5	1
Behavior	103	5	3	3
Other	8	1	1	0
Total	224	9	9	4

3.2.2. Aspects of Dog Welfare by Scientific Domain

As shown in Table 7, behavioral issues (Behavior) were the most often addressed dog welfare issues in publications from animal behavior science (*Behavior*; 67.2% of 58), as well as from animal welfare science and social science (*Welfare*, 78.6%; *Social*, 64.1%). Unsurprisingly, health issues (Health) ranked high in medical science and veterinary science publications (*Medical* 75.0%; *Veterinary* 47.5%), while behavioral welfare issues (Behavior) also featured prominently in publications from veterinary science, (*Veterinary* 34.3%), showing the diverse approach in veterinary science to dog welfare in the context of the human–dog bond.

Table 7. Welfare domains in human–dog bond articles with a primary or secondary focus on dog welfare by scientific domain (2012–2023; n = 246).

	Behavior	*Welfare*	*Medical*	*Social*	*Veterinary*
Nutrition	1	0	1	4	9
Environment	6	1	1	2	7
Health	9	0	27	7	47
Behavior	39	11	5	25	34
Other	3	2	2	1	2
Total	58	14	36	39	99

3.2.3. Behavioral Aspects of Dog Welfare by Scientific Domain

The analysis presented in Section 3.2.2. showed that behavioral interactions and their associated effects on dog welfare (i.e., welfare domain Behavior) were the most common aspect of dog welfare addressed in our selection of publications on the human–dog bond. We followed up on this finding with a detailed look at the various types of behavioral interaction issues and we identified the eighteen distinct ones listed in Table 8a.

Table 8. (a) Behavioral interaction effects on dog welfare (welfare domain Behavior) in human–dog bond articles with a primary or secondary focus on dog welfare by scientific domain (2012–2023; n = 114). (b). Sources of evidence for the analysis presented in (a).

(a)

Science Domain:	*Behavior*	*Welfare*	*Medical*	*Social*	*Veterinary*	*Total*
Behavioral interactional effects associated with:						
Abuse	0	1	0	0	0	1
Attachment	5	0	0	4	0	9
Synchrony	1	0	0	1	0	2
COVID-19 pandemic	1	0	0	0	1	2
Dog characteristics	2	1	3	2	0	8
Dog emotion	5	0	0	1	2	8
Dog preferences	2	0	0	0	0	2
Dog-related injuries	0	0	1	0	0	1
Dog training	3	1	1	2	4	11
Ethical/moral issues	1	0	0	1	0	2
Intraspecies interactions	6	0	0	0	0	6
Owner characteristics	7	3	0	13	11	34
Physical activity	0	0	0	0	0	2
Population management	1	0	0	0	0	1
Post-conflict behavior	0	0	0	1	0	1
Problem behaviors	5	3	0	0	8	16
Stress levels	0	1	0	0	0	1
Veterinary visits	0	1	0	0	5	6
Total	39	11	5	25	34	114

Table 8. *Cont.*

(b)					
Scientific Domain	*Behavior*	*Welfare*	*Medical*	*Social*	*Veterinary*
Abuse	-	McMillan et al. [24]	-	-	-
Attachment	Carreiro et al. [25]; Riggio et al. [26]; Saavedra-Aracena et al. [27]; Sipple et al. [28]; Thielke and Udell [29]	-	-	Lewis [30]; Solomon et al. [31]; Payne et al. [32]; Konok et al. [33];	
Behavioral synchrony	Duranton et al. [34]	-	-	Duranton et al. [35]	-
COVID-19 pandemic	Brand et al. [36]	-	-	-	Sherwell et al. [37]
Dog characteristics	Samet et al. [38]; Protopopova et al. [39]	Döring et al. [40]	Lee et al. [41]; Smith [42]; Scotney and Clay [43]	Chira et al. [7]; Yamasaki [44]	-
Dog emotion	Hakanen et al. [45]; Savalli et al. [46]; Lenkei et al. [47]; Arahori et al. [48]; Kurachi et al. [49]	-	-	McMillan [50]	Qiasvand et al. [51]; Ballantyne [52]
Dog preferences	Bhattacharjee et al. [53]; Duranton et al. [54]	-	-	-	-
Dog-related human injuries	-	-	Schurer et al. [55]	-	-
Dog training	D'Angelo et al. [56]; Vieira de Castro et al. [57]; Harris et al. [58]	Vitulli et al. [59]	Alers & Simpson [60]	Smith et al. [61]; Pręgowski [23]	Townsend et al. [62]; Learn et al. [63]; LaFollette et al. [64]; Masson et al. [65];
Ethical/moral issues	NG et al. [66]	-	-	Benz-Schwarzburg et al. [67]	-
Intraspecies interaction	Cheng et al. [68]; Uccheddu et al. [69]; Cimarelli et al. [70]; Mariti et al. [71]; Romero et al. [72]; Mariti et al. [73]	-	-	-	-
Owner/handler characteristics	Brubaker and Udell [74]; Holland et al. [75]; Stevens et al. [76]; Holland [16]; Philpotts et al. [77]; González-Ramírez [78]; Rehn et al. [79]	Reese [80]; Rossi and Maia [81]; Mehrkam et al. [82]	-	Ferrell and Crowley [83]; Giraudet et al. [84]; Karvinen and Rhodes [85]; Włodarczyk [86]; Bouma et al. [17]; Westgarth et al. [87]; Maharaj et al. [88]; Dodman et al. [89]; Davis et al. [90]; Hall et al. [91]; Bathurst and Lunghofer [92]; Sirois [93]; Alcaidinho et al. [94]	The Veterinary Record [95]; Powell et al. [96]; Pirrone [97]; Laurence et al. [98]; Hall et al. [99]; Nardoia et al. [100]; Strickler [101]; McGreevy et al. [102]; Diverio et al. [103]; Herron et al. [104]; Tiplady et al. [105];
Physical activity	-	-	-	-	Väätäjä et al. [106]; Yuma et al. [107]
Population management	Ma et al. [108]	-	-	-	-
Post-conflict behavior	-	-	-	Cavalli et al. [109]	-
Problem behavior	Feuerbacher and Muir [110]; Gates et al. [111]; Wormald et al. [112]; Thielke and Udell [113]; Mills et al. [114]	Stephens-Lewis et al. [115]; Bräm Dubé et al. [116]; Mehrkam et al. [117]	-	-	Powell et al. [118]; Normando et al. [119]; Teixeira and Hall [120]; Taylor et al. [121]; Buller and Ballantine [122]; Dinwoodie et al. [123]; Canejo-Teixeira et al. [124]; Rajapaksha [125]
Stress levels	-	Koda et al. [126]	-	-	Riggio et al. [127]
Veterinary visits	-	Mariti et al. [128]	-	-	Helsly et al. [129]; Kogan et al. [130]; Csoltova et al. [131]; Döring et al. [132]; Martin et al. [133]

With 29.8% of the total, owner characteristics and their interactional effects on dog welfare were the most often addressed behavioral dog welfare issue across scientific domains (four out of five scientific domains reporting), followed by problem behaviors (14.0%; three of five scientific domains). Dog training was another relatively common behavioral dog welfare issue (9.5%; five of five scientific domains), followed by interactional effects related to attachment to owner (7.9%; two of five scientific domains), dog characteristics (7.0%;

four of five scientific domains), and dog emotion (7.0%; three of five scientific domains). Intraspecies interactions (5.3%) and veterinary visits (5.3) were specific to scientific domains with, respectively, one and two scientific domains reporting on these behavioral interactional dog welfare issues. Table 8b shows the sources of evidence for the above analysis.

4. Discussion

4.1. The Five-Domain Model of Dog Welfare and the Sentient Dog

As the Behavior domain was the most frequently addressed welfare domain in the selected literature of this scoping review with a focus on dog welfare, we proceeded with a more detailed review of the publications in this area. The behavioral interaction domain of Mellor et al.'s five-domain model of animal welfare (Behavior) breaks down positive human attributes and behaviors into attitude, voice, aptitude, and handling/controlling. For attitude, confidence, caring, sensitivity, patience, kindness, and empathy are listed as positive influences on the human–animal interaction, resulting in a bonded, alert, and responsive animal ready to explore novel events. A calm, clear, and encouraging human voice adds to that positive mix, and so does an experienced and skilled aptitude on the part of the human partner. Handling and controlling of the animal by the human partner should be skillful and gentle, based on an insightful mix of firmness and restraint, and a focus on rewards. Deficiencies on the part of the human in those attributes and behaviors can result in the animal being anxious and insecure in the interaction with the human and can trigger fear and even panic in the animal, as well as helplessness and an avoidance of novel events [20]. In terms of the operational definition of the human–dog bond that we advance in this paper, we propose that with the right attributes and behaviors on the part of the human, the likelihood that the dog will be ready to share feelings, interests, and experiences with the human partner (see Section 1.2.1) and establish a bond will be greatly increased.

Groetzinger Strickler [101] states that in recent years, the relationship between pet and owner has changed significantly. Mirroring empirical and theoretical developments in the study of animal behavior as a whole [9,11] (see also Section 1.2.1), professionals including veterinarians and trainers, as well as owners, are more likely than before to perceive companion animals as thinking, feeling beings. According to Groetzinger Strickler, this allows for a transition in training and care from one of poor and inadequate behavioral welfare to an approach that acts on the full potential of the animal. Our own perspective here acknowledges the science, which shows that many animals that were previously depicted as mere stimulus response automata are indeed thinking and feeling beings, capable of making their own decisions while navigating through life. We think that this new understanding of animal behavior both maps onto and enriches the five-domain model of animal welfare, especially as it applies, but is not limited to, the human–dog bond.

4.1.1. The Pre-Acquisition Phase and Dog Welfare

Many of the articles that we reviewed for this presentation discuss or investigate behavioral elements of Mellor et al.'s five-domain model. Several authors stress that the foundation for a close bond between a human and a dog starts at the time the dog enters the human's life, or even before that, during the phase the human contemplates bringing a dog into her life. Holland and colleagues [16,75] warn against impulse acquisitions without any research into the background of the dog or its breed. These authors state that acquiring a dog out of a desire to help a vulnerable animal, including adopting a dog from a shelter, does not necessarily guarantee the establishment of a bond favorable to the welfare of the dog, unless and until the human contributes the right mix of attitudes and behaviors expected to positively affect the dog's welfare.

Research by Bouma et al. [134] shows that people who frequently read books about owning dogs and who often talked about this with others were more likely to make an informed decision about getting a dog, compared to people who visited websites offering or selling dogs and as such, were more prone to impulse buying. Pirrone et al. [97] stress that potential owners should see the puppy they intend to buy while it is with its mother, as this can provide useful information on the behavior of the mother and can help to predict the behavior of the puppy as an adult. Diverio et al. [103] show that there can be a gap between what people imagine as the ideal dog and their actual dog, and that adequate education of potential dog owners about the specifics of the breed and background of the dog, and of the positive effects of training activities on the dog's behavior, is important for establishing a bond that optimizes dog welfare.

Herron at al [104] point out that newly adopted shelter dogs often experience separation anxiety, and that pre-adoption counseling can be helpful to inform the new owner about effective prevention tools to use in the home to minimize the development of separation anxiety. Reese [80] suggests, for shelter dog adoptions to be successful, matching discussions with a potential owner before showing any dogs are important, as they can serve as a guide to show only those dogs that in terms of temperament and behavior can be predicted to be a good match with the new owner. Taken together, these studies show the responsibilities of potential and new dog owners for doing their part in laying the foundation from which a bond that optimizes dog welfare can develop.

4.1.2. Owner Characteristics and Dog Welfare

The articles we reviewed for this section present a variety of approaches to the role of owner characteristics in the human–dog bond and dog welfare. Different aspects of the Behavior domain of dog welfare are implied, associated with different contexts and perspectives. Brubaker and Udell [74] found inspiration in human developmental science for their approach to the human–dog bond. Developmental science shows a link between parenting style (and parenting practice) and social developmental outcomes in children and adolescents [135]. In some sociocultural groups, especially in WEIRD societies, authoritative parenting, which gives children some input in the interaction and respects their relative autonomy, tends to result in positive social developmental outcomes, such as positive peer relations and self-confidence and self-esteem in social situations. Authoritarian parenting, based on less or no child input in the interaction, and, especially, rejecting/neglecting parenting, have been associated with less positive social developmental outcomes (Ibid.). Brubacker and Udell showed that authoritative owners with high expectations and high responsiveness tended to have dogs that were highly social and sensitive to social context and good at solving an experimental problem task. In contrast, dogs with authoritarian owners (high expectations and low responsiveness) tended to show less positive social behavior and problem-solving skills. These authors suggest that a highly social and cognitively well-functioning dog can be seen as a "happy" dog, and that their research shows that the style in which owners interact with their dog can have a direct effect on the dog's welfare.

González-Ramírez [78] showed that compatibility in energy levels, temperament, and daily activity between owners and their dogs was associated with less aggressive and fearful behaviors and higher trainability scores in dogs, compared to less compatible human–dog dyads. Powell [118] found that owner conscientiousness, extraversion and openness, and the quality of the bond with their dog (defined here as "attachment"), were positive factors in the dog's response to clinical behavioral intervention. Stevens et al. [76] provide evidence that owners who scored higher on cognitive measures were more likely to have their dogs complete an obedience training program. Karvinen and Rhodes [85] showed that owners who train their dog in agility engage in more physical activity with their dogs but less without their dog, compared to other dog owners. As physical activity is generally beneficial for a healthy dog, engaging in agility can indirectly have a positive effect on dog welfare.

We detected a growing interest in personality profiles and mental health of owners and other humans in the close social circle of a dog and the possible links to the quality of the human–dog bond and dog welfare. Dodman et al. [89] found a significant correlation between moderate depression in male owners and the use of aversive and confrontational dog training techniques. Tiplady et al. [105] report on the effects of domestic violence on dog welfare. Dogs that were the target of redirected violence were more likely to be owned by women rather than men, children, or both partners. These authors also found that people experiencing domestic violence are often unwilling to confide in veterinarians or seek help from animal shelters. Hall [91] studied dogs in families with children with neuro-developmental disorders and with neurotypical children. She found that harsh contact and rough and tumble play with children with neuro-developmental disorders, and having to cope with child meltdowns and tantrums, were negative factors in terms of the welfare of dogs in such families. Hall suggests a safe haven for the dog to escape to, parent's awareness of stress signs, and child education in dog interaction to help limit negative effects on dog welfare.

Ferrell and Crowley [83] studied emotional support dogs' interactions with their human beneficiaries and report that approximately one in seven ESA dogs in their study may not have been receiving consistent quality care. Moreover, explicitly discussing animal welfare was not associated with actual welfare items except for adequate shelter. While cautioning that their findings are preliminary, these authors suggest that oversight of ESAs may need to include an assessment of caretaking behaviors and a determination of how to best meet the needs of both the beneficiary and the ESA dog.

Giraudet [84] suggests that dogs may show greater levels of stress in the presence of children, and that the welfare of assistance and therapy dogs who may interact with children remains underexplored. She suggests that for children, the benefits of interacting with dogs may outweigh the risks, but that this is not necessarily the case for dogs. Older children and adolescents may have interactions with dogs that have a more positive effect on dog welfare. Bathurst and Lunghofer [92], for example, report on 'Lifetime Bonds', a program in Chicago that teams up at-risk youth, particularly adolescent males, and at-risk dogs impounded as victims of cruelty and neglect. Both youths and dogs help each other in this program, with the youths learning about the responsibilities of taking care of another sentient being, and the dogs having a chance to overcome some of the effects of their violent past through the care and training provided by their young partners.

Behavior problems of dogs are an important dog welfare concern, as they are one of the most common causes of relinquishment to shelters and a common reason for euthanasia [136]. Anxious dogs can be more vulnerable to disease and, through aggressive behavior, can become a public health concern (Ibid.). Several authors reviewed here discuss the role of owner characteristics in dog behavior problems and ways to increase owner knowledge about the causes and interventions associated with problem behavior. Philpotts [77] argues that while improving an owner's knowledge through an education intervention has the potential to improve dog welfare, the complexity of dog welfare and dog ownership requires significantly more informed input from cross-disciplinary and boundary-crossing research on dog welfare in the design and execution of such education intervention. Westgarth et al. [87] express a similar view while discussing responsible dog ownership and ways to promote it. These authors argue that telling owners that they should be responsible is of limited use for promoting behavior change if it is not accompanied by targeted education on the dog's role within the family and wider society.

McGreevy et al. [102] argue for an applied science of dogmanship. They emphasize the role veterinary behavioral medicine can play in such a science, using information technology tools such as a computer or personal electronic device-based interactive doglogbook and dogmanship coaching tool, based on a detailed dog–human interaction ethogram. They suggest these tools can be combined with real-time measures of heart rate, balance and movement to give biofeedback, as the user develops timing, consistency, and calmness, toward an ideal interaction pattern that protects and enhances dog welfare. In a similar vein, Alcaidinho et al. [94] report on a pilot study on whether the use of a specially designed smartphone application registering owner and dog interaction can increase the perceived strength of the bond between owners and newly adopted dogs from a California shelter, with the goal of reducing returns to the shelter. And in the United Kingdom, a smartphone application that rewards owners for walking their dog and taking care of its wellbeing with points that can be swapped for vouchers, gained 50,000 registered users since its launch in March 2022 [95].

4.1.3. Problem Behaviors as a Dog Welfare Issue

Problem behavior is discussed extensively in the literature on the human–dog bond reviewed here, and it is often linked to the welfare of both dogs and their human partners [122]. There is no one-size-fits-all definition for problem behavior, as what is perceived as problematic dog behavior for one dog owner may not be a problem for another. We found some reoccurring themes in the literature, however, including aggressive behavior and separation anxiety. This latter term, borrowed from human psychology, refers to often destructive behavior exhibited by a dog when left at home with the owner away and no other humans present. Several authors emphasize that problem behavior can be linked to the specific functioning, or malfunctioning, of a given human–dog bond. This implies that in terms of remedying the problem behavior, the behavior and characteristics of both the human and the dog need to be considered. The first author has experience with this reality, going back many years when he worked as a professional dog trainer in a large boarding and breeding facility in Germany. My (PV) task was to correct problem behavior in dogs boarded at the facility where I worked, ranging from serious issues such as aggression toward other dogs or people, to more mundane issues, such as excessive leash pulling when walked, or ignoring commands and generally not being a good canine citizen in the opinion of the owner. I would work intensively with such dogs for about 6 weeks, at the end of which I would demonstrate the behavior of the dog to its owner in the expectation that the dog's behavior was now acceptable to the owner. In some cases, this was not the end of it, however, as the dog would be returned to us by the owner with the same problem behavior occurring again. This is when I learned that to remedy problem behavior in dogs, both the owner and the dog need to be trained.

Veterinary behavioral scientist Daniel Mills and colleagues [114] propose eight overlapping dimensions to be characteristic to the human–dog bond: content of interactions, diversity of interactions, reciprocity versus complementarity of interactions, quality of interactions, frequency and patterning of interactions, intimacy, cognitive perspective of interactions, and multidimensional qualities. They link these eight dimensions to (1) the emotional involvement/dependency, (2) common interest, and (3) working partnership that make up a human–dog bond. We see significant similarity here to our own view of the human–dog bond as one of shared feelings, interests, and experiences. From this conceptual approach to the human–dog bond, Mills and colleagues suggest that approaches to problem behavior in dogs should consider the bidirectionality of the human–dog bond, including the expectations and biases that both partners may bring to the bond. This, of course, is a more eloquent way of stating what the first author learned as a professional dog trainer all those years ago: it usually takes two, the owner and the dog, to remedy dog problem behavior.

Stephen-Lewis et al. [115] make a similar case for the need to consider both owner and dog when dog problem behavior occurs in their discussion of reactivity in dogs, which may be a problem to some owners and less of a problem to others. They propose that dog reactivity is influenced by canine characteristics, human expectations, and human capabilities, each of which features multiple subfactors, such as culture and lifestyle. Canejo-Teixeira [124] distinguishes between functional and dysfunctional dyads in the context of dog aggression and other problem behavior. Human members of functional human–dog dyads are described as being responsible for the dog's welfare, providing the necessary care, and avoiding situations of risk. Humans in dysfunctional dyads are described as showing the opposite characteristics, and as not always being aware that their behavior may be placing themselves and/or their dog at risk. Taken together, each of these authors underlines the need to consider the bidirectionality and overall functioning of the human–dog bond when dog problem behavior occurs.

A study by Dinwoodie and colleagues [123] provides a good insight into various types of problem behavior in a large sample survey of 4114 dogs of mixed and pure breeds submitted by 2480 dog owners. Male and female, mostly neutered, dogs were equally represented in the sample. Problem behavior was reported in 85% of the survey responses and included anxiety and fear, aggression, excessive barking, house soiling, destructive behavior, and a range of other issues. The survey focused on dog characteristics and found that age, neutered status, origin, and lineage were all correlated with problem behavior. Rajapaksha [125] reflects on age as a dog characteristic in problem behavior and suggests that problem behavior is more difficult to diagnose in older dogs than in younger dogs. The reason for this is that many degenerative disease conditions in older dogs are reflected as a change in behavior. Rajapaksha recommends detailed clinical examinations for older dogs and owner education and behavioral enrichment as measures to improve the welfare of older dogs.

While framing dog characteristics in the human–dog bond context, research by Dubé [116] showed that problem behavior was more common in dogs described as highly sensitive and when there was a mismatch between owner personality and dog "personality". Gates et al. [111] report on problem behavior in dogs adopted from shelters. In their sample, most dogs were reported to exhibit problem behavior, including aggression toward people or dogs, destructive behavior in the home, and excessively high energy. Interestingly, most new owners showed little concern about the problem behavior, but these authors do recommend support programs for adopters to increase adoption satisfaction.

While not apparent in the Gates et al. study, problem behavior can result in adopted dogs being returned to a shelter, or to dogs being relinquished to a shelter in the first place. Powel and colleagues [118] suggest that problem behavior is in fact one of the leading causes of dog relinquishment. They surveyed owner perception of problem behavior of relinquishing and non-relinquishing owners to see whether there were differences between these two types of owners in how problem behavior was perceived. Relinquishing owners were found to be significantly less likely to report problem behavior compared to the matched sample of non-relinquishing owners. Powel et al. suggest that if the relinquishing owners were indeed answering honestly, they might have been less informed about normative dog behavior and did not always recognize their dog's behavior as a problem.

Several authors in our review discuss separation anxiety, including in dogs newly adopted from shelters, and possible interventions for this problem behavior. One strategy that has been suggested is for owners to show little or no excitement during arrivals and departures from the home. Recent research by Teixiera and Hall [120], however, showed that this strategy had no effect in mitigating separation anxiety in newly adopted dogs. Feuerbacher and Muir [110] tried to use the return of the owner as a reward for teaching a desired behavior to dogs, and the ability to stay alone without showing signs of separation anxiety increased over baseline, but none of the dogs in the study was able to stay alone for very long.

4.1.4. Dog Training as a Dog Welfare Issue

Dog training is related to dog welfare in multiple ways, including through the goals of the training as well as through its execution. The literature on dog training that we identified as part of our search for articles on the human–dog bond and dog welfare deals exclusively with dog training in WEIRD societies, highlighting the need for comparative and ethnographic work on dog training and dog welfare from other parts of the world. Training goals can be indirectly related to dog welfare, as illustrated by d'Angelo et al. [56], who report on a training program in Italy with the goal of increasing the chances of shelter dogs being adopted. The program is described as having met this goal, as adult and older dogs that underwent a 4-month good canine citizen training program were more likely to be adopted than an age-matched sample of untrained shelter dogs. In case the adoptions of trained shelter dogs indeed resulted in a new "forever home" for these dogs, it can be argued that their welfare was indirectly served by the training program.

Vitulli et al. [59] report on another shelter dog training program in Italy that is guided by a systematic pre-training assessment of the dogs to be included in the training program. Commenting that shelters are often lacking the financial and staff resources to manage training programs, the authors propose students as a source of labor for such programs. In Italy, secondary education students can take anthrozoology courses that are focused on dog training methods respectful of dogs and the human–dog bond. As the mission of many of these courses is to complement in-class learning with a practicum, Vitulli et al. see this as an opportunity for shelters to team up with schools to have students conduct their practical learning through participation in a shelter dog training program.

The Washington Humane Society (WHS) teamed up with the Walter Reed National Military Medical Center (WRNMMC) in a shelter dog training program. Alers and Simpson [60] report on this program, in which soldiers recovering at WRNMMC learned and applied positive reinforcement training to dogs waiting for adoption at WHS. Similarly, as in the program discussed by Vitulli et al. [59], the dogs in the training program were prescreened on health and behavior. Both dogs and soldiers were said to benefit from the program, as it increased the chances for the dogs to be adopted, and the soldiers developed new skills, built positive bonds with the dogs, and continued to serve their community.

The reference to "positive reinforcement" in the above section brings us to several articles from our selection here that we consider to be especially relevant and thought provoking in terms of the relationship between dog welfare and training. In a published research proposal focused on the training of working dogs, Vieira de Castro et al. [57] present an overview of traditional dog training methods based on the behaviorist principles of classical conditioning and operant conditioning. Classical conditioning methods include a conditioned punisher such as when an initially neutral stimulus (e.g., the word 'No!') is paired with a punishing stimulus, e.g., a slap. After repeated pairing of these two stimuli, the word No! can stand alone in achieving the effect of stopping any undesired behavior the dog may be engaged in. The second classical conditioning method is referred to as a conditioned reinforcer, where a previously neutral stimulus, e.g., a clicker, is repeatedly paired with a reward, e.g., a food reward. After repeated pairing of these two stimuli, the clicker can stand alone in reinforcing a desired (to be trained) behavior that the dog is engaged in. Operant conditioning provides four training methods: positive punishment, negative reinforcement, positive reinforcement, and negative punishments. Positive punishment is used to try to stop an undesirable behavior, for example by jerking on the dog's leash or by yelling at the dog. Negative reinforcement involves stopping a stimulus that is being perceived as unpleasant by the dog (e.g., the vibrations caused by an e-collar [65]) as soon as the dogs shifts to or shows a desired behavior. Positive reinforcement involves providing the dog with a reward, e.g., a dog treat, as soon as the dogs exhibits the desired behavior. Negative punishment refers to removing a stimulus perceived as pleasant by the dog after it shows an undesirable behavior, e.g., a time-out session in a dog crate.

Trainers engaged in these techniques often refer to purportedly genetically predisposed temperament traits in the dog, such as prey drive (shown, for example, by eagerness to chase down and retrieve a toy), or food drive, shown by interest in obtaining dog treats or other foods. In case of positive reinforcement, a toy or treat can be the reward, or can become a conditioned reinforcer. And in the case of conditioned punishment and positive punishment, specific behaviors of the handler can be the punishers. As with the pigeons in BF Skinner's operant conditioning chambers, there seems to be no need for the dog to do much else than to focus on the incoming rewards or punishers and to behave accordingly as it is driven along by its drives.

Pręgowski [23] critically reviews traditional and more recent dog training techniques. He comments that dog training practices tend to rely either on the traditional approach of positioning the owner/handler as the dominant "leader of the pack", resulting in a discipline heavy approach, or on behaviorism, with currently more of an emphasis on positive reinforcement than on punishment. Pręgowski comments that what is left out of the picture of this otherwise coherent and force-free positive reinforcement approach is the well-being of the trained dog, including the dog's immanent needs, especially psychological needs, such as attention and bonding.

Smith et al. [61] used an ethnographic approach to study the training of police dogs in the United Kingdom. These authors suggested that in working dog training, the dogs tend to become "instruments of human work whose capabilities and subjectivities are left unexplored". The central aim of their study was to obtain a better insight into the degree to which this assertion holds in police dog training, and whether and how this type of training creates a bond between dog and human. The police instructors in the study shared that they saw the bond between dog and handler as the central pillar in the training relationship. Through a mix of interviews and observations and the survey of video and photo materials, Smith and colleagues showed how interests, feelings, and experiences were shared in the training through "interactions between voices (praising, commanding, warning, types of barks and growls), sensing and touching bodies (hands, noses, teeth, fur, licks, and bites), material objects (toys, bite sleeves, leashes, ground, and kennel)". The trainers commented that they thought that their canine partners had a better quality of life than the average pet dog resting at home, but with reference to the work of Bekoff [12], the authors state that this view omits the consideration of other aspects of a dog's life, such as play, including with other dogs, and reproduction and care of offspring.

Finally, in this discussion of articles on dog training and dog welfare included in our scoping review, LaFollette et al. [64] report on a study that investigated the relationships among training methods, posttraumatic stress disorder (PTSD) severity, service dog behavior, and the veteran–service dog bond in a program matching military veterans with service dogs. The veteran used multiple training methods, and positive reinforcement or bond-based training methods were associated with reporting more positive outcomes, while positive punishment was associated with more negative outcomes. LaFollette and colleagues suggest that education about training methods could be beneficial for service dog efficacy and welfare.

4.1.5. Attachment to Owner as a Dog Welfare Issue

Earlier in this scoping review, we discussed the adaptation of the concept of parenting style to the study of the human–dog bond and dog welfare. Inherent to this approach is the view that in the human–dog bond, the human is the "adult", and the dog the immature, dependent, partner. Attachment theory is another approach from human developmental science that has been adapted to the study of the human–dog bond and dog welfare. We identified several attachment publications in our search, and we discuss them here. Before discussing individual publications, it is useful to review the history of attachment research. Inspired by the ethologist Konrad Lorenz' work on imprinting in gosling, John Bowlby [137] developed attachment theory based on the premise that in species where infants are fully dependent on their parents for survival, including the human species, newborns are biologically predisposed to establish a close bond with their caregivers, thus promoting their chances of survival and, ultimately, reproduction at sexual maturity.

The psychologist Mary Ainsworth developed a standardized test titled 'The Strange Situation' (SST) for measuring the quality of attachment to the primary caregiver of human infants between 1 and 2 years of age. The SST is purported to identify a secure attachment pattern and distinct patterns of insecure attachment. Like parenting style/parenting practice theory and research in developmental science, attachment theory has been further developed through longitudinal research that has linked infant attachment patterns as measured with SST with later social developmental outcomes, including, in the case of secure attachment, positive peer relations in childhood and adolescence, and even positive romantic relationships in adulthood.

The SST standardized test of attachment in human infants has been adapted for use with the human–dog dyad to measure the attachment of dog to owner, and in some cases, owner to dog and dog to dog in cohabiting dog pairs. Several human–dog bond studies refer to the adapted SST by its original name, while other studies refer to it as 'The Secure Base Test' (SBT) [31,32]. Other studies have developed surveys to be completed by dog owners to assess attachment of the dog to the owner without employing the SST [26].

Carreiro and colleagues [25] studied the correlation between SST attachment and sleep patterns in mature dogs. They found that secure attachment in dogs was associated with more time spent in NREM sleep. They refer to a previous EEG study on pet dogs that showed that participation in a negative social interaction, including separation from the owner, was associated with a decrease in NREM duration following the negative interaction compared to following a positive social interaction. Commenting on these findings, Carreiro and colleagues suggest that securely attached dogs might have had a more stable inner state due to the secure base provided by the owner. Thielke and Udell [29] compared dog attachment to shelter and foster care staff with the dog's performance on cognitive tasks. They report that secure attachments were associated with higher persistence in cognitive tasks in dogs, and with less survey-rated neurotic behavior, compared to insecurely attached dogs.

Saavedra-Aracena and colleagues [27] asked the question of whether attachment status would be associated with roaming behavior in owned but free-ranging dogs. The study was conducted at Navarino Island in Southern Chile, where 30% of owned dogs roam free. They found that owners of free-ranging dogs are less likely to represent a secure base to the dogs as measured through SST and complementary surveys. Following these findings the authors argue for the need for educational campaigns to foster responsible dog ownership. Sipple et al. [28] found comparable attachment patterns between dogs and their owners as in similar studies, but no evidence for attachment between cohabiting dog pairs. These authors suggest that bonds formed among adult dogs likely serve a different function than those between dogs and their owners. Konok and colleagues [33] found a correlation between insecure avoidant attachment in owners and separation anxiety in their dogs. They suggest that avoidant owners may not provide a secure base for their dog when needed, and as a result, the dog may develop separation anxiety.

Lewis [30] critiques the application of attachment theory and methods to the study of the human–dog bond and dog welfare. Lewis argues that this approach infantilizes mature animals and proposes instead that dogs form mature social bonds with their guardians and that separation anxiety is the result either of the frustration of mature adult group behaviors, or an overdependency fostered by the guardian. Lewis suggests that social bonds are adaptive, as they maximize survival and reproductive fitness and, through social buffering, can ameliorate the physiological response to acute stressors. Lewis adds that it is reasonable to infer that domestic dogs are emotionally and socially mature individuals with adult social skills specifically adapted to human social groups. Importantly, Lewis argues that separation anxiety can arise when a dog's natural social behaviors are thwarted when left alone in the home, resulting in frustration of normal adult social behavior. Such frustration may also occur when dogs are left alone without prior rigorous exercise or other suitable sensory stimulation. In sum, such frustration of normal adult social behavior when left alone at home is fundamentally different from a subjective feeling of anxiety in the absence of an attachment figure. Lewis concludes that from the perspective of dog welfare, it is necessary that dogs are no longer viewed as immature, infantile individuals, but rather as mature sentient beings with a predisposition to bond with humans.

4.1.6. Dog Characteristics as a Dog Welfare Issue

Like owner characteristics, dog characteristics affect the human–dog bond, but our review suggests that more needs to be learned about how this works. Samet and her colleagues noted that survey studies seeking to define the human–dog bond commonly do so without specific questions about the dog's investment in the bond. Samet et al. define the human–dog bond as "the unique, dynamic and reciprocated relationship between a person and dog, one in which each member can influence the other's psychological and physiological state". As a first step to develop more dog-centered questions for survey research on the human–dog bond, Samet et al. conducted a series of semi-structured interviews asking dog owners and handlers to comment on their perception of a dog's place in the human–dog bond. Themes that emerged included 'adaptation', 'understanding of a dog's preferences, likes, and dislikes', and 'affirmation'. Subthemes included 'boundaries' and 'expectations' (within adaptation), 'excitement', 'proximity', 'affection', and 'recall' (within affirmation) [38].

Protopopova and colleagues investigated to what extent behaviors of shelter dogs may affect their length of stay at the shelter. Controlling for morphological preferences, they found that leaning or rubbing on the kennel wall, facing away from the front of the kennel, and standing, were all associated with an increase in length of stay compared to dogs who did not exhibit these behaviors. No association between length of time at the shelter and consistent behavioral changes was found. The authors conclude that their findings can help shelters to focus their behavioral modification efforts on behaviors likely to influence adopters' choices [39].

Döring and colleagues report on rehoming of laboratory beagles in Germany. They cite 'The European Directive 2010/63/EU (EU, 2010; recital no. 26)' that states as follows: "(...) animals such as dogs and cats should be allowed to be rehomed in families since there is a high level of public concern about the fate of such animals". The adopted laboratory beagles, most of whom had never known life outside their laboratory kennel, showed desired behavior in their new homes within 6–12 weeks and thus proved to be highly adaptable. Nine dogs were returned, resulting in a 94% adoption success rate. This study showed that the rehoming of laboratory dogs presents a valuable alternative to euthanasia [40].

Lee and colleagues report on owner-reported characteristics of older dogs. This research is part of the Dog Aging Project. Owners reported that older dogs were less active than younger dogs; rural dogs were more active than suburban and urban dogs, especially at younger ages; and larger dogs were more active than smaller dogs. Somewhat surprisingly, older owners were found to have more active dogs than younger owners [41]. Related to Lee et al.'s study, Yamasaki reports on efforts by dog welfare organizations to motivate people to adopt older dogs. The approach involves a narrative that describes older dogs as potentially damaged but always resilient, deserving of care and still capable of a meaningful life [44].

4.1.7. Dog Emotion as a Dog Welfare Issue

The emotions of dogs, especially negative emotions, were linked to dog welfare by several authors. Hakanen and colleagues (2020) investigated correlates of non-social fear in dogs. Non-social fear is an important welfare issue, as it causes distress in fearful dogs. It was found that less socialization early in life, inexperienced owners, living without other dogs present, urban environments, and less frequent participation in activities or training were all factors predictive of non-social fear. The authors suggest that several of these factors can be improved upon by changes in owner behavior [45]. A survey study conducted in Brazil showed that negative emotional activation in pet dogs was associated with single-dog households, with being neutered, and with being owned by women. Mixed breeds, which account for most of the pet dogs in Brazil, showed higher levels in both negative and positive emotional activation compared to purebred dogs [46].

Lenkei et al. surveyed personality traits of owners and their dogs and observed separation behavior with an outdoor test. Dogs with lenient owners were more likely to bark than to whine during the test, and the authors concluded that the owner's attitude toward the dog can be related to the dog's frustration-related separation behavior [47].

A survey administered in Tokyo, Osaka, and Sendai in Japan comprising owners of 262 dogs revealed that more than half of the owners reported that their dog showed anxiety-related behavior, and more than 20% of the owners were concerned about it. Triggers for anxiety-related behavior reported by owners included 'separation' and 'subject', 'storm' and 'fireworks', 'storm' and 'sound', and 'fireworks' and 'sound' [49]. A mixed-methods study in Iran used surveys and a behavioral test to investigate stress and fear-related factors and behavioral problems in dogs. The results showed that neuroticism and fear caused by other dogs and humans were commonly associated with problem behavior, followed by separation from the owner. Small dogs were more likely to show problem behavior linked to separation and fear caused by other dogs than larger dogs. Keeping dogs indoors with limited access to a yard was also found to be associated with problem behavior [51].

4.2. *Human–Dog Bond Research across the Globe*

We found that the number of annual publications on the human–dog bond increased significantly over the period 2012–2022, and that this upward trend continued in 2023. We also found that research on the human–dog bond takes place around the globe, with significant concentrations of this work in Europe and North America. We recognize that it is possible that publications represented in the database that we searched for this review are biased in origin toward work from Western, Educated, Industrialized, Rich, Democratic (WEIRD) societies. As such, our findings concerning the geographical distribution of work on the human–dog bond are preliminary. More work should be done using other sources to uncover research on the realities of dog welfare and the human–dog bond in non-WEIRD countries.

4.2.1. Focus on Dog Welfare in Context

We found that Europe had the highest proportion of publications on the human–dog bond with a primary focus on dog welfare, followed by North America and Asia. Science is never conducted in a societal or political vacuum, and future research should study local animal welfare legislation as a societal context that could motivate or necessitate research on dog welfare. Moreover, in democracies, legislation derives from societal needs and sentiments, and as such, the presence or absence of animal welfare legislation specific to dogs can give insight into a democratic society's perception of the place of dogs within the society.

We conducted some preliminary research on animal welfare legislation, using the World Animal Protection website https://api.worldanimalprotection.org/ accessed on 23 May 2024, and we found that, for example, Austria and The Netherlands, like other EU member states, ban modifying surgery on dogs, such as the cropping of ears and docking of tails in certain breeds, as well as the use of E-collars and electric fences and prong collars. Austria's animal welfare laws specifically mention the mental wellbeing of animals, and in The Netherlands, dog welfare is being lobbied for by a 112-year-old organization for the protection of dogs (Koninklijke Honden Bescherming; https://hondenbescherming.nl/ accessed on 23 May 2024). The US is one of the pioneers in animal welfare legislation with the 1966 Animal Welfare Act, but much of this legislation is focused on farm- and industry-related animal practices. The use of E-collars and prong collars has not been banned in the US, nor has the practice of modifying surgeries on dogs.

4.2.2. Dog Welfare and the Function of Dogs

Our review suggests that function matters in terms of dog welfare. For example, a study conducted in 124 globally distributed societies showed that human–dog bonds were closer, and positive care of the dogs increased, in the case of herding dogs, while the opposite was found for hunting dogs [7]. Research on culturally specific functions of dogs has the potential to tell us much about cultural diversity in the expression of the human–dog bond and how it relates to dog welfare.

In WEIRD societies with breed associations and kennel clubs in place, there is a need, in our view, for breed-specific investigations of dog welfare as an aspect of the bond between owners and handlers of the breed. The way in which the original function of a breed is maintained (or not) through specialized training and specific breeding requirements, such as done within the member states of the WUSV (Weltunion der Vereine für Deutsche Schäferhunde; World Union of German Shepherd Dog Club Associations), also deserves attention. Does formal adherence to a functional breed standard affect the welfare of the dogs, and, if so, in what way, and how is it related to the nature of the bond between owners and handlers of the breed? We think that these are questions worthy of pursuing in future research.

4.2.3. Reciprocity and Individual Differences

Studies reviewed here illustrate reciprocity between humans and dogs in the establishment and sustenance of the human–dog bond, and they shed a light on how that interplay affects dog welfare. The shared feelings, interests and experiences that build the human–dog bond vary with the input of both partners, and there is a need for research on how the interplay of the abilities, predispositions, and characteristics of the partners affects the bond and dog welfare. There is no generic dog as much as there is no generic human, and to better understand the reciprocity within the human–dog bond, there is a need for more attention to individual differences such as 'personality' factors in members of both species. We argue for a relational approach that does justice to uncovering universal patterns of interaction as well as to the variations thereof.

4.2.4. Crossing Disciplinary Boundaries

This scoping review illustrates that the human–dog bond is a topic of interest across a range of scientific disciplines. Perhaps this should not come as a surprise, considering that dogs feature in so many different aspects of human life and human culture. Quite a few of the studies that we reviewed here were the result of cross-disciplinary collaboration, and we think that such collaboration will be important to take the research on dog welfare and the human–dog bond to the next phase.

5. Conclusions

This scoping review shows that during the past decade, 44.8% of publications on the human–dog bond did not mention dog welfare. This finding can be interpreted from either a 'the glass is half empty' or a 'the glass is half full' perspective. We opt for the latter, as this scoping review also shows that the proportion of publications with a primary focus on dog welfare more than doubled over the period 2018–2022 compared to the period 2012–2017 (11.8% vs. 24.2%). In addition, the number of annual publications on the human–dog bond, irrespective of the presence or absence of a focus on dog welfare, increased significantly during the latter half of the past decade, with the upward trend continuing in 2023. This scoping review also suggests that more than in previous years, studies are considering dogs as deserving partners capable of complex interactions with humans that go far beyond mere responses to reward or punishment. Context matters, however, and the main limitation of this scoping review is that it presents findings that are mostly restricted to research conducted in WEIRD societies. There certainly is a need for research on dog welfare and the human–dog bond in non-WEIRD societies. In addition, the work included in this review is primarily focused on pet dogs, and we need to know much more about dog welfare in working dogs and service and assistance dogs as it relates to the bonds of these active dogs with their owners, handlers, and beneficiaries.

In closing, we suggest that the ethologist Nico Tinbergen's eminently useful four questions for the study of behavior can help guide us in our thinking about where to take the research on dog welfare and the human–dog bond next. Answers to questions like 'What enhances, protects, and sustains dog welfare in the human dog bond?' (proximate causation); 'How does the human–dog bond develop and how does its development affect dog welfare?' (ontogeny); 'What is the function of the human–dog bond and how does function affect dog welfare?' (function); and 'How did the ability of dogs and humans to form close bonds evolve?' (ultimate causation) could tell us a lot about our dogs as well as about ourselves. Much remains to be done.

Supplementary Materials: The following supporting information can be downloaded at: https://www.mdpi.com/article/10.3390/ani14131985/s1, Selected articles on the human–dog bond without a focus on dog welfare.

Author Contributions: Conceptualization, P.V.; methodology, P.V.; analyses, P.V., C.A.M. and L.Q.; writing—original draft preparation, P.V.; writing—review and editing, P.V., C.A.M., L.Q. and S.J.T. All authors have read and agreed to the published version of the manuscript.

Funding: This research received no external funding.

Institutional Review Board Statement: Not applicable.

Informed Consent Statement: Not applicable.

Data Availability Statement: Not applicable.

Acknowledgments: Thank you, Jackson Junior vom Fleischerheim ('JJ') and Rose Alpha von Wilhendorf ('Rose'), trusted German Shepherd Dogs, for your companionship during the writing of this manuscript.

Conflicts of Interest: The authors declare no conflicts of interest.

References

1. Pengelly, M. Trump VP Contender Kristi Noem Writes of Killing Dog–and Goat–in New Book. *The Guardian*, 26 April 2024.
2. Sykes, N.; Beirne, P.; Horowitz, A.; Jones, I.; Kalof, L.; Karlsson, E.; King, T.; Litwak, H.; McDonald, R.A.; Murphy, L.J.; et al. Humanity's Best Friend: A Dog-Centric Approach to Addressing Global Challenges. *Animals* **2020**, *10*, 502. [CrossRef] [PubMed]
3. Strauss, E.G.; McCune, S.; MacLean, E.; Fine, A. Our Canine Connection: The History, Benefits and Future of Human-Dog Interactions. *Front. Vet. Sci.* **2021**, *8*, 784491. [CrossRef] [PubMed]
4. Koler-Matznick, J.; Adair, K.; Whittbecker, A. *Dawn of the Dog: The Genesis of a Natural Species*; Cynology Press: Mumbai, India, 2016.
5. Boros, M.; Magyari, L.; Morvai, B.; Hernández-Pérez, R.; Dror, S.; Andics, A. Neural evidence for referential understanding of object words in dogs. *Curr. Biol.* **2024**, *34*, 1750–1754.e1754. [CrossRef] [PubMed]
6. Prichard, A.; Cook, P.F.; Spivak, M.; Chhibber, R.; Berns, G.S. Awake fMRI Reveals Brain Regions for Novel Word Detection in Dogs. *Front. Neurosci.* **2018**, *12*, 737. [CrossRef] [PubMed]
7. Chira, A.M.; Kirby, K.; Epperlein, T.; Brauer, J. Function predicts how people treat their dogs in a global sample. *Sci. Rep.* **2023**, *13*, 4954. [CrossRef] [PubMed]
8. Belsare, A.; Vanak, A.T. Modelling the challenges of managing free-ranging dog populations. *Sci. Rep.* **2020**, *10*, 18874. [CrossRef] [PubMed]
9. de Waal, F.B.M.; Andrews, K. The question of animal emotions. *Science* **2022**, *375*, 1351–1352. [CrossRef] [PubMed]
10. Rowan, A.; D'Silva, J.; Duncan, I.; Palmer, N. Animal sentience: History, science, and politics. *Anim. Sentience* **2021**, *6*, 1. [CrossRef]
11. de Waal, F. *Mama's Last Hug: Animal Emotions and What They Tell Us about Ourselves*; W. W. Norton & Company: New York, NY, USA, 2019; p. viii, 326.
12. Bekoff, M.; Goodall, J. *The Emotional Lives of Animals: A Leading Scientist Explores Animal Joy, Sorrow, and Empathy—and Why They Matter*; New World Library: Novato, CA, USA, 2024.
13. Andrews, K. "All animals are conscious": Shifting the null hypothesis in consciousness science. *Mind Lang.* **2024**, *39*, 415–433. [CrossRef]
14. Dutrow, E.V.; Serpell, J.A.; Ostrander, E.A. Domestic dog lineages reveal genetic drivers of behavioral diversification. *Cell* **2022**, *185*, 4737–4755.e18. [CrossRef]
15. Morrill, K.; Hekman, J.; Li, X.; McClure, J.; Logan, B.; Goodman, L.; Gao, M.; Dong, Y.; Alonso, M.; Carmichael, E.; et al. Ancestry-inclusive dog genomics challenges popular breed stereotypes. *Science* **2022**, *376*, eabk0639. [CrossRef]
16. Holland, K.E. Acquiring a Pet Dog: A Review of Factors Affecting the Decision-Making of Prospective Dog Owners. *Animals* **2019**, *9*, 124. [CrossRef]
17. Bouma, E.M.C.; Vink, L.M.; Dijkstra, A. Social-Cognitive Processes Before Dog Acquisition Associated with Future Relationship Satisfaction of Dog Owners and Canine Behavior Problems. *Anthrozoös* **2020**, 33, 659–672. [CrossRef]
18. Horowitz, A. Considering the "Dog" in Dog–Human Interaction. *Front. Vet. Sci.* **2021**, *8*, 642821. [CrossRef] [PubMed]

19. Bekoff, M.; Pierce, J. *Unleashing Your Dog: A Field Guide to Giving Your Canine Companion the Best Life Possible*; New World Library: Novato, CA, USA, 2019.
20. Mellor, D.J.; Beausoleil, N.J.; Littlewood, K.E.; McLean, A.N.; McGreevy, P.D.; Jones, B.; Wilkins, C. The 2020 Five Domains Model: Including Human-Animal Interactions in Assessments of Animal Welfare. *Animals* **2020**, *10*, 1870. [CrossRef] [PubMed]
21. Tricco, A.C.; Lillie, E.; Zarin, W.; O'Brien, K.K.; Colquhoun, H.; Levac, D.; Moher, D.; Peters, M.D.J.; Horsley, T.; Weeks, L.; et al. PRISMA Extension for Scoping Reviews (PRISMA-ScR): Checklist and Explanation. *Ann. Intern. Med.* **2018**, *169*, 467–473. [CrossRef]
22. Horowitz, A. *Inside of a Dog: What Dogs See, Smell, and Know*; Scribner: New York, NY, USA, 2009.
23. Pręgowski, M.P. Your Dog is Your Teacher: Contemporary Dog Training Beyond Radical Behaviorism. *Soc. Anim.* **2015**, *23*, 525–543. [CrossRef]
24. McMillan, F.D.; Duffy, D.L.; Zawistowski, S.L.; Serpell, J.A. Behavioral and psychological characteristics of canine victims of abuse. *J. Appl. Anim. Welf. Sci.* **2015**, *18*, 92–111. [CrossRef]
25. Carreiro, C.; Reicher, V.; Kis, A.; Gacsi, M. Attachment towards the Owner Is Associated with Spontaneous Sleep EEG Parameters in Family Dogs. *Animals* **2022**, *12*, 895. [CrossRef]
26. Riggio, G.; Noom, M.; Gazzano, A.; Mariti, C. Development of the Dog Attachment Insecurity Screening Inventory (D-AISI): A Pilot Study on a Sample of Female Owners. *Animals* **2021**, *11*, 3381. [CrossRef]
27. Saavedra-Aracena, L.; Grimm-Seyfarth, A.; Schüttler, E. Do dog-human bonds influence movements of free-ranging dogs in wilderness? *Appl. Anim. Behav. Sci.* **2021**, *241*, 105358. [CrossRef]
28. Sipple, N.; Thielke, L.; Smith, A.; Vitale, K.R.; Udell, M.A.R. Intraspecific and Interspecific Attachment between Cohabitant Dogs and Human Caregivers. *Integr. Comp. Biol.* **2021**, *61*, 132–139. [CrossRef] [PubMed]
29. Thielke, L.E.; Udell, M.A.R. Evaluating Cognitive and Behavioral Outcomes in Conjunction with the Secure Base Effect for Dogs in Shelter and Foster Environments. *Animals* **2019**, *9*, 932. [CrossRef] [PubMed]
30. Lewis, A. Infantilizing Companion Animals through Attachment Theory: Why Shift to Behavioral Ecology-Based Paradigms for Welfare. *Soc. Anim.* **2020**, *31*, 178–195. [CrossRef]
31. Solomon, J.; Beetz, A.; Schoberl, I.; Gee, N.; Kotrschal, K. Attachment security in companion dogs: Adaptation of Ainsworth's strange situation and classification procedures to dogs and their human caregivers. *Attach. Hum. Dev.* **2019**, *21*, 389–417. [CrossRef] [PubMed]
32. Payne, E.; Bennett, P.C.; McGreevy, P.D. Current perspectives on attachment and bonding in the dog-human dyad. *Psychol. Res. Behav. Manag.* **2015**, *8*, 71–79. [CrossRef] [PubMed]
33. Konok, V.; Kosztolanyi, A.; Rainer, W.; Mutschler, B.; Halsband, U.; Miklosi, A. Influence of owners' attachment style and personality on their dogs' (*Canis familiaris*) separation-related disorder. *PLoS ONE* **2015**, *10*, e0118375. [CrossRef] [PubMed]
34. Duranton, C.; Bedossa, T.; Gaunet, F. Interspecific behavioural synchronization: Dogs exhibit locomotor synchrony with humans. *Sci. Rep.* **2017**, *7*, 12384. [CrossRef] [PubMed]
35. Duranton, C.; Bedossa, T.; Gaunet, F. When walking in an outside area, shelter dogs (*Canis familiaris*) synchronize activity with their caregivers but do not remain as close to them as do pet dogs. *J. Comp. Psychol.* **2019**, *133*, 397–405. [CrossRef]
36. Brand, C.L.; O'Neill, D.G.; Belshaw, Z.; Pegram, C.L.; Stevens, K.B.; Packer, R.M.A. Pandemic Puppies: Demographic Characteristics, Health and Early Life Experiences of Puppies Acquired during the 2020 Phase of the COVID-19 Pandemic in the UK. *Animals* **2022**, *12*, 629. [CrossRef]
37. Sherwell, E.G.; Panteli, E.; Krulik, T.; Dilley, A.; Root-Gutteridge, H.; Mills, D.S. Changes in Dog Behaviour Associated with the COVID-19 Lockdown, Pre-Existing Separation-Related Problems and Alterations in Owner Behaviour. *Vet. Sci.* **2023**, *10*, 195. [CrossRef] [PubMed]
38. Samet, L.E.; Vaterlaws-Whiteside, H.; Harvey, N.D.; Upjohn, M.M.; Casey, R.A. Exploring and Developing the Questions Used to Measure the Human-Dog Bond: New and Existing Themes. *Animals* **2022**, *12*, 805. [CrossRef]
39. Protopopova, A.; Mehrkam, L.R.; Boggess, M.M.; Wynne, C.D. In-kennel behavior predicts length of stay in shelter dogs. *PLoS ONE* **2014**, *9*, e114319. [CrossRef]
40. Doring, D.; Nick, O.; Bauer, A.; Kuchenhoff, H.; Erhard, M.H. Behavior of laboratory dogs before and after rehoming in private homes. *ALTEX-Altern. Anim. Exp.* **2017**, *34*, 133–147. [CrossRef] [PubMed]
41. Lee, H.; Collins, D.; Creevy, K.E.; Promislow, D.E.L.; Dog Aging Project, C. Age and Physical Activity Levels in Companion Dogs: Results From the Dog Aging Project. *J. Gerontol. A Biol. Sci. Med. Sci.* **2022**, *77*, 1986–1993. [CrossRef] [PubMed]
42. Smith, T.D. The threads that bind us. *Anat. Rec.* **2021**, *304*, 206–209. [CrossRef] [PubMed]
43. Scotney, R.; Clay, L. Welfare of dogs and humans in animal-assisted interventions. In *Animal-Assisted Interventions for Health and Human Service Professionals*; Nova Science Publishers, Inc.: Hauppauge, NY, USA, 2020; pp. 139–164.
44. Yamasaki, J. Old friends make the best friends: A counter-narrative of aging for people and pets. *J. Aging Stud.* **2020**, *54*, 100873. [CrossRef] [PubMed]
45. Hakanen, E.; Mikkola, S.; Salonen, M.; Puurunen, J.; Sulkama, S.; Araujo, C.; Lohi, H. Active and social life is associated with lower non-social fearfulness in pet dogs. *Sci. Rep.* **2020**, *10*, 13774. [CrossRef]
46. Savalli, C.; Albuquerque, N.; Vasconcellos, A.S.; Ramos, D.; de Mello, F.T.; Mills, D.S. Assessment of emotional predisposition in dogs using PANAS (Positive and Negative Activation Scale) and associated relationships in a sample of dogs from Brazil. *Sci. Rep.* **2019**, *9*, 18386. [CrossRef]

47. Lenkei, R.; Alvarez Gomez, S.; Pongracz, P. Fear vs. frustration-Possible factors behind canine separation related behaviour. *Behav. Process.* **2018**, *157*, 115–124. [CrossRef]
48. Arahori, M.; Kuroshima, H.; Hori, Y.; Takagi, S.; Chijiiwa, H.; Fujita, K. Owners' view of their pets' emotions, intellect, and mutual relationship: Cats and dogs compared. *Behav. Process.* **2017**, *141*, 316–321. [CrossRef] [PubMed]
49. Kurachi, T.; Irimajiri, M.; Mizuta, Y.; Satoh, T. Dogs predisposed to anxiety disorders and related factors in Japan. *Appl. Anim. Behav. Sci.* **2017**, *196*, 69–75. [CrossRef]
50. McMillan, F.D. The psychobiology of social pain: Evidence for a neurocognitive overlap with physical pain and welfare implications for social animals with special attention to the domestic dog (*Canis familiaris*). *Physiol. Behav.* **2016**, *167*, 154–171. [CrossRef] [PubMed]
51. Qiasvand, M.A.; Alimohammadi, S.; Hassanpour, S. Study of fear and fear evoking stimuli in a population of domestic dogs in Iran: A questionnaire-based study. *Vet. Ital.* **2022**, *58*, 161–171. [CrossRef]
52. Ballantyne, K.C. Separation, Confinement, or Noises: What Is Scaring That Dog? *Vet. Clin. Small Anim. Pract.* **2018**, *48*, 367–386. [CrossRef]
53. Bhattacharjee, D.; Sau, S.; Das, J.; Bhadra, A. Free-ranging dogs prefer petting over food in repeated interactions with unfamiliar humans. *J. Exp. Biol.* **2017**, *220*, 4654–4660. [CrossRef] [PubMed]
54. Duranton, C.; Bedossa, T.; Gaunet, F. Do shelter dogs engage in social referencing with their caregiver in an approach paradigm? An exploratory study. *Appl. Anim. Behav. Sci.* **2017**, *189*, 57–65. [CrossRef]
55. Schurer, J.M.; McKenzie, C.; Okemow, C.; Viveros-Guzman, A.; Beatch, H.; Jenkins, E.J. Who Let the Dogs Out? Communicating First Nations Perspectives on a Canine Veterinary Intervention Through Digital Storytelling. *Ecohealth* **2015**, *12*, 592–601. [CrossRef] [PubMed]
56. d'Angelo, D.; Sacchettino, L.; Quaranta, A.; Visone, M.; Avallone, L.; Gatta, C.; Napolitano, F. The Potential Impact of a Dog Training Program on the Animal Adoptions in an Italian Shelter. *Animals* **2022**, *12*, 1759. [CrossRef]
57. Vieira de Castro, A.C.; Araujo, A.; Fonseca, A.; Olsson, I.A.S. Improving dog training methods: Efficacy and efficiency of reward and mixed training methods. *PLoS ONE* **2021**, *16*, e0247321. [CrossRef]
58. Harris, L.; Durston, T.; Flatman, J.; Kelly, D.; Moat, M.; Mohammed, R.; Smith, T.; Wickes, M.; Upjohn, M.; Casey, R. Impact of Socio-Economic Status on Accessibility of Dog Training Classes. *Animals* **2019**, *9*, 849. [CrossRef] [PubMed]
59. Vitulli, V.; Zanin, L.; Trentini, R.; Lucidi, P. Anthrozoology in Action: Performing Cognitive Training Paths in a Garden Shelter to Make Dogs More Suitable as Pets. *J. Appl. Anim. Welf. Sci.* **2020**, *23*, 29–40. [CrossRef] [PubMed]
60. Alers, E.V.; Simpson, K.M. Reclaiming identity through service to dogs in need. *U.S. Army Med. Dep. J.* **2012**, 70–73.
61. Smith, H.; Miele, M.; Charles, N.; Fox, R. Becoming with a police dog: Training technologies for bonding. *Trans. Inst. Br. Geogr.* **2021**, *46*, 478–494. [CrossRef]
62. Townsend, L.; Dixon, L.; Buckley, L. Lead pulling as a welfare concern in pet dogs: What can veterinary professionals learn from current research? *Vet. Rec.* **2022**, *191*, e1627. [CrossRef]
63. Learn, A.; Radosta, L.; Pike, A. Preliminary assessment of differences in completeness of house-training between dogs based on size. *J. Vet. Behav.* **2020**, *35*, 19–26. [CrossRef]
64. LaFollette, M.R.; Rodriguez, K.E.; Ogata, N.; O'Haire, M.E. Military Veterans and Their PTSD Service Dogs: Associations Between Training Methods, PTSD Severity, Dog Behavior, and the Human-Animal Bond. *Front. Vet. Sci.* **2019**, *6*, 23. [CrossRef] [PubMed]
65. Masson, S.; de la Vega, S.; Gazzano, A.; Mariti, C.; Pereira, G.D.G.; Halsberghe, C.; Muser Leyvraz, A.; McPeake, K.; Schoening, B. Electronic training devices: Discussion on the pros and cons of their use in dogs as a basis for the position statement of the European Society of Veterinary Clinical Ethology. *J. Vet. Behav.* **2018**, *25*, 71–75. [CrossRef]
66. Ng, Z.Y.; Albright, J.D.; Fine, A.H.; Peralta, J.M. Our Ethical and Moral Responsibility. In *Handbook on Animal-Assisted Therapy*; Academic Press: Cambridge, MA, USA, 2019; pp. 175–198.
67. Benz-Schwarzburg, J.; Monso, S.; Huber, L. How Dogs Perceive Humans and How Humans Should Treat Their Pet Dogs: Linking Cognition With Ethics. *Front. Psychol.* **2020**, *11*, 584037. [CrossRef]
68. Cheng, W.K.; Leong, W.C.; Tan, J.S.; Hong, Z.W.; Chen, Y.L. Affective Recommender System for Pet Social Network. *Sensors* **2022**, *22*, 6759. [CrossRef]
69. Uccheddu, S.; Ronconi, L.; Albertini, M.; Coren, S.; Da Graca Pereira, G.; De Cataldo, L.; Haverbeke, A.; Mills, D.S.; Pierantoni, L.; Riemer, S.; et al. Domestic dogs (*Canis familiaris*) grieve over the loss of a conspecific. *Sci. Rep.* **2022**, *12*, 1920. [CrossRef]
70. Cimarelli, G.; Marshall-Pescini, S.; Range, F.; Viranyi, Z. Pet dogs' relationships vary rather individually than according to partner's species. *Sci. Rep.* **2019**, *9*, 3437. [CrossRef]
71. Mariti, C.; Carlone, B.; Votta, E.; Ricci, E.; Sighieri, C.; Gazzano, A. Intraspecific relationships in adult domestic dogs (*Canis familiaris*) living in the same household: A comparison of the relationship with the mother and an unrelated older female dog. *Appl. Anim. Behav. Sci.* **2017**, *194*, 62–66. [CrossRef]
72. Romero, T.; Nagasawa, M.; Mogi, K.; Hasegawa, T.; Kikusui, T. Oxytocin promotes social bonding in dogs. *Proc. Natl. Acad. Sci. USA* **2014**, *111*, 9085–9090. [CrossRef]
73. Mariti, C.; Carlone, B.; Ricci, E.; Sighieri, C.; Gazzano, A. Intraspecific attachment in adult domestic dogs (*Canis familiaris*): Preliminary results. *Appl. Anim. Behav. Sci.* **2014**, *152*, 64–72. [CrossRef]
74. Brubaker, L.; Udell, M.A.R. Does Pet Parenting Style predict the social and problem-solving behavior of pet dogs (Canis lupus familiaris)? *Anim. Cogn.* **2023**, *26*, 345–356. [CrossRef]

75. Holland, K.E.; Mead, R.; Casey, R.A.; Upjohn, M.M.; Christley, R.M. "Don't Bring Me a Dog...I'll Just Keep It": Understanding Unplanned Dog Acquisitions Amongst a Sample of Dog Owners Attending Canine Health and Welfare Community Events in the United Kingdom. *Animals* **2021**, *11*, 605. [CrossRef]
76. Stevens, J.R.; Wolff, L.M.; Bosworth, M.; Morstad, J. Dog and owner characteristics predict training success. *Anim. Cogn.* **2021**, *24*, 219–230. [CrossRef]
77. Philpotts, I.; Dillon, J.; Rooney, N. Improving the Welfare of Companion Dogs-Is Owner Education the Solution? *Animals* **2019**, *9*, 662. [CrossRef]
78. Gonzalez-Ramirez, M.T. Compatibility between Humans and Their Dogs: Benefits for Both. *Animals* **2019**, *9*, 674. [CrossRef]
79. Rehn, T.; Lindholm, U.; Keeling, L.; Forkman, B. I like my dog, does my dog like me? *Appl. Anim. Behav. Sci.* **2014**, *150*, 65–73. [CrossRef]
80. Reese, L.A. Make Me a Match: Prevalence and Outcomes Associated with Matching Programs in Dog Adoptions. *J. Appl. Anim. Welf. Sci.* **2021**, *24*, 16–28. [CrossRef] [PubMed]
81. Rossi, A.P.; Maia, C.M. Owners Frequently Report that They Reward Behaviors of Dogs by Petting and Praising, Especially When Dogs Respond Correctly to Commands and Play with Their Toys. *J. Appl. Anim. Welf. Sci.* **2020**, *23*, 402–409. [CrossRef] [PubMed]
82. Mehrkam, L.R.; Verdi, N.T.; Wynne, C.D. Human interaction as environmental enrichment for pair-housed wolves and wolf-dog crosses. *J. Appl. Anim. Welf. Sci.* **2014**, *17*, 43–58. [CrossRef] [PubMed]
83. Ferrell, J.; Crowley, S.L. Emotional Support Animal Partnerships: Behavior, Welfare, and Clinical Involvement. *Anthrozoös* **2023**, *36*, 471–487. [CrossRef]
84. Giraudet, C.S.E.; Liu, K.; McElligott, A.G.; Cobb, M. Are children and dogs best friends? A scoping review to explore the positive and negative effects of child-dog interactions. *PeerJ* **2022**, *10*, e14532. [CrossRef] [PubMed]
85. Karvinen, K.H.; Rhodes, R.E. Association Between Participation in Dog Agility and Physical Activity of Dog Owners. *Anthrozoös* **2021**, *34*, 217–231. [CrossRef]
86. Włodarczyk, J. "My dog and I, we need the park": More-than-human agency and the emergence of dog parks in Poland, 2015–2020. *Cult. Geogr.* **2021**, *28*, 495–511. [CrossRef]
87. Westgarth, C.; Christley, R.M.; Marvin, G.; Perkins, E. The Responsible Dog Owner: The Construction of Responsibility. *Anthrozoös* **2019**, *32*, 631–646. [CrossRef]
88. Maharaj, N.; Kazanjian, A.; Borgen, W. Investing in human–animal bonds: What is the psychological return on such investment? *Loisir Société/Soc. Leis.* **2018**, *41*, 393–407. [CrossRef]
89. Dodman, N.H.; Brown, D.C.; Serpell, J.A. Associations between owner personality and psychological status and the prevalence of canine behavior problems. *PLoS ONE* **2018**, *13*, e0192846. [CrossRef] [PubMed]
90. Davis, W.E.; Hicks, J.A.; Foster, C.L.; Holub, M.K.; Krecek, R.C.; Richburg, A.W. Attitudes and Motivations of Owners Who Enroll Pets in Pet Life-Care Centers. *Anthrozoös* **2018**, *31*, 211–219. [CrossRef]
91. Hall, S.S.; Wright, H.F.; Mills, D.S. Parent perceptions of the quality of life of pet dogs living with neuro-typically developing and neuro-atypically developing children: An exploratory study. *PLoS ONE* **2017**, *12*, e0185300. [CrossRef] [PubMed]
92. Bathurst, C.L.; Lunghofer, L. Lifetime bonds: At-risk youth and at-risk dogs helping one another. In *Men and Their Dogs: A New Understanding of Man's Best Friend*; Springer International Publishing: Berlin/Heidelberg, Germany, 2016; pp. 133–150.
93. Sirois, L.M. Recovering for the loss of a beloved pet: Rethinking the legal classification of companion animals and the requirements for loss of companionship tort damages. *Univ. Pa. Law Rev.* **2015**, *163*, 1199–1239.
94. Alcaidinho, J.; Valentin, G.; Tai, S.; Nguyen, B.; Sanders, K.; Jackson, M.; Gilbert, E.; Starner, T. Leveraging Mobile Technology to Increase the Permanent Adoption of Shelter Dogs. In Proceedings of the 17th International Conference on Human-Computer Interaction with Mobile Devices and Services, Copenhagen, Denmark, 24–27 August 2015; pp. 463–469.
95. App that rewards owners for good dog care. *Vet. Rec.* **2023**, *192*, 63. [CrossRef] [PubMed]
96. Powell, L.; Stefanovski, D.; Siracusa, C.; Serpell, J. Owner Personality, Owner-Dog Attachment, and Canine Demographics Influence Treatment Outcomes in Canine Behavioral Medicine Cases. *Front. Vet. Sci.* **2020**, *7*, 630931. [CrossRef] [PubMed]
97. Pirrone, F. The importance of educating prospective dog owners on best practice for acquiring a puppy. *Vet. Rec.* **2020**, *187*, 109–111. [CrossRef]
98. Laurence, C.; Burrows, H.; Zulch, H. Educating owners isn't enough to prevent behavioural problems. *Vet. Rec.* **2020**, *187*, 237–238. [CrossRef] [PubMed]
99. Hall, S.S.; Finka, L.; Mills, D.S. A systematic scoping review: What is the risk from child-dog interactions to dog's quality of life? *J. Vet. Behav.* **2019**, *33*, 16–26. [CrossRef]
100. Nardoia, M.; Arena, L.; Berteselli, G.; Migliaccio, P.; Valerii, L.; Di Giustino, L.; Dalla Villa, P. Development of a questionnaire to evaluate the occupational stress in dog's shelter operators. *Vet. Ital.* **2019**, *55*, 5–14. [CrossRef]
101. Strickler, B.G. Helping Pet Owners Change Pet Behaviors: An Overview of the Science. *Vet. Clin. Small Anim. Pract.* **2018**, *48*, 419–431. [CrossRef]
102. McGreevy, P.; Starling, M.; Payne, E.; Bennett, P. Defining and measuring dogmanship: A new multidisciplinary science to improve understanding of human-dog interactions. *Vet. J.* **2017**, *229*, 1–5. [CrossRef]
103. Diverio, S.; Boccini, B.; Menchetti, L.; Bennett, P.C. The Italian perception of the ideal companion dog. *J. Vet. Behav.* **2016**, *12*, 27–35. [CrossRef]

104. Herron, M.E.; Lord, L.K.; Husseini, S.E. Effects of preadoption counseling on the prevention of separation anxiety in newly adopted shelter dogs. *J. Vet. Behav.* **2014**, *9*, 13–21. [CrossRef]
105. Tiplady, C.M.; Walsh, D.B.; Phillips, C.J. Intimate partner violence and companion animal welfare. *Aust. Vet. J.* **2012**, *90*, 48–53. [CrossRef]
106. Vaataja, H.; Majaranta, P.; Cardo, A.V.; Isokoski, P.; Somppi, S.; Vehkaoja, A.; Vainio, O.; Surakka, V. The Interplay Between Affect, Dog's Physical Activity and Dog-Owner Relationship. *Front. Vet. Sci.* **2021**, *8*, 673407. [CrossRef] [PubMed]
107. Yuma, P.; Fowler, J.; Duerr, F.; Kogan, L.; Stockman, J.; Graham, D.J.; Duncan, C. Promoting Outdoor Physical Activity for People and Pets: Opportunities for Veterinarians to Engage in Public Health. *Top. Companion Anim. Med.* **2019**, *34*, 18–21. [CrossRef] [PubMed]
108. Ma, G.C.; Withers, A.M.; Spencer, J.; Norris, J.M.; Ward, M.P. Evaluation of a Dog Population Management Intervention: Measuring Indicators of Impact. *Animals* **2020**, *10*, 1061. [CrossRef] [PubMed]
109. Cavalli, C.; Dzik, V.; Carballo, F.; Bentosela, M. Post-Conflict Affiliative Behaviors Towards Humans in Domestic Dogs (*Canis familiaris*). *Int. J. Comp. Psychol.* **2016**, *29*, 1–13. [CrossRef]
110. Feuerbacher, E.N.; Muir, K.L. Using Owner Return as a Reinforcer to Operantly Treat Separation-Related Problem Behavior in Dogs. *Animals* **2020**, *10*, 1110. [CrossRef]
111. Gates, M.C.; Zito, S.; Thomas, J.; Dale, A. Post-Adoption Problem Behaviours in Adolescent and Adult Dogs Rehomed through a New Zealand Animal Shelter. *Animals* **2018**, *8*, 93. [CrossRef] [PubMed]
112. Wormald, D.; Lawrence, A.J.; Carter, G.; Fisher, A.D. Reduced heart rate variability in pet dogs affected by anxiety-related behaviour problems. *Physiol. Behav.* **2017**, *168*, 122–127. [CrossRef] [PubMed]
113. Thielke, L.E.; Udell, M.A. The role of oxytocin in relationships between dogs and humans and potential applications for the treatment of separation anxiety in dogs. *Biol. Rev. Camb. Philos. Soc.* **2017**, *92*, 378–388. [CrossRef] [PubMed]
114. Mills, D.; van der Zee, E.; Zulch, H. When the Bond Goes Wrong. In *The Social Dog*; Academic Press: Cambridge, MA, USA, 2014; pp. 223–245.
115. Stephens-Lewis, D.; Johnson, A.; Turley, N.; Naydorf-Hannis, R.; Scurlock-Evans, L.; Schenke, K.C. Understanding Canine 'Reactivity': Species-Specific Behaviour or Human Inconvenience? *J. Appl. Anim. Welf. Sci.* **2022**, *27*, 546–560. [CrossRef] [PubMed]
116. Bram Dube, M.; Asher, L.; Wurbel, H.; Riemer, S.; Melotti, L. Parallels in the interactive effect of highly sensitive personality and social factors on behaviour problems in dogs and humans. *Sci. Rep.* **2020**, *10*, 5288. [CrossRef] [PubMed]
117. Mehrkam, L.R.; Perez, B.C.; Self, V.N.; Vollmer, T.R.; Dorey, N.R. Functional analysis and operant treatment of food guarding in a pet dog. *J. Appl. Behav. Anal.* **2020**, *53*, 2139–2150. [CrossRef] [PubMed]
118. Powell, L.; Duffy, D.L.; Kruger, K.A.; Watson, B.; Serpell, J.A. Relinquishing Owners Underestimate Their Dog's Behavioral Problems: Deception or Lack of Knowledge? *Front. Vet. Sci.* **2021**, *8*, 734973. [CrossRef] [PubMed]
119. Normando, S.; Bertomoro, F.; Bonetti, O. Satisfaction and satisfaction affecting problem behavior in different types of adopted dogs. *J. Vet. Med. Sci.* **2021**, *83*, 566–572. [CrossRef] [PubMed]
120. Teixeira, A.R.; Hall, N.J. Effect of greeting and departure interactions on the development of increased separation-related behaviors in newly adopted adult dogs. *J. Vet. Behav.* **2021**, *41*, 22–32. [CrossRef]
121. Taylor, S.; Webb, L.; Montrose, V.T.; Williams, J. The behavioral and physiological effects of dog appeasing pheromone on canine behavior during separation from the owner. *J. Vet. Behav.* **2020**, *40*, 36–42. [CrossRef]
122. Buller, K.; Ballantyne, K.C. Living with and loving a pet with behavioral problems: Pet owners' experiences. *J. Vet. Behav.* **2020**, *37*, 41–47. [CrossRef]
123. Dinwoodie, I.R.; Dwyer, B.; Zottola, V.; Gleason, D.; Dodman, N.H. Demographics and comorbidity of behavior problems in dogs. *J. Vet. Behav.* **2019**, *32*, 62–71. [CrossRef]
124. Canejo-Teixeira, R.; Neto, I.; Baptista, L.V.; Niza, M. Identification of dysfunctional human-dog dyads through dog ownership histories. *Open Vet. J.* **2019**, *9*, 140–146. [CrossRef] [PubMed]
125. Rajapaksha, E. Special Considerations for Diagnosing Behavior Problems in Older Pets. *Vet. Clin. Small Anim. Pract.* **2018**, *48*, 443–456. [CrossRef] [PubMed]
126. Koda, N.; Watanabe, G.; Miyaji, Y.; Ishida, A.; Miyaji, C. Stress levels in dogs, and its recognition by their handlers, during animal-assisted therapy in a prison. *Anim. Welf.* **2015**, *24*, 203–209. [CrossRef]
127. Riggio, G.; Borrelli, C.; Campera, M.; Gazzano, A.; Mariti, C. Physiological Indicators of Acute and Chronic Stress in Securely and Insecurely Attached Dogs Undergoing a Strange Situation Procedure (SSP): Preliminary Results. *Vet. Sci.* **2022**, *9*, 519. [CrossRef] [PubMed]
128. Mariti, C.; Pierantoni, L.; Sighieri, C.; Gazzano, A. Guardians' Perceptions of Dogs' Welfare and Behaviors Related to Visiting the Veterinary Clinic. *J. Appl. Anim. Welf. Sci.* **2017**, *20*, 24–33. [CrossRef]
129. Helsly, M.; Priymenko, N.; Girault, C.; Duranton, C.; Gaunet, F. Dog behaviours in veterinary consultations: Part II. The relationship between the behaviours of dogs and their owners. *Vet. J.* **2022**, *281*, 105789. [CrossRef] [PubMed]
130. Kogan, L.R.; Hellyer, P.W.; Rishniw, M.; Schoenfeld-Tacher, R. Veterinary Behavior: Assessment of Veterinarians' Training, Experience, and Comfort Level with Cases. *J. Vet. Med. Educ.* **2020**, *47*, 158–169. [CrossRef]
131. Csoltova, E.; Martineau, M.; Boissy, A.; Gilbert, C. Behavioral and physiological reactions in dogs to a veterinary examination: Owner-dog interactions improve canine well-being. *Physiol. Behav.* **2017**, *177*, 270–281. [CrossRef]

132. Döring, D.; Haberland, B.E.; Ossig, A.; Küchenhoff, H.; Dobenecker, B.; Hack, R.; Schmidt, J.; Erhard, M.H. Behavior of laboratory beagles towards humans: Assessment in an encounter test and a simulation of experimental situations. *J. Vet. Behav.* **2014**, *9*, 295–303. [CrossRef]
133. Martin, K.M.; Martin, D.; Shaw, J.K. Small animal behavioral triage: A guide for practitioners. *Vet. Clin. Small Anim. Pract.* **2014**, *44*, 379–399. [CrossRef]
134. Bouma, E.M.C.; Vink, L.M.; Dijkstra, A. Expectations versus Reality: Long-Term Research on the Dog-Owner Relationship. *Animals* **2020**, *10*, 772. [CrossRef]
135. Darling, N.; Steinberg, L. Parenting style as context: An integrative model. *Psychol. Bull.* **1993**, *113*, 487–496. [CrossRef]
136. Love, hope and charity: How low-cost veterinary clinics make a difference. *Vet. Rec.* **2020**, *186*, 655. [CrossRef]
137. Bowlby, J. *Attachment and Loss*; Basic Books: New York, NY, USA, 1969.

MDPI
Grosspeteranlage 5
4052 Basel
Switzerland
Tel.: +41 61 683 77 34
www.mdpi.com

Animals Editorial Office
E-mail: animals@mdpi.com
www.mdpi.com/journal/animals

www.ingramcontent.com/pod-product-compliance
Lightning Source LLC
Chambersburg PA
CBHW061553220926
51850CB00005B/321
9783725816934